Biofuels and Bioenergy

Biofuels and Bioenergy

Edited by **Robbie Larkin**

SYRAWOOD
PUBLISHING HOUSE
New York

Published by Syrawood Publishing House,
750 Third Avenue, 9th Floor,
New York, NY 10017, USA
www.syrawoodpublishinghouse.com

Biofuels and Bioenergy
Edited by Robbie Larkin

International Standard Book Number: 978-1-68286-248-3 (Hardback)

Printed in China.

Contents

Preface

The purpose of the book is to provide a glimpse into the dynamics and to present opinions and studies of some of the scientists engaged in the development of new ideas in the field from very different standpoints. This book will prove useful to students and researchers owing to its high content quality.

Biofuels are rapidly emerging as a significant source of alternative energy around various parts of the world. They are produced with the help of plant extracts, biomass, agricultural wastes, etc. This innovative and comprehensive book integrates the well-developed theory and practical applications of bioenergy and biofuels with concepts such as biomass pretreatment, enzyme production, biodegradable wastes, biodiesel, biocatalytic conversion, etc. A number of latest researches have been included to keep the readers up-to-date with the important concepts and advancements in this area of study. Those in search of information to further their knowledge about biofuels and bioenergy will be greatly assisted by this book.

At the end, I would like to appreciate all the efforts made by the authors in completing their chapters professionally. I express my deepest gratitude to all of them for contributing to this book by sharing their valuable works. A special thanks to my family and friends for their constant support in this journey.

Editor

1

Bioenergy and African transformation

Lee R Lynd[1†], Mariam Sow[2†], Annie FA Chimphango[3], Luis AB Cortez[4], Carlos H Brito Cruz[5,6], Mosad Elmissiry[2], Mark Laser[1], Ibrahim A Mayaki[2], Marcia AFD Moraes[7], Luiz AH Nogueira[4], Gideon M Wolfaardt[8,9], Jeremy Woods[10] and Willem H van Zyl[8*]

Abstract

Among the world's continents, Africa has the highest incidence of food insecurity and poverty and the highest rates of population growth. Yet Africa also has the most arable land, the lowest crop yields, and by far the most plentiful land resources relative to energy demand. It is thus of interest to examine the potential of expanded modern bioenergy production in Africa. Here we consider bioenergy as an enabler for development, and provide an overview of modern bioenergy technologies with a comment on application in an Africa context. Experience with bioenergy in Africa offers evidence of social benefits and also some important lessons. In Brazil, social development, agricultural development and food security, and bioenergy development have been synergistic rather than antagonistic. Realizing similar success in African countries will require clear vision, good governance, and adaptation of technologies, knowledge, and business models to myriad local circumstances. Strategies for integrated production of food crops, livestock, and bioenergy are potentially attractive and offer an alternative to an agricultural model featuring specialized land use. If done thoughtfully, there is considerable evidence that food security and economic development in Africa can be addressed more effectively with modern bioenergy than without it. Modern bioenergy can be an agent of African transformation, with potential social benefits accruing to multiple sectors and extending well beyond energy supply per se. Potential negative impacts also cut across sectors. Thus, institutionally inclusive multi-sector legislative structures will be more effective at maximizing the social benefits of bioenergy compared to institutionally exclusive, single-sector structures.

Introduction

Universal access to affordable, reliable, and modern energy services is and will increasingly be required for growth and development across Africa. As such, energy provision will be a central pillar in national and regional industrialization policy and strategies. In turn, delivering energy services is a critical component of the advancement of agriculture as a basis for a broad and inclusive socio-economic growth and development strategy. In this regard, bioenergy is already playing a central role in food production and provision and is considered in most developed countries as one among several routes for diversification of energy sources. Its role might be more crucial in Sub-Saharan Africa, where so many are entirely dependent on access to land and its products, which include traditional forms of bioenergy, to survive.

With annual gross domestic product (GDP) growth rates reaching 5% during the past decade, more than twice that of the 1980s and 1990s, Africa has become one of the fastest growing continents. However, this growth has not been equally distributed and, despite substantial progress made in creating skills and jobs, poverty and food insecurity are still widespread. According to the most recent estimates available, 47% of the population of Sub-Saharan Africa lives on less than \$1.25 per day, and 27% are hungry or undernourished [1]. 43% of Africans have no access to electricity, and this percentage rises to 80% in rural areas [2]. The situation in some African countries is much worse. The challenge of addressing these issues is further heightened by population demographics featuring two-thirds of the population below 25 years of age, most of whom are unemployed. According to the UN Population Division, "the largest regional percentage increase in population between 2013 and 2050 will be in Africa, whose population can be expected to at least double and increase

* Correspondence: whvz@sun.ac.za
†Equal contributors
8Department of Microbiology, University of Stellenbosch, Stellenbosch, South Africa
Full list of author information is available at the end of the article

from 1.3 billion to about 2.3 billion, with a further 1.8 billion increase between 2050 and 2100. That projection, however, depends on the assumption that sub-Saharan Africa's total fertility rate (the average number of children per woman) will decline from 5.1 to approximately 3.0 by 2050" [3], which is yet to be supported by data.

The New Partnership for Africa's Development (NEPAD) Agency, along with regional organizations, believes that innovative approaches beyond business-as-usual should be undertaken to address Africa's multiple, interconnected challenges. Such approaches are adopted through the transformation agenda designed and implemented by the continental and regional bodies, and include among others: 1) the Comprehensive Africa Agriculture Development Programme (CAADP) Framework, 2) the Program for Infrastructure Development in Africa (PIDA), and more recently 3) the Rural Futures Program [4]. These programs are about fostering transformation. Such a transformation has been defined as "a people-centered approach based on equity and inclusiveness where rural men and women can develop their potential and reach their aspirations including income security, whilst securing environmental sustainability and where all territories in a country can express their development potential and none of them are persistently marginalized" [4]. This innovative approach is based on three basic principles: economic profitability, social equity, and environmental sustainability. Well-designed and implemented bioenergy strategies can contribute substantially to this transformation goal. In particular, modern bioenergy brings a distinctive set of attributes such that the range of development approaches and outcomes with bioenergy is substantially expanded, and can in some cases be improved, as compared to the case without bioenergy.

In considering the many intricacies and challenges associated with bioenergy and development in Africa, it is important to not lose sight of the obvious: bioenergy provides a route for Africans, from the most vulnerable to the wealthiest, to obtain critically needed energy from a resource in which the continent is rich, that is, land. To equal the land area of Africa, one can add that of China, India, Europe, and the United States - which together represent just under half the world's population. Africa has the most arable land of any continent, a substantial fraction of land well suited for production of rain-fed crops that is not currently cultivated, and the lowest per hectare crop yields in the world [5]. The potential to increase the production and harvest of biomass for both food and energy is thus very large. With land per capita above the global average and by far the lowest per capita primary energy use in the world, Africa's land resources are uniquely plentiful relative to demand for energy (Figure 1). Africa's singularly high ratio of bioenergy potential compared to current demand may of course change somewhat in the course of future development, and this will be important to consider.

Translating this potential into reality requires that daunting challenges be overcome, including those that have limited development in the agricultural sector for decades, such as widespread lack of agricultural extension, degraded soils, poorly developed infrastructure, conflict and poor governance, and complications associated with land tenure. Also critical is the availability of water resources and competing demands for land use including food and fiber crops, pasture, timber, and the whole range of forest products which constitute a substantial component of local populations' food security and well-being in terms of health. Ultimately, bioenergy cannot solve Africa's longstanding problems by itself and must be seen as one tool among many in the context of a systemic approach.

Bioenergy production requires land, and is thus inextricably linked with social development, agriculture, and environmental quality. These linkages increase the complexity of analysis and deployment of bioenergy, and can result in undesirable consequences if managed poorly. If managed well, they also have potential to greatly multiply the benefits beyond energy provision per se. Illustrative of the potential for bioenergy to be a double-edged sword, a 2011 working paper prepared by Practical Action Consulting [9] observes that biofuels development has the potential to produce a paradigm shift in agriculture, industrial, and rural development in Africa, while simultaneously providing opportunities to significantly increase energy self-sufficiency. However, the working paper also notes that ineffective policies risk displacing indigenous populations, agricultural productivity, and ecosystems for crops that may, in some cases, fail.

There is thus both a moral imperative to consider and beneficially deploy bioenergy to address critical issues on the African continent at the same time that there is potential to deploy bioenergy in harmful ways. Clear vision, strong policies, and good governance will likely be essential in order for the potential of bioenergy to be realized, and they represent an urgent need. Our objective in writing this paper is to be responsive to this need.

Bioenergy as a potential enabler of development

As noted by the United Nations Development Program (sustainable energy) "Energy is central to sustainable development and poverty reduction efforts. It affects all aspects of development – social, economic, and environmental – including livelihoods, access to water, agricultural productivity, health, population levels, education, and gender-related issues. None of the Millennium Development Goals (MDGs) can be met without major improvement in the quality and quantity of energy services in developing countries" [10].

Continents	Europe	Asia	North America	North America	Africa	Global Avg.
Total Land (ha/capita)	1.3	1.1	7.1	4.6	3.0	2.2
Primary Energy Use (TJ/capita/yr)	0.172	0.055	0.364	0.062	0.017	0.079
Land/Energy Use (Ha/(TJ/yr))	8	20	20	74	176	28

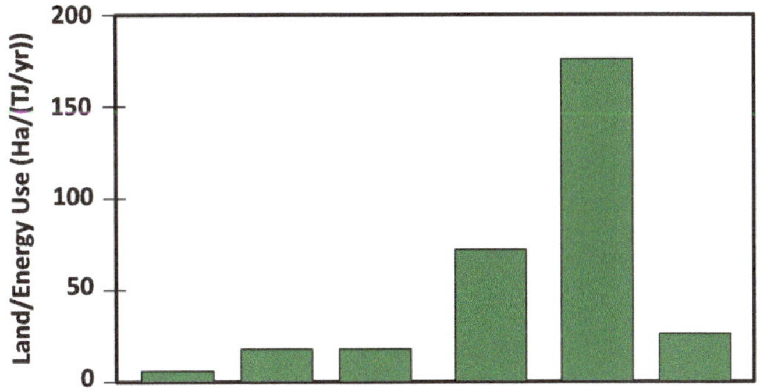

Figure 1 Comparative land resources and energy demand. Sources: population [6]; energy [7]; land [8].

Several potential contributions of bioenergy to development are listed by Lynd and Woods [11], including employment; development of marketable and transferable skills for the rural population; introduction of agricultural infrastructure and know-how; improved balance of payments and currency valuation; energy democratization, self-sufficiency, and availability for agricultural machinery and processing; and an economically rewarding way to regenerate Africa's vast areas of degraded land. A substantial literature points to disproportionately large benefits to the rural poor from agricultural development as compared to other kinds of development [12-14].

A comprehensive study of 15 small-scale bioenergy projects in 12 countries, 5 from Africa [15], drew preliminary lessons and conclusions as follows:

- Natural resource efficiency is possible in small-scale bioenergy initiatives.
- Local and productive energy end uses develop virtuous circles.
- Where fossil energy prices dominate, partial insulation is an option.
- Longer term planning and regulation has a crucial role if small-scale bioenergy projects are to succeed.
- Flexibility and diversity can also reduce producer risk.
- Collaboration in the market chain is key at start-up.
- Long local market chains spread out the benefits.
- Moving bioenergy resources up the energy ladder adds value.
- Any new activity raising demand will raise prices, even those for wastes.
- Cases do not appear to show local staple food security to be affected.
- Small-scale bioenergy initiatives can offer new choices in rural communities.

Experience with bioenergy in Africa, including positive as well as cautionary examples, is presented in the section entitled Experience with bioenergy in Africa. As considered in more detail in the section entitled The Brazilian experience, Brazil provides a prominent example of simultaneous and apparently synergistic advancement of large-scale bioenergy production, food security, and economic well-being.

As a consequence of the continent's very large land area, some of the most remote places on earth are in Africa. African agricultural producers far from ports and trade centers face the "double penalty" of lower prices for their products and higher costs for fuel and other inputs. In the 40 years preceding 2010, per capita world food production grew 17%, while in Africa it fell 10%, as population growth outstripped agricultural output [16]. One of the big problems faced by African farmers is the steep cost of transport, which means that African farmers pay two to six times the global cost of fertilizers [16]. Local production of bioenergy (heat, electricity, and biofuels for transport) to power farm machinery, dry and safely store crops, and enable transportation of goods to market could substantially alleviate this double penalty. It is notable in this context that diesel engines used in tractors and trucks can be powered by established biofuels, including not only biodiesel but also ethanol in the form of "E95" (personal communication, Jonas Stomborg, Scania).

Losses in the food supply chain, both in quantity and quality, exacerbate chronic food insecurity and malnutrition in Africa. The Food and Agriculture Organization (FAO)

[17] estimates close to one-third of the world food supply to be lost in the supply chain. These losses occur at every step of the food supply chain, including harvesting, processing, preservation, storage, transportation, and cooking. Poor access to energy is among the most important factors responsible for these limitations. By improving such access, bioenergy development could play a crucial role in preventing crop and food losses.

A multitude of factors conspire to make it difficult for African farmers to sell crops competitively into world markets, as elaborated in compelling detail by Thurow and Kilman [18]. North America and Europe export large amounts of subsidized food at prices difficult for African farmers to compete with. However, these regions do not export biofuels and are unlikely to do so in the future, and exporting heat and electricity is not feasible. Thus, energy provides a potential catalyst for socioeconomic advancement in Africa that is largely independent of several important factors that have made this difficult in the case of food production.

Government subsidies, international trade agreements, and other factors have led to relatively stable markets for producers and supply for consumers in developed countries. The consumer in the developed world, where distance between producer and the table has little impact, seldom notices regional droughts and transient decreases in production. In contrast, their counterparts in the developing world are much more vulnerable to even slight fluctuations in weather patterns or factors such as the availability of transport, fuel, and electricity. Typically, in years of abundance they do not have sufficient markets for their produce nor the means to store their produce, consequently leading to widespread spoilage and falling producer prices. But on multiple occasions oversupply has been followed by famine and skyrocketing prices in less than a year, with Ethiopia in 2003 and 2004 a notable example [18].

The precarious nature of food supply in Africa has often led to dependence on foreign aid. Yet the drivers for transformation on the African continent cannot be based on policies and regulations designed for the market-based Western economies. They also cannot be dictated by the food versus fuel debate that takes place in countries where food waste occurs not because of lack of transport infrastructure or storage facilities, but because of excess and consumer preferences, thus primarily at the retail and consumer level.

Any bioenergy strategy must be reconciled with the potential for collision between bioenergy feedstocks and food on a continent where an alarming fraction of the population is undernourished. Advancing bioenergy at the expense of food security is an unacceptably bad trade for Africa. There is increasing acceptance that bioenergy production and food security need not be in competition and could be complementary [11,19-24], but that is not the same as saying that food-fuel competition will not happen. Commenting on biofuels and local food security in developing countries, Locke and Henley [25] observe that

- Few studies use or attempt to measure the balance of all four pillars of availability, access, utilization, and stability of food.
- Available evidence does not provide a robust basis for a strong statement about the impact of biofuel projects on local food security in developing countries.
- The impact of biofuel feedstocks on food security may be similar to that of other commercial crops. It is not necessarily the fact that it is a biofuel feedstock that matters. What seems to matter is the production model used; the timing of impact measurement; the profitability of production; and the terms and conditions under which entitlements to land, wages, and prices are defined and productivity is raised.

Evaluating the effect of bioenergy on indicators of food security is somewhat different from evaluating the impacts of bioenergy on the causal factors that give rise to food insecurity, which include poverty, lack of economic development, and also physical, institutional, and market infrastructure [26]. Both evaluative frameworks are important, with the potential benefits of bioenergy likely more apparent in the latter.

Bioenergy is prominently featured in low-carbon global energy scenarios, for example, representing an average of 25% of primary energy supply in five scenarios compiled by Dale *et al.* [27]. Africa, today a small contributor to greenhouse gas emissions, has in many locations abundant resources to develop low-carbon bioenergy without having to compete with an established fossil energy infrastructure. Being the last continent to develop an economy based on fossil resources is unlikely to be a wise strategy for Africa. If unwisely deployed, bioenergy could make adaptive responses to climate change more difficult in Africa and elsewhere [28]. However, bioenergy can be an asset for such responses if wisely deployed. At a continental scale, substantive impacts from climate change are expected on Africa's cropping systems, with severe high temperature episodes and increasing frequency and severity of droughts and floods potentially causing catastrophic failures in production [29]. Indeed, yields in many important staple crops, such as maize, rice, and wheat, in Africa are increasingly volatile and in a number of cases in decline [30]. At a local level, predicting the consequences of climate change remains highly uncertain [29]. Bioenergy systems should therefore be deployed in ways that support resilience (economic and climatic) in African food cropping by, for example, enabling economically productive novel crop rotations and cropping patterns to combat increasing levels of pests and diseases in both food crops and forestry systems [31,32] and alternative markets during times of oversupply [26].

UNEP has estimated that more than a quarter of the African continent is at present in the process of becoming useless for cultivation due to degradation [33]. Cultivation of perennial grasses, which are potential bioenergy feedstocks, is well established as a means to increase soil carbon stocks and restore degraded land [34-36]. However, this subject has in general received more study in temperate climates than under conditions typical in Africa.

In pursuit of maximizing the development benefits of bioenergy, it is important to consider the entire bioenergy supply chain. At the front end, the availability of land and means by which land is accessed are critical [25]. At the back end, the extent to which bioenergy products are - or are not - aligned with and used to address high-priority social needs is equally important. We note in this context that electricity, cooking fuel, and fuel for agricultural machinery are key needs in many parts of Africa, whereas the need for fuel for light-duty vehicles is often less critical. In situations where bioenergy can provide previously missing links that enable new value chains, there is potential for large and indeed transformative development benefits.

Bioenergy overview

There are a substantial number of bioenergy feedstocks, conversion processes, and products, as summarized in Tables 1 and 2 and reviewed in more detail elsewhere [37,38]. Established combinations include:

- Woody cellulosic biomass undergoes combustion to produce electricity and heat.
- Starch- and sugar-rich crops undergo fermentation to produce ethanol.
- Oil seeds undergo pressing and transesterification to produce biodiesel.

Processes based on grains, sugarcane, or palm oil achieve rather high per hectare fuel productivity. However, this parameter is generally lower for fuels from oil seeds, which are in many cases coproducts of animal feed production. Fossil fuel displacement ratios, as well as greenhouse gas emission reductions, are generally high for processes based on sugarcane, cellulosic feedstocks, and oil-rich crops, and positive but moderate for bioenergy production from grains. Processes based on cellulosic feedstocks offer broad site range, potential for high per hectare yields, and low feedstock purchase cost. In addition, there is well-documented potential for environmental benefits from incorporating perennial grasses into agricultural landscapes with respect to soil fertility and land reclamation, water quality, and wildlife habitat [34,35,56-58]. While cellulosic feedstocks are widely thought to offer great promise for the future, conversion technology to liquid fuels is still under development and is not yet widely applied.

The potential of drought-resistant plants in regions with lower precipitation should also be considered. For example, agave plants are drawing attention as a prospective feedstock for biofuel production because of their ability to grow in dry climates, high biomass yield, and high concentrations of soluble sugar content [59]. A recent life cycle analysis of the potential of these succulent plants as a feedstock for first-generation biofuel production suggests that they show much promise with minimal impact on food production or pressure on water resources [60]. Traditionally, agaves are commercially cultivated primarily as a fiber source, often in arid, warm regions; some can tolerate temperatures of up to 65°C [61] and are therefore a good feedstock candidate for second-generation biofuels in an African context, where residues could potentially be further processed in small-scale operations for heat or electricity generation. Another intriguing aspect of some of the agaves is their response to increases in CO_2 concentration. Graham and Nobel [62] performed long-term experiments that showed a greater than 100% increase in water-use efficiency and a significant increase in dry mass production when the CO_2 concentration was doubled.

Compounding new technology risks with risks likely inherent in many African applications - for example, those involving infrastructure, business models, and governance - is unlikely to be a good strategy. As a result, a strong argument can be made for deploying established bioenergy technology in an African context. At the same time, improvements in technology for both biomass production and conversion may make possible more beneficial and widespread application in the future. Considering these two factors together, it is important to employ meritorious, current bioenergy technologies in ways that enable rather than impede deployment of future technologies, and to develop and deploy future processes in ways that expand rather than contract opportunities for early adopters and investors [63].

An illustrative and potentially important example is the possible progression from established processing of sugarcane to not-yet-established cellulosic biofuel technology. Sugarcane processing to ethanol, often accompanied by electricity and/or sugar, produces fuel competitive with global petroleum prices, has a very positive ratio of fossil fuel displacement: fossil fuel input, high fuel yields per hectare, and generally positive sustainability metrics

Table 1 Bioenergy feedstocks

Crop category	Example	Industry status	Land, environment, and energy
Starch-rich[1]	Maize, wheat, sorghum	About 50 billion L ethanol in the US based on maize	Typically grown on high-quality cropland with substantial fertilizer input. Fossil energy displacement ratio 1.3 to 1.7. 4,000 L ethanol/ha in the U.S.
Sugar-rich[2]	Sugarcane, sugar beets	About 23 billion L ethanol in Brazil based on sugarcane	Grown primarily on former pastureland in Brazil. Agrichemical inputs less than maize. Fossil energy displacement ratio about 8 to 10. 6,700 L/ha in Brazil today, could be substantially higher with conversion of cellulosics, energy cane.
Oil-rich[3]	Rapeseed, soy, sunflower, palm oil	About 23 billion L produced worldwide, most in the EU, US, and Brazil	Rapeseed, soy generally grown on cropland. Most palm oil plantations are on former forests. Fossil energy displacement ratio 2 to 2.5 for rapeseed and soy, about 4 to 8 for palm. 530 L/ha for soy in Argentina; 3,600 L/ha for palm in Malaysia.
Cellulosic[4]	Grass, trees, various wastes	331 TWh electricity globally. Liquid fuel capacity about 175 million L worldwide	Could in principle grow on land unsuitable for crops. Potential environmental benefits when incorporated into agricultural landscapes. Fossil energy displacement ratio somewhat speculative for liquid fuel production but expected to be similar to sugarcane. Over 7500 L/ha based on miscanthus yields in the US (25 tonnes/ha), 75 US gal/ton.

[1]Starch-rich crops: annual production [39]; fossil energy displacement [40]; corn yield [41]; dry mill yield [42].
[2]Sugar-rich crops: annual production [43]; fossil energy displacement and ethanol land yield [44].
[3]Oil-rich crops: annual production [7]; fossil energy displacement [45,46]; soy oil yield [47]; palm oil yield [48].
[4]Cellulosic crops: global electricity [49]; global cellulosic biofuel capacity [37]; current miscanthus yields [50].

[64,65]. Lignocellulose is present in sugarcane in about a 2:1 ratio relative to sugar. Converting the lignocellulose as well as the sucrose fractions in sugarcane would substantially increase yields of energy and revenue per ton, and growing "energy cane" with reduced sugar content would have the multiplicative effect of increasing tons per hectare. Once conversion of the lignocellulose component of sugarcane is established, this would enable conversion of other cellulosic crops, for example, those with a higher tolerance to drought, that could be grown where sugarcane cannot. Thus, there is a continuous and potentially advantageous path from fermenting only the soluble sugars present in cane to also fermenting cellulosic residues once the required conversion technology is available.

Table 2 Modern bioenergy conversion technology summary

Category	Products	Technology
Non-biological		
Combustion[1]	Electricity, heat	Mature. Electricity generation rather capital intensive (about $1,900 - $4,300/installed kW).
Gasification[2]	Electricity (via gas turbines) or synthetic gasoline and diesel (e.g., Fischer-Tropsch)	Limited commercial application. Often highly capital intensive (about $375/L annual capacity for coal liquefaction in South Africa).
Pyrolysis[3]	"Biocrude", a mixture of liquid-phase organics	Limited commercial application. $2/L annual capacity for production of naptha and diesel.
Pressing and transesterification[4]	Biodiesel from oil-rich crops	Mature. Relatively simple, low capital ($0.33/L installed capacity for biodiesel in Europe).
Biological		
Fermentation of starch and sugars	Ethanol, potentially many other molecules	Mature for ethanol production. Capital cost[5] about $1.20/L installed capacity for ethanol with cogeneration in Brazil, about $2/L installed capacity for maize ethanol in the US.
Anaerobic digestion	Methane	Rather mature. Can be applied to both liquid and solid wastes. Many thousand small-scale digesters operative, particularly in China and Germany.
Lignocellulose hydrolysis and fermentation	Ethanol, potentially many other molecules	Not mature. Hydrolysis can be accomplished via acid or enzymes. Several fermentation options and configurations.

[1]Combustion capital costs: [51].
[2]Gasification capital costs: [52].
[3]Pyrolysis capital costs: [53].
[4]Biodiesel capital costs: [54].
[5]Sugarcane ethanol capital costs: [55].

Bioenergy can and is being produced over a broad range of scales from village-scale digesters and biodiesel refining operations to industrial-scale facilities which produce up to a half billion liters per year of fuel and process up to over five thousand dry metric tons per day of feedstock. Large-scale facilities require large land areas as well as technological expertise and capital not available within many African communities. At the same time, high efficiency and financial viability are often easier to achieve at a larger scale as compared to smaller scale and scattered markets with low purchasing power of the populations. This conundrum remains to be resolved, and is likely to be fertile ground for creative approaches that are tailored to location-specific circumstances and will likely evolve over time. The Brazilian experience suggests (see the section later in the paper) that broadly distributed social benefits and large-scale efficient bioenergy production need not be mutually exclusive.

Experience with bioenergy in Africa

In 1990, Africa's primary energy consumption had reached 16 EJ, less than 5% of the global energy demand, of which bioenergy provided 60%. By 2010, its primary energy consumption had risen to 28 EJ, slightly more than 5% of the global demand, with bioenergy providing about half of this for the continent as a whole and much larger shares in some regions [66]. Africa's dependence on traditional forms of biomass for energy has not diminished and is not predicted to do so in the foreseeable future (Figure 2).

Biomass has been and remains the main source of energy for many people in Africa both in rural and in urban areas. For Sub-Saharan Africa (excluding South Africa), over 80% of the total energy supply for heating, cooking, and processing of agricultural produce is derived from biomass, such as fuel wood and agricultural residues [66,67]. In most of this region's cities, where the population is still booming, the majority of households are dependent on wood energy more than on any other sources for such purposes. Cooking on open fires is highly energy inefficient and also poses a major public health problem; an estimated four thousand Africans die prematurely every day from household smoke pollution [67,68]. Demand for wood for cooking, particularly when converted to charcoal to sell into urban markets, can exceed supply, resulting in environmental degradation in addition to serious health impacts [67,68]. By contrast, modern bioenergy involves using higher efficiency technology to produce fuels, electricity, and heat.

Africa is looking for more efficient and affordable household energy sources that can enhance rural development and reduce the burden on women to provide the energy needs of their households while combating deforestation, land degradation, and desertification. In this context, there have been various bioenergy initiatives implemented to increase access of rural and peri-urban populations to clean and sustainable energy and modern bioenergy sources. These initiatives have targeted both the demand and supply sides. Projects can be categorized as follows:

1. Increasing access to traditional sources of energy such as wood and charcoal in a more sustainable manner through reforestation and investments in energy production plantations while increasing diversification of products and income opportunities on the end-user side, and the use of efficient conversion technologies such as improved cookstoves. Examples include projects funded by the World Bank in the Democratic Republic of Congo, and in Malawi with the Jatropha, Neem, and Moringa project [67,69,70].

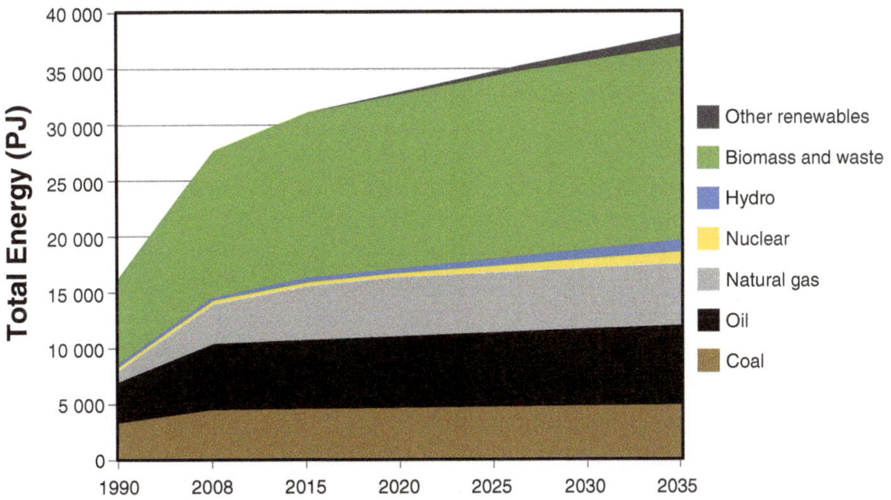

Figure 2 Total primary energy demand for energy sources on the African continent, 1990 to 2035 [66].

2. Using agricultural residues, municipal waste, and non-food crops, hence avoiding competition with food crops. Such energy sources are not fully developed and constitute a promising avenue, as demonstrated through several experiences in various regions of the continent. Country-specific projects include those in Senegal, Ghana, Kenya, Uganda, Tanzania, and Malawi, as presented in Table 3.
3. Using liquid biofuels such as ethanol and biodiesel and the corresponding technologies for conversion and utilization to substitute the traditional sources and conversion technologies. This is the case in the Ethiopian government-led project, but also in several other Southern and East Africa countries including Madagascar, Mauritius, South Africa, Zambia, and Malawi, to name a few. Examples of these options and related initiatives are summarized in Table 3.

Diaz-Chavez [20] reported a detailed study of biofuel development and potential in African countries selected to represent different regions: Senegal, Mali, Kenya, Tanzania, Mozambique, and Zambia. This study concluded that Africa has potential to meet both its food and fuel needs from biomass, neither of which occurs today, and that biofuel production could help unlock Southern Africa's latent potential and positively increase food production if it brings investment in land, infrastructure, and human resources. Further conclusions, illustrative of both potential and challenges, included the following:

- Yields of currently cultivated land in the less developed countries could be tripled by using improved management practices, potentially freeing up more land for biofuel production.
- It is estimated that the area under sugarcane in the region could be doubled without reducing food or destroying valuable habitats.
- Mozambique has immense agricultural potential, with an estimated 36 million ha of arable land of which only 10% is presently in productive use.
- Negative impacts have occurred in some areas (not whole countries), such as displacement, and these should not only be avoided but legally penalized.
- The capacity to implement and monitor needed policies is limited in some countries.
- Bioenergy projects in Africa have not been without challenges related to feedstock production, technology, and social factors such as consumer preferences and institutional coordination. In particular:
- There is a constraint of reliable feedstock supply under circumstances that achieve low agricultural yields today. Given the low and/or volatile level of yield for many crops - most of which are rain fed with low access to quality inputs and equipment - bioenergy projects have suffered from irregular feedstock provision in terms of quality and quantity, making the availability of bioenergy products unstable and unpredictable. When feedstocks are derived from non-food crops for which a research gap remains to be filled, for example, jatropha or other tree crops, the situation has often been particularly challenging. Under such circumstances, price stability and confidence of consumers are easily eroded, and the new adopters shift back very quickly to traditional biomass sources of energy and equipment, for which sources of supply are well established. The myth that some favored new crops, such as jatropha, would be immediately commercially productive on marginal land is now realized to be predominantly false [9].
- Consumer preferences are difficult to shift to new technologies in cases where the energy density and efficiency of new biomass-derived products is lower than that of well-established products. On the other hand, ease of handling, including safety and cleanliness, have been found to be a significant factor for adopting liquid-based biofuels such as ethanol for cooking [74].
- Experience in many African countries reveals that price incentives have not been sufficient for adoption of biofuels given the lower energy density of the new product (briquettes for instance) compared to charcoal. Under such circumstances, more research is needed in order to improve the efficiency of these new technologies.
- Isolated projects, even those with tangible outcomes, have in some cases not proved sustainable or conducive to a qualitative transformation process. This has been the case in a number of projects conducted by external partners with weak involvement of government and national stakeholders. Furthermore, many projects still need to be scaled up for a real impact on a large fraction of the population.
- Institutional constraints must also be faced in terms of coordination and synergies to be built among government units. Agriculture, environment, and energy departments rarely work together to discuss and design bioenergy strategy frameworks and harmonized policies and regulations. Private sector participation is also at its early stage, as most projects are initiated by non-governmental organizations (NGOs) and international partners.

Although modern bioenergy industries are emerging in several African countries, in particular, where an incentive exists for blending ethanol with gasoline, most of them still lack the capacity to develop an economically viable and sustainable bioenergy industry. However,

Table 3 Examples of bioenergy initivatives in Africa

Country	Initiative	Opportunities/comments
Ethiopia	National biogas program, which plans to build 14,000 domestic biogas digesters [71]. A 5% blending of petrol and ethanol since 2008.	Under the national biomass program, a 4-year demonstration project has demonstrated notable benefits of replacing fuelwood (currently 29%) and kerosene (42%) with ethanol stoves; notably reduced foreign exchange to import kerosene, reduced distance traveled to collect firewood by 73%, and improved indoor air quality [15].
Ghana	Jatropha oil for mixing with diesel (70% plant oil/30% diesel) to fuel butter processing equipment, and as a kerosene substitute for use in lanterns [72].	Village-level biofuel production. *Note:* Jatropha has been planted in a number of other African countries such as Malawi and Mozambique (see below) as well as Mali [15]. In South Africa currently only allowed for experimentation [73].
Kenya, Tanzania, and Uganda	Afforestation for sustainable charcoal production [74].	Charcoal making supports about 500,000 full-time and part-time charcoal producers. Wood fuel demand is double the supply, with forest cover decrease by 2% annually, thus incentive for tree planting. Charcoal remains preferred choice over briquettes despite higher price and more pollution. *Note:* See also initiatives in Senegal [15].
Madagascar	Ethanol as a household fuel and alternative sources of energy to relieve the pressure on forest resources and reduce childhood mortality [75].	Identified need for a regulation, Government support and optimization identified as key requirements for success.
	Gel fuel to replace charcoal as a cooking fuel in urban areas [69].	Identified need for economic sustainability.
Malawi	Restoration and commercial use of tree crops, including marginal lands [70].	Potential for integrating various tree species to increase crop yield, rehabilitate degraded land, and improve the soil fertility. Products are used as bio fertilizer and green charcoal.
Mauritius	Cogeneration, primarily using bagasse, renders sugar industry electricity self-sufficient, with estimates that excess bagasse-derived power accounts for 30% of total electricity demand in the country [76].	Life cycle analysis shows that despite potential negative consequences such as high water consumption and eutrophication, benefits include lower GHG emissions and acidification; probably the only stable alternative to 100% coal imports.
Mozambique	Initiated in 2004, biofuel production originally dominated by small-scale farmers, now by foreign commercial investors [77].	Originally the focus was primarily on jatropha biodiesel, now there is increased emphasis on bioethanol derived from sugarcane and sorghum.
South Africa	Mandatory blending of petrol and diesel with biofuels as follows: 5% minimum concentration for biodiesel blending, and permitted range for bioethanol blending from 2% to 10% v/v [78]. Target date of 1 October 2015.	South African Airways plans 50% use of aviation biofuels by 2020. Energy crops include sweet sorghum and sugarcane [79,80]. Renewable energy feed-in tariff implemented to establish energy prices including a profit margin to attract developers to invest [81].
Tanzania	Sisal biogas. Conventionally only 4% of the plant (fiber) has been used to make items such as ropes and carpets. Two projects to date resulted in improved efficiency for biogas and biofertilizer production; current electricity output is150 kW with plans to expand to other estates for a total of 6 MW [15].	A private company without external support leads this initiative, which led to an 80% increase in the number of children attending school, while access to health care also improved as a result of the energy supplied to schools and hospitals.
Zimbabwe	Planned current 5% blending of ethanol in petrol to 15% [82].	The technical feasibility and potential were demonstrated when the commercial producer reached maximum generation capacity of 18 MWe. About 8 MWe is used for sugarcane ethanol, leaving 10 MWe surplus.
	Jatropha cultivation for biodiesel [83].	Objective is to produce biodiesel to meet 10% import substitution (approximately 100 million L per year) from jatropha, using an existing facility operating on cotton and sunflower seeds.

opportunities exist as several regional economic communities have defined very clear strategies that need substantial support to be adapted and implemented in a comprehensive manner at the national level. This is the case, for example, of the West African Economic and Monetary Union, which has adopted a bioenergy strategy since 2008 [84]. One of the main drivers of bioenergy development in this region resides in reversing the trend of desertification and land degradation, and developing sustainable energy sources for cooking, heating, and food processing. Therefore, the key strategies aimed at providing alternative fuels can be expected to benefit from reliance on a combination of feedstocks provided through reforestation with fast-growing and adapted species that can be harvested sustainably and processed into cleaner fuels. In areas where reforestation is not possible, bioenergy development has been encouraged through multicropping systems and careful management of water resources [84].

The Brazilian experience

Brazil's modern bioenergy industry, one of the two largest in the world in absolute terms, is by far the largest in terms of fractional energy supply, and is the foremost example of bioenergy deployed in a developing country context. Soils and climates in much of Africa have similarities to those in Brazil, and Africa and South America are widely recognized as the continents with the greatest potential to increase modern bioenergy production [85]. Over the last three decades, Brazil saw marked increases in social development (minimum wage increase, poverty and hunger reduction), went from being a small player in international agriculture to the largest exporter in the world (number one in soybeans, beef, chicken, oranges, and coffee), and became energy independent with a large contribution from modern bioenergy (Table 4). There is substantial evidence that the emergence of Brazil's bioenergy industry positively impacted simultaneous advances in social development and agriculture. Brazil's bioenergy experience is thus of distinctive relevance to Africa.

However, we acknowledge at the outset the tremendous diversity of circumstances on the African continent, and that the Brazilian bioenergy model will in most cases require some adaptation to these circumstances. We note that development of bioenergy in Brazil has until recently targeted national markets, which for some African countries are small and/or otherwise impractical to rely on. As well, the expansion of Brazilian bioenergy production seen since 1980 began with already established industrial production of both sugar and ethanol, thereby providing a foundation of expertise and purchasing power that are present in some but by no means all African countries.

Sugarcane has been cultivated in Brazil since the sixteenth century and has always represented an important economic activity. In 1931, aiming to reduce dependence on imported liquid fuels and absorb the excess production of the sugar industry, the Brazilian government implemented a compulsory blend of at least 5% anhydrous ethanol in gasoline. During the period from 1931 to 1975, an average of 7.5% of gasoline demand was met by ethanol. In order to further reduce oil imports and increase energy security, the Brazilian government created the National Alcohol Program (Proálcool) in 1975. This program has evolved since then, with ethanol reaching price parity with gasoline on a BTU basis in about 2005 [65]. A particularly significant development was the introduction of flex-fuel cars, able to use any blend of gasoline (E25) and hydrous ethanol. Flex-fuel cars currently represent 95% of sales of new cars, and pure ethanol can be used by 12.7 million Brazilian vehicles representing 47% of the national fleet [92]. Ethanol currently provides about 50% of light-duty fuel and 25% of total road transport fuel in Brazil, with biodiesel production about one-tenth that of ethanol [91]. However, the growth of ethanol production in Brazil has stalled in recent years due to government policies that maintain lower-than-market gasoline prices [93]. Ethanol production as practiced today in Brazil has generally positive sustainability indicators, notably including life cycle greenhouse gas emissions on the order of 10% of a gasoline base case [94].

As in many other countries, Brazilian mills processing sugarcane use bagasse to produceheat and electricity. Increasingly, surplus electricity is sold to the grid. Today bagasse is the second leading source of energy for electricity generation in Brazil after hydropower [91]. The progressive introduction of more efficient cogeneration systems allowed surplus electricity per ton of sugarcane processed to increase from approximately 20 kWh to up to 140 kWh in the most efficient mills, with room for

Table 4 Summary of Brazil's advances in social, agricultural, and energy sectors: 1980 to 2010

Sector	Social			Agricultural	Energy		
Index	Minimum wage ($US\$_{2010}$/month)	Population out of poverty (%)	Global hunger Index (GHI)[a]	Exports (million $US\$_{2010}$)	Net imported (% supply)	Renewable (% supply)[b]	Biofuels (% liquid fuel supply)
	[86]	[87]	[88,89]	[90]	[91]		
1980	207	67	10.4 (1981)	24,700	42%	46%	8%
2010	298	90	4.0 (2011)	62,100	10%	47%	27%

[a]The Global Hunger Index is used to evaluate the hunger situation by countries, considering: a) the undernourished population as a percentage of the total population, b) the prevalence of underweight children under the age of 5, and c) the under-5 mortality rate. Values less than 4.9 reflect "low hunger", values between 5 and 9.9 reflect "moderate hunger", and values between 10 and 19.9 indicate "serious hunger". The worst global hunger scores in 2011 were ascribed to Burundi and the Democratic Republic of Congo, with scores of 37.9 and 39, respectively.

[b]The share of renewable energy supply remained about constant, but shifted from wood fuel used in households for cooking to liquid biofuels used in the transportation sector. In this period the total energy supply increased 234% (115 to 269 Mtoe) [91].

further improvement to reach to about 200 kWh via integrated biomass gasification and combined cycles [95]. The electricity produced in Brazil from bagasse in 2012, 25 TWh, represents 5.6% of the electricity consumption in Brazil [96]. The installed power generating capacity of cogeneration systems in Brazilian mills, 9.3 GW, is a third of the 28 GW of installed capacity in the 47 Sub-Saharan African countries excluding South Africa [97]. Development of electrical generating capacity from bagasse in Brazil is a relatively recent event, occurring entirely within the last decade. As previously mentioned in Table 3, cogeneration from bagasse in Mauritius is extensive.

It is interesting to stress the relevance of yields improvement and densification to reduce the land requirement for agriculture, including bioenergy production, in Brazil [98]. In recent decades, the sugarcane yield (tons/hectare) grew at a cumulative average annual rate of 1.4% and the process yield (liters ethanol/ton) grew at an average rate of 1.6%, resulting in an average annual increase of 3.1% in ethanol production per hectare. Thanks to these gains, the area currently dedicated to the cultivation of sugarcane for ethanol production is 38% of the area that would have been required to obtain such production with the yields observed when Proálcool began. Almost all of the 4.8 Mha used to produce ethanol in Brazil, representing about 1.3% of the total area of rural properties, is former pasture land. Over the lifetime of the Proálcool program, pasture land devoted to beef production has decreased by 10%, but beef production has more than tripled as a result of both higher stocking densities (head/ha) as well as higher animal performance (kg beef/head/year). Roughly threefold yield gains have also been observed over this period for grains and maize [99]. As shown in Figure 3, Brazil has achieved both food and gasoline independence, whereas substantial reliance on imports is observed for several African countries with substantial land resources.

There are about 400,000 direct jobs specifically related to ethanol production in Brazil, excluding the workers associated with sugar production [100]. Under current conditions, the production of bioethanol per unit of energy produced, compared with mineral carbon, hydroelectricity, and oil, requires, respectively, 38, 50, and 152 times more human labor [44]. About 81.4% of the employees work under a formal labor contract, compared to about 40% in the Brazilian agricultural sector as a whole. Formal work relations assure legislatively mandated rights such as retirement and annual paid vacations, unemployment insurance, extra monthly wages per year, health programs, and improved work conditions. Cooperative relations with workers' unions where sugarcane mills operate has fostered, among other benefits, reduction of illiteracy and increase of years attending school, and a reduction in underage workers (from 15.3% in 1981 to less than 0.3% in 2009 [100].

In a detailed analysis of the socio-economic impacts caused by the expansion of sugarcane cultivation, Assato and Moraes [101] studied the results of establishing sugarcane processing plants in two municipalities, Nova Alvorada do Sul and Rio Brilhante. They found an increase in the aggregate income which boosted local markets, as evidenced by an increase in the number of shops and services as well as a more active real estate sector. They also noted that jobs which derived from the expansion of the sugarcane industry, and from other industries related to this activity, have played a key role in retaining and attracting residents, thus reducing rural exodus and contributing to increased population in the two towns they analyzed. These towns feature a large number of surrounding rural settlements, in which crops are cultivated that existed before the arrival of the sugarcane industry. Assato and Moraes observed that the income (often subsistence initially) of family farms in these settlements was supplemented with the wages from the jobs created by the sugarcane industry either in the ethanol plants or in the sugarcane fields. A significant portion of family farmers reported improvement in their quality of life due to the social programs offered by the sugarcane industry-related companies and due to opportunities for (re)training, employment and education, especially for children. Data collected from interviews indicated improved education during the period after the installation of the sugarcane industry. The authors conclude that introduction of sugarcane culture created jobs that led to an increase in aggregate income of the municipalities, and through multiplier effects enabled improved indicators of health, education, and quality of life.

The question of how the Brazilian agricultural sector would have developed without the simultaneous rapid growth of the bioenergy industry is complex and would likely benefit from more study. Although bioenergy development was not a primary cause of the growth of Brazil's agricultural sector, it has likely been an accelerating factor in light of contributions to the development of rural communities and human resources together with improvements in logistics and trading infrastructure. Social development, agricultural development and food security, and bioenergy development in Brazil have been synergistic rather than antagonistic.

Important lessons from the Brazilian bioenergy experience of potential relevance in the African context include the following:

1. It is valuable for bioenergy feedstocks to be well known in agricultural terms, taking into consideration regional factors. Support by

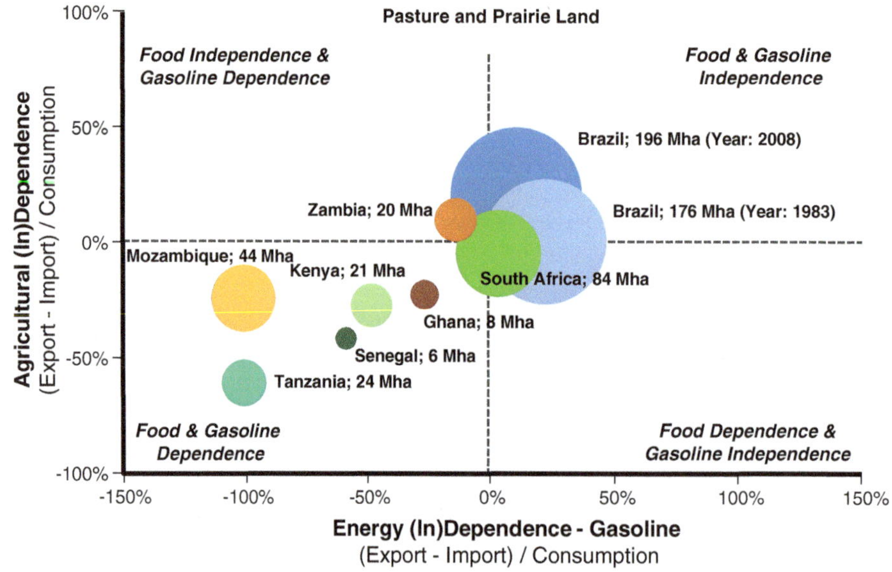

Figure 3 Agriculture (in)dependence and gasoline (in)dependence and pastures/prairies area [89,91].

breeding programs, built on a foundation of good germ plasm, is essential.

2. Selling into multiple product markets (for example, food, fuel, electricity) has been advantageous in Brazil.
3. Bioenergy production chains should score well in terms of life cycle indicators, which are generally fostered by efficient use of land, water, and energy.
4. The state and its agencies have fundamental roles in fostering sound biofuel programs by assessing/creating/monitoring/enforcing the conditions for production/use, preferentially within a clear legal and normative framework. Important tasks include defining fuel (and blends) specifications, setting mandatory blending levels and implementing the program, and establishing a balanced tax regime taking into account appropriate externalities. These tasks are complex and demand both a technical background and negotiation among stakeholders, who frequently present contradictory perceptions and aims.
5. Social benefits should be explicitly considered within an integrated framework that also considers commercial viability, and are generally fostered by efficient production chains (point 3).

The evolution of African agriculture

Van Kuelen and Schiere [102] suggest a scheme for the evolution of agriculture, focusing on mixed farming systems. Borrowing heavily from the four-stage progression they outline, we adapt this scheme here to describe agriculture in general and present attributes of each developmental stage.

As depicted in Figure 4, increasing population and resource pressure drives agriculture through a progression of modes from expansive/long fallow, to low external input/highly integrated, to high external input/specialized, to new conservation agriculture featuring extensive integration and high knowledge intensity. Agricultural integration, involving the exchange of material and energy between various agricultural activities and in particular crop and livestock production, plays a central role in this progression.

Most of Africa is supported by low input agriculture. Integration is practiced widely in some locations, for example, raising animals and crops on the same land in different parts of the year. However, the scope for integration can be restricted somewhat by very small farm sizes, for example, one or two hectares. Although much of the world's effort to increase food productivity is focused on specialized agriculture with high inputs, 50% of the world's food production and 70% of the world's people are supported by mixed crop-livestock agricultural systems featuring a significant level of integration, and much of this agriculture involves low inputs [103]. Just as cell phones proliferated within Africa, bypassing the need to build a network of wires and poles, we see potential - and many benefits - to Africa progressing from the low external input, often integrated, mode to an African brand of new conservation agriculture bypassing some aspects of the high input/specialized mode. Realizing this potential is a challenge for policy makers, as we partially address in the section entitled Future directions.

Much has been written about bioenergy production from food crops grown outside of Africa leading to higher food prices and compromised food security [104-106].

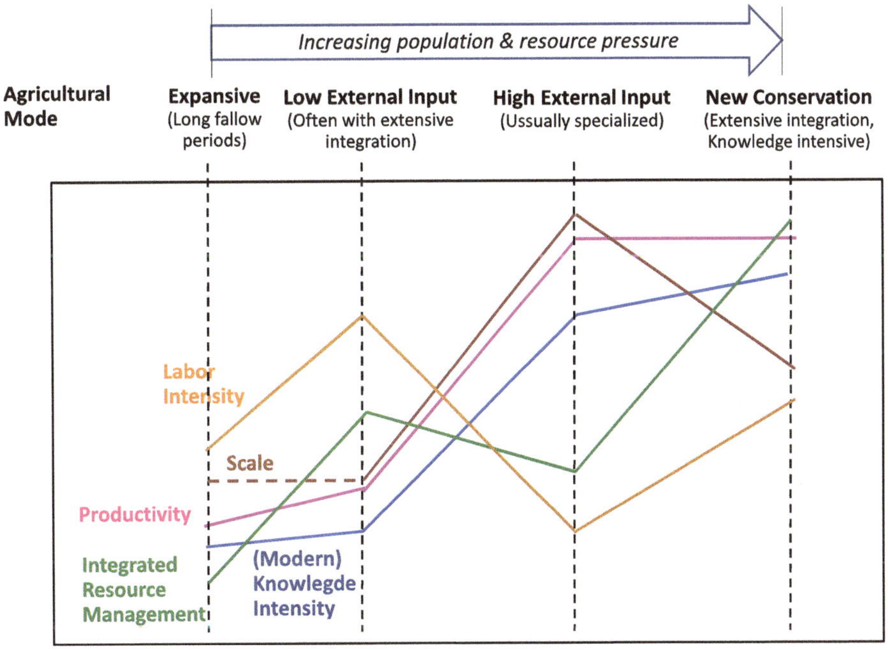

Figure 4 Evolution of agriculture.

Considerably less attention has been given to impacts of modern bioenergy production in Africa and, in particular, the potential benefits of such production with respect to food security. In fact, very few African examples of modern bioenergy production have been in existence at either small or large commercial scales over a long enough duration for sufficient data to be available to draw robust conclusions. In the Scurlock *et al.* [107] analysis of the relatively large-scale Triangle sugarcane ethanol plant in Zimbabwe, mainly benign and positive impacts on sugarcane production and productivity were found by the implementation of an ethanol plant annexed to a sugar mill.

Perhaps more speculatively (and we acknowledge, controversially), it is possible to foresee an important role for biofuels in supporting resilience in food cropping, as opposed to the competitive outcome with food provision and access that is most often assumed. Here we speculate what might have happened over the last decade if Zambia, and indeed South Africa, had implemented a large-scale biofuel production program based on the use of maize as the primary feedstock. Crop production in Sub-Saharan Africa can be described as a boom, but under-supply, cycle that can lead to bouts of severe undernourishment. For example, in Zambia during 2010 and 2011, as a result of adverse climate conditions the maize crop failed, and Jayne [108] states that "the government of Zambia spent 2-3% of GDP stabilising food prices. In 2012, better climatic conditions returned and a 1.5 million tonne maize surplus was generated. However, as the country only had the capacity to export 70 000 tonnes per month to other countries, it would have taken '20 months to export the surplus by which time (as a result of a lack of storage infrastructure) most would be unsuitable for human consumption. "Similar cycles are seen throughout the continent. And yet, if Zambia had a biofuel industry capable of using all or part of the grain surplus, an economic take-off would have been available supporting the development of the production and storage infrastructure, and during times of crop failure the remaining crop could be diverted back to human food markets. In this way the maize supply chain could become more resilient to climate shocks.

Future directions

Modern bioenergy can be an agent of African transformation, with potential social benefits accruing to multiple sectors and extending well beyond energy supply per se. Potential negative impacts also cut across sectors. Thus, institutionally inclusive multi-sector legislative structures will be more effective at maximizing the social benefits of bioenergy compared to institutionally exclusive, single-sector structures. This critical point is articulated well by the 2011 Practical Action working paper [9]:

> "the role of government is to provide stimulus for private investment and initiatives, as well as promote effective regulation, monitoring and co- ordination of the biofuels sector. The particular multifaceted opportunity that liquid biofuels offers for Africa demands a new type of public, private and governmental engagement and

> integration, which may be very beneficial for Africa's overall growth and development. Given the complexity of the different policy objectives, and the many unknowns, the industry is still more likely to succeed within a purpose built legislative structure, than within the current inadequate and/or conflicting frameworks. Subsequently, working with all relevant ministries and aligning policy within a clear dedicated biofuels policy is the best way to achieve sustainable results".

Conceptual models for modern bioenergy deployment in Africa may be thought of along an axis defined by the extent of social engagement. At one end of this axis, which we term the "low social engagement model", bioenergy feedstock production can be imagined in areas that are unoccupied and unused or nearly so, hence destined to consumers located outside of the area, that is, urban, regional, or export markets. At the other end, termed here the "high social engagement model", feedstock production can be imagined in areas that are occupied and used to a considerable extent. In this case the business model can be either a cash feedstock crop or a local feedstock for local bioenergy development.

Pursuing deployment according to the low social engagement model is certainly simpler, may be beneficial in some instances, and could be a step in a sequence of actions leading to realization of development objectives. However, the potential development benefits of the high social engagement model are likely to be substantially greater. We note that occupied areas capable of growing energy crops are far more plentiful in Africa compared to unoccupied areas, which are in most cases degraded, dry, and/or landlocked and remote. Although difficult to quantify objectively, we offer the impression that considerably more effort has gone into analysis along the lines of "How much bioenergy could be produced once food needs were provided for?" as compared to "How much more food security and other social benefits could be realized with bioenergy than without it?" In the context of African development, we find the latter question to be considerably more compelling.

Although there is widespread awareness of a stunning gap between the actual and potential output of Africa's land resources [14], and some important initial exploration has occurred (see the section Experience with bioenergy in Africa), there is much more to be done in the area of analyzing integrated scenarios featuring increased production of food and bioenergy. Table 5 presents a framework wherein the "What is?" and "What could be?" questions are examined from the point of view of geography, land management, society, environment, and synthesis, culminating in a vision for multiply beneficial land use.

The spatial scale chosen for analysis will impact the execution and outcome of efforts to develop a vision for multiply beneficial land use featuring production of both food and energy from the land. Analysis at the national or multi-national level will be informative with respect to operative federal policies and regulation, aggregated impacts, and consideration of integrated strategies and benefits at a high level. Analysis at the level of the feedstock catchment area for a single potential production facility will be particularly informative with respect to local circumstances, objectives, and benefits, and will be more relevant to potential projects. For many purposes analysis at both levels will be needed. Visions developed in different locations will likely have some features in common, but will also reflect the tremendous diversity of circumstances across the African continent.

Once a vision for multiply beneficial land use is developed, regardless of scale, the next step is to ask "What needs to be done to close the gap between *what is* and *what could be*?" The answers will in general be location-specific, and will usually involve a variety of players including communities, companies, federal and local governments, and NGOs. In many cases it will be useful to target the simultaneous realization of two goals: 1) sustainable and widely distributed social benefits and 2) commercial viability. Given this duality, there is significant scope for creative partnerships between the public, private, and NGO sectors. The impetus for such partnerships can be expected to result from further analysis of multiply beneficial land use.

Pursuing social benefits and commercial viability within the context of the high social engagement model for African bioenergy development might proceed via the following steps:

1. Develop a multiply beneficial land use vision and strategy along with appropriate and inclusive land tenure systems (see above).
2. Provide - either by government, companies, or public-private partnerships - adequate incentives in terms of facilitating access to input and output markets and mitigating investment risk for smallholder farmers to increase food and non-food crop yields. Note that several-fold increases have been observed to result from simple extension measures [109,110].
3. Investment would be gathered - by a company, co-op, or public-private partnership - to build a bioenergy conversion facility with bioenergy feedstocks planted on land made available by and fostered by an enabling environment and adequate incentives.
4. Monitor and optimize social benefits and environmental impact.
5. Share best practices within and across regions.

Table 5 Framework for development of a vision for multiply beneficial land use

Domain	What is?	What could be?
a. Society	Wealth generation/distribution, and access to capital; supply and demand of food, water, fodder, and energy; land ownership and occupation	Define needs and aspirations based on community and stakeholder input at relevant scales.
b. Geography	Precipitation, temperature, soil texture, irrigation potential.	Define potential yields of food crops, pasture, and energy crops.
c. Land management	Land cover, use, and disturbances; current crop yields	Define how management would have to optimize the potential defined in domain b based on the needs and aspirations defined in domain a.
d. Environment	Inventory C and N? Flows, ecosystem services, soil and air quality, water quality and access.	Evaluate the changes in domain c with respect to environmental objectives; propose strategies to mitigate any conflicts.
e. Synthesis		Considering all aspects, develop a vision for multiply beneficial land use responsive to social and economic priorities featuring production of food and bioenergy without compromising water and other natural resources, and catalyzed by responsible investment.

As elaborated conceptually in the section Bioenergy as a potential enabler of development, and supported by experience in both Africa and Brazil, as we have discussed, we see strong evidence that the benefits of proceeding to step 4 can be substantially greater than those achieved by stopping at step 2. That is, we think it is very likely that measures to advance food security and bioenergy development can be a substantially more effective development strategy when pursued together than either can alone.

In many examples of bioenergy deployment in developing countries, social consequences have been an afterthought rather than an integral part of project planning. Even when pursued in this mode, it appears that impacts of bioenergy on food security and economic development have in some cases been demonstrably positive, with the experience in Brazil being a prominent example. Still, some projects are more beneficial than others, there are examples of projects that have had negative impacts, and even projects with positive impacts for a majority are likely to have negative impacts on a minority that would be desirable to mitigate [25]. To the extent that development objectives become integral to project planning, the magnitude, probability, and distribution of anticipated social benefits from bioenergy rise markedly. Developing and implementing policies and institutional structures that foster such integration is challenging and very much a work in progress. Notwithstanding, the potential of bioenergy to positively impact Africa's pressing challenges requires that it be urgently considered and advanced.

Conclusions

Africa has the highest incidence of food insecurity and poverty and the highest rates of population growth, but it also has the most arable land, the lowest crop yields, and by far the most plentiful land resources relative to energy demand. In Brazil, social development, agricultural development and food security, and the development of modern bioenergy have been synergistic rather than antagonistic. Achieving such synergies in African countries will require clear vision, good governance, and adaptation of technologies, knowledge, and business models to myriad local circumstances. Strategies for integrated production of food crops, livestock, and bioenergy are potentially attractive and offer an alternative to an agricultural model featuring specialized land use. Modern bioenergy can be an agent of African transformation, with potential social benefits accruing to multiple sectors and extending well beyond energy supply per se. Potential negative impacts also cut across sectors. Thus, institutionally inclusive multi-sector legislative structures will be more effective at maximizing the social benefits of bioenergy compared to institutionally exclusive, single-sector structures. Innovative business models (such as public-private partnerships) aimed at maximizing social benefits are also promising. If done thoughtfully, there is considerable evidence that food security and economic development in Africa can be addressed more effectively with modern bioenergy than without it. This review is relevant to economic development, and in particular rural development, in African countries and poor countries elsewhere. Our findings are significant because they point to opportunities for development that are not fully realized, and because they highlight potential positive outcomes in domains wherein the impact of bioenergy has often been assumed to be negative.

Abbreviations

BTU: British thermal unit; C: carbon; CAADP: Comprehensive Africa Agriculture Development Programme (framework); CO_2: carbon dioxide; E25: blend of 25% ethanol and 75% gasoline; EJ: exajoule; EU: European Union; FAO: Food and Agriculture Organization (United Nations); gal: gallon; GDP: gross domestic product; GHG: greenhouse gas; GHI: Global Hunger Index; GW: gigawatt; ha: hectare; kW: kilowatt; kWh: kilowatt-hour; L: liter; Mha: million hectare; MDGs: Millenium Development Goals; Mtoe: megaton oil equivalent; MW: megawatt; MWe: megawatts of electricity; N: nitrogen;

NEPAD: New Partnership for Africa's Development; NGO: non-governmental organization; PIDA: Program for Infrastructure Development in Africa; TWh: terawatt-hour; v/v: volume per volume; UN: United Nations; US: United States.

Competing interests

LRL and WHVZ are involved with private sector activities in the biofuel sector. All other authors declare that they have no competing interests.

Authors' contributions

LRL was the primary author of the manuscript and had lead responsibility for developing Figures 1 and 4 and Table 5. MS was an active coauthor, played a lead role in providing information and perspectives on development in Africa, and played a co-lead role in editing the manuscript. AC had co-lead responsibility for developing Table 3 and provided information and perspectives on bioenergy deployment in Africa and its relationship to development. LABC and CHBC provided information and perspectives on bioenergy deployment in Brazil and its relationship to development, and also played a lead role in discussions leading to initiation of this manuscript. ME and IM provided information and perspectives on bioenergy deployment in Africa and its relationship to development. ML had lead responsibility for developing Tables 1 and 2. MAFDM had primary responsibility for the text discussing the social impacts of bioenergy in the section on the Brazilian experience. LAHN had lead responsibility for developing Table 4 and Figure 3, as well as much of the text in the section on the Brazilian experience. GMW contributed a discussion of water-efficient crops, had co-lead responsibility for developing Table 3, and provided information and perspectives on bioenergy deployment in Africa. JW contributed a discussion of climate change, provided information and perspectives on bioenergy deployment in Africa including potential for enhancing the resilience of food supplies, and played a co-lead role in editing the manuscript. WHVZ provided information and perspectives on bioenergy deployment in Africa and played a lead role in discussions leading to initiation of this manuscript. All coauthors reviewed the manuscript and responses to reviewer comments. All authors read and approved the final manuscript.

Acknowledgement

Partial support was provided by grant 2012/00282-3 from the Sao Paulo Research Foundation, the NEPAD-FAO project under the EU-FAO Global Governance for Hunger Reduction Programme, and the Link Foundation (support for LRL).

Author details

[1]Thayer School of Engineering, Dartmouth College, Hanover, NH, USA. [2]New Partnership for Africa's Development (NEPAD), Johannesburg, South Africa. [3]Department of Process Engineering, University of Stellenbosch, Stellenbosch, South Africa. [4]Faculty of Agricultural Engineering, University of Campinas, Campinas, Brazil. [5]São Paulo Research Foundation São Paulo, São Paulo, Brazil. [6]Physics Institute, University of Campinas, Campinas, Brazil. [7]Department of Applied Economics, University of São Paulo, ESALQ, Piracicaba, Brazil. [8]Department of Microbiology, University of Stellenbosch, Stellenbosch, South Africa. [9]Water Institute, University of Stellenbosch, Stellenbosch, South Africa. [10]Centre for Environmental Policy, Imperial College London, London, UK.

References

1. United Nations: The millennium development goals report 2012. New York; 2012.
2. International Energy Agency. World energy outlook 2011. Paris: IEA Publishing; 2011.
3. Haub C. Fact sheet: world population trends. Washington, DC: Population Reference Bureau; 2012.
4. New Partnership for Africa's Development (NEPAD): NEPAD programmes. http://www.nepad.org/nepad-programmes.
5. Bank W. World Development Report 2008: Agriculture for development. Washington, DC; 2007.
6. United States Census Bureau: International Programs. http://www.census.gov/population/international/.
7. United States Energy Information Administration (EIA): International Energy Statistics. http://www.eia.gov/cfapps/ipdbproject/IEDIndex3.cfm?tid=44&pid=44&aid=2.
8. Graphic Maps: World atlas. http://www.worldatlas.com/.
9. Practical Action Consulting: in selected African countries. Policy Innovation Systems for Clean Energy Security. Research for Development (R4D), Edinburgh; 2011.
10. UNDP: United Nations Development Programme: Sustainable energy. http://www.undp.org/content/undp/en/home/ourwork/environmentandenergy/focus_areas/sustainable-energy.html.
11. Lynd LR, Woods J. A new hope for Africa. Nature. 2011;474:S20–1.
12. Christaensen L, Demery L. The evolving role of agriculture in poverty reduction: an empirical perspective. J Dev Econ. 2011;96:239–54.
13. Ligon E, Sadoulet E: Estimating the effects of aggregate agricultural growth on the distribution of expenditures. Background paper for the World Development Report 2008. 2007.
14. United Nations Development Project (UNDP). African human development report 2012: towards a food secure future. New York, USA: United Nations Publication; 2012.
15. Practical Action Consulting. Small-scale bioenergy initiatives: brief description and preliminary lessons on livelihood impacts from case studies in Asia, Latin America and Africa. Rome: PISCES and FAO; 2009.
16. Accenture, 2013. Food for thought: unlocking the economic potential of sub-Saharan Africa by addressing food security. http://www.accenture.com/za-en/Pages/insight-unlocking-sub-saharan-africa-food-security.aspx Accessed 18 March 2014.
17. Food and Agriculture Organisation (FAO). Global food losses and food waste: extent, causes and prevention. 2011. http://www.fao.org/docrep/014/mb060e/mb060e00.pdf Accessed 21 March 2014.
18. Thurow R, Kilman S. Enough: why the world's poor starve in an age of plenty. New York, US: Public Affairs; 2009.
19. Achterbosch T, Meijerink G, Slingerland M, Smeets E. 2013. Combining bioenergy production and food security. NL Agency, Ministry of Economic Affairs. http://edepot.wur.nl/260061.
20. Diaz-Chavez R. Mapping food and bioenergy in Africa: a report prepared for FARA. Ghana: Forum for Agricultural Research in Africa; 2010.
21. Hamelinck, C. Biofuels and food: risks and opportunities. http://www.ecofys.com/files/files/ecofys-2013-biofuels-and-food-security.pdf.
22. Langeveld JWA, Dixon J, van Keulen H, Quist-Wessel PMF. Analyzing the effect of biofuel expansion on land use in major producing countries: evidence of increased multiple cropping. Biofuels, Bioprod Bioref. 2014;8(1):49–58.
23. Lotze C, von Lampe HM, Kyle P, Fujimori S, Havlik P, van Meijl H, et al. Impacts of increased bioenergy demand on global food markets: an AgMIP economic model intercomparison. Agric Econ. 2014;45:1–14.
24. Rosillo-Calle F, Johnson F. Food versus fuel. London: ZED Books; 2010.
25. Locke A, Henley G. Biofuels and food security: what does the evidence say? Overseas Development Institute. 2014. http://www.odi.org.uk/publications/7485-biofuels-food-security-food-prices-evidence.
26. Osseweijer P, Watson HK, Johnson FX, Batistella M, Cortez LAB, Lynd LR, et al. Bioenergy and food security. In: Bioenergy: bridging the gaps. Paris. France: SCOPE; In press.
27. Dale BE, Anderson JE, Brown RC, Csonka S, Dale VH, Herwick G, et al. Take a closer look: biofuels can support environmental, economic and social goals. Environ Sci Technol. 2014;48:7200–3.
28. Smith P, Bustamante M, Ahammad H, Clark H, Dong H, Elsiddig EA, et al. IPCC. Bioenergy: climate effects, mitigation options, potential and sustainability implications, Appendix to Chapter 11 (AFOLU) Final Draft. IPCC WGIII AR5; Geneva, 2014.
29. Porter, JR., et al. Food security and food production systems. In WGII AR5. IPCC. Chap 7; 2014
30. Ray DK, Ramankutty N, Mueller ND, West PC, Foley JA. Recent patterns of crop yield growth and stagnation. Nature Commun. 2012; 3
31. Long SP, Karp A, Buckeridge MS, Davis SC, Moore PH, Moose SP, et al. Crop feedstocks for biofuels and bioenergy, vol. 2. Paris, France: Bioenergy: bridging the gaps. SCOPE; In press.
32. Woods J, Lynd LR, Laser M, Batistella M, de Castro VD, Kline K, et al. Land and bioenergy. In: Bioenergy: bridging the gaps, vol. 9. Paris, France: SCOPE; In press.
33. FAO. Land and environmental degradation and desertification in Africa. FAO Corporate Document Repository. http://www.fao.org/docrep/x5318e/x5318e00.HTM.

34. Anderson-Teixeira KJ, Masters MD, Black CK, Zeri M, Hussain MZ, Bernacchi CJ, et al. Altered below ground carbon cycling following land use change to perennial energy crops. Ecosystems. 2013; doi:10.1007/s10021-012-9628-x.
35. Jordan N, Boody G, Broussard W, Glover JD, Keeney D, McCown BH, et al. Sustainable development of the agricultural bio-economy. Science. 2007;316:1570–1.
36. Lal R. Soil carbon sequestration impacts on global climate change and food security. Science. 2004;304(5677):1623–7.
37. Biofuels for transportation: global potential and implications for sustainable agriculture and energy in the 21st century. World Watch Institute, Washington, DC; 2006.
38. International Energy Agency: Status of advanced biofuels demonstration facilities in 2012. A report to Bioenergy Task 39. http://demoplants.bioenergy2020.eu/files/Demoplants_Report_Final.pdf.
39. Renewable Fuels Association: Statistics. http://www.ethanolrfa.org/pages/statistics.
40. Hammerschlag R. Ethanol's energy return on investment: a survey of the literature 1990 - present. Environ Sci Technol. 2006;40:1744–50.
41. National Agricultural Statistics Service: Statistics by subject. http://www.nass.usda.gov/Statistics_by_Subject/index.php.
42. Mueller S. National dry mill corn ethanol survey. Biotechnol Lett. 2008;2010(32):1261–4.
43. UNICA (Sugar Cane Industry Association): UNICA data. http://www.unicadata.com.br/.
44. Goldemberg J. The Brazilian biofuels industry. Biotechnology for Biofuels. 2008;1:1–7.
45. Souza SP, Pacca S, Avila MT, Borges JLB. Greenhouse gas emissions and energy balance of palm oil biofuel. Renew Energy. 2010;35:2552–61.
46. Smith EG, Janzen H, Newlands NK. Energy balances of biodiesel production from soybean and canola in Canada. Can J Plant Sci. 2007;87:793–801.
47. Panichelli L, Dauriat A, Gnansounou E. Life cycle assessment of soybean-based biodiesel in Argentina for export. Int J Life Cycle Assess. 2009;14:144–59.
48. Malaysian Palm Oil Board: Statistics. http://www.palmoilworld.org/PDFs/Overview-2011.pdf.
49. International Energy Agency: World energy outlook 2012, renewable energy outlook. http://www.worldenergyoutlook.org/media/weowebsite/2012/WEO2012_Renewables.pdf.
50. United States Department of Agriculture: Planting and managing giant Miscanthus as a biomass energy crop. http://www.plant-materials.nrcs.usda.gov/pubs/flpmstn10548.pdf.
51. International Renewable Energy Agency: Renewable energy technologies: cost analysis series. Volume 1: Power Sector Issue 5/5; 2012. https://www.irena.org/DocumentDownloads/Publications/RE_Technologies_Cost_Analysis-WIND_POWER.pdf.
52. World Coal Institute. Coal: liquid fuels. UK: World Coal Association; 2006.
53. Wright MW, Satrio JA, Brown RC, Daugaard DE, Hsu DD. Techno-economic analysis of biomass fast pyrolysis to transportation fuels. Golden, CO: National Renewable Energy Laboratory; 2010. http://www.nrel.gov/docs/fy11osti/46586.pdf.
54. Charles C, Gerasimchuk I, Bridle R, Moerenhout T, Asmelash E, Laan T. Biofuels - at what cost? A review of costs and benefits of EU biofuel policies. Winnipeg, Manitoba, Canada: International Institute for Sustainable Development; 2013.
55. Dias M, Junqueira TL, Cavalett O, Cunha MP, Jesus CDF, Rossell CEV, et al. Integrated versus stand-alone second generation ethanol production from sugarcane bagasse and trash. Bioresour Technol. 2012;103:152–61.
56. Gopalakrishnan G, Negri M, Snyder S. Redesigning agricultural landscapes for sustainability using bioenergy crops: quantifying the tradeoffs between agriculture, energy and the environment. Aspects of Applied Biology, Biomass and Energy Crops IV. 2011;112:139–46.
57. Quinkenstein A, Wőlleke J, Böhm C, Grünewald H, Freese D, Schneider B, et al. Ecological benefits of the alley cropping agroforestry system in sensitive regions of Europe. Environ Sci Pol. 2009;12:1112–21.
58. Skenhall SA, Berndes G, Woods J. Integration of bioenergy systems into UK agriculture - new options for management of nitrogen flows. Biomass Bioenergy. 2013;54:219–26.
59. Somerville C, Youngs H, Taylor C, Davis SC, Long SP. Feedstocks for lignocellulosic biofuels. Science. 2010;329:790–2.
60. Yan X, Tan DKY, Oliver R, Inderwild OR, Smith JAC, King DA. Life cycle energy and greenhouse gas analysis for agave-derived bioethanol. Energy Environ Sci. 2011;4(9):3110–21.
61. Lujan R, Lledıas F, Martınez L, Barreto R, Cassab G, Nieto-Sotelo J. Small heat-shock proteins and leaf cooling capacity account for the unusual heat tolerance of the central spike leaves in Agave tequilana var. Weber. Plant Cell Environ. 2009;32:1791–803.
62. Graham EA, Nobel PS. Long-term effects of a doubled atmospheric CO_2 concentration on the CAM species Agave deserti. J Exp Bot. 1996;47:61–9.
63. Lynd L, Aziz R, Cruz C, Chimphango A, Cortez L, Faaij A, et al. A global conversation about energy from biomass: the continental conventions of the global sustainable bioenergy project. Interface Focus. 2011;1:271–9.
64. Cortez LAB: Sugarcane ethanol: R&D for productivity and sustainability. Blucher, Editora Edgard Blucher Ltda, Sao Paulo, 2010.
65. Nogueira LHA. Sugarcane-based ethanol: energy for sustainable development. Rio de Janeiro: BNDES; 2008.
66. IRENA, DFBZ, 2013. Biomass potential in Africa. Report Authors: K. Stecher, A. Brosowski, D. Thrän. The International Renewable Energy Agency (IRENA), Abu Dhabi. http://www.irena.org/DocumentDownloads/Publications/IRENA-DBFZ_Biomass%20Potential%20in%20Africa.pdf.
67. World Bank. Wood-based biomass, energy development, for sub-Saharan Africa - issues and approaches. 2011. http://www-wds.worldbank.org/external/default/WDSContentServer/WDSP/IB/2012/12/27/000386194_20121227062926/Rendered/PDF/NonAsciiFileName0.pdf.
68. World Bank. Household cookstoves, environment, health, and climate change: a new look at an old problem. Washington, D.C. 20433 USA: The World Bank Group; 2013. http://climatechange.worldbank.org/sites/default/files/documents/Household%20Cookstoves-web.pdf.
69. Kambewa P, Chiwaula L. Biomass energy use in Malawi. A background paper prepared for the International Institute for Environment and Development (IIED) for an international ESPA workshop on biomass energy. Zomba, Malawi. Parliament House Hotel, Edinburgh, Chancellor College; 2010.
70. Sibale B, Kafakoma R, Shaba A, Macqueen D. Trees on-farm: removing the obstacles to enterprise. A review of current climate-smart tree-based experiences in Malawi. London: International Institute for Environment and Development; 2013.
71. Lakew H. Ethiopian energy sector review for up to 2008. Addis Ababa: Forum for Environment, No 1, Forum for environment; 2010. p. 79–104
72. Mensah S. Extraction and use of jatropha oil by a village women's group to power shea butter processing equipment. In: Karlsson G, Banda K, editors. Biofuels for rural development and empowerment of women. Leusden, The Netherlands: ENERGIA & IUCN; 2009. p. 13–7.
73. Blanchard R, Richardson DM, O'Farrell PJ, Maltitz GPV: Biofuels and biodiversity in South Africa. S Afr J Sci. 2011:107:Art.#186 (1–8).
74. EEPAfrica.org. Policy brief: Analyzing briquette markets in Tanzania, Kenya and Uganda: a comparative study. Southern and East Africa: Energy and Environment Partnership; 2013.
75. Pswarayi-Riddihough I. Ethanol as a household fuel in Madagascar. Economic assessment, and review of African lessons for scaling-up. World Bank, Washington, DC: Health Benefits; 2011.
76. Janssen R, Rutz D, Janssen R, Rutz D. Bioenergy for sustainable development in Africa. Dordrecht Heidelberg London New York: Springer; 2012.
77. Schut M, Slingerland M, Locke A. Biofuel developments in Mozambique. Update and analysis of policy, potential and reality. Energy Policy. 2010;38:5151–65.
78. Department of Energy. South African Government Gazette No. 35623, 23 August 2012. Pretoria: Department of Energy in Republic of South Africa; 2012.
79. Sapp M: IDC to invest $606 million for biofuel production. Biofuels Digest; 2012—this reference is an internet article, not a journal article, so volume and page number are not applicable. Suggest citing http://www.biofuelsdigest.com/bdigest/2012/08/27/idc-to-invest-606-million-for-biofuel-production/ as URL with date of access.
80. Biofuels Digest: South Africa's FE&AP plans $373 M investment in sweet sorghum ethanol. 2012—this reference is an internet article, not a journal article, so publisher location is not applicable. Suggest citing http://www.biofuelsdigest.com/bdigest/2012/01/26/south-africasfeap-plans-373m-investment-in-sweet-sorghum-ethanol/ as URL with date of access and Biofuels Digest as the publisher.
81. Odeku K, Meyer E, Miraka O, Jetsaolo J. Implementing a renewable energy feed-in tariff in South Africa: the beginning of a new dawn. Sustainable Development Law & Policy. 2011;2:45–9.
82. Green Fuel ethanol, power venture forges ahead in Zimbabwe. Biofuels Digest. http://www.biofuelsdigest.com/bdigest/2012/01/20/green-fuelethanol-power-venture-forges-ahead-in-zimbabwe/ January 20, 2012.

83. Gandure S. Women's roles in the national jatropha-growing project. In: Karlsson G, Banda K, editors. Biofuels for sustainable rural development and empowerment of women. The Netherlands: ENERGIA & IUCN Leusden; 2009. p. 44–9.
84. UNF. Sustainable bioenergy development in UEMOA member countries. New York, USA: United Nations Foundation; 2008. http://www.unfoundation.org/. http://www.globalproblems-globalsolutions-files.org/gpgs_files/pdf/UNF_Bioenergy/UNF_Bioenergy_full_report.pdf [Accessed 18 March 2014.
85. Smeets EMW, Faaij APC, Lewandowski IM, Turkenburg WC. A bottom-up assessment and review of global bio-energy potentials to 2050. Prog Energy Combust Sci. 2007;33:56–106.
86. IBGE. Anuário Estatístico do Brasil 2011. Brasília: Instituto Brasileiro de Geografia e Estatística; 2012.
87. Rocha S. Pobreza no Brasil: a evolução de longo prazo (1970–2011). Rio de Janeiro: XXV Fórum Nacional, Instituto Nacional de Altos Estudos; 2013.
88. IFPRI. A global hunger index: measurement concept, ranking of countries, and trends. Washington, DC: International Food Policy Research Institute; 2012.
89. USDA. International Food Security Assessment 2012–2022. Economic Research Service, United States Department of Agriculture: Washington; 2012.
90. FAO: Food and Agriculture Organization of the United Nations. FAOSTAT. http://faostat3.fao.org/home/index.html#HOME.
91. EPE. National energy balance 2012, Empresa de Pesquisa Energética. Brasília: Ministério de Minas e Energia; 2012.
92. ANFAVEA. Statistical yearbook of Brazilian automotive industry. São Paulo: National Association of Automotive Industry; 2012.
93. Angelo C. Growth of ethanol fuel stalls in Brazil. Nature. 2012;491:646–7.
94. Macedo IC. Sugar cane's energy: twelve studies on Brazilian sugar cane agribusiness and its sustainability. São Paulo: UNICA - Sao Paulo Sugar Cane Agroindustry Union; 2005.
95. Seabra JEA, Macedo IC. Comparative analysis for power generation and ethanol production from sugarcane residual biomass in Brazil. Energy Policy. 2011;39:421–8.
96. MME. National Energy Balance. Brasília: Ministry of Mines and Energy; 2012.
97. World Bank. Energy in Africa: Overview. 2012. http://web.worldbank.org/WBSITE/EXTERNAL/COUNTRIES/AFRICAEXT/EXTAFRREGTOPENERGY/0,,menuPK:717332~pagePK:51065911~piPK:64171006~theSitePK:717306,00.html.
98. Lapola DM, Martinelli LA, Peres CA, Ometto JPHB, Ferreira ME, Nobre CA, et al. Pervasive transition of the Brazilian land-use system. Nature climate change, vol. 4, January 2014, doi:10.1038/nclimate2056.
99. CONAB. Companhia Nacional de Abastecimento. Séries Históricas. 2012. Available in: http://www.conab.gov.br/conteudos.php?a=1252&t=2 Accessed 20 June 2013.
100. Moraes MAFD, Zilberman D. Production of ethanol from sugarcane in Brazil. From state intervention to a free market. Cham: Springer International Publishing; 2014.
101. Assato MM, de Moraes MAFD. Impactos sócio-econômicos da expansão do setor bioenergético no Estado do Mato Grosso do Sul: os casos dos municípios de Nova Alvorada do Sul e Rio Brilhante. São Paulo: Corecon; 2011.
102. Van Kuelen H, Schiere H. Crop-livestock systems: old wine in new bottles? Brisbane, Australia: Proceedings of the 4th International Crop Science Congress; 2004.
103. Herrero M, Thornton PK, Notenbaert AM, Wood S, Msangi S, Freeman HA, et al. Smart investments in sustainable food production: revisiting mixed crop-livestock systems. Science. 2010;327:822–4.
104. Baffes J, Dennis A. Long-term drivers of food prices. Policy Research Working Paper 6455. Washington, DC: World Bank; 2013. http://papers.ssrn.com/sol3/papers.cfm?abstract_id=2268316##.
105. HLPE. Biofuels and food security. A report by the High Level Panel of Experts on Food Security and Nutrition of the Committee on World Food Security. Rome: Food and Agriculture Organization of the United Nations; 2013.
106. Mitchell D. A note on rising food prices. Washington, DC: World Bank; 2008.
107. Scurlock JMO, Rosenschein A, Hall DO. Fuelling the future: power alcohol in Zimbabwe. Nairobi, Kenya: ACTS Press; 1991.
108. Jayne TS. Managing food price instability in East and Southern Africa. Global Food Security. 2012;1:143–9. http://dx.doi.org/10.1016/j.gfs.2012.10.002.
109. Thurow R. The last hunger season: a year in an African farm community on the brink of change. New York, US: Public Affairs; 2012.
110. Wilson, K. One billion hungry. Can we feed the world? Facts and Figures. Agriculture for Impact. http://ag4impact.org/wp-content/uploads/2012/10/Facts-and-Figures-One-Billion-Hungry_3Nov2014.pdf. Accessed 18 March 2014.

2

Biofilm formation by designed co-cultures of *Caldicellulosiruptor* species as a means to improve hydrogen productivity

Sudhanshu S Pawar[1*], Thitiwut Vongkumpeang[1], Carl Grey[2] and Ed WJ van Niel[1]

Abstract

Background: *Caldicellulosiruptor* species have gained a reputation as being among the best microorganisms to produce hydrogen (H_2) due to possession of a combination of appropriate features. However, due to their low volumetric H_2 productivities (Q_{H2}), *Caldicellulosiruptor* species cannot be considered for any viable biohydrogen production process yet. In this study, we evaluate biofilm forming potential of pure and co-cultures of *Caldicellulosiruptor saccharolyticus* and *Caldicellulosiruptor owensensis* in continuously stirred tank reactors (CSTR) and up-flow anaerobic (UA) reactors. We also evaluate biofilms as a means to retain biomass in the reactor and its influence on Q_{H2}. Moreover, we explore the factors influencing the formation of biofilm.

Results: Co-cultures of *C. saccharolyticus* and *C. owensensis* form substantially more biofilm than formed by *C. owensensis* alone. Biofilms improved substrate conversion in both of the reactor systems, but improved the Q_{H2} only in the UA reactor. When grown in the presence of each other's culture supernatant, both *C. saccharolyticus* and *C. owensensis* were positively influenced on their individual growth and H_2 production. Unlike the CSTR, UA reactors allowed retention of *C. saccharolyticus* and *C. owensensis* when subjected to very high substrate loading rates. In the UA reactor, maximum Q_{H2} (approximately 20 mmol $\cdot L^{-1} \cdot h^{-1}$) was obtained only with granular sludge as the carrier material. In the CSTR, stirring negatively affected biofilm formation. Whereas, a clear correlation was observed between elevated (>40 μM) intracellular levels of the secondary messenger bis-(3′-5′)-cyclic dimeric guanosine monophosphate (c-di-GMP) and biofilm formation.

Conclusions: In co-cultures *C. saccharolyticus* fortified the trade of biofilm formation by *C. owensensis*, which was mediated by elevated levels of c-di-GMP in *C. owensensis*. These biofilms were effective in retaining biomass of both species in the reactor and improving Q_{H2} in a UA reactor using granular sludge as the carrier material. This concept forms a basis for further optimizing the Q_{H2} at laboratory scale and beyond.

Keywords: Caldicellulosiruptor saccharolyticus, *Caldicellulosiruptor owensensis*, Biohydrogen, Co-culture, c-di-GMP, UA reactor, CSTR, Volumetric H_2 productivity

Introduction

Amid the findings of vast reserves of shale oil and convenient negligence towards its (alleged) side-effects on the environment, the utopian world of 'hydrogen economy' still looks distant. One of the key bottlenecks is the unavailability of economical and eco-friendly ways of hydrogen production. Credible research is underway for developing sustainable processes producing hydrogen through electrolysis of water using wind and solar power [1]. However, more alternatives are needed to complement these technologies. In this respect, fermentative hydrogen (biohydrogen) production at a higher temperature, thermophilic biohydrogen production, using renewable biomass can be a viable option.

Caldicellulosiruptor species belong to a group of extremely thermophilic obligate anaerobes, which possess a natural ability to produce hydrogen from a wide range of mono-, di-, and oligo-saccharides and raw materials [2-6]. In addition to this, various other beneficial metabolic features enable the genus *Caldicellulosiruptor* as

* Correspondence: sudhanshu.pawar@tmb.lth.se
[1]Division of Applied Microbiology, Lund University, Getingevägen 60, PO Box 124, SE-221 00 Lund, Sweden
Full list of author information is available at the end of the article

one of the best, yet not ideal, groups of bacteria with the natural ability to produce H_2 [7]. Within this genus, *Caldicellulosiruptor saccharolyticus* and *Caldicellulosiruptor owensensis* are two of the best-studied species, both known to produce H_2 near the theoretical maximum of 4 mol · mol^{-1} [8,9].

However, increasing Q_{H2} (volumetric H_2 productivity; mmol · L^{-1} · h^{-1}) is one of the major challenges in developing a cost-effective biohydrogen process with *Caldicellulosiruptor* species. The Q_{H2} depends on various factors such as cell density, extent of substrate conversion, and reactor configuration. The cell density can be increased by retaining more cells through different approaches, such as immobilization, cell entrapment, or cell retention. However immobilized or trapped cells can face mass transfer issues [10]. In contrast, biofilms, are well-organized structures, and are inherent to cell retention [11,12]. Moreover, biofilms generally follow 'feed-and-bleed' cycles allowing cell growth, which can be significant for growth-dependent product formation [11]. Among *Caldicellulosiruptor* species, *C. owensensis* has been previously reported to form biofilms [13] mainly by flocculating at the bottom of the reactor. However, no further information could be found regarding the factor(s) leading to biofilm formation by *C. owensensis* [13]. Bis-(3′-5′)-cyclic dimeric guanosine monophosphate (c-di-GMP) has been recognized as a ubiquitous secondary messenger in bacteria with multilayer control, i.e. at transcriptional, translational, and posttranslational level [14,15]. The c-di-GMP is synthesized using two molecules of guanosine-5′-triphosphate (GTP) by the enzyme diguanylate cyclase (DGC) and is hydrolyzed by the enzyme phosphodiesterase (PDE) [15]. Numerous studies have proven that high intracellular levels of c-di-GMP promote expression of extracellular-matrix related components needed for biofilm formation [14-16].

So far, most of the research pertaining to biohydrogen has been performed to investigate the physiological properties of H_2-producing microbes. These studies have mainly been performed in continuously stirred tank reactors (CSTR). However, CSTR systems do not allow cell retention. Hence, it is of paramount importance to evaluate alternative reactor types that can help retain the biomass. Several different reactor types, such as packed bed reactor [13], membrane bioreactor [17], anaerobic sequencing blanket reactor [18], trickle bed reactor [19], and up-flow anaerobic (UA) reactor [20] aiding cell retention have been reported to produce H_2 at higher rates. In fact, UA reactors are widely exploited for studies pertaining to biogas production. Their medium recirculation loop aids in achieving higher substrate conversion and also allows cells to adhere to the biofilms flocculated at the bottom of reactor. On the other hand, in case of the CSTR, using carriers has been reported to increase Q_{H2} by several folds [21].

In this study we aimed to evaluate the biofilm forming potential of *C. saccharolyticus* and *C. owensensis* in pure culture, and also evaluate whether *C. owensensis* through biofilm formation aids *C. saccharolyticus* when cultivated in co-cultures. Furthermore, we report the intracellular levels of c-di-GMP in both the organisms and its relationship with biofilm formation. We also evaluate the potential of UA reactors in improving Q_{H2} compared to CSTRs and whether carrier materials affect retaining the biomass and improving Q_{H2}.

Material and methods

Microorganism and its maintenance

C. saccharolyticus DSM 8903 and *C. owensensis* DSM 13100 were purchased from the Deutsche Sammlung von Mikroorganismen (DSM) und Zellkulturen (Braunschweig, Germany). Routine subcultures and maintenance were conducted in 250 mL serum bottles containing 50 mL of a modified DSM 640 medium [22] unless stated otherwise. Anoxic solutions of glucose, cysteine · HCl, and magnesium sulphate were autoclaved (1.5 atm, at 120°C for 20 minutes) separately and added to the sterile medium at the required concentration. A 1,000× concentrated vitamins solution was prepared as described previously [8] and used in the growth medium at 1× concentration as a replacement for yeast extract. A 1,000× concentrated trace element solution was prepared as described previously [23].

Fermentation setup and culture medium

To study the effect of any excretion of *C. saccharolyticus* on the growth of *C. owensensis* and vice versa, batch cultures of each were performed in biological duplicates and previously collected cell-free culture supernatant of one organism was added into the batch medium of another prior to inoculation. The volume of supernatant added in each respective case was equivalent to that of containing 1 g cell dry weight (CDW) of the respective organism.

To study the effect of different reactor systems on biofilm formation and cell retention, *C. saccharolyticus* and *C. owensensis* were cultivated independently (pure culture) or together (co-culture) in two different reactor systems: continuously stirred tank reactor (CSTR) and up-flow anaerobic (UA) reactor (Table 1). To allow for biofilm formation and/or cell retention, co-cultures of *C. saccharolyticus* and *C. owensensis* were performed in both the reactor systems with K1-carriers (Catalogue # K1, AnoxKaldnes AB, Lund, Sweden). K1-carrier is made of polyethylene in a tube-like structure (length, 7.2 mm; diameter, 9.1 mm with an internal cross and 18 external fins. In the case of the CSTR, co-cultures were performed with or without stirring, however, the pure cultures were only performed without stirring but with the K1-carriers (Table 1). In the case of the UA reactor, the co-cultures were performed with and without using the

Table 1 Various cultivation conditions applied during this study

Name	Cultivation condition
Case A	Co-culture in CSTR* without stirring with carriers
Case B	Co-culture in CSTR with stirring without carriers
Case C	*C. saccharolyticus* without stirring with carriers
Case D	*C. owensensis* without stirring with carriers
Case E	Co-culture in UA** reactor with sludge
Case F	*C. saccharolyticus* in UA reactor with sludge
Case G	*C. owensensis* in UA reactor with sludge
Case H	Co-culture in UA reactor without sludge with carriers
Case I	Co-culture in UA reactor without sludge without carriers

*CSTR, continuously stirred tank reactor; **UA, up-flow anaerobic.

granular sludge as the packed bed, however, the pure cultures were performed only with granular sludge (Table 1).

All experiments were conducted in a jacketed, 3 L (CSTR) or 1 L (UA), equipped with an ADI 1025 Bio-Console and an ADI 1010 Bio-Controller (Applikon, Schiedam, The Netherlands) at a working volume of 1 L (CSTR) or 0.85 L (UA), either in batch or continuous mode. The water height in the UA reactor was maintained at approximately 20 cm. The pH was maintained at 6.5 ± 0.1 at 70°C by automatic titration with 4 M NaOH. The temperature was thermostatically kept at 70 ± 1°C. In case of the CSTR, a condenser with 5°C cooling water was fitted to the bioreactor's headplate and the stirring was kept at 250 rpm unless specified otherwise. The UA reactor's top was fitted with a rubber cork inserted with a collection tube releasing the flue gas out of the reactor. During batch cultivations, culture samples were collected at different time intervals for monitoring growth, and the culture supernatant was collected for analysis of glucose, acetic acid, lactic acid, propionic acid, and ethanol. Gas samples were collected from the headspace to analyze levels of H_2 and CO_2. During continuous cultures, samples for c-di-GMP were collected at steady state. Batch cultures were performed in two independent biological replicates, whereas, for continuous cultures steady states were obtained in technical duplicates.

All the reactors were autoclaved with a base medium (BM) containing per litre of demineralized water: KH_2PO_4 0.75 g; $K_2HPO_4 \cdot 2H_2O$ 1.5 g; NH_4Cl 0.9 g; yeast extract 1.0 g; resazurin 1 mg; 1000 × modified SL-10 1 mL. Solutions of glucose, 10 $g \cdot L^{-1}$ for CSTRs (Case A, B, C, and D) and 20 $g \cdot L^{-1}$ for UA reactors (Case E. F, G, H, and I), cysteine · HCl, 0.25 $g \cdot L^{-1}$, and $MgSO_4 \cdot 6H_2O$, 0.5 $g \cdot L^{-1}$ were autoclaved and added separately prior to inoculation. UA reactors containing 250 g of granular sludge as a carrier material (Case E, F, and G) were autoclaved twice to eliminate the risk of methanogenic or hydrogenogenic contaminants. Autoclaving conditions did not affect the shape or the integrity of the granules. Gas samples were regularly taken from the headspace of UA reactors to detect any traces of methane. Carriers were autoclaved separately and were added prior to inoculation. The granular sludge was obtained from methanogenic reactors treating municipal waste water under mesophilic conditions. The granules of anaerobic sludge were circular in shape, measuring about 2 mm in diameter. Inocula for each organism were prepared through a succession of at least three sub-cultivations prior to inoculation. In the case of co-cultures, inocula of each organism were grown separately.

For continuous cultivations, the bioreactor started to be fed with fresh medium at the end of the logarithmic growth phase of the batch culture. Glucose was used as a primary substrate in all continuous experiments at an initial concentration of 10 $g \cdot L^{-1}$. Steady states were assessed after at least five volume changes based on the criteria of constant H_2 and CO_2 production rates and constant biomass concentration.

Analytical methods

Headspace samples were analyzed for CO_2, H_2, and CH_4 by gas chromatography, using a dual channel Micro-GC (CP-4900; Varian, Micro gas chromatography, Middelburg, The Netherlands), as previously described [8]. The results were analyzed with a Galaxie Chromatography Workstation (version 1.9.3.2, Middelburg, The Netherlands). The optical density of the culture was measured at 620 nm (OD_{620}) using a U-1100 spectrophotometer (Hitachi, Tokyo, Japan). CDW was determined by filtration as previously described [24]. Glucose, acetate, lactate, propionate, and ethanol were analyzed by HPLC (Waters, Milford, Massachusetts, United States) on an Aminex HPX-87H ion exchange column (Bio-Rad, Hercules, United States) at 45°C, with 5 mM H_2SO_4 (0.6 $ml \cdot min^{-1}$) as the mobile phase. The column was equipped with a refractive index detector (RID-6A; Shimadzu, Kyoto, Japan).

Scanning electron microscopy of biofilm samples

Biofilm samples were scraped from the pH probe and/or carrier at the end of the cultivation (Case A) and were immediately stored overnight in glutaraldehyde solution (2 to 3%) to allow fixation. The samples were then stored with sodium cacodylate buffer (about pH 7) until further use. A few hours prior to scanning electron microscopy (SEM) imaging, samples were dried by first washing with ethanol solutions from 50 to 100% in series and then subjecting to 'critical point drying' using liquid CO_2. Subsequently, the dry biofilm samples were then glued on a stub and were sputter coated with gold/palladium alloy and finally viewed under the SEM (Hitachi SU3500, Hitachi, Japan).

Determination of intracellular levels of c-di-GMP

During batch cultivations, 5 mL of culture samples were collected in quadruplets when cultures reached stationary phase. Similarly, in continuous cultures, 5 mL of culture samples were collected in quadruplets at steady state obtained under various conditions. All samples were collected on ice and were centrifuged immediately at 4000 rpm in a swinging bucket rotor at 4°C and were subsequently processed for the extraction of c-di-GMP. The extraction was performed as described by [25] with the exception that in the final step, samples were dried by incubating overnight at approximately 50°C.

The quantification of c-di-GMP was performed as previously described [25] but with the following modifications. The LC-separation were performed using isocratic conditions, 3.5% MeOH (A) and 96.5% 10 mM ammonium acetate in 0.1% acetic acid (B) at 400 μL/min for 6.5 min. The internal standard, xanthosine 5'- monophosphate (XMP), eluted after 3.1 min and c-di-GMP at 4.7 min. A wash program was run every 16 samples to ensure a robust analysis, in which 90% A was applied for 15 min before equilibrating the column for 20 min using the isocratic conditions. Standards, seven levels, ranging from 10 nM to 10 μM were included in the beginning and end of the sequence. The detection was performed using an Orbitrap-Velos Pro mass spectrometer (Thermo Scientific, Waltham, MA, USA) using the electrospray ionization (ESI) in positive mode. Two scan events were applied: ion trap (ITMS) for quantification, including selected reaction monitoring (SRM) on XMP (m/z 347/153 between 0 and 4 min) and c-di-GMP (m/z 691/540 between 4 and 6.5 min) and orbitrap fullscan (FTMS) for accurate mass identification, using a resolution of 30000.

Bioinformatics analysis for genes related to bis-(3′-5′)-cyclic dimeric guanosine monophosphate

Genomes of *C. saccharolyticus* and *C. owensensis* were analyzed to locate genes coding for DGC and PDE. All the information regarding genome sequences and corresponding annotations were retrieved from the Integrated Microbial Genomes (IMG, Berkeley, United States).

Population dynamics in the biofilm samples of co-cultures using qPCR

During all the co-culture experiments, 2 mL of culture samples were collected and immediately centrifuged and the cell pellets were stored at −20°C until further use. Similarly, sufficient amounts of biofilm samples were collected from the pH probe and from the reactor wall after the cultivations were ceased. The genomic DNA from the samples were extracted using Invitrogen's (Carlsbad, United States) EasyDNA genomic DNA extraction kit (Catalogue number K1800-01) as per manufacturer's protocol and stored at −20°C until further use.

To determine the relative presence of *C. saccharolyticus* and *C. owensensis* in the co-cultures, quantitative PCR (qPCR) assays were performed as described below. The 16S rDNA sequence was used as target for identification and quantification of each species. To design specific primers (Table 2), dissimilar regions were identified between target sequences using various sequence alignment tools available in the computer software BioEdit (Ibis Biosciences, Carlsbad, California, United States, 92008). PCR amplification and detection were performed in a LightCycler® Nano instrument (Roche Diagnostics, Mannheim, Germany). The PCR assay mixture (20 μL) contained: 1 × ExTaq buffer, 1U TaKaRa ExTaq HS DNA polymerase, 4.5 mM $MgCl_2$, 0.2 mM dNTP (all from Th. Geyer GmbH, Renningen, Germany), 2 μg BSA, 1 × Eva green solution (Bioline GmbH, Luckenwalde, Germany), forward and reverse primers (each 0.5 μM, Table 2) and 4 μL of DNA template. For *C. saccharolyticus* the qPCR amplification protocol started with an initial denaturation at 95°C for 180 seconds, followed by 45 cycles of denaturation at 95°C for 10 seconds, annealing at 67°C for 10 seconds, and elongation and fluorescence acquisition at 72°C for 25 seconds. To confirm the absence of unspecific products, melting-curve analysis was performed as follows: heating at 60°C for 60 seconds followed by an increase in temperature by 0.1°C/s up to 97°C. Similar assays were performed for *C. owensensis*; albeit by changing the annealing temperature to 60°C. Quantification was performed using the method of absolute quantification with the help of LightCycler Nano software version 1.1. Pure genomic DNA samples (2.4 to 48 ng/μL) of each species were used in each run of the LightCycler Nano to establish a standard curve. Each run consisted of a blank assay with a PCR mixture containing dH2O instead of DNA template. It also consisted of a negative control assay with a PCR mixture containing the primers designed for one of the organisms from the pair of *Caldicellulosiruptor* species used in this study and genomic DNA of the other as a template and vice versa. For a particular sample, the DNA

Table 2 Primers used in this study

Organism (Locus tag)	Primer	Sequence	Product (bp)
C. saccharolyticus (Csac_R0001)*	F_R0001	GGTGCGTAGG CGGCTATGCG	448
	R_R0001	CCCACCCTTTC GGGCAGGTC	
C. owensensis (Calow_R0003)	F_R0003	GCTAAGCGGA TGGGGGAAACT	582
	R_R0003	CTGGCAGTGTT GAACGC	

*Primers for *C. saccharolyticus* were obtained from a previous study [31].

concentration of each species was added together and then their relative fractions were determined.

Calculations

The Q_{H2} (mmol · L^{-1} · h^{-1}) and cumulative H_2 formation (CHF, mmol · L^{-1}) were calculated in two different ways, depending on the experimental design. All calculations were based on the ideal gas law and the H_2 and CO_2 concentrations in the headspace. For the cultures in the CSTR, the calculations were based on the flow rate of the influent N_2 gas and the percentages of H_2 and CO_2 in the effluent gas, as no other gases were detected. Thus, Q_{H2} and CHF were calculated based on hydrogen concentration in the effluent gas and the flow rate of the effluent gas. For the experiments performed in the UA reactor, the Q_{H2} was assumed to be twice the respective acetate productivities based on the stoichiometry [26]. Product yields were calculated by determining moles of products formed per mole of glucose consumed. Biomass yield was calculated as moles of biomass formed per mole of glucose consumed. Carbon and redox balances were calculated as described previously [9].

Results

Results obtained from continuously stirred tank reactors

Pure cultures in batch mode were tested for the influence of excretory metabolites from one species to another. For this reason, the supernatant of one organism was added to the reactor of the other prior to inoculation. As a control, both organisms were also grown in pure culture in absence of each other's supernatant. Batch cultures of both *C. saccharolyticus* and *C. owensensis* displayed significantly shorter lag phases when grown in the presence of each other's supernatant rather than in absence of it (Figure 1A and B). Moreover, when exposed to each other's supernatant the cultures accumulated higher amounts of H_2 and biomass, and were less prone to cell lysis in the stationary phase (Figure 1A and B). These are clear indications that both species might influence each other when in co-culture.

To evaluate the biofilm-forming potential and its effect on biomass retention, Q_{H2}, substrate conversion rate, and lactate formation by *C. saccharolyticus* and *C. owensensis*, experiments were performed in the CSTR with or without K1-carriers (Cases A to D, Table 1). In continuous cultures performed in the CSTR, maximum Q_{H2} and maximum substrate conversion were obtained in Case A, whereas, maximum lactate productivity was observed in Case D (Figure 2A, B and D). Cultures of Case A and D sustained growth at higher dilution rate, d (h^{-1}), than those of Case B and C (Figure 2C). In case of Q_{H2}, no particular trend was observed for Case A with increasing d (h^{-1}), whereas, for Case B and C the Q_{H2} increased until d = 0.2 h^{-1} and then decreased. For Case D, Q_{H2} increased until d = 0.3 h^{-1} and then slightly decreased. The hydrogen yield was at its theoretical maximum only at low d (0.03 to 0.05 h^{-1}). Generally, for all the continuous cultures performed in the CSTR, the H_2 yield decreased with increasing d (h^{-1}), with the exception of Case A where it slightly increased at d >0.3 h^{-1} (Figure 2A). For Case A, the substrate conversion rate (SCR) increased with increasing substrate loading rate (SLR). For Cases B and C, the SCR increased with increasing SLR until d = 0.2 h^{-1} and then dropped. Similarly, for Case D, the SCR increased with increasing SLR until d = 0.3 h^{-1} and then decreased. For all the continuous cultures performed in the CSTR

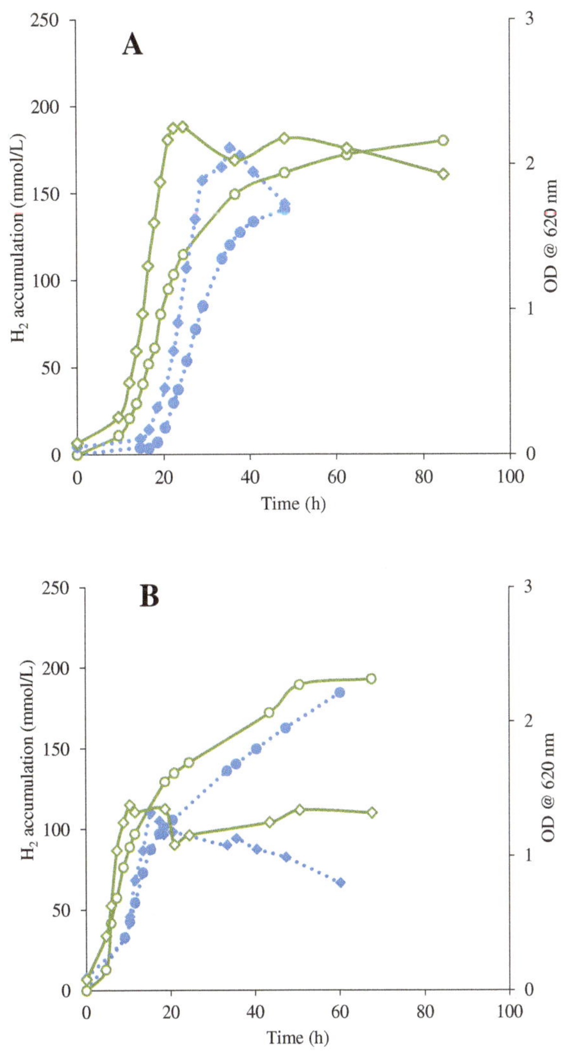

Figure 1 Growth and H_2 accumulation by *C. saccharolyticus* and *C. owensensis* in pH-controlled batch fermentations. Presence (solid green line) and absence (dotted blue line) of each other's supernatant, *C. saccharolyticus* **(A)** and *C. owensensis* **(B)**. Optical density (OD) measured at 620 nm when grown with supernatant (open diamond) and without supernatant (filled diamond); H_2 accumulation when grown with supernatant (open circles) and without supernatant (filled circles).

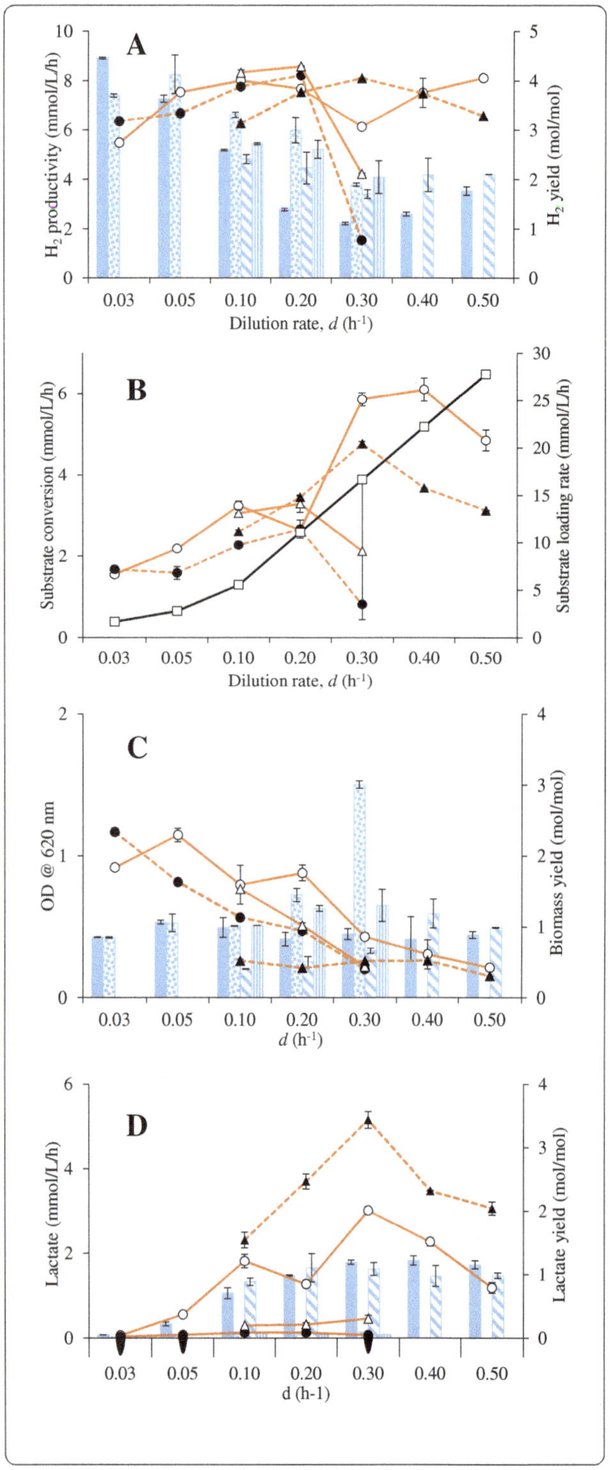

Figure 2 Results of the continuous cultures of *C. saccharolyticus* and *C. owensensis* performed in the continuously stirred tank reactor (CSTR). (A) Q_{H2}, line graph (mmol · L^{-1} · h^{-1}) and H_2 yield, bar graph (mol · mol^{-1}); **(B)** substrate conversion rate (mmol · L^{-1} · h^{-1}); **(C)** Optical density (OD) measured at 620 nm, line graph, and biomass yield, bar graph (mol/mol) from planktonic phase; and **(D)** lactate productivity (mmol · L^{-1} · h^{-1}), line graph, and lactate yield (mol · mol^{-1}), bar graph. Case A (open circles, filled bar); Case B (filled circles, dotted bar); Case C (open triangles, bar with vertical lines); and Case D (filled triangles, bar with skewed lines). Substrate loading rate, solid black line with open squares.

(Cases A to D) at $0.05 > d > 0.4\ h^{-1}$, in most cases the SLR was always higher than the SCR (Figure 2B).

For all the continuous cultures performed in the CSTR, the planktonic biomass concentration generally decreased with increasing d (h^{-1}) (Figure 2C). At any particular d (h^{-1}), Case A generally accumulated more planktonic biomass than Cases B, C, or D. Considering the pure cultures, *C. saccharolyticus* (Case C) showed higher biomass concentration compared to *C. owensensis* (Case D). Surprisingly, for Case B, the biomass yield suddenly increased at 0.3 h^{-1}, but was non-existent at higher d due to washout. No particular trend was observed in biomass yields with increasing d (h^{-1}) for Case A, C, or D. The cultures of Cases A and D could not sustain growth at d >0.5 h^{-1}, whereas, cultures of Case B and C washed out at d >0.3 h^{-1} (Figure 2C). Of the co-cultures, lactate production was only observed when the culture was not stirred (Case A), and increased with the d until 0.3 h^{-1} where it decreased thereafter. Similarly, for the pure cultures, only *C. owensensis* (Case D) produced significant amounts of lactate, which increased with the d until 0.3 h^{-1} and decreased thereafter. A similar trend was observed with the lactate yield for Cases A and D. Overall, the CSTR appeared to be an inappropriate system with respect to achieving higher SCR and Q_{H2}. Therefore, another reactor type was used for further studies.

Results obtained from continuous cultures in the up-flow anaerobic reactor

Again, to evaluate the biofilm-forming potential and its effect on biomass retention, Q_{H2}, substrate conversion rate, and lactate formation by *C. saccharolyticus* and *C. owensensis*, experiments were performed in a UA reactor with either granular sludge or K1-carriers as carrier materials (Cases E to H, Table 1), or without any carrier (Case I, Table 1). The highest Q_{H2} (approximately 20 mmol · L^{-1} · h^{-1}) was obtained in a co-culture with granular sludge at a d = 1.25 h^{-1} (Case E, Figure 3A). The Q_{H2} of this culture increased steadily with increasing d (h^{-1}) and was higher than any other culture performed in the UA reactor at any particular d (h^{-1}). Other co-cultures, with and without K1-carriers, produced H_2 at significantly lower rates, but without any particular trend with increasing d (h^{-1}). On the other hand, the pure cultures of both organisms in the presence of granular sludge (Case F and G) produced H_2 at higher rates than the co-cultures without granular sludge (Case H and I, Figure 3A). Among these pure cultures no significant differences were observed in Q_{H2} at any d (h^{-1}) except at 0.8 and 1.0 h^{-1}, where *C. owensensis* (Case G) displayed a slightly higher Q_{H2} (Figure 3A). The H_2 yields were the highest for the co-culture with granular sludge compared to all other cultures at any particular d (h^{-1}) and generally varied between 2 and 3.3 mol of H_2/ mol of glucose consumed (Figure 3A). The SCR in the UA reactor with granular sludge (Case E, F, and G) generally increased with the SLR (at d ≤0.8 h^{-1}) (Figure 3B). Even though cultures with granular sludge (Case E, F, and G) survived SLR values up to 140 mmol · L^{-1} · h^{-1}, none of them displayed SCR more than 10 mmol · L^{-1} · h^{-1}. At d >0.1 h^{-1}, cultures without granular sludge (Cases H and I) could not sustain growth at SLR values beyond approximately 90 mmol · L^{-1} · h^{-1} and generally displayed much lower SCR compared to cultures with granular sludge (Case E, F and G, Figure 3B).

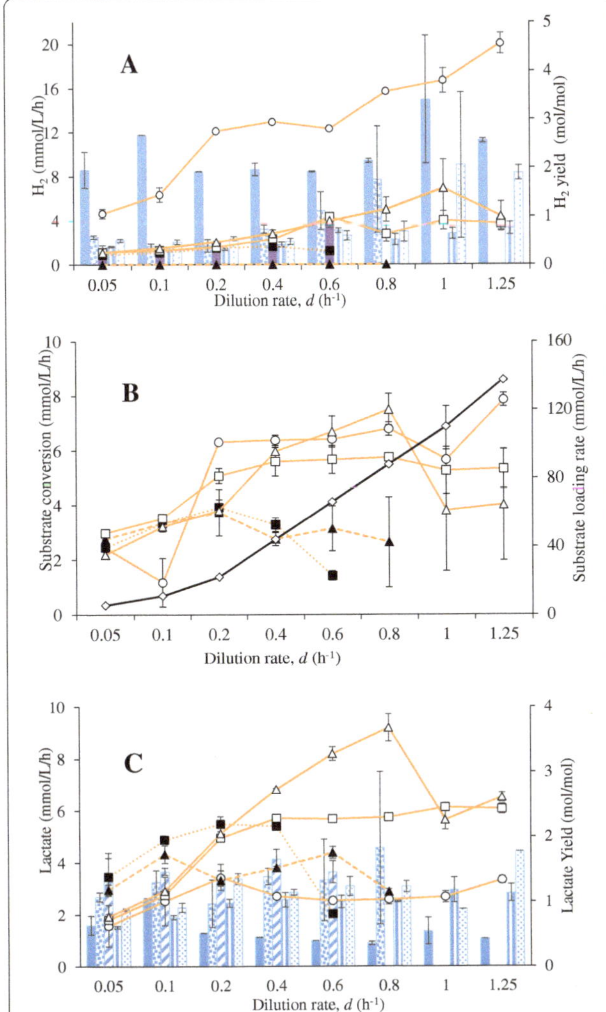

Figure 3 Results of the continuous cultures of *C. saccharolyticus* and *C. owensensis* performed in the up-flow anaerobic (UA) reactors. (A) Q_{H2}, line graph (mmol · L^{-1} · h^{-1}) and H_2 yield, bar graph (mol · mol^{-1}); **(B)** substrate conversion rate and substrate loading rate (mmol · L^{-1} · h^{-1}); and **(C)** lactate productivity (mmol · L^{-1} · h^{-1}), line graph and, lactate yield (mol · mol^{-1}), bar graph. Case E (open circles, filled bar); Case F (open squares, open bar); Case G (open triangles, bar with vertical lines); Case H (filled triangles, dotted bar); and Case I (filled squares, bar with horizontal lines). Substrate loading rate, solid line with open squares.

All liquid samples withdrawn from the granular sludge containing cultures (Case E, F, and G) contained sludge granules, which made it difficult to determine the planktonic biomass concentration, thus no reliable data could be obtained. On the other hand, planktonic biomass concentration in cultures without granular sludge was very low (data not shown), as is evident from the low SCR values obtained in these cultures (Case H and I, Figure 3B).

The highest lactate productivity was observed in the *C. owensensis* culture with granular sludge (Case G, Figure 3C). At d >0.2 h^{-1}, both the pure cultures with granular sludge (Case F and G) displayed higher lactate productivity than the co-culture with (Case E) or without sludge (Case H and I). Of these co-cultures, the one without granular sludge (Case H and I) produced lactate at higher rates than the one with granular sludge (Case E). The lactate yields were lowest for the co-culture with granular sludge (Case E) at any particular d (h^{-1}). No significant differences in lactate yield were observed among the other cultures (Case F, G, H, and I).

Biofilm formation by *Caldicellulosiruptor* species

No biofilm was observed during any of the batch cultures performed. In the continuous cultures, at d >0.2 h^{-1} a substantial amount of flocculation was observed at the bottom of the CSTR in the co-culture when stirring was not applied (Case A, Additional files 1 and 2). In addition, in this culture at d >0.2 h^{-1}, biofilm was also observed on the reactor walls, pH probe, and K1-carriers. In contrast, when stirring was applied (Case B), no biofilm was observed. Among the pure cultures, no biofilm was observed on the reactor wall, pH probe, or K1-carriers in either of the Cases C and D. However, a biofilm in the form of flocculation of cells was observed in the *C. owensensis* culture for the entire duration (Case D). When viewed under SEM, the biofilm growing on the pH probe of the CSTR with co-culture (Case A) revealed distinct cells attached to each other with visible fibre-like structures (Figure 4). Two different kinds of cell structures were observed, one as rod-shaped and unicellular form with dimensions 0.2 to 0.4 μm by 3 to 4 μm, whereas the other in a chain-like,

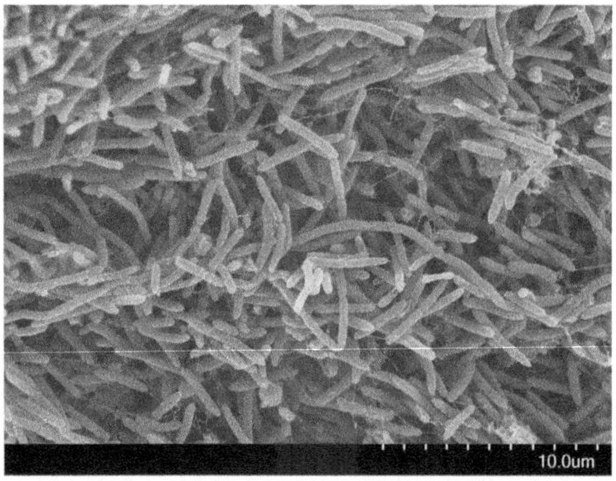

Figure 4 SEM image of a biofilm obtained from the pH probe from the co-culture (Case A).

multi-cellular structure with similar width (0.2 to 0.4 μm) but variable length depending on the number of cells in a chain (Figure 4).

The co-culture with sludge (Case E) displayed significant flocculation and biofilm on the reactor wall which was especially pronounced at d >0.2 h^{-1}. Among the pure cultures, the *C. owensensis* culture with sludge (Case G) also displayed significant flocculation atop the sludge bed but hardly any biofilm was observed on the reactor walls. The co-cultures without sludge also displayed traces of biofilm on the reactor wall (Case H and I), however, no significant biofilm was observed on the K1-carriers (Case H).

Intracellular levels of bis-(3′-5′)-cyclic dimeric guanosine monophosphate

The genomes of *C. saccharolyticus* and *C. owensensis* contain multiple genes coding for diguanylate cyclase (DGC), and phosphodiesterase (PDE) (Additional file 3). In batch cultures of *C. saccharolyticus* cells contained very low c-di-GMP levels compared to those observed in cells of *C. owensensis* (Figure 5). Interestingly, when grown in the presence of each other's supernatant, cells of *C. saccharolyticus* accumulated higher levels of c-di-GMP compared to those cells grown without the supernatant of a *C. owensensis* culture (Figure 5). In contrast, the opposite trend was observed for *C. owensensis*. In continuous cultures, the co-culture without stirring (Case A) accumulated very low (<20 μM) levels of c-di-GMP at d ≤0.2 h^{-1}. However, at d ≥0.2 h^{-1} the same culture accumulated at least 5 to 10-fold higher levels of c-di-GMP, albeit with no particular trend. Interestingly, in the co-culture without stirring (Case A), the levels of c-di-GMP appear to have increased when levels of residual sugar increased beyond 2 $g \cdot L^{-1}$ (Figure 6), without any particular pattern. In contrast, the co-culture with stirring accumulated very low (>30 μM) levels of c-di-GMP regardless of the d (h^{-1}). Among the pure cultures, cells of *C. owensensis* (Case D) accumulated similar levels to those observed in the co-culture without stirring (Case A) at d ≥0.2 h^{-1}, but approximately 10-fold higher levels than those observed in cells of *C. saccharolyticus* (Case C, Figure 5).

Among the UA cultures, the co-culture without K1-carriers (Case I), except for d 0.2 and 0.4 h^{-1}, cells accumulated very low (<30 μM) c-di-GMP levels. The co-culture with K1-carriers (Case H) contained very low (<30 μM) c-di-GMP levels regardless of the d (h^{-1}) (Figure 5). No samples were collected from cultures performed with sludge (Case E, F, and G) due to contaminations from granular sludge.

Population dynamics in co-cultures of *C. saccharolyticus* and *C. owensensis*

In the co-culture without stirring performed in the CSTR (Case A), the biofilm on the pH probe consisted of *C. saccharolyticus* and *C. owensensis* in about a 1:1 ratio (Figure 7). However, in the same culture, the biofilm on the K1-carriers contained about 10 to 12 times more cells of *C. owensensis* than cells of *C. saccharolyticus*. Similarly, in the co-culture performed in the UA reactor (Case H), the biofilm on the K1-carriers contained the cell ratio of about 10:1 for *C. owensensis* compared to *C. saccharolyticus* (Figure 7). No results could be obtained with samples collected from planktonic cells in any of the cultures, possibly due to the low target DNA concentration.

Discussion

Effect of biofilm formation on Q_{H2}, substrate conversion, and lactate formation

In a techno-economic analysis of a representative biohydrogen process, low Q_{H2} has been identified as a key bottleneck for making the process economically viable [27]. This study reports a higher Q_{H2} (approximately. 20 $mmol \cdot L^{-1} \cdot h^{-1}$, Case E) than most of the previously obtained values in continuous cultures of *Caldicellulosiruptor* species [28], but which is still about an order of magnitude lower than the maximum Q_{H2} ever reported for thermophilic hydrogen producers [20]. Nevertheless, the highest maximum Q_{H2} in both these studies were obtained at very high d (>1.0 h^{-1}), which may not be ideal for reasonable process economics [27]. Thus, further investigations are needed to determine the implications of high d (h^{-1}) on a biohydrogen process.

Numerous studies have asserted that biofilm formation improves substrate conversion leading to increased Q_{H2} [20,21,29]. Similarly, in this study, formation of biofilm by co-cultures of *C. saccharolyticus* and *C. owensensis* improved the substrate conversion in the CSTR as well

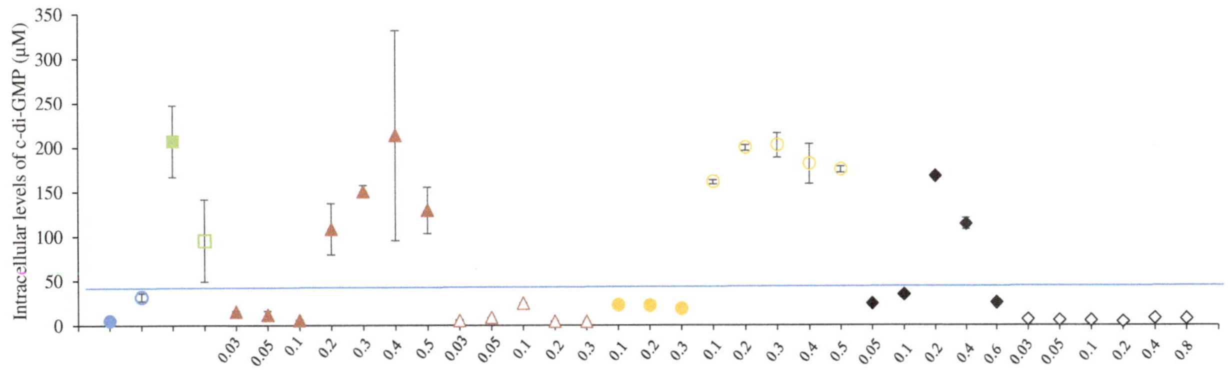

Figure 5 Intracellular levels of c-di-GMP in batch and continuous cultures performed in CSTR and UA reactors. Batch cultures without supernatant: *C. saccharolyticus* (filled circle, green), *C. owensensis* (filled square, green); batch cultures with each other's supernatant: *C. saccharolyticus* (open circle, blue), *C. owensensis* (open square, green); Continuous cultures: Case A (filled triangle, red); Case B (open triangle, red); Case C (filled circle, yellow); Case D (open circle, yellow); Case I (filled diamond, black); and Case H (open diamond, black). For continuous cultures, the values on X-axis represent d (h^{-1}) at which the sample was collected.

as the UA reactor (Case A and E). However, it had a varied effect on Q_{H2}. In the UA reactor biofilm formation indeed improved Q_{H2}. In the CSTR, however, improved substrate conversion was accompanied by an increase in lactate production (Case A), which consequently subdued Q_{H2}. This abnormality of the CSTR accumulating relatively higher amounts of reduced by-products, such as lactate and ethanol, than UA reactors (Case A and E) was also observed in a similar study comparing conversion of wheat straw hydrolysate using mixed culture in CSTR and UA reactors [30]. In the present study, the aforementioned abnormality may have occurred due to the presence of a higher proportion of *C. owensensis* compared to *C. saccharolyticus* in the planktonic phase at high d (>0.2 h^{-1}) in the CSTR. This hypothesis is supported by the fact that *C. owensensis* produced higher amounts of lactate than *C. saccharolyticus* regardless of the reactor system (Figure 2D and 3C), and that unlike the CSTR, the UA reactors inherently allow biomass retention, thus perhaps a higher fraction of cells of *C. saccharolyticus* were retained in the UA reactor compared to the CSTR when operated at higher d (h^{-1}).

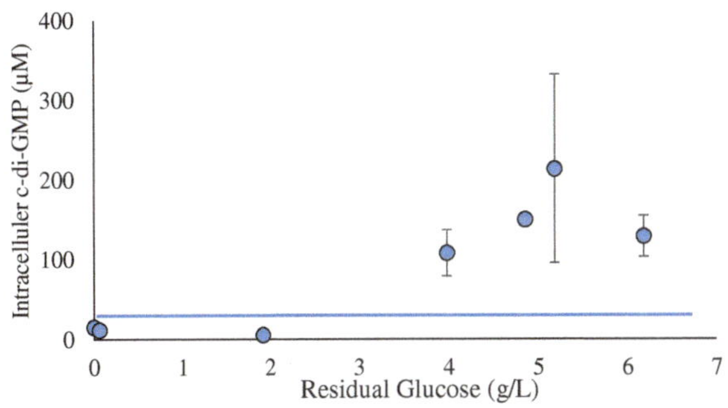

Figure 6 Correlation between intracellular c-di-GMP levels and residual sugar concentration in the co-culture (Case A).

Designed co-cultures versus pure cultures

Regardless of the reactor system used, the co-cultures converted higher amounts of substrate and, in the UA, displayed higher Q_{H2} than the pure culture of each species. This is in agreement with previous studies, where designed co-cultures of *C. saccharolyticus* and *Caldicellulosiruptor kristjanssonii* showed higher H_2 yields than their pure cultures [31]. Similarly, a co-culture of *Clostridium thermocellum* JN4 and *Thermoanaerobacterium thermosaccharolyticum* GD17 reported two-fold higher Q_{H2} than either of their pure cultures [32], even though they are of different genus.

Both *Caldicellulosiruptor* species performed better in batch growth in the presence of each other's supernatant, which clearly indicate that both species excrete compounds positively affecting the other one. A similar observation has been made for *C. saccharolyticus* excreting compound(s) that boosted the growth of *C.*

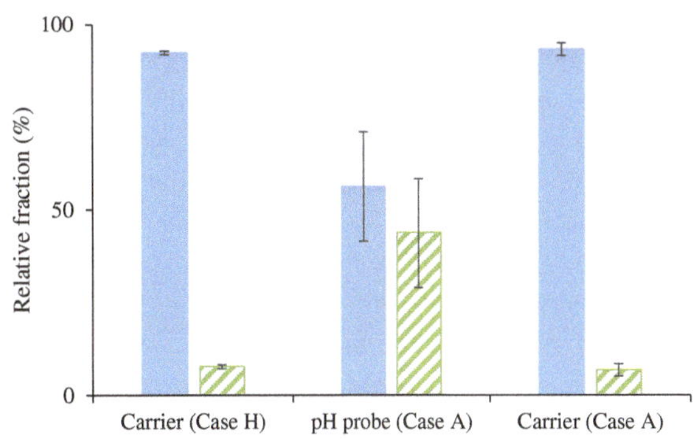

Figure 7 Fraction of *C. saccharolyticus* and *C. owensensis* in biofilm samples (Case A and H). *C. owensensis* (filled, blue) and *C. saccharolyticus* (horizontal lines, green), values on X-axis represent the source of the biofilm sample with respect to reactor system and the carrier.

kristjanssonii [31]. In fact, co-culture *C. saccharolyticus* boosted the growth performance of *C. kristjanssonii,* which can be interpreted as altruistic behaviour [31]. In the current study, a similar behaviour was seen with *C. saccharolyticus* fortifying *C. owensensis'* ability to form biofilm. On its turn, *C. owensensis* showed altruistic behaviour by aiding *C. saccharolyticus* to take part in the biofilm formation (Figure 6). This phenomenon is explained by 'kin selection theory' [33], according to which closely related species help each other to reproduce to pass its own genes on to next generation, even if indirectly. According to Hamilton's rule, higher relatedness (r) between the species, higher fitness benefit (b) to the beneficiary, and lower fitness cost (c) to the altruist will ensure better cooperation ($r \times b - c > 0$) [33]. This may explain why the co-culture of *C. saccharolyticus* and *C. kristjanssonii* reported higher H_2 yields [31] than any of the mixed cultures consisting of microorganisms of various genera ever reported. Indeed, another study argues simply that higher cooperation can be expected between highly related species [34].

Among the pure cultures, both *C. saccharolyticus* and *C. owensensis* produced higher amounts of lactate than previously reported studies [8,9] performed in similar conditions, except that stirring was not applied for the cultures in this study. Obviously, the non-stirring condition led to oversaturation of H_2 and CO_2 in the culture, leading to a shift in the metabolism [35,36]. Finally, the observation of an unusual increase in biomass yield in the pure culture of *C. saccharolyticus* (Case C) near its critical d (0.3 h^{-1}) can be attributed to relatively higher energy spent by the culture on cell growth than product formation, as a reaction to wash-out conditions at a high d (h^{-1}). A similar observation was reported in a previous study performed with *C. saccharolyticus* [23]. As far as we know, this has not been described before in the literature, and a clear rationale behind this phenomenon is lacking.

Effect of reactor system and culture conditions

In UA reactors, only granular sludge provided a supporting bed to the flocculating biofilms of *C. owensensis* and *C. saccharolyticus*. This explains the very low Q_{H2} observed in the UA reactor without granular sludge. Similar results were obtained in a previous study performed with *Thermoanaerobacterium thermosaccharolyticum* PSU-2 [20]. However, despite its benefits, the risk of contamination from hydrogenotrophic methanogens threatens the stability of UA reactors when granular sludge is used. It could be that porous glass beads may be a viable alternative carrier. A recent study reported an increase in Q_{H2} and H_2 yield by 70% and 30%, respectively, when cells of *Thermotoga neapolitana* were immobilized on porous glass beads in a CSTR [37].

Although, higher Q_{H2} (>15 $mmol \cdot L^{-1} \cdot h^{-1}$) is desirable for better process economics, a higher H_2 yield (>3 $mol \cdot mol^{-1}$) can certainly contribute to improving the process economics when relatively expensive raw materials are used. In that respect, when the results obtained in this study are compared, UA reactors appear to offer a process alternative to achieve high Q_{H2} and yield (Figure 8). The CSTR, on the other hand, seems to have a boundary value around 10 $mmol \cdot L^{-1} \cdot h^{-1}$ for Q_{H2} regardless of the H_2 yield (Figure 8).

The UA reactor allowed d (h^{-1}) well beyond the maximum specific growth rates of *C. saccharolyticus* and *C. owensensis* in pure and co-cultures, underlining the ability of UA reactors to retain the biomass of these species.

Biofilm and intracellular levels of bis-(3′-5′)-cyclic dimeric guanosine monophosphate

A clear correlation was observed between the high intracellular c-di-GMP levels (>40 μM) and the stage of a particular culture initiating a biofilm. Although the samples were collected from planktonic biomass and not the biofilm itself, since the biofilms go through feed-and-bleed cycles, the planktonic cells can be assumed to be representative of the cells in the biofilm. Conversely, in the absence of any biofilm, very low c-di-GMP levels were observed when stirring was applied in continuous cultures in the CSTR (Case B). However, batch cultures of *C. owensensis* accumulated high levels of c-di-GMP but no biofilm was observed, perhaps due to the stirring. Moreover, c-di-GMP levels in co-culture performed without stirring (Case A) increased as the concentration of residual sugar increased beyond 2 $g \cdot L^{-1}$ (Figure 6). This may be because of a combination of the fact that the flocculating cells of *C. owensensis* at the bottom of the CSTR did not have access to the influent feed being dropped from the top of the CSTR, and that cells of *C.*

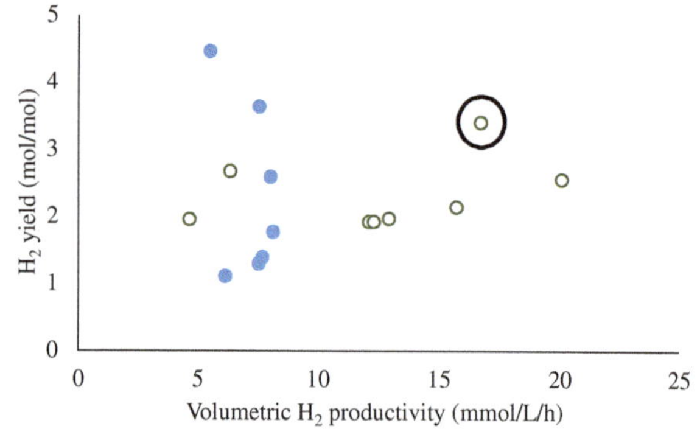

Figure 8 The correlation between Q_{H2} and H_2 yield in co-cultures (Case A and E). Q_{H2} ($mmol \cdot L^{-1} \cdot h^{-1}$), H_2 yield ($mol \cdot mol^{-1}$). Case A (filled circle, blue); Case E (open circle, red). The encircled data point represent the best case scenario where both Q_{H2} and H_2 yield are reasonably high.

saccharolyticus dominating the planktonic phase consumed most of the substrate until a d of 0.1 h^{-1}, after which the residual concentration increased beyond 2 g · L^{-1} (Figure 6). Beyond that point the glucose gradient may have reached *C. owensensis* at the bottom, allowing the development of biofilms at d ≥0.2 h^{-1}. Thus, it can be argued that if the co-cultures were performed at high substrate concentration, biofilm could have been obtained even at d <0.2 h^{-1}. This knowledge may help in achieving SLRs as well as biofilms at low d (h^{-1}), similar to those obtained at high d (h^{-1}) in this study. However, the vulnerability of *C. saccharolyticus* to high osmotic pressure limits the option of performing cultures using feed with high substrate concentration [22]. Alternatively, a reactor system such as a UA reactor which feeds the influent from bottom may also be more appropriate, as shown in the present study.

Although, *C. saccharolyticus* possesses genes required for the synthesis of c-di-GMP, its intracellular levels are well below the critical level (40 μM). This perhaps explains the inability of *C. saccharolyticus* to form biofilm independent of *C. owensensis*. Arguably, overexpression of DGC may elevate the levels of c-di-GMP in *C. saccharolyticus*, allowing biofilm formation. Thus, encouraging *C. saccharolyticus* to form biofilms on its own may provide a better alternative to its co-culture with *C. owensensis*, considering the propensity of the latter to produce lactate and ethanol.

Conclusions

Only when grown together in co-culture do, *C. saccharolyticus* and *C. owensensis* form substantial amounts of biofilm, improving substrate conversion and Q_{H2}. Thus, such a constructed co-culture is an effective means to be exploited in any bioreactor designed for biomass retention, such as UA reactors. Indeed, UA reactors allow retention of *C. saccharolyticus* and *C. owensensis* when subjected to very high substrate loading rates, improving substrate conversion, and Q_{H2}. Granular sludge showed superior support to biofilm formation in UA reactors. However, as sludge can be a potential source of methanogenic contaminants, it either needs proper pre-treatment, or more suitable alternatives should be found. Elevated intracellular levels of c-di-GMP are clearly linked to biofilm formation by *C. saccharolyticus* and *C. owensensis*. The maximum Q_{H2} obtained in this study was obtained at very high d (h^{-1}) which may not be ideal for a reasonable process economics. Alternatively, a biofilm forming pure or co-cultures of *Caldicellulosiruptor* species, which can withstand feed containing high substrate concentrations, can be operated at a reasonably low d (h^{-1}), which will allow similar substrate loading rates to that obtained in this study at high d (h^{-1}). The way forward for industrial application is to further exploit the concept of this designed co-culture in UA-type reactors using granular sludge-type of carriers for obtaining higher volumetric hydrogen productivities.

Additional files

Additional file 1: The planktonic biomass in the co-culture without stirring (Case A). The boxes filled with different colours represent a particular d (h^{-1}).

Additional file 2: A short film showing the biofilm in action (Case A).

Additional file 3: Table S1. Genes related to c-di-GMP synthesis and hydrolysis in *C. saccharolyticus* and *C. owensensis*.

Abbreviations

c-di-GMP: bis-(3′-5′)-cyclic dimeric guanosine monophosphate; CDW: Cell dry weight; CHF: Cumulative H_2 formation (mmol · L^{-1}); CSTR: Continuously stirred tank reactor; d: Dilution rate (h^{-1}); DSM: Deutsche Sammlung von Mikroorganismen; DGC: diguanylate cyclase; PDE: phosphodiesterase; XMP: xanthosine 5'- monophosphate; ESI: electrospray ionization; SRM: selected reaction monitoring; Q_{H2}: Volumetric H_2 productivity (mmol · L^{-1} · h^{-1}); SCR: Substrate conversion rate (mmol · L^{-1} · h^{-1}); SLR: Substrate loading rate (mmol · L^{-1} · h^{-1}); UA: Up-flow anaerobic.

Competing interests

The authors declare that they have no competing interests.

Authors' contributions

SSP planned the content of the article and planned and performed the experiments. TV assisted SSP in some of the batch and continuous cultures. CG optimized and performed the analysis of c-di-GMP and also contributed with writing related to c-di-GMP. EvN was involved in the planning of the experiments and supervised the process. EvN also critically reviewed the text. All the authors read and approved the final manuscript.

Acknowledgements

SSP acknowledges support from the Swedish research council (VR). We thank Valentine Nkongndem Nkemka for providing the UA reactor with granular sludge. We thank Anox-Kaldnes AB, Lund, Sweden, for generously providing the K1-carriers used in this study. We are grateful to Ola Gustafsson for his expert advice and co-operation during the SEM analysis of the biofilm samples. We also thank Linda Jansson and Johannes Hedman for their expert advice regarding qPCR work.

Author details

[1]Division of Applied Microbiology, Lund University, Getingevägen 60, PO Box 124, SE-221 00 Lund, Sweden. [2]Department of Biotechnology, Lund University, Getingevägen 60, PO Box 124221 00 Lund, Sweden.

References

1. Delucchi MA, Jacobson MZ. Providing all global energy with wind, water, and solar power, part II: reliability, system and transmission costs, and policies. Energy Policy. 2011;39:1170–90.
2. Pawar SS, Nkemka VN, Zeidan AA, Murto M, van Niel EWJ. Biohydrogen production from wheat straw hydrolysate using *Caldicellulosiruptor saccharolyticus* followed by biogas production in a two-step uncoupled process. Int J Hydrog Energy. 2013;38:9121–30.
3. de Vrije T, Bakker RR, Budde MAW, Lai MH, Mars AE, Claassen PAM. Efficient hydrogen production from the lignocellulosic energy crop Miscanthus by the extreme thermophilic bacteria *Caldicellulosiruptor saccharolyticus* and *Thermotoga neapolitana*. Biotechnol Biofuels. 2009;2:12.
4. de Vrije T, Budde MAW, Lips SJ, Bakker RR, Mars AE, Claassen PAM. Hydrogen production from carrot pulp by the extreme thermophiles *Caldicellulosiruptor saccharolyticus* and *Thermotoga neapolitana*. Int J Hydrog Energy. 2010;35:13206–13.

5. van de Werken HJG, Verhaart MRA, VanFossen AL, Willquist K, Lewis DL, Nichols JD, et al. Hydrogenomics of the extremely thermophilic bacterium *Caldicellulosiruptor saccharolyticus.* Appl Environ Microbiol. 2008;74:6720–9.
6. Ivanova G, Rákhely G, Kovács KL. Thermophilic biohydrogen production from energy plants by *Caldicellulosiruptor saccharolyticus* and comparison with related studies. Int J Hydrog Energy. 2009;34:3659–70.
7. Pawar SS, van Niel EWJ. Thermophilic biohydrogen production: how far are we? Appl Microbiol Biotechnol. 2013;97:7999–8009.
8. Zeidan AA, van Niel EWJ. A quantitative analysis of hydrogen production efficiency of the extreme thermophile *Caldicellulosiruptor owensensis* OLT. Int J Hydrogen Energy. 2010;35:1128–37.
9. de Vrije T, Mars AE, Budde MAW, Lai MH, Dijkema C, de Waard P. Glycolytic pathway and hydrogen yield studies of the extreme thermophile *Caldicellulosiruptor saccharolyticus.* Appl Microbiol Biotechnol. 2007;74:1358–67.
10. Kumar N, Das D. Continuous hydrogen production by immobilized *Enterobacter cloacae* IIT-BT 08 using lignocellulosic materials as solid matrices. Enzyme Microb Technol. 2001;29:280–7.
11. Karatan E, Watnick P. Signals, regulatory networks, and materials that build and break bacterial biofilms. Microbiol Mol Biol Rev. 2009;73:310–47.
12. Dufour D, Leung V, Lévesque CM. Bacterial biofilm: structure, function, and antimicrobial resistance. Endod Topics. 2010;22:2–16.
13. Peintner C, Zeidan AA, Schnitzhofer W. Bioreactor systems for thermophilic fermentative hydrogen production: evaluation and comparison of appropriate systems. J Cleaner Prod. 2010;18 Suppl 1:S15–22.
14. Hengge R. Principles of c-di-GMP signalling in bacteria. Nat Rev Microbiol. 2009;7:263–73.
15. Jenal U, Malone J. Mechanisms of Cyclic-di-GMP Signaling in bacteria. Annu Rev Genet. 2006;40:385–407.
16. Pérez-Mendoza D, Coulthurst SJ, Sanjuán J, Salmond GPC. N-Acetylglucosamine-dependent biofilm formation in *Pectobacterium atrosepticum* is cryptic and activated by elevated c-di-GMP levels. Microbiology. 2011;157:3340–8.
17. Kim M-S, Lee D-Y, Kim D-H. Continuous hydrogen production from tofu processing waste using anaerobic mixed microflora under thermophilic conditions. Int J Hydrog Energy. 2011;36:8712–8.
18. Prasertsan P, O-Thong S, Birkeland N-K. Optimization and microbial community analysis for production of biohydrogen from palm oil mill effluent by thermophilic fermentative process. Int J Hydrog Energy. 2009;34:7448–59.
19. van Groenestijn JW, Geelhoed JS, Goorissen HP, Meesters KPM, Stams AJM, Claassen PAM. Performance and population analysis of a non-sterile trickle bed reactor inoculated with *Caldicellulosiruptor saccharolyticus*, a thermophilic hydrogen producer. Biotechnol Bioeng. 2009;102:1361–7.
20. O-Thong S, Prasertsan P, Karakashev D, Angelidaki I. High-rate continuous hydrogen production by *Thermoanaerobacterium thermosaccharolyticum* PSU-2 immobilized on heat-pretreated methanogenic granules. Int J Hydrog Energy. 2008;33:6498–508.
21. Koskinen PEP, Lay C-H, Puhakka JA, Lin P-J, Wu S-Y, Örlygsson J, et al. High-efficiency hydrogen production by an anaerobic, thermophilic enrichment culture from an Icelandic hot spring. Biotechnol Bioeng. 2008;101:665–78.
22. Willquist K, Claassen PAM, van Niel EWJ. Evaluation of the influence of CO_2 on hydrogen production in *Caldicellulosiruptor saccharolyticus.* Int J Hydrog Energy. 2009;34:4718–26.
23. Pawar SS, Van Niel EWJ. Evaluation of assimilatory sulphur metabolism in *Caldicellulosiruptor saccharolyticus.* Bioresour Technol. 2014;169:677–85.
24. Willquist K, van Niel EWJ. Lactate formation in *Caldicellulosiruptor saccharolyticus* is regulated by the energy carriers pyrophosphate and ATP. Metab Eng. 2010;12:282–90.
25. Spangler C, Böhm A, Jenal U, Seifert R, Kaever V. A liquid chromatography-coupled tandem mass spectrometry method for quantitation of cyclic di-guanosine monophosphate. J Microbiol Methods. 2010;81:226–31.
26. Shen N, Zhang F, Song X-N, Wang Y-S, Zeng RJ. Why is the ratio of H_2/acetate over 2 in glucose fermentation by *Caldicellulosiruptor saccharolyticus*? Int J Hydrog Energy. 2013;38:11241–7.
27. Ljunggren M, Wallberg O, Zacchi G. Techno-economic comparison of a biological hydrogen process and a 2nd generation ethanol process using barley straw as feedstock. Bioresour Technol. 2011;102:9524–31.
28. Bielen AAM, Verhaart MRA, van der Oost J, Kengen SVM. Biohydrogen production by the thermophilic bacterium *Caldicellulosiruptor saccharolyticus*: current status and perspectives. Life. 2013;3:52–85.
29. Kim JO, Kim YH, Ryu JY, Song BK, Kim IH, Yeom SH. Immobilization methods for continuous hydrogen gas production biofilm formation versus granulation. Process Biochem. 2005;40:1331–7.
30. Kongjan P, Angelidaki I. Extreme thermophilic biohydrogen production from wheat straw hydrolysate using mixed culture fermentation: effect of reactor configuration. Bioresour Technol. 2010;101:7789–96.
31. Zeidan A, Rådström P, van Niel E. Stable coexistence of two *Caldicellulosiruptor* species in a de novo constructed hydrogen-producing co-culture. Microb Cell Fact. 2010;9:102.
32. Liu Y, Yu P, Song X, Qu Y. Hydrogen production from cellulose by co-culture of *Clostridium thermocellum* JN4 and *Thermoanaerobacterium thermosaccharolyticum* GD17. Int J Hydrog Energy. 2008;33:2927–33.
33. Hamilton WD. The genetical evolution of social behaviour. I J Theor Biol. 1964;7:1–16.
34. West SA, Griffin AS, Gardner A, Diggle SP. Social evolution theory for microorganisms. Nat Rev Microiol. 2006;4:597–607.
35. van Niel EWJ, Claassen PAM, Stams AJM. Substrate and product inhibition of hydrogen production by the extreme thermophile, *Caldicellulosiruptor saccharolyticus.* Biotechnol Bioeng. 2003;81:255–62.
36. Willquist K, Pawar SS, van Niel EWJ. Reassessment of hydrogen tolerance in *Caldicellulosiruptor saccharolyticus.* Microb Cell Fact. 2011;10:111.
37. Ngo TA, Bui HTV. Biohydrogen production using immobilized cells of hyperthermophilic eubacterium *Thermotoga neapolitana* on porous glass beads. J Technol Innov Renewable Energy. 2013;2:231–8.

3

High titer and yield ethanol production from undetoxified whole slurry of Douglas-fir forest residue using pH profiling in SPORL

Jinlan Cheng[1,2], Shao-Yuan Leu[3,2], JY Zhu[2*] and Rolland Gleisner[2]

Abstract

Background: Forest residue is one of the most cost-effective feedstock for biofuel production. It has relatively high bulk density and can be harvested year round, advantageous for reducing transportation cost and eliminating onsite storage. However, forest residues, especially those from softwood species, are highly recalcitrant to biochemical conversion. A severe pretreatment for removing this recalcitrance can result in increased sugar degradation to inhibitors and hence cause difficulties in fermentation at high solid loadings. Here, we presented high titer ethanol production from Douglas-fir forest residue without detoxification. The strong recalcitrance of the Douglas-fir residue was removed by sulfite pretreatment to overcome the recalcitrance of lignocelluloses (SPORL). Sugar degradation to inhibitors was substantially reduced using a novel approach of "pH profiling" by delaying acid application in pretreatment, which facilitated the simultaneous enzymatic saccharification and fermentation of undetoxified whole slurry at a solid loading of 21%.

Results: "pH profiling" reduced furan production by approximately 70% in using SPORL pretreating Douglas-fir forest residue (FS-10) comparing with the control run while without sacrificing enzymatic saccharification of the resultant substrate. pH profiling also reduced carbohydrate degradation. The improved carbohydrate yield in pretreated solids and reduced fermentation inhibitors with pH profiling resulted in a terminal ethanol titer of 48.9 ± 1.4 g/L and yield of 297 ± 9 L/tonne FS-10, which are substantially higher, i.e., by 27% in titer and by 38% in yield, than those of a control SPORL run without pH profiling.

Conclusions: Economical and large-volume production of commodity biofuels requires the utilization of feedstocks with low value (therefore low cost) and sustainably producible in large quantities, such as forest residues. However, most existing pretreatment technologies cannot remove the strong recalcitrance of forest residues to produce practically fermentable high titer sugars. Here, we demonstrated a commercially scalable and efficient technology capable of removing the strong recalcitrant nature of forest residues using "pH profiling" together with "low temperature SPORL". The resultant pretreated whole slurry of a Douglas-fir forest residue using this technology can be easily processed at high solids of 21% without detoxification to achieve a high ethanol yield of 297 L/tonne at 48.9 g/L.

Keywords: Forest residue, Enzymatic hydrolysis and fermentation, High solids processing, Fermentation inhibitors, High-titer biofuel

* Correspondence: jzhu@fs.fed.us
[2]USDA Forest Service, Forest Products Laboratory, Madison, WI 53719, USA
Full list of author information is available at the end of the article

Background

Lignocellulosic biomass, as a structural material, has natural resistance to enzymatic deconstruction for production of fermentable sugars. Pretreatment, a step to remove this recalcitrance, increases the cellulose accessibility to cellulase for efficient saccharification of polysaccharides in lignocelluloses. Most promising and commonly practiced acidic pretreatments, such as dilute acid [1,2], sulfite pretreatment to overcome recalcitrance of lignocellulose (SPORL) [2,3], organosolv [4], and SO_2-catalyzed steam explosion [5,6], however, can degrade sugars to undesirable compounds such as furans. These pretreatments also convert acetyl groups on the hemicellulose backbone into acetic acid. The undesirable sugar degradation products and acetic acid are inhibitive to many microbes and catalysts, such as *Saccharomyces cerevisiae*, during conversion of sugars to biofuel and bioproducts. Removal of the inhibitive compounds through detoxification steps is possible, but at additional costs, which negatively affects production economics.

Reduced sugar degradation during pretreatment is desirable for high solid processing without detoxification to achieve high biofuel titer. Low-temperature pretreatment can reduce inhibitor formation. An early study arbitrary reduced the reaction duration to result in a lower pretreatment severity to reduce degradation [7], but at the expense of lower sugar yield. Additional processing steps such as alkali extraction, disk refining, and xylanase supplementation were necessary to maintain enzymatic saccharification efficiency. Recently, we conducted low-temperature pretreatment while maintaining reaction severity measured by a combined hydrolysis factor (CHF) based on hemicellulose dissolution [8]. We used a constant CHF to determine the extended reaction time at a reduced pretreatment temperature. We were able to reduce furan formation by approximately 50% in pretreating Douglas-fir wood chips while maintaining similar high (>90%) cellulose enzymatic saccharification efficiency by reducing the pretreatment temperature from 180°C to 165°C [8]. Further reduction in furan formation is possible by using an even lower temperature but requires a much longer pretreatment time, which reduces production capacity [9].

Forest residue is one of the most cost-effective feedstock for biofuel production based on a recent study by the US National Academy of Sciences [10]. It can be sustainably produced in large quantities in North America and various regions of the globe [11-13]. Furthermore, it has relatively high bulk density and can be harvested year round to eliminate onsite storage, both of which are advantages over agriculture residue and herbaceous biomass for improving the supply chain logistics and reducing transportation costs [14,15]. However, forest residues have high lignin content due to the presence of bark and juvenile wood which can increase recalcitrance to biochemical conversion. A severe pretreatment required to remove this recalcitrance can produce increased sugar degradation to furan even using the low-temperature pretreatment strategy [16]. Because low furan formation can not only reduce sugar loss but also facilitate fermentation to increase fermentation yield especially at high solid loadings [17], improving pretreatment with low furan formation is always desirable and needed.

Here, we demonstrated the "pH profiling" concept in SPORL [3] to substantially reduce furan formation to achieve high titer and high ethanol yield from SPORL pretreated whole slurry of a Douglas-fir forest residue without solid and liquid separation, solid washing, and detoxification. In this concept, the application of acid such as SO_2 or H_2SO_4 was delayed (the amount of acid applied was unchanged) to purposely control the time-dependent pH profile during pretreatment as shown by the insert plot in Figure 1. The concept is based on two fundamental understandings: (1) the role of acid and sulfite on hemicellulose dissolution [2], sugar degradation, delignification, and sulfonation [18], and deacetylation [19] and (2) the effects of hemicellulose dissolution and delignification on improving cellulose accessibility and cellulose enzymatic saccharification [20-24]; specifically, the quantitative contributions of delignification and hemicellulose removal to enzymatic digestibility of softwood species and forest residues [8,20]. The improved delignification due to increased pH in the early stage of pretreatment compensated for the reduced hemicellulose dissolution as a result of delayed acid application in a "pH profiling" pretreatment. Therefore, equivalent enzymatic saccharification can be achieved but at much low sugar degradation and furan formation. SPORL was chosen for demonstrating the "pH profiling" concept because delignification can be achieved at acidic conditions using sulfite which can reduce the pH range and therefore acid application in addition to the robust performance of SPORL for bioconversion of softwoods and forest residues [16,25]. The concept is similar to two-stage sulfite pulping [26-28] but with the purpose to reduce sugar degradation rather to improve pulp yield. When implementing "pH profiling" for pretreatment, lower chemical loadings, higher temperatures, and shorter residence time than those in sulfite pulping are used as demonstrated here. Multiple or continuous acid application can also be implemented.

Results and discussions

Upgrade of Douglas-fir forest residue by physical fractionation

The results from the present study indicated that fractionation was effective to reduce bark content of the Douglas-fir residue harvested using grinding. Visual

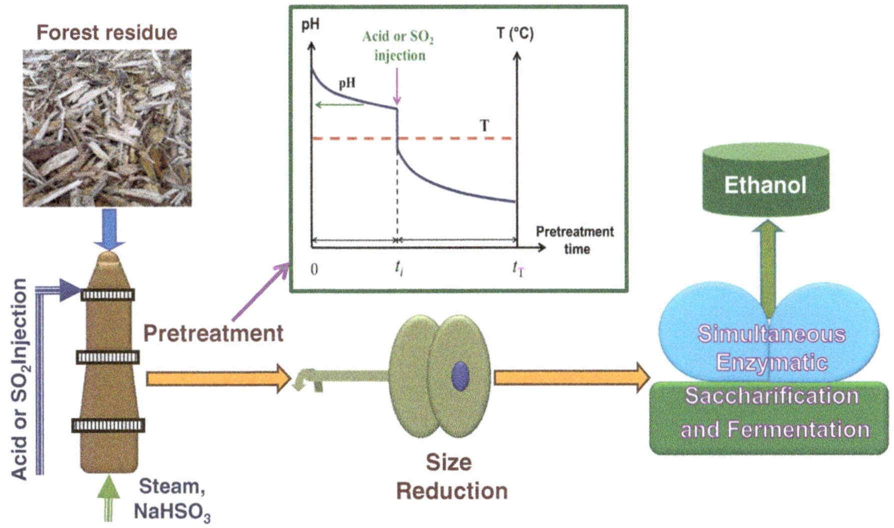

Figure 1 A schematic experimental block flow diagram with the illustration of the "pH profiling" concept in SPORL for high titer and yield ethanol production without detoxification from a Douglas-fir forest residue.

observation indicted that the small fractions had darker colors and contained more bark than the large fractions (Additional file 1: Figure S1). The smallest fraction (<3.2 mm) was only approximately 5% of the total mass (Additional file 1: Figure S2a) but contained 20% of the total bark in the harvested residue (Additional file 1: Figure S2b). Therefore, rejecting the smallest fraction (<3.2 mm) through screening can selectively reduce the bark and therefore lignin content to result in an upgraded residue. The resultant Douglas-fir forest residue labeled as FS-10 had a bark content of 4.7% compared with 5.9% in the as-harvested residue to result in a lower lignin and higher carbohydrate contents of 29.3% and 56.4%, respectively, than the 30.5% and 50.3% of the as-harvested residue (Table 1). Upgrading through screening increased the carbohydrate content by 12%.

FS-10 cell wall modification by pretreatments with and without pH profiling

Four SPORL pretreatments of Douglas-fir forest residue FS-10 were conducted with one control, i.e., no active pH profiling, and three pH profiling runs. The key wood component yields from the pretreated washed solids and pretreatment hydrolysates (spent liquors) are listed in Table 2. The sample labels txxA4B12 represents acid application delay time t_i = xx, sulfuric acid loading of 4 mL/L and sodium bisulfite charge on wood 12 wt.%, respectively. Therefore, t0A4B12 is the control run and the rests are pH profiling runs. As expected, the pH profiling runs increased delignification due to delayed acid application, i.e., lignin yield in washed solids was reduced from 171 g/kg for the control run to approximately 130 g/kg for the pH profiling runs. pH profiling reduced hemicellulose removal. Xylan and mannan yields were increased from 14 and 16 to approximately 21 and 24 g/kg on average of three pH profiling runs, respectively. pH profiling also slightly increased glucan yield from the washed solids which resulted in a lower glucose yield in the spent liquor. The increased carbohydrate yields in the solid substrates potentially can result in increased monomeric sugar recovery through enzymatic saccharification, beneficial to increasing overall sugar yield. pH profiling also reduced monomeric xylose and mannose yields in the spent liquor. However, xylose yield as percentage of dissolved xylan was remained approximately 40% while mannose yield as percentage of dissolved mannan was reduced from approximately 67% to 45% perhaps due to the incomplete hydrolysis as a result of delayed acid application. A substantial amount of dissolved xylan and mannan are expected in the form of oligo-xylose and oligo-mannose, respectively, in the pretreatment spent liquor. The reduction in furan formation by pH profiling was apparent (Table 2) as further discussed in the next section. However, the delay time in acid injection among the three

Table 1 Chemical compositions of the as harvested and screening upgraded (FS-10) Douglas-fir forest residue

Untreated forest residue	Bark (%)	Glucan (%)	Xylan (%)	Mannan (%)	K Lignin (%)	G + X + M[a] (%)
As harvested	5.9	38.4	4.4	7.5	30.5	50.3
Screening upgraded (FS-10)	4.7	41.0	5.7	9.7	29.3	56.4

[a]Major carbohydrate = glucan (G) + xylan (X) + mannan (M).

Table 2 Comparisons of yields (per kg of FS-10) of key wood components in the recovered solids and liquid hydrolysate from SPORL pretreatments at 165°C for 75 min with and without pH profiling

	Washed solids (g)					Pretreatment hydrolysate (spent liquor) (g) (only monomeric sugars were reported)							
Pretreated sample[a]	Glucan	Xylan	Mannan	K lignin	Solids yields[b]	Glucose as glucan	Xylose as xylan	Mannose as mannan	K lignin[c]	Furfural as pentosan	HMF as hexosan	Yield[d]	Total yield
t0A4B12	388.4	13.7	15.8	170.6	612.7	21.8	17.7	54.6	120.3	5.0	9.3	228.7	841.4
t25A4B12	405.6	22.7	23.7	130.1	603.9	10.0	13.5	32.8	160.8	2.2	4.0	222.9	826.8
t35A4B12	410.7	20.4	21.8	137.7	616.5	15.7	15.1	38.2	153.2	2.0	3.8	228.0	844.5
t45A4B12	406.5	21.0	25.7	131.9	622.3	9.1	14.2	30.9	159.0	2.0	2.9	218.1	840.4

[a]txx stands for acid injection time in min, A4 stands for 4 mL of sulfuric acid in 1 L of initial pretreatment solution, and B12 stands for 12 wt.% of sodium bisulfite charge on wood.
[b]As measured after disk milling.
[c]Based on balance of lignin.
[d]Sum of listed pretreatment hydrolysate components.

pH profiling runs had minimal effects on component mass yields.

Effects of pH profiling on furan and acetic acid formation

To illustrate the effectiveness of pH profiling in reducing the formation of fermentation inhibitors such as furan and acetic acid, the hydroxymethyl furfural (HMF), furfural, and acetic acid concentrations in the pretreatment spent liquor was plotted against the acid injection delay time t_i as shown in Figure 2. HMF, furfural, and acetic acid concentrations in the pretreatment spent liquor were reduced from approximately 2.5, 1.2, and 5.3 g/L, respectively, for the control run to approximately 0.8, 0.5, and 3.5 g/L, or by approximately 70%, 60%, and 35% when using the pH profiling technique. The reductions of furan formation resulted from improved carbohydrate preservation (Table 2) due to delayed acid application. The delayed acid application might also reduce the degree of deacetylation reaction

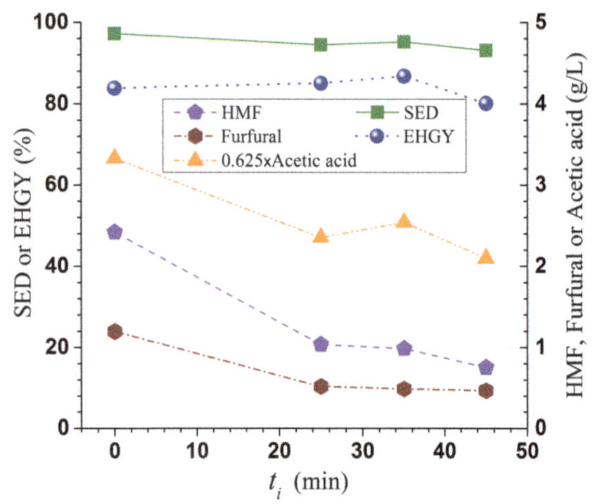

Figure 2 Effects of acid injection delay time t_i in pretreating a Douglas-fir forest residue on resultant substrate enzymatic digestibility (SED), enzymatic hydrolysis glucose yield (EHGY), HMF, furfural, and acetic acid formation.

to result in low acetic acid production. The results also suggested that acid injection time t_i did not substantially affect the reductions in furan and acetic acid formation for the studied t_i range between 25 to 45 min.

Effect of pH profiling on enzymatic saccharification of pretreated FS-10 solids

The effects of pH profiling on pretreated FS-10 substrate enzymatic digestibility (SED), defined as the percentage of substrate glucan enzymatically saccharified to glucose, and enzymatic hydrolysis glucose yield (EHGY), defined as the percentage of wood glucan recovered as glucose through enzymatic hydrolysis alone, can be observed from Figure 2. It appears that SED was not negatively affected by pH profiling considering that the measurement errors in enzymatic hydrolysis were approximately 2% (not shown for clarity). The results suggested that improved delignification in the pH profiling runs compensated for the reduced hemicellulose removal (Table 2) to achieve similar SED. Similarly, EHGY was not substantially affected by pH profiling. The increased glucan recovery from the solid substrates as discussed previously (Table 2) compensated for the slight reduction in SED by pH profiling to maintain the same level of EHGY. The time-dependent SED data showed that pH profiling reduced the rate of saccharification, but the terminal saccharification efficiency after 72 h was approximately the same as that of the control run (Figure 3).

Comparisons of high solid fermentation among control and pH profile runs

Quasi-simultaneous enzymatic saccharification and fermentation (Q-SSF) of the pretreated whole slurries from the control and two pH profiling runs were conducted at 21 wt.% total solid loading (approximately water insoluble solid loading of 15 wt.%). The low-temperature pretreatment of 165°C allowed Q-SSF without detoxification even for the control run, perhaps due to sufficient yeast loading at initial optical density (600 nm) of 5 (Figure 4a) [17,29]. The reduced furan formation in the two pH profile runs

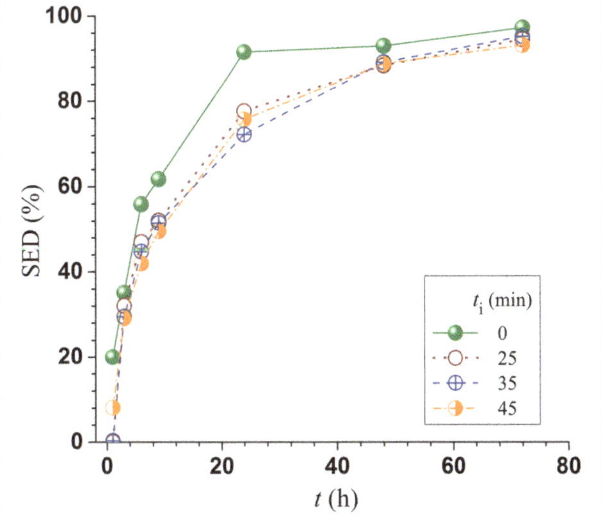

Figure 3 Effect of acid injection delay time t_i in pretreating a Douglas-fir forest residue on time-dependent resultant substrate enzymatic digestibility (SED).

facilitated fermentation as can be seen from the increased glucose consumption and ethanol productivity in the first 72 h (Table 3), which resulted in higher final terminal ethanol concentrations and yields (Table 3, Figure 4a). The improved carbohydrate preservation on solids (Table 2) also contributed to the higher terminal ethanol titer and yield for the two pH profiling runs. This can be seen from the higher initial glucose concentration for the pH profiling run with acid injection delay time $t_i = 25$ min (Figure 4b), though the increase in initial glucose concentration for the pH profiling run with acid injection delay time $t_i = 45$ min was insignificant (Figure 4b) due to slightly reduced saccharification efficiency (Figure 3). The continuous enzymatic saccharification throughout the entire Q-SSF process as can be seen from the increase in glucose concentration even after 144 h of fermentation suggests the importance of improving carbohydrate preservation in solids to ethanol production. A terminal ethanol titer of 48.9 g/L and yield 0.494 g/g sugar FS-10 was achieved for the pH profiling run with acid application delay time $t_i = 25$ min compared with 38.6 g/L and yield of 0.390 g/g sugar for the control run (Table 3), or an increase over 25% in titer and 27% in yield. It was noticed in Figure 4a that the control run had a very high standard deviation at the peak ethanol concentration or 120 h. Examining the time-dependent ethanol concentration data for the two duplicate fermentation runs (not shown) indicated that run II had a large spike at 120 h followed by a sudden decrease to the same ethanol level in run I. However, both runs showed very similar time-dependent glucose consumption behavior with no abrupt changes. Furthermore, glucose was nearly consumed after 120 h, suggesting no support for the ethanol spike observed in run II. We believe that the large standard deviation in ethanol at 120 h was due to sampling in a nonuniform slurry. Using the HMF concentration in the control run as the inhibitor tolerance limit, an estimated solid loading of approximately 38% can be implemented using pH profiling pretreated whole slurry of FS-10 without detoxification. Considering the effect of other inhibitors on fermentation, a realistic solid loading of approximately 30% may be implemented. This is a substantial advantage in terms of achieving high ethanol titer and reducing water usage in processing. With reduced inhibitor concentration, one can also reduce yeast loading to achieve good ethanol productivity as we demonstrated previously [17].

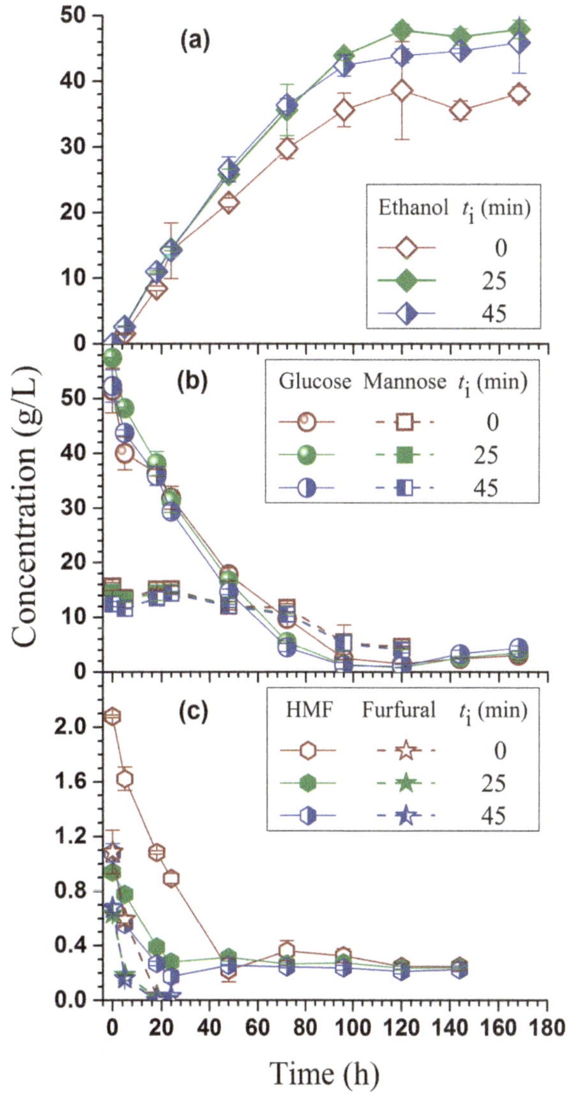

Figure 4 Comparisons of time-dependent (a) ethanol production, (b) sugar consumptions, (c) and furan metabolization during high solid fermentation of the SPORL pretreated Douglas-fir forest residue with and without pH profiling.

Table 3 Comparisons of fermentation performance among pretreatments with and without pH profiling.

	Control ($t_i = 0$)	pH profiling ($t_i = 25$ min)	pH profiling ($t_i = 45$ min)
Average fermentation performance measure (g/L/h)			
Glucose consumption (72 h)	−0.54 ± 0.06	−0.70 ± 0.07	−0.64 ± 0.05
Ethanol productivity (72 h)	0.42 ± 0.03	0.50 ± 0.02	0.51 ± 0.02
HMF metabolization (24 h)	−0.05 ± 0.002	−0.03 ± 0.002	−0.02 ± 0.006
Maximal ethanol production			
Terminal ethanol concentration (g/L)	38.6 ± 7.5	48.9 ± 1.4	45.9 ± 4.6
Ethanol yield (g g/sugar)[a]	0.390 ± 0.076	0.494 ± 0.014	0.460 ± 0.046
Ethanol yield (L/tonne wood)[b]	215 ± 42	297 ± 9	259 ± 26
Ethanol yield (% theoretical)[c]	52.8 ± 10.3	73.1 ± 2.1	63.7 ± 6.4

[a]Based on the total of glucan, mannan, and xylan in the pretreated-solids and glucose, mannose, and xylose in the pretreatment spent liquor.
[b]Calculated from measured amount of ethanol yield from the amount of pretreated whole slurry used in fermentation and the yield of whole slurry from pretreatment.
[c]Theoretical yield (406 L tonne wood^{-1}) based on total glucan, mannan, and xylan in the untreated forest residue of FS10.

The acid injection time did not substantially affect glucose consumption in fermentation (Figure 4b, Table 3) because both runs produced similar level of inhibitors (Figure 2, Table 2). Similarly, acid injection time did not substantially affect mannose consumption either (Figure 4b). Mannose consumption was slow in the first 24 h most likely due to the presence of inhibitors. Mannose was not completely consumed after 120 h fermentation. Xylose fermentation was negligible with xylose consumption of only approximately 10% (not shown) though the strain YRH400 is capable of fermenting xylose, in agreement with a previous study [25] using the same strain fermenting a whole slurry of lodgepole pine with a low xylose concentration. The strain YRH400 was able to metabolize both HMF and furfural (Figure 4c), in agreement with a previous study [25]. However, a longer metabolization time was needed for the control run than those for the two pH profiling runs (Figure 4c).

When comparing the two fermentation runs using pH profiling pretreated whole slurry of FS-10, a longer acid application delay time of $t_i = 45$ min slightly reduced terminal ethanol concentration (45.9 g/L) as well as ethanol yield (0.460 g/g sugar) compared with those achieved from $t_i = 25$ min run of 48.9 g/L and (0.494 g/g sugar), respectively, (Table 3). This is in agreement with the slightly reduced SED and EHGY (Figure 2). The ethanol yield and titer from pH profiling run with $t_i = 45$ min run were still respectively higher than those from the control run, i.e., 38.6 g/L and 0.390 g/g sugar.

Overall mass balance in ethanol production with pH profiling

An overall mass balance for the pH profiling run with injection delay time $t_i = 25$ was conducted. Component yields from both the washed solids and pretreatment spent liquor were separately determined based on composition analyses of samples of washed solids and liquor (Figure 5). Actual liquor and solid separation was not conducted as shown in Figure 1, rather the whole slurries were used for simultaneous enzymatic saccharification and fermentation in ethanol production. Approximately 56% of the FS-10 lignin was solubilized to become lignosulfonate that can be recovered as a valuable co-product [17,25]. Substantial amounts of mannan (25%) and xylan (40%) were retained on the washed solids. Glucan dissolution was negligible. Percentage of major carbohydrates (glucan, mannan, and xylan) retained on solids was 80% compared with 74% for the control run. Final ethanol yield was 234 ± 6.7 kg (297 ± 9 L)/tonne FS-10 at 48.9 g/L, equivalent to 73.1 ± 2.1% theoretical ethanol yield based on the glucan, xylan, and mannan contents of FS-10. Comparing with 169 ± 32 kg (215 ± 42 L)/tonne FS-10 at 38.6 g/L, pH profiling increased ethanol yield by 38%.

Conclusions

pH profiling in SPORL effectively reduced furan formation by as much as 70% depending on acid delay time when pretreating a Douglas-fir forest residue at a moderate temperature of 165°C for 75 min. pH profiling also reduced carbohydrate degradation. The improved carbohydrate yield in pretreated solids and reduced fermentation inhibitors with pH profiling enabled fermentation at 21% solids to result in a terminal ethanol titer of 48.9 ± 1.4 g/L and yield of 297 ± 9 L/tonne FS-10, which are substantially higher, i.e., by 27% in titer and by 38% in yield, than those of a control SPORL run without pH profiling. Based on the amount of furan formation, the estimated potential solid loading over 35% can be efficiently fermented without detoxification. SPORL was developed based on sulfite pulping which reduced risk for commercial scale-up. The demonstrated robust performance for high titer and high yield ethanol production with detoxification from one of the most recalcitrant feedstock—Douglas-fir forest residue—has commercial significance.

Materials and methods

Materials

Douglas-fir forest residue was collected from roadside piles resulting from a regeneration harvest in a Douglas-fir stand on Mosby Creek owned by Weyerhaeuser Company southeast of Cottage Grove in Lane County, OR. The residues were ground on January 8, 2013 by a Peterson horizontal drum-fixed hammer grinder (4710 Horizontal Grinder) using a combination of 76 and 102 mm

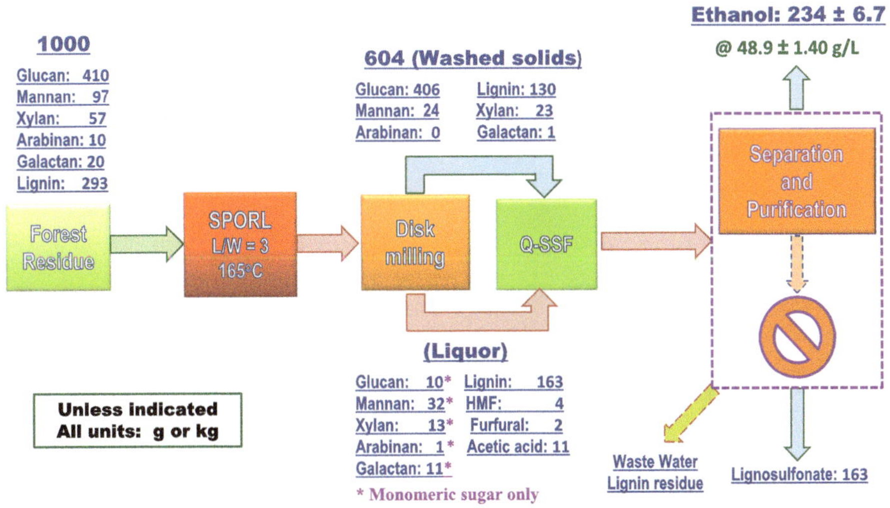

Figure 5 An overall mass balance for high titer and yield ethanol production from a Douglas-fir forest residue pretreated by SPORL with "pH profiling" and acid injection delay time $t_i = 25$ min.

grates. The harvested residues were shipped to Weyerhaeuser Company at Federal Way, WA. The moisture content was 43.9% measured at arrival. The collected residues were screened using a gyratory screen (Black-Clawson) equipped with a 1.75-inch diameter round hole punched plate top deck to remove oversized particles and a 1/8-inch clear opening woven wire bottom screen (6 wires/inch mesh) to remove fines. The screen reject fines were 7.6%. Physical fractionation through screening was proven effective to reduce ash and bark content to upgrade forest residue harvested using a chipping method [30]. The oversized particles were hammer milled at West Salem Equipment, which resulted near-zero oversized particles and 14.9% fines of the 9.8% initial screen oversize particles. The total rejection of fines was therefore at 9% with near zero-rejection of oversize particles. The accept residue or FS10 was then air-dried to moisture content of 15% and shipped to the USDA Forest Products Laboratory, Madison, WI. The chemical composition of FS10 is listed in Table 1.

The bark contents of the large fractions (>25.4 mm) were determined gravimetrically after manually separating the bark from the bark-free wood. Because of difficulties in separating bark from wood for the small fractions (<25.4 mm), the bark contents of the small fractions were calculated using the measured lignin and glucan contents of bark, bark-free wood, and the forest residue fractions as described in our previous study [30].

Commercial cellulase enzymes Cellic® CTec3 (abbreviated CTec3) were generously provided by Novozymes North America (Franklinton, North Carolina, USA). The cellulase activity was 217 filter paper unit (FPU)/mL as calibrated by a literature method [31]. Sodium acetate buffer, sulfuric acid, and sodium bisulfite were used as received from Sigma-Aldrich (St. Louis, MO). All chemicals were ACS reagent grade.

S. cerevisiae YRH400 is an engineered fungal strain for xylose fermentation [32]. The strain was grown at 30°C for 2 days on YPD agar plates containing 10 g/L yeast extract, 20 g/L peptone, 20 g/L glucose, and 20 g/L agar. A colony from the plate was transferred by loop to liquid YPD medium in a flask and cultured overnight at 30°C with agitation at 90 rpm on a shaking bed incubator (Thermo Fisher Scientific, Model 4450, Waltham, MA). The yeast seed concentration was monitored using optical density at 600 nm by a UV-vis spectrometer (Model 8453, UV-visible spectroscopy system, Agilent Technologies, Palo Alto, CA). The cultured medium was used to inoculate the fermentation culture.

Pretreatment and pH profiling

SPORL pretreatments of Douglas-fir forest residue FS-10 with or without pH profiling were conducted in a 23 L rotating laboratory wood pulping digester using a dilute solution of sodium sulfite as described elsewhere [3,25]. In commercial practice, sulfite solution is prepared by bubbling SO_2 into a hydroxide solution of desired pH [9]. Using pH profiling, the amount of SO_2 applied initially should be based on the desired initial pH whether in the alkaline (pH 8–12), neutral (pH = 6–8), or bisulfite range (pH = 3–5). To simplify laboratory practice, aqueous sodium bisulfite solution of pH 4.0 was used together with sulfuric acid to adjust the pH of the sulfite solution. In the control SPORL run, sodium bisulfite together with sulfuric acid were all applied at the beginning ($t_i = 0$) to a desired initial solution pH of approximately 1.8 when measured at room temperature (Table 4). No additional acid was applied later in pretreatment. For the three

Table 4 List of pretreatment conditions of FS10 at 165°C using SPORL with and without pH profiling

Run label[a]	t_T (min)	Initial pH	Sulfuric acid at $t = 0$ (wt.%)	Sodium bisulfite at $t = 0$ (wt.%)	t_i (min)	Sulfuric acid at t_i (wt.%)	Final pH	L/W (L/kg)
t0A4B12	75	1.79	2.2	12	0	0	1.45	3
t25A4B12	75	4.06	0	12	25	2.2	1.72	3
t35A4B12	75	4.06	0	12	35	2.2	1.40	3
t45A4B12	75	4.06	0	12	45	2.2	1.66	3

[a]txx stands for acid injection time in min, A4 stands for 4 mL of sulfuric acid in 1 L of initial pretreatment solution, and B12 stands for 12 wt.% of sodium bisulfite charge on wood.

pH profiling runs, the same amount of sodium bisulfite ($B = 12$ wt%) and sulfuric acid ($A = 2.2$ wt%; or 4 mL/L) on oven dry wood base as those of the control run were used, but acid was applied at varied delayed times of t_i during pretreatment (insert in Figure 1) through injection. The total pretreatment duration ($t_T = 75$ min), temperature ($T = 165$°C), and liquid volume to wood mass ratio ($L/W = 3$ L/kg) were also identical for all four runs with and without pH profiling (Table 4), so that fair comparisons can be made to demonstrate the advantages of the pH profiling concept. This set of pretreatment conditions was chosen based on previous SPORL optimization study using softwood [25] as well as a similar Douglas-fir forest residue [16].

At the end of the each pretreatment, an aliquot of spent liquor was taken for chemical composition analysis. The solids and spent liquor (neutralized) were together fed to a laboratory disk mill (Andritz Sprout-Bauer Atmospheric Refiner, Springfield, OH) for size reduction as described previously [33]. The disk plates had a pattern of D2B-505 and the disk plate gap was set at 1.0 mm. A sample of mill solids was washed for chemical composition analysis, yield determination, and enzymatic hydrolysis. The resultant FS-10 whole slurries, i.e., the complete mixture of disk-milled pretreated solids and spent liquor, were neutralized again to pH approximately 6.2 with solid lime for saccharification and fermentation.

Enzymatic hydrolysis

Enzymatic hydrolyses of the washed solids were conducted at 2% (w/v) in 50 mL of 50 mM acetate buffer (pH 5.5) on a shake/incubator (Thermo Fisher Scientific, Model 4450, Waltham, MA) at 50°C and 200 rpm. An elevated pH of 5.5, higher than the commonly used pH 4.8–5.0, can significantly reduce nonproductive cellulase binding to lignin leading to enhanced lignocellulose saccharification [34-36]. The CTec3 loading was 15 FPU/g glucan. Aliquots of 1 mL enzymatic hydrolysate were taken periodically for glucose analysis after centrifugation at 13,000 *g* for 5 min. Each data point is the average of two analyses. The mean values and standard deviations (as error bars) from replicate runs were reported in plots.

Quasi-simultaneous enzymatic saccharification and fermentation

Q-SSF of the pretreated FS-10 whole slurry was carried out in 250-mL Erlenmeyer flasks using a shaker/incubator (Thermo Fisher Scientific, Model 4450, Waltham, MA). Acetic acid/sodium acetate buffer (50 mM) of pH 6.0 was added into each of the pH-adjusted pretreated FS-10 whole slurry to conduct enzymatic saccharification using CTec3 at 24 FPU/g glucan at a total solid loadings of 21%. An elevated pH of 6.0, higher than the commonly used pH of 4.8–5.0, and lignosulfonate in the SPORL pretreatment spent liquor reduced nonproductive cellulase binding to lignin and enhanced lignocellulose saccharification [34-36]. The use of a higher CTec3 loading than that used for enzymatic hydrolysis was to facilitate solid liquefaction at high solids as Q-SSF was conducted on shaking bed without shear mixing. Liquefaction of the solids was conducted at 50°C and 200 rpm. The mixture was then cooled down to 35°C, and the shaker speed was reduced to 90 rpm and inoculated with 1 mL of yeast seed. The initial optical density at 600 nm of the yeast for all fermentation experiments was estimated at 5. No nutrients were applied during fermentation. Samples of the fermentation broth were taken periodically for analysis of monosaccharides, inhibitors, and ethanol. Reported results were the average of duplicate analyses. Replicate fermentation runs were conducted to ensure experimental repeatability. The standard deviations were used as error bars in plotting.

Analytical methods

The chemical compositions of the untreated and pretreated lignocelluloses were analyzed as described previously [37]. All lignocellulosic samples were Wiley milled (Model No. 2, Arthur Thomas Co, Philadelphia, PA, USA) to 20 mesh (~1 mm) and hydrolyzed in two stages using sulfuric acid of 72% (v/v) at 30°C for 1 h and 3.6% (v/v) at 120°C for 1 h. Carbohydrates of the hydrolysates were analyzed by high-performance anion exchange chromatography with pulsed amperometric detection (ICS-5000, Dionex). Klason lignin was quantified gravimetrically [38]. For fast analysis, glucose in the enzymatic hydrolysates were measured using a commercial glucose analyzer (YSI 2700S, YSI Inc., Yellow Springs, OH, USA).

Monosaccharides (glucose, mannose, xylose, arabinose, and galactose) in the enzymatic hydrolysates and fermentation broths were determined using a HPLC system (Ultimate 3000, Thermo Scientific, Sunnyvale, CA) equipped with an RI (RI-101) and UV (VWD-3400RS) detector and BioRad Aminex HPX-87P column (300 × 7.8 mm) operated at 80°C. Double distilled water (d.d.w.) was used as eluent at a flow of 0.6 mL/min. Inhibitors (acetic acid, furfural and HMF) and ethanol were measured by the same HPLC system equipped with BioRad Aminex HPX-87H column (300 × 7.8 mm) operated at 60°C. Dilute sulfuric acid solution of 5 mM was used as eluent at a flow rate of 0.6 mL/min. All sample injection volumes were 20 μL. Samples were diluted in deionized water and filtered by a 0.22-μm filter prior to injection.

Additional files

Additional file 1 Figure S1: Photos of different fractions of the as harvested Douglas-fir forest residue. **(a)** <3.2 mm. **(b)** 4.8–6.4 mm. **(c)** 9.5–12.7 mm. **(d)** 19.1–22.2 mm. **(e)** 28.6–31.8 mm. **(f)** The wet as-harvested Douglas-fir forest residue. **Figure S2:** Fractional (particle size) mass distributions of the as harvested Douglas-fir forest residue. **(a)** Oven dry and wet mass. **(b)** Bark mass.

Abbreviations

SPORL: sulfite pretreatment to overcome the recalcitrance of lignocelluloses; CHF: combined hydrolysis factor that accounts for pretreatment temperature, time, and chemical concentrations; SED: substrate enzymatic digestibility defined as percent of solid substrate glucan enzymatically saccharified into glucose; EHGY: enzymatic hydrolysis glucose yield defined as enzymatic glucose yield as percent of glucan in the untreated lignocellulose; Q-SSF: quasi-simultaneous enzymatic saccharification and fermentation; FPU: filter paper unit.

Competing interests

Zhu and Gleisner are co-inventors of a US patent application on SPORL.

Authors' contributions

JC, SYL, and RG conducted the pretreatment experiments. JC conducted enzymatic saccharification and fermentation and all analytical works. JYZ conceived the pH profile concept and drafted and revised the manuscript. RG implemented the pH profiling concept in initial experiments. All authors read and approved the final manuscript.

Authors' information

Cheng was a visiting scientist and Leu was a postdoctoral fellow at the USDA Forest Service (USDA-FS), Forest Products Lab (FPL). Gleisner and Zhu are technical staff at the USDA-FS-FPL. Zhu is the Grantee of the Fulbright-Aalto University Distinguished Chair in Energy and Sustainable Use of Natural Resources (2015-2016), Aalto University, Helsinki, Finland. He serves on the editorial boards of several technical journals.

Acknowledgements

This work is supported by the Agriculture and Food Research Initiative Competitive grant (no. 2011-68005-30416), USDA National Institute of Food and Agriculture (NIFA) through the Northwest Advanced Renewables Alliance (NARA). The authors thank Novozymes North America for providing the cellulase enzymes, Drs. Ronald Hector and Bruce Dien for providing the YRH400 strain, Fred Matt of USDA-FS-FPL for conducting chemical composition analysis, Diane Dietrich USDA-FS-FPL for assisting HPLC analyses, Drs. Xuejun Pan and Troy Runge of University of Wisconsin-Madison for allowing us access to their lab equipment, and Gevan Marrs of Weyerhaeuser Company and John Sessions of Oregon State University and the NARA feedstock team for providing the Douglas-fir forest residue. Jiangsu Government Scholarship for Overseas Studies provided partial financial support for the visiting appointment of Dr. Jinlan Cheng at the USDA Forest Service, Forest Products Laboratory (FPL), Madison, WI.

Author details

[1]Jiangsu Provincial Key Lab of Pulp and Paper Science and Technology, Nanjing Forestry University, Nanjing, China. [2]USDA Forest Service, Forest Products Laboratory, Madison, WI 53719, USA. [3]Department of Civil and Environmental Engineering, Hong Kong Polytechnic University, Kowloon, Hong Kong.

References

1. Larsson S, Palmqvist E, Hahn-Hagerdal B, Tengborg C, Stenberg K, Zacchi G, et al. The generation of fermentation inhibitors during dilute acid hydrolysis of softwood. Enzyme Microb Technol. 1999;24:151–9.
2. Zhu W, Houtman CJ, Zhu JY, Gleisner R, Chen KF. Quantitative predictions of bioconversion of aspen by dilute acid and SPORL pretreatments using a unified combined hydrolysis factor (CHF). Process Biochem. 2012;47:785–91.
3. Zhu JY, Pan XJ, Wang GS, Gleisner R. Sulfite pretreatment (SPORL) for robust enzymatic saccharification of spruce and red pine. Bioresour Technol. 2009;100(8):2411–8.
4. Pan XJ, Gilkes N, Kadla J, Pye K, Saka S, Gregg D, et al. Bioconversion of hybrid poplar to ethanol and co-products using an organosolv fractionation process: optimization of process yields. Biotechnol Bioeng. 2006;94(5):851–61.
5. Cantarella M, Cantarella L, Gallifuoco A, Spera A, Alfani F. Effect of inhibitors released during steam-explosion treatment of poplar wood on subsequent enzymatic hydrolysis and SSF. Biotechnol Prog. 2004;20(1):200–6.
6. Soderstrom J, Galbe M, Zacchi G. Effects of washing on yield in one- and two-step steam pretreatment of softwood for production of ethanol. Biotechnol Prog. 2004;20:744–9.
7. Chen X, Shekiro J, Franden MA, Wang W, Zhang M, Kuhn E, et al. The impacts of deacetylation prior to dilute acid pretreatment on the bioethanol process. Biotechnol Biofuels. 2012;5:8.
8. Zhang C, Houtman CJ, Zhu JY. Using low temperature to balance enzymatic saccharification and furan formation in SPORL pretreatment of Douglas-fir. Process Biochem. 2014;49:466–73.
9. Zhu JY, Chandra MS, Gu F, Gleisner R, Reiner R, Sessions J, et al. Using sulfite chemistry for robust bioconversion of Douglas-fir forest residue to bioethanol at high titer and lignosulfonate: a pilot-scale evaluation. Bioresour Technol. 2015;179:390–7.
10. National Research Council: Renewable Fuel Standard: Potential Economic and Environmental Effects of U.S. Biofuel Policy. http://www.nap.edu/openbook.php?record_id=13105 2011, ISBN 978-0-309-18751-0.
11. Perlack RD, Stokes BJ: U.S. Billion-Ton Update: Biomass Supply for a Bioenergy and Bioproducts Industry. DOE Report DE-AC05-00R22725, Oakridge National Laboratory, 2011.
12. Kirschbaum MUF. To sink or burn? A discussion of the potential contributions of forests to greenhouse gas balances through storing carbon or providing biofuels. Biomass Bioenerg. 2003;24(4–5):297–310.
13. Gan J, Smith CT. Availability of logging residues and potential for electricity production and carbon displacement in the USA. Biomass Bioenerg. 2006;30(12):1011–20.
14. Stephen JD, Mabee WE, Saddler JN. Biomass logistics as a determinant of second-generation biofuel facility scale, location, and technology selection. Biofpr. 2010;4:503–18.
15. Zhu JY, Zhuang XS. Conceptual net energy output for biofuel production from lignocellulosic biomass through biorefining. Prog Energ Combustion Sci. 2012;38(4):583–9.
16. Leu S-Y, Gleisner R, Zhu JY, Sessions J, Marrs G. Robust enzymatic saccharification of a Douglas-fir forest harvest residue by SPORL. Biomass Bioenerg. 2013;59:393–401.
17. Zhou H, Leu S-Y, Wu X, Zhu JY, Gleisner R, Yang D, et al. Comparisons of high titer ethanol production and lignosulfonate properties by SPORL pretreatment of lodgepole pine at two temperatures. RSC Adv. 2014;4:27033–8.
18. Ingruber O: Sulfite science, part I: sulfite pulping cooking liquor and the four bases. In: Sulfite Science and Technology. Edited by Ingruber O, Kocurek M,

Wong A, vol. 4, 3rd edition. Atlanta: The joint textbook Committee of the Paper Industry, TAPPI/CPPA; 1985: 3–23.

19. Tunc MS, Van Heiningen ARP. Hemicellulose extraction of mixed southern hardwood with water at 150°C: effect of time. Industr Eng Chem Res. 2008;47(18):7031–7.
20. Leu SY, Zhu JY. Substrate-related factors affecting enzymatic saccharification of lignocelluloses: our recent understanding. Bioenerg Res. 2013;6(2):405–15.
21. Wang QQ, He Z, Zhu Z, Zhang Y-HP, Ni Y, Luo XL, et al. Evaluations of cellulose accessibilities of lignocelluloses by solute exclusion and protein adsorption techniques. Biotechnol Bioeng. 2012;109(2):381–9.
22. Yang B, Wyman CE. Effect of xylan and lignin removal by batch and flowthrough pretreatment on the enzymatic digestibility of corn stover cellulose. Biotechnol Bioeng. 2004;86(1):88–95.
23. Rollin JA, Zhu Z, Sathitsuksanoh N, Zhang Y-HP. Increasing cellulose accessibility is more important than removing lignin: a comparison of cellulose solvent-based lignocellulose fractionation and soaking in aqueous ammonia. Biotechnol Bioeng. 2011;108(1):22–30.
24. Jeoh T, Ishizawa CI, Davis MF, Himmel ME, Adney WS, Johnson DK. Cellulase digestibility of pretreated biomass is limited by cellulose accessibility. Biotechnol Bioeng. 2007;98(1):112–22.
25. Zhou H, Zhu JY, Luo X, Leu S-Y, Wu X, Gleisner R, et al. Bioconversion of beetle-killed lodgepole pine using SPORL: process scale-up design, lignin coproduct, and high solids fermentation without detoxification. Ind Eng Chem Res. 2013;52(45):16057–65.
26. Croon I: The flexibility of sodium-based two-stage neutral-acid sulfite pulping. In: Pulp Paper Mag Canada, February 1965:T71-T76.
27. Lagergren S: The Stora pulping process and its implications for European paper makers. Svensk Papperstidning 1964, 67(238–243).
28. Sanyer N, Keller EL, Chidester GH. Multistage sulfite pulping of jack pine, balsam fir, spruce, oak, and sweetgum. TAPPI. 1962;45(2):90–104.
29. Zhou H, Zhu JY, Gleisner R, Qiu X, Horn E, Negron J: Pilot-scale demonstration of SPORL for bioconversion of lodgepole pine to bio-ethanol and lignosulfonate. Holzforschung. 2015; 69, DOI: 10.1515/hf-2014-0332.
30. Zhang C, Zhu JY, Gleisner R, Sessions J. Fractionation of forest residues of Douglas-fir for fermentable sugar production by SPORL pretreatment. Bioenergy Res. 2012;5(4):978–88.
31. Wood TM, Bhat M: Methods for measuring cellulase activities. In: In: Colowick SP, Kaplan NO, editors Methods in Enzymology, Vol 160, Biomass (Part A, Cellulose and Hemicellulose) Vol editors:Wood WA, Kellogg ST. New York: Academic Press, Inc, 1988: 87–112.
32. Hector RE, Dien BS, Cotta MA, Qureshi N. Engineering industrial Saccharomyces cerevisiae strains for xylose fermentation and comparison for switchgrass conversion. J Ind Microbiol Biotechnol. 2011;38(9):1193–202.
33. Zhu JY, Wang GS, Pan XJ, Gleisner R. Specific surface to evaluate the efficiencies of milling and pretreatment of wood for enzymatic saccharification. Chem Eng Sci. 2009;64(3):474–85.
34. Lan TQ, Lou H, Zhu JY. Enzymatic saccharification of lignocelluloses should be conducted at elevated pH 5.2–6.2. Bioenerg Res. 2013;6(2):476–85.
35. Lou H, Zhu JY, Lan TQ, Lai H, Qiu X. pH-induced lignin surface modification to reduce nonspecific cellulase binding and enhance enzymatic saccharification of lignocelluloses. ChemSusChem. 2013;6(5):919–27.
36. Wang ZJ, Lan TQ, Zhu JY. Lignosulfonate and elevated pH can enhance enzymatic saccharification of lignocelluloses. Biotechnol Biofuels. 2013;6:9.
37. Luo X, Gleisner R, Tian S, Negron J, Horn E, Pan XJ, et al. Evaluation of mountain beetle infested lodgepole pine for cellulosic ethanol production by SPORL pretreatment. Ind Eng Chem Res. 2010;49(17):8258–66.
38. Dence CW: The determination of lignin. In: Methods in lignin chemistry. Edited by Lin SY, Dence CW. Berlin: Springer-Verlag; 1992: 33–61.

4

Influence of bark on fuel ethanol production from steam-pretreated spruce

Balázs Frankó*, Mats Galbe and Ola Wallberg

Abstract

Background: Bark and bark-containing forest residues have the potential for utilization as raw material for lignocellulosic ethanol production due to their abundance and low cost. However, the different physical properties and chemical composition of bark compared to the conventionally used wood chips may influence the spruce-to-ethanol bioconversion process. This study assesses the impact of bark on the overall bioconversion in two process configurations, separate hydrolysis and fermentation (SHF) and simultaneous saccharification and fermentation (SSF), utilizing steam-pretreated spruce bark and wood mixtures.

Results: Mixtures of different proportions of spruce bark and wood chips were subjected to SO_2-catalyzed steam pretreatment at 210°C for five minutes, which has been shown to be effective for the pretreatment of spruce wood chips. The final ethanol concentration was the highest without bark and decreased significantly with increasing proportions of bark in both process configurations. However, this decrease cannot be attributed solely to the lower availability of the carbohydrates in mixtures containing bark, as the ethanol yield also decreased, from 85 to 59% in SSF and from 84 to 51% in SHF, as the mass fraction of bark was increased from 0 to 100%.

Conclusions: The results show that it was significantly more difficult to hydrolyse spruce bark to monomeric sugars than wood chips. Bark had an adverse effect on the whole bioconversion process due to its lower enzymatic hydrolyzability. On the other hand, bark inclusion had no detrimental effect on the fermentability of steam-pretreated spruce wood and bark mixtures. It was also observed that lower amounts of inhibitory degradation products were formed during the steam pretreatment of spruce bark than during the steam pretreatment of wood chips.

Keywords: Ethanol, Softwood, Spruce, Bark, Steam pretreatment, SSF, SHF

Background

The driving force behind the exploitation of renewable energy sources is the necessity to shift from a fossil-fuel dependent economy to one based on renewable resources. Biomass can be used to efficiently produce renewable liquid or gaseous fuels, providing alternatives to fossil fuels [1]. Ethanol, for instance, is already being produced from sugar and starch crops, and is used worldwide, as a consequence of policies promoting ethanol production [2-5]. However, the controversy of ethanol production from sugar and starch crops (first-generation ethanol) has led to the development of technologies employing lignocellulosic biomass as raw material [6-8].

The utilization of lignocellulosic biomass to produce ethanol provides an alternative to sugar and starch crops. However, the additional cost of the lignocellulosic ethanol production process, resulting from the necessity of pretreatment and enzymes for lignocellulosic biomass refining, has led to limited profitability in comparison with sugar- and starch-based ethanol production [9]. Thus, there is a need to further decrease the production cost of lignocellulosic ethanol in order for it to become competitive with the first-generation ethanol [10].

One possible means of cost reduction is to utilize abundant low-cost lignocellulosic raw materials such as bark [11]. Bark and bark-containing forest residues could serve as a potential feedstock for ethanol production, although bark is considered to be an inferior raw material to higher-value wood chips due to its composition. Compared with spruce wood chips, spruce bark has a lower content of carbohydrates, and contains significantly more extractives and ash [12]. The lower content of carbohydrates results in decreased sugar concentration after hydrolysis, and thus

* Correspondence: Balazs.Franko@chemeng.lth.se
Department of Chemical Engineering, Lund University, PO Box 124, Getingevägen 60, Lund SE-221 00, Sweden

a lower ethanol concentration after fermentation. The pre-hydrolyzates obtained from pretreated bark can contain elevated amounts of water-soluble extractives and poly-phenols, which may have inhibitory effects on fermenting microorganisms and cellulolytic enzymes [13-15]. Therefore, the combined utilization of bark and wood chips for ethanol production might pose an even greater challenge than the use of softwood chips, which is already demanding. In this case, not only the inherent recalcitrance of the material must be overcome, but also the problems resulting from the significant difference in the physical properties and chemical composition of bark and wood chips. However, if these limitations could be overcome, then abundantly available, low-value forestry residues could be exploited, and existing spruce-to-ethanol production processes could be simplified by not having to debark the material. Ultimately, it is likely that forestry residues available for bioethanol production will contain varying amounts of bark, and it is therefore important to investigate the effects of bark on production processes previously optimized for wood chips only.

Previous studies have mainly focused on the effects of bark on fermentability. Boussaid *et al.* found that including bark led to decreased fermentability of pre-hydrolyzates prepared under low-severity pretreatment conditions, while pre-hydrolyzates prepared under higher severity conditions could be fermented comparably well to ethanol when 9% bark was included [13]. Similar results were obtained by Robinson *et al.*, who found that up to 30% bark, on a dry basis, had negligible effects on the fermentability of pre-hydrolyzates obtained from SO_2-catalyzed steam-exploded Douglas fir whitewood [16]. Although the enzymatic hydrolyzability of bark has been investigated previously [17], there are few reports on the influence of bark on enzymatic hydrolysis and the overall ethanol yield of the ethanol bioconversion process when performed at higher water-insoluble solids (WIS) content [18]. Enzymatic hydrolysis and fermentation must be performed at a higher WIS loading in order to increase the ethanol concentration after fermentation, which is essential to reduce the cost of distillation and thus the marginal production cost production [19].

The aim of the present study was to assess the feasibility of utilizing bark together with spruce wood chips for the fermentative conversion of biomass to ethanol at 10% WIS content, using a commercial enzyme cocktail to hydrolyse the steam-pretreated material, and an industrial strain of *Saccharomyces cerevisiae* as the fermenting microorganism. Spruce bark mixed with wood chips at different ratios ranging from 0 to 100% were subjected to SO_2-catalyzed steam pretreatment at 210°C for five minutes, which has previously been shown to be effective for spruce wood chips [20]. The effects of bark inclusion on the spruce-to-ethanol bioconversion process were investigated by performing separate hydrolysis and fermentation (SHF) and simultaneous saccharification and fermentation (SSF) of the steam-pretreated wood and bark mixtures.

Results and discussion

Steam pretreatment of wood and bark mixtures

The composition of the spruce wood chips and bark is given in Table 1. The content of carbohydrates, with the exception of arabinan, was lower in the bark than in the wood chips. The amount of hexose sugars available in the bark feedstock was only about half of that in the wood feedstock. Even though only neutral carbohydrates were analyzed, spruce bark also contains a significant amount of other polysaccharides, such as pectin [21,22]. The bark contained significantly more extractives and ash than were found in the wood chips. The content of acid-soluble lignin may be slightly overestimated for both feedstocks due to possible interference from other non-lignin components [23].

The compositions of the water-insoluble solid fractions of the steam-pretreated materials were determined, and the results are presented in Table 2. As a result of the lower glucan content of the bark feedstock, the glucan content of the steam-pretreated mixtures decreased with increasing proportions of bark (Table 2). Detectable amounts of sugars originating from the hemicellulose were observed in the solid fraction after pretreatment of 100% bark. Steam pretreatment dissolved most of the hemicelluloses in all other steam-pretreated materials. This could be due to the higher recalcitrance of bark or the possible neutralization of the SO_2 added to the raw material by the higher ash and extractives content of the bark feedstock. This indicates that bark requires more SO_2 or higher severity steam pretreatment conditions to dissolve hemicellulose to the same extent as in wood chips. Interestingly, the acid-insoluble lignin content of

Table 1 Composition of the spruce wood and bark feedstocks as a percentage of dry matter (% of DM)

Feedstock	Carbohydrates						Lignin		Extractives[c]	Ash
	Glucan	Xylan	Galactan	Arabinan	Mannan	Sum of neutral carbohydrates	ASL[a]	AIL[b]		
Wood	42.4	5.6	1.3	0.7	9.9	59.9	7.6	26.2	3.3	0.2
Bark	23.1	3.6	0.8	4.3	3.4	35.2	13.3	20.5	28.2	2.2

[a]Acid-soluble lignin.
[b]Acid-insoluble lignin.
[c]Water and ethanol extractives.

Table 2 Composition of the water-insoluble fraction of steam-pretreated wood and bark mixtures as a percentage of dry matter (% of DM)

Bark content (% of DM)	Carbohydrates						Lignin		Ash
	Glucan	Xylan	Galactan	Arabinan	Mannan	Sum of neutral carbohydrates	ASL[a]	AIL[b]	
0	54.5	n.d.	n.d.	n.d.	0.7	55.2	3.1	42.9	0.3
10	51.8	n.d.	n.d.	n.d.	0.5	52.3	2.9	43.1	0.5
30	48.4	n.d.	n.d.	n.d.	0.5	48.9	2.7	46.3	0.8
50	44.5	1.4	n.d.	n.d.	0.4	46.3	2.8	48.5	1.2
100	36.7	2.1	0.3	0.2	0.8	40.1	3.6	48.8	2.5

[a]Acid-soluble lignin.
[b]Acid-insoluble lignin.
n.d. Not detected.

the water-insoluble fractions increased as the bark content increased in the feedstock (Table 2), although the acid-insoluble lignin content was found to be higher in the wood chips than in the bark (Table 1). This has been found in previous studies reporting that bark contains water-soluble phenolic compounds that can condense with lignin during acid-catalyzed steam pretreatment and appear as acid insoluble lignin in the subsequent compositional analysis [18,24]. This phenomenon could play a significant part in the structural changes of the bark during the acid-catalyzed steam pretreatment carried out at the optimal condition for the wood chips.

The composition of liquid fractions obtained from the steam-pretreated materials is presented in Table 3. As a consequence of the lower carbohydrate content of bark, the concentration of total sugars (expressed in monomeric form) in the liquid fraction decreased with increasing proportions of bark, with the exception of arabinose due to the higher arabinan content of bark. As it can be seen in Figure 1, the addition of bark seemed to have a negligible effect on the overall recovery of glucose (94 to 96% for all steam-pretreated materials) and mannose (80 to 82% for all steam-pretreated materials) in the steam pretreatment step, although a lower proportion of sugars was dissolved in monomeric form in the liquid fraction as bark was added. This confirms the hypothesis that the pretreatment conditions previously found to be optimal for spruce wood chips may be too mild to overcome the recalcitrance of bark.

The amount of degradation products generated during steam pretreatment is a function of the severity of the pretreatment and the concentration of carbohydrates present in the feedstock. Therefore, the most likely explanations of the decreasing concentrations (Table 3), and also the amount of all measured inhibitors expressed as grams of inhibitors formed per 100 g dry raw material (Figure 2) with increasing bark inclusion, are the lower carbohydrate content and the higher recalcitrance of the bark feedstock.

Fermentability of pre-hydrolyzates

In order to evaluate the effect of the inhibitory compounds present in the liquid fractions obtained from the steam-pretreated materials, the pre-hydrolyzates were subjected to a fermentation test. As shown in Figure 3, all the pre-hydrolyzates showed similar high degrees of fermentability to ethanol, and the ethanol yields were in the same range as in the control solution, which contained only pure monomeric glucose and mannose.

Although bark contained more extractives than wood chips (Table 1), these had no detrimental effect on the fermentability of the pre-hydrolyzates. Boussaid *et al.* reported decreased fermentability of pre-hydrolyzates following low-severity steam pretreatment of bark-containing softwood, and attributed it to extractives, which were recovered in the water-soluble fraction [13]. However, additional acid hydrolysis of the pre-hydrolyzates increased the ethanol yield significantly, and they believed this to be due to the condensation and polymerization of water-soluble phenolic compounds. Moreover, they also showed that pre-hydrolyzates obtained from steam pretreatment at a

Table 3 Composition of the liquid fraction of the steam-pretreated wood and bark mixtures

Bark content (% of dry matter)	Total sugars (expressed as monomeric sugar) (g/L)					Inhibitors (g/L)				
	Glucose	Xylose	Galactose	Arabinose	Mannose	HMF[a]	Furfural	Formic acid	Acetic acid	Levulinic acid
0	26.5	9.7	5.1	2.7	21.7	2.0	1.3	1.5	1.9	1.0
10	21.8	9.2	4.0	3.3	20.7	1.8	1.1	1.3	1.7	0.7
30	20.1	9.2	4.2	4.8	17.8	1.4	0.8	0.7	1.4	0.4
50	18.3	8.0	4.1	5.9	13.7	1.0	0.7	0.4	1.2	0.3
100	18.2	6.0	4.3	10.0	5.6	0.4	0.4	0.4	0.6	0.2

[a]5-Hydroxymethylfurfural.

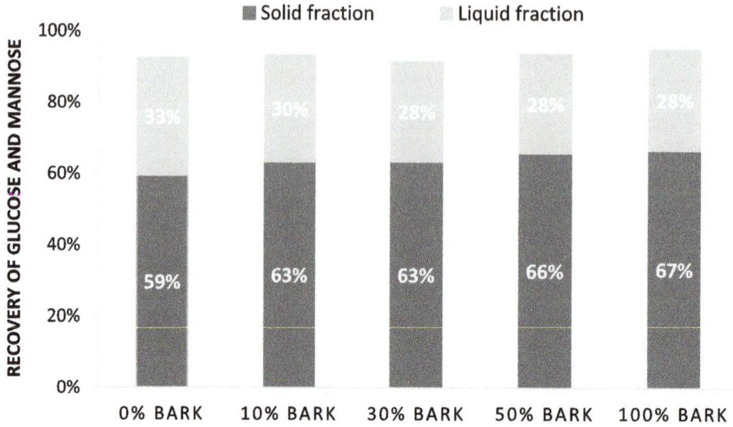

Figure 1 Recovery of glucose and mannose after steam pretreatment of wood and bark mixtures. Recovery expressed as percentage of the theoretical based on the glucan and mannan content of the raw materials.

severity factor above three (as defined by Overend *et al.* [25]) fermented well to ethanol despite the inclusion of 9% bark. This indicates that the relatively high severity of the steam pretreatment applied in our study also made the condensation and polymerization of water-soluble phenolic compounds possible, hence possibly eliminating their inhibitory effect. These results support previous findings that lower amounts of inhibitory compounds are generally formed during steam pretreatment of softwood bark than in the case of wood chips, and as a consequence, prehydrolyzates of steam-pretreated spruce bark can be fermented comparably well into ethanol [12,16,18].

Separate hydrolysis and fermentation of wood and bark mixtures

Previous studies have mostly been devoted to the investigation of the fermentability of steam-pretreated softwood bark [12,13,16,17], while little has been reported on the effect of including bark on the enzymatic hydrolyzability of steam-pretreated softwoods. Furthermore, most previous studies have been performed at lower WIS contents [17,18]. The implementation of enzymatic hydrolysis and fermentation, either separately or simultaneously, at a higher WIS content is driven by the possible energy savings in the distillation step [26]. In order to investigate the effects of bark on enzymatic hydrolyzability and fermentability of the hydrolyzed pretreated materials separately, SHF experiments were performed at 10% WIS content. The major advantage of SHF is that hydrolysis and fermentation can be carried out under their optimal conditions. However, SHF in general requires longer overall process time in comparison with SSF [27], and the end-product inhibition of enzymes by glucose and cellobiose results in a reduced rate of saccharification [28].

Figure 4 shows the concentration profiles for glucose during the enzymatic hydrolysis of steam-pretreated wood and bark mixtures and final glucose yields. The highest final glucose concentration (80.6 g/L) was achieved with

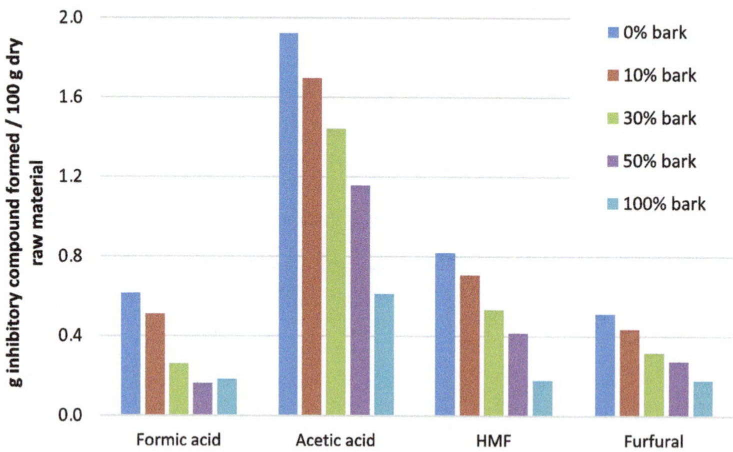

Figure 2 The amounts of inhibitory compounds formed during steam pretreatment of wood and bark mixtures. HMF, 5-hydroxymethylfurfural.

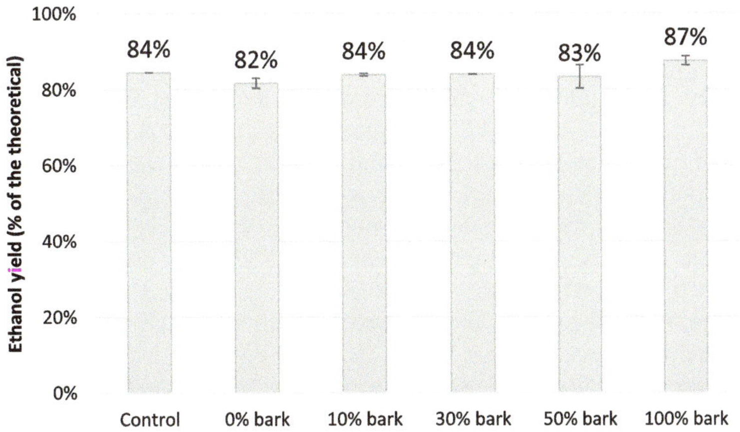

Figure 3 Ethanol yields of pre-hydrolyzate fermentation. Ethanol yields expressed as percentage of the theoretical based on glucose and mannose. Fermentation was performed with *Saccharomyces cerevisiae* at 30°C, pH 5.5 for 24 hours.

no bark addition, and decreased to 34 g/L with 100% bark. Furthermore, the highest glucose yield (90% based on all available glucose) and the highest rate of hydrolysis were obtained when no bark was added. The significant extractives content of bark can be a possible explanation for the lower glucose yields with bark addition. Phenolic compounds, either in monomeric or oligomeric form, deriving from bark can have inhibitory effect on the enzymes [14,15,29], which makes the enzymatic hydrolysis of bark more challenging than in the case of spruce wood chips. As can be seen in Figure 4, the glucose yield decreased significantly with increasing proportions of bark, and reached 53% at 100% bark. However, the same trend of decreasing hydrolyzability was also observed with increasing proportions of bark at 5% WIS loading, both on whole slurry and on washed fibre (data not shown). This suggests that the underlying reason for the lower enzymatic hydrolyzability is not the inhibitory effect of phenolic compounds in the liquid phase, but rather the higher recalcitrance or the structural changes caused by the relocation of bark extractives during acid-catalyzed steam pretreatment.

Previous studies have reported higher sugar yields from bark-containing softwood or bark as the only raw material. However, these enzymatic hydrolysis experiments were performed at a lower WIS content. For instance, Kemppainen *et al.* reported between 70 and 80% yields, depending on the steam pretreatment conditions, after 48 hours of enzymatic hydrolysis of spruce bark at 1% dry matter content [18]. A glucose yield of 79.6% was obtained by Robinson *et al.* in the enzymatic hydrolysis of softwood containing bark pretreated with steam and an additional alkaline peroxide treatment [17]. Another possible way can be to use additional accessory enzymes in order to increase the yield of the enzymatic hydrolysis of bark. For instance, 24% improvement was observed in hydrolysis of spruce bark

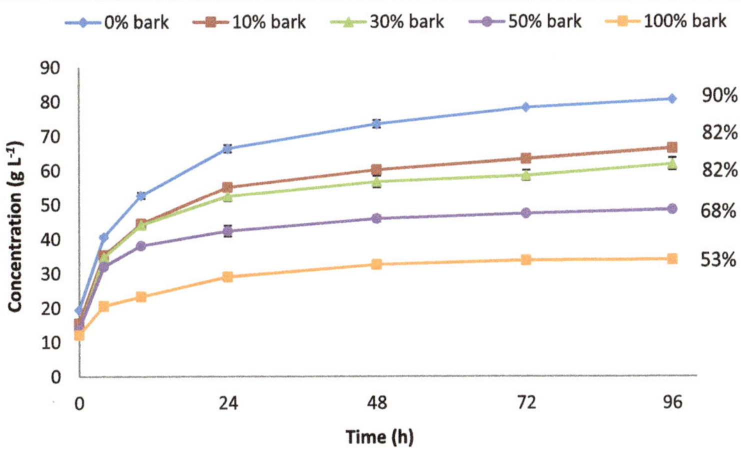

Figure 4 Concentration profiles for glucose during enzymatic hydrolysis and the final glucose yields. Enzymatic hydrolysis of steam-pretreated wood and bark mixtures was performed at 10% WIS loading, 45°C, pH 5 for 96 hours using 20 FPU/g WIS Cellic CTec3. Final glucose yields expressed as percentage of the theoretical based on all available glucose. FPU, filter paper unit; WIS, water-insoluble solids.

after 48 hours when pectinase enzymes were used as a supplementation to the cellulolytic enzymes [18]. Due to the presence of pectin in bark, the pectinase activity of the applied enzyme cocktail may also significantly affect the hydrolysis yields.

After the removal of the solid fraction of enzymatically hydrolyzed mixtures by filtration, the liquid fractions were subjected to fermentation. Figure 5 shows the concentration profiles for total hexose sugars and the ethanol produced during fermentation, together with the final ethanol yields (percentage of the theoretical maximum stoichiometric yield based on all available hexose sugars). The ethanol concentrations reached their maximum values after 24 hours in all cases. However, the final ethanol concentrations were considerably lower as the amount of bark was increased, due to the lower concentration of hexose sugars available for ethanol production in the enzymatically hydrolyzed bark-containing mixtures. As can be seen in Figure 5, almost all hexose sugars were consumed by the yeast and fermented into ethanol. The ethanol yields were largely unaffected by the addition of bark, and were in the same range; above 90%.

Table 4A shows the final concentrations of the substrates and products, together with the ethanol yield and the initial volumetric ethanol productivity in the SHF experiments. The volumetric ethanol productivity during the first two hours increased with bark inclusion, from 1.7 g/L·h to 3.2 g/L·h as the bark content was increased from 0 to 100%. A possible explanation for the higher initial volumetric ethanol productivity could be the lower concentration of 5-hydroxymethylfurfural (HMF) and furfural in the mixtures with higher bark content (Table 3). HMF and furfural are known to be inhibitory to yeast [30], and the fermentability of pre-hydrolyzates is significantly decreased with increasing concentrations of HMF and furfural. Taherzadeh *et al.* found that the rate of fermentation decreased considerably when the combined amount of HMF and furfural exceeded approximately 2 g/L [12]. In the present study, the HMF and furfural concentrations were highest when no bark was added to the spruce chips (1.6 g/L and 1.0 g/L, respectively), which were completely detoxified by the yeast in the first four hours of fermentation. As a consequence of the inhibitory effect of these degradation products, lower initial volumetric ethanol productivity was observed with the pretreated mixtures containing wood chips, than that for 100% bark, which contained the lowest concentrations of HMF and furfural.

The results of the SHF experiments showed that bark had no detrimental effects on the fermentability, which is in agreement with the results of the pre-hydrolyzates fermentation. However, the addition of bark has an adverse effect on enzymatic hydrolyzability, which limits the amount of ethanol that can be produced by the yeast in the bioconversion process.

Simultaneous saccharification and fermentation of wood and bark mixtures

In the SSF process configuration, enzymatic hydrolysis and fermentation are performed simultaneously in the same vessel, and the end-product inhibition during hydrolysis is minimized by the continuous conversion of glucose to ethanol by the fermenting microorganism [31]. However, enzymatic hydrolysis and fermentation are performed under sub-optimal conditions. In the present study, SSF was performed at 10% WIS content to assess the effect of including bark on both fermentability and enzymatic hydrolyzability. Figure 6 shows the concentration profiles for total hexose sugars and the

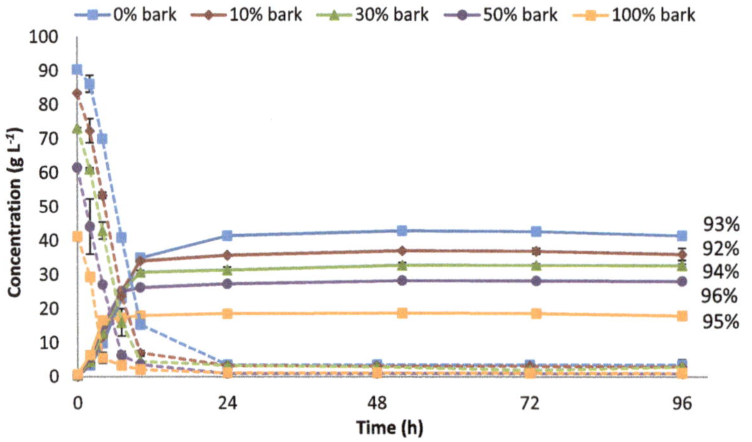

Figure 5 Concentration profiles for total hexose sugars (dashed lines) and ethanol (solid lines) during fermentation and the final ethanol yields. Fermentation of the supernatants obtained from the enzymatic hydrolysis of steam-pretreated wood and bark mixtures was performed at 10% WIS loading, 30°C, pH 5 for 96 hours using Ethanol Red yeast (5 g/L). Final ethanol yields expressed as percentage of the theoretical based on all available hexose sugars.

Table 4 Substrate, product and by-product concentrations and yields obtained from SHF (A) and SSF (B) experiments

A)

	Bark content (% of DM)	0	10	30	50	100
Enzymatic hydrolysis step in SHF	Glucose concentration (g/L)	80.6 ± 0.1	66.5 ± 0.7	61.8 ± 1.8	48.5 ± 0.5	34.0 ± 0.5
	Glucose yield (% of the theoretical)	89.8 ± 0.3	81.6 ± 1.0	81.9 ± 3.3	67.9 ± 1.1	53.3 ± 0.9
Fermentation step in SHF	Total hexose sugar concentration (g/L)	3.5 ± 0	3.1 ± 0	2.9 ± 2.1	1.0 ± 0.7	1.1 ± 0
	Glycerol concentration (g/L)	5.2 ± 0	4.5 ± 0.1	4.2 ± 0.1	3.6 ± 0	3.0 ± 0.1
	Volumetric ethanol productivity[a] (g/L·h)	1.7 ± 0	2.0 ± 0	2.3 ± 1.2	2.8 ± 0.1	3.2 ± 0
	Ethanol concentration (g/L)	41.5 ± 0.8	36.1 ± 1.8	32.8 ± 0.4	28.2 ± 0.1	18.1 ± 0
	Ethanol yield (% of the theoretical)	93.1 ± 0.1	92.0 ± 0	93.8 ± 0.8	96.2 ± 0.9	95.0 ± 0.2
SHF	Overall ethanol yield (% of the theoretical)	83.6	75.1	76.8	65.3	50.7

B)

	Bark content (% of DM)	0	10	30	50	100
SSF	Total hexose sugar concentration (g/L)	4.2 ± 0.3	1.9 ± 0.1	2.2 ± 0.1	2.4 ± 0.1	3.3 ± 0.1
	Glycerol concentration (g/L)	4.8 ± 0.1	4.2 ± 0.2	3.8 ± 0.1	3.4 ± 0	3.6 ± 1.4
	Volumetric ethanol productivity[a] (g/L·h)	1.7 ± 0.1	1.8 ± 0.3	2.5 ± 0	2.9 ± 0.1	4.0 ± 0.1
	Ethanol concentration (g/L)	45.8 ± 0.8	39.3 ± 0.9	34.5 ± 0.4	29.4 ± 0.1	20.9 ± 0
	Overall ethanol yield (% of the theoretical)	85.4 ± 1.9	81.1 ± 2.3	77.5 ± 1.3	70.5 ± 0.4	59.3 ± 0.1

[a]Calculated for the first two hours.
SSF and SHF experiments of steam-pretreated wood and bark mixtures were performed at 10% WIS loading, pH 5 using Cellic CTec3 enzyme cocktail (20 FPU/g WIS) and an industrial *S. cerevisiae* strain, Ethanol Red (5 g/L). DM, dry matter; SHF, separate hydrolysis and fermentation; SSF, simultaneous saccharification and fermentation; WIS, water-insoluble solids.

ethanol produced during SSF, together with the ethanol yields obtained (percentage of the theoretical based on all available hexose sugars).

As can be seen in Figure 6, the highest final ethanol concentration (46 g/L) was obtained when no bark was added, and it decreased significantly with increasing additions of bark. However, this decrease cannot be attributed solely to the lower amount of carbohydrates available in the bark-containing mixtures, as the ethanol yield also decreased, from 85 to 59%, as the proportion of bark was increased from 0 to 100%.

As can be seen in Table 4B, the volumetric ethanol productivity during the first two hours of SSF increased from 1.7 to 4.0 g/L·h as the amount of bark was increased from 0 to 100%. Furthermore, accumulation of the hexose sugars was also observed during the same time period in all SSF experiments, with the exception of 100% bark, where no accumulation of hexose sugars

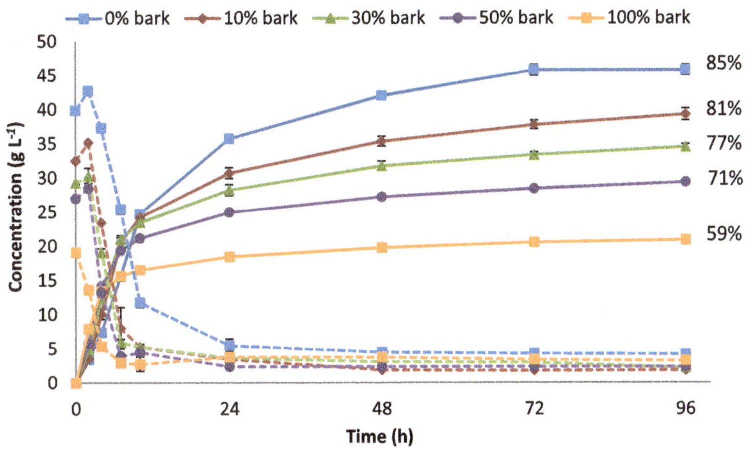

Figure 6 Concentration profiles for total hexose sugars (dashed lines) and ethanol (solid lines) during SSF and the final ethanol yields. SSF of the steam-pretreated wood and bark mixtures was performed at 10% WIS loading, 35°C, pH 5 for 96 hours using 5 g/L Ethanol Red yeast and 20 FPU/g WIS Cellic CTec3. Final ethanol yields expressed as percentage of the theoretical based on all available hexose sugars. FPU, filter paper unit; SSF, simultaneous saccharification and fermentation; WIS, water-insoluble solids.

occurred during SSF (Figure 6). Similarly to the SHF experiments, the lower concentrations of inhibitory compounds with increasing amounts of bark contributed to higher initial volumetric ethanol productivities than without bark. This indicates that the fermentability was not affected negatively by the inclusion of bark. This is also confirmed by that the addition of bark had no noticeable negative effects on the sugar utilization of the yeast in the SSF experiments (Figure 6). All the available glucose and mannose were consumed by the yeast, and only galactose was detected at low concentrations after 96 hours. Glycerol was the main by-product produced by the yeast in all cases (Table 4), and there was no significant difference in the yield of glycerol based on all available hexose sugars (approximately 4% in all SSF experiments).

The ethanol yield obtained in the SSF experiments with no bark was in the same range as reported for spruce chips in previous studies [32,33], while Kemppainen *et al.* reported an ethanol yield of 66.4% of the theoretical in SSF of sequentially hot-water extracted and steam-pretreated spruce bark [18]. This higher yield might be explained by the removal of the extractives from the bark prior steam pretreatment. Moreover, the six-hour pre-hydrolysis applied before SSF, and the possible structural differences between industrial bark and the freshly processed bark used in the present study, should not be neglected.

Comparing the two process configurations, it is apparent that SSF was superior to SHF in all cases, since SSF resulted in higher overall yields regardless of the bark content (Table 4). It is also evident that the decreased enzymatic hydrolyzability of bark is a decisive factor behind the declining ethanol yields, with increasing amounts of bark in both process configurations. Bark was found to be significantly more difficult to hydrolyse to monomeric sugars than wood chips. Lower amounts of monomeric sugars were recovered in the liquid fraction after steam pretreatment, and lower yields were observed in the enzymatic hydrolysis step with increasing bark content. Although, it appears that bark might require more severe steam pretreatment to overcome its inherent recalcitrance, the possible unfavorable structural changes in the steam-pretreated bark could also hamper the enzymatic hydrolysis. The relocation of extractives and bark lignin during the acid-catalyzed steam pretreatment might reduce the accessibility for the enzymes to cellulose, which can result in lower enzymatic hydrolyzability. Delignification methods might also be an alternative to achieve higher yields in enzymatic hydrolysis, and thus provide more sugars for ethanol fermentation; however, chemical delignification operations are expensive and would constitute an additional burden on the already sensitive economics of second-generation ethanol production [34]. Thus, further research is needed to improve the enzymatic hydrolyzability of bark in order to achieve higher yields at lower enzyme dosages.

Conclusions

The effect of including bark in the spruce-to-ethanol production process has been assessed. The results showed that adding bark had no detrimental effects on the fermentability of steam-pretreated spruce bark and wood mixtures, and it was observed that lower amounts of degradation products were formed during the steam pretreatment of spruce bark than spruce wood chips. However, the addition of bark had an adverse effect on the whole bioconversion process due to the low hydrolyzability of bark. This was reflected by the decreasing overall ethanol yield with increasing proportions of bark in both process configurations. SSF proved to be more efficient than SHF for all wood and bark mixtures, since this process configuration resulted in higher overall yields, regardless of bark content.

Materials and methods

Materials

Fresh spruce, *Picea abies*, was debarked and kindly provided by a local sawmill (ATA Timber Widtskövle AB, Everöd, Sweden), together with the bark fraction. The bark and the bark-free chipped wood were further chipped using a knife mill (Retsch GmbH, Haan, Germany) and sieved in order to obtain the fraction with a size range between 2 and 10 mm. The spruce had a dry matter content of 40%, while the bark had a somewhat lower dry matter content of 35%. The raw materials were stored in plastic bags at 4°C until used.

The enzyme preparation used was Cellic CTec3, kindly provided by Novozymes A/S (Bagsværd, Denmark). The yeast used in both the SSF and SHF experiments was Ethanol Red, kindly provided by Leaf Technologies (Marcq-en-Baroeul Cedex, France). The yeast used to determine the fermentability of the pre-hydrolyzates was prepared on an agar plate from ordinary baker's yeast, *Saccharomyces cerevisiae*, produced by Jästbolaget (Rotebo, Sweden). Vitahop, kindly provided by BetaTec (Schwabach, Germany), was used in the SSF and SHF experiments to avoid bacterial contamination. All chemicals used were of reagent grade quality.

Feedstock preparation and steam pretreatment

The bark and wood fractions were mixed to obtain batches containing 0, 10, 30, 50 and 100% bark on a dry weight basis. Each batch had a total dry weight of 700 g. The mixtures were impregnated with gaseous SO_2 (2.5% w/w, based on the water content of the mixtures) in tightly sealed plastic bags for 20 minutes at room temperature, and then subjected to steam pretreatment.

Steam pretreatment was performed at 210°C for five minutes in a 10 L reactor (Process- & Industriteknik AB, Kristianstad, Sweden), as described previously by Palmqvist *et al.* [35]. The pretreated materials were stored at 4°C before subsequent analysis and experiments.

Fermentation of pre-hydrolyzates

Yeast that was used to evaluate the fermentability of pre-hydrolyzates was aerobically cultivated. The inoculum culture was prepared by adding yeast cells, previously grown on a YPG agar plate (10 g/L yeast extract, 20 g/L peptone, 20 g/L glucose and 15 g/L agar) for three days at 30°C, to two 250 mL Erlenmeyer flasks, together with 70 mL of an aqueous solution containing 23.8 g/L glucose, 10.8 g/L $(NH_4)_2SO_4$, 5.0 g/L KH_2PO_4 and 1.1 g/L $MgSO_4{\cdot}7\ H_2O$. The solution also contained 14.4 mL/L trace element solution and 1.4 mL/L vitamin solution, prepared according to Taherzadeh *et al.* [36]. The pH was adjusted to pH 5 with 2.5 M NaOH solution. The Erlenmeyer flasks were sealed with cotton plugs and incubated at 30°C on a rotary shaker (Adolf Kühner AG, Basel, Switzerland) for 20 hours. Aerobic batch propagation was performed in a 2 L LABFORS fermentor (Infors AG, Bottmingen, Switzerland) at 30°C for 24 hours, with a working volume of 1 L. Propagation was initiated by the addition of 140 mL inoculum cultures to an autoclaved medium containing 20 g/L glucose, 22.5 g/L $(NH_4)_2SO_4$, 10.5 g/L KH_2PO_4 and 2.2 g/L $MgSO_4{\cdot}7\ H_2O$, 60.0 g/L trace element solution and 6.0 g/L vitamin solution. The aeration rate was 1 L/min, corresponding to 1 vvm (gas volume flow per unit working volume per minute). The stirrer speed was 700 rpm and the pH was maintained at pH 5 with 2.5 M NaOH solution. The dissolved oxygen concentration was monitored continuously with an O_2-sensor (Mettler-Toledo GmbH, Urdorf, Switzerland). When all the sugars had been consumed as indicated by the O_2-sensor, cultivation was stopped and the cells were harvested by centrifugation in 700 mL bottles at 3,600 × g for 10 minutes. The supernatant was discarded and the dry matter content of the harvested cells was determined. The time between cell harvesting and the initialization of the fermentation tests was less than two hours.

Fermentation tests were carried out on the pre-hydrolyzates to assess their fermentability and the extent of inhibition by the compounds formed during steam pretreatment. Pre-hydrolyzates were obtained from the steam-pretreated materials by vacuum filtration using grade five filter paper (Munktell Filter AB, Falun, Sweden). The pre-hydrolyzates were then diluted with deionized water to obtain an equivalent solids concentration (the concentration of inhibitors corresponding to an SSF with a certain WIS load) corresponding to a WIS load of 10% mass fraction. The initial concentrations of fermentable sugars were adjusted to 30 g/L glucose and 20 g/L mannose in order to obtain comparable fermentation results. A reference solution was prepared with the same sugar concentrations to serve as a control. Fermentation was performed anaerobically on a rotary shaker in shake flasks with a working volume of 100 mL, containing 0.5 g/L $(NH_4)_2HPO_4$, 0.025 g/L $MgSO_4{\cdot}7\ H_2O$ and 0.2 mL/L Vitahop. Fermentation tests were conducted at 30°C and pH 5.5 for 24 hours with a yeast concentration of 5 g/L. The fermentation experiments were performed in duplicate.

Separate hydrolysis and fermentation

Enzymatic hydrolysis of the whole pretreated slurry was performed in 2 L LABFORS bioreactors with a working weight of 1.2 kg. A WIS load of 10% mass fraction and Cellic CTec3 enzyme cocktail at a load of 20 FPU/g WIS based on the final weight, were applied. The hydrolysis experiments were performed at 45°C, with a stirring rate of 400 rpm, at pH 5 maintained with 2.5 M NaOH solution. After 96 hours of enzymatic hydrolysis the supernatants were separated by vacuum filtration using grade five filter paper (Munktell Filter AB). The supernatants obtained from the duplicates were mixed and stored at −20°C prior to fermentation.

Fermentation of the supernatant was performed in 2 L LABFORS bioreactors with a working weight of 0.55 kg. Ethanol Red yeast was added at a dry weight concentration of 5 g/L based on the final weight. The supernatant was supplemented with $(NH_4)_2HPO_4$ solution at a concentration of 0.5 g/L and 0.125 mL/L Vitahop. Fermentation was carried out at 30°C, with a stirring rate of 250 rpm for 96 hours, at pH 5 maintained with 2.5 M NaOH solution. All experiments were performed in duplicate.

Simultaneous saccharification and fermentation

The SSF experiments using the whole pretreated slurry were performed in sterilized 2 L LABFORS bioreactors with a working weight of 1 kg. A WIS load of 10% mass fraction, the Cellic CTec3 enzyme cocktail at a load of 20 FPU/g WIS and Ethanol Red yeast at a dry weight concentration of 5 g/L based on the final amount, were applied. The experiments were carried out at 35°C, with a stirring rate of 400 rpm for 96 hours, at pH 5 maintained with 2.5 M NaOH solution. The SSF media were supplemented with $(NH_4)_2HPO_4$ solution at a concentration of 0.5 g/L and 0.125 mL/L Vitahop. All experiments were performed in duplicate.

Analysis

The total solids content of biomass materials and total dissolved solids content of liquid samples were determined according to the National Renewable Energy Laboratory (NREL) standardized laboratory analytical procedure [37].

The structural carbohydrates, lignin, extractives and ash content of the solid fractions and the composition of the liquid fractions were determined according to NREL standardized laboratory analytical procedures [38-41].

All samples obtained from experiments or compositional analysis were centrifuged in 2 mL Eppendorf tubes at 16,000 × g for 10 minutes. The supernatant was filtered using 0.2 μm syringe filters (GVS Filter Technology Inc., Indiana, United States), and filtered samples were stored at −20°C prior to high-performance liquid chromatography (HPLC) analysis. Sugars, ethanol, organic acids and other by-products were analyzed using a Shimadzu LC-20 AD HPLC system equipped with a Shimadzu RID 10A refractive index detector (Shimadzu Corporation, Kyoto, Japan). Monomeric sugars were quantified with isocratic ion-exchange chromatography using an Aminex HPX-87P column with a De-Ashing Bio-Rad micro-guard column at 85°C (both from Bio-Rad Laboratories, Hercules, California, United States) using reagent grade water as the mobile phase at a flow rate of 0.5 mL/min. Ethanol, organic acids and other by-products were determined using an Aminex HPX-87H chromatography column with a Cation-H Bio-Rad micro-guard column at 50°C (Bio-Rad Laboratories, Hercules, California, United States), with a mobile phase of 5 mM sulfuric acid at a flow rate of 0.5 mL/min.

Yield calculations

The glucose yield in the enzymatic hydrolysis experiments was calculated on the basis of total available glucose in the liquid and the solid fraction of the steam-pretreated materials. The theoretical amount of glucose released during enzymatic hydrolysis is 1.11 times the amount of glucan in the solid fraction of the steam-pretreated materials (due to the addition of water in hydrolysis). The ethanol yield is expressed as a percentage of the theoretical stoichiometric ethanol yield (0.51 g/g), based on total available hexose sugars, namely glucose, mannose and galactose, in the solid and/or the liquid fraction of the steam-pretreated materials.

Abbreviations

AIL: Acid-insoluble lignin; ASL: Acid-soluble lignin; DM: Dry matter; FPU: Filter paper unit; HMF: Hydroxymethylfurfural; HPLC: High-performance liquid chromatography; NREL: National Renewable Energy Laboratory;
SHF: Separate hydrolysis and fermentation; SSF: Simultaneous saccharification and fermentation; WIS: Water-insoluble solids; YPG: Yeast peptone glucose.

Competing interests

The authors declare that they have no competing interests.

Authors' contributions

BF, MG and OW designed and coordinated the study. BF carried out the experiments, analyzed the results and prepared the manuscript. MG and OW reviewed the manuscript. All authors read and approved the final manuscript.

Acknowledgements

The authors gratefully acknowledge the Swedish Energy Agency, Energimyndigheten, for financial support (grant number: 35355-1).

References

1. Himmel ME, Ding S-Y, Johnson DK, Adney WS, Nimlos MR, Brady JW, et al. Biomass recalcitrance: engineering plants and enzymes for biofuels production. Science. 2007;315(5813):804–7.
2. Balan V, Chiaramonti D, Kumar S. Review of US and EU initiatives toward development, demonstration, and commercialization of lignocellulosic biofuels. Biofuels Bioprod Biorefin. 2013;7(6):732–59.
3. The US Energy Independence and Security Act of 2007. Pub L. No. 110-140, 121 Stat. 1492, HR6 (Jan 4, 2007). http://www.gpo.gov/fdsys/pkg/BILLS-110hr6enr/pdf/BILLS-110hr6enr.pdf. Accessed 18 Jan 2015.
4. Directive 2009/28/EC of the European Parliament and of the council of 23 April 2009 on the promotion of the use of energy from renewable sources and amending and subsequently repealing Directives 2001/77/EC and 2003/30/EC. OJ. L. No. 140, p. 16–62. (April 23, 2009).
5. Goldemberg J. Ethanol for a sustainable energy future. Science. 2007;315 (5813):808–10.
6. Groom MJ, Gray EM, Townsend PA. Biofuels and biodiversity: principles for creating better policies for biofuel production. Conserv Biol. 2008;22(3):602–9.
7. Williams PRD, Inman D, Aden A, Heath GA. Environmental and sustainability factors associated with next-generation biofuels in the U.S.: what do we really know? Environ Sci Technol. 2009;43(13):4763–75.
8. Searchinger T, Heimlich R, Houghton RA, Dong F, Elobeid A, Fabiosa J, et al. Use of U.S. croplands for biofuels increases greenhouse gases through emissions from land-use change. Science. 2008;319(5867):1238–40.
9. Wyman CE. What is (and is not) vital to advancing cellulosic ethanol. Trends Biotechnol. 2007;25(4):153–7.
10. Stephen JD, Mabee WE, Saddler JN. Will second-generation ethanol be able to compete with first-generation ethanol? Opportunities for cost reduction. Biofuels Bioprod Biorefin. 2012;6(2):159–76.
11. von Sivers M, Zacchi G. Ethanol from lignocellulosics: a review of the economy. Bioresource Technol. 1996;56:131–40.
12. Taherzadeh MJ, Eklund R, Gustafsson L, Niklasson C, Lidén G. Characterization and fermentation of dilute-acid hydrolyzates from wood. Ind Eng Chem Res. 1997;36(11):4659–65.
13. Boussaid A, Cai Y, Robinson J, Gregg DJ, Nguyen Q, Saddler JN. Sugar recovery and fermentability of hemicellulose hydrolysates from steam-exploded softwoods containing bark. Biotechnol Prog. 2001;17(5):887–92.
14. Ximenes E, Kim Y, Mosier N, Dien B, Ladisch M. Deactivation of cellulases by phenols. Enzyme Microb Tech. 2011;48(1):54–60.
15. Kim Y, Ximenes E, Mosier NS, Ladisch MR. Soluble inhibitors/deactivators of cellulase enzymes from lignocellulosic biomass. Enzyme Microb Tech. 2011;48(4–5):408–15.
16. Robinson J, Keating J, Boussaid A, Mansfield S, Saddler J. The influence of bark on the fermentation of Douglas-fir whitewood pre-hydrolysates. Appl Microbiol Biotechnol. 2002;59(4–5):443–8.
17. Robinson J, Keating JD, Mansfield SD, Saddler JN. The fermentability of concentrated softwood-derived hemicellulose fractions with and without supplemental cellulose hydrolysates. Enzyme Microb Tech. 2003;33(6):757–65.
18. Kemppainen K, Inkinen J, Uusitalo J, Nakari-Setälä T, Siika-aho M. Hot water extraction and steam explosion as pretreatments for ethanol production from spruce bark. Bioresource Technol. 2012;117:131–9.
19. Zacchi G, Axelsson A. Economic evaluation of preconcentration in production of ethanol from dilute sugar solutions. Biotechnol Bioeng. 1989;34:223–33.
20. Stenberg K, Tengborg C, Galbe M, Zacchi G. Optimisation of steam pretreatment of SO2-impregnated mixed softwoods for ethanol production. J Chem Technol Biotechnol. 1998;71(4):299–308.
21. Kemppainen K, Siika-aho M, Pattathil S, Giovando S, Kruus K. Spruce bark as an industrial source of condensed tannins and non-cellulosic sugars. Ind Crops Prod. 2014;52:158–68.
22. Krogell J, Holmbom B, Pranovich A, Hemming J, Willför S. Extraction and chemical characterization of Norway Soruce inner and outer bark. Nord Pulp Pap Res J. 2012;27(1):6–17.

23. Sluiter JB, Ruiz RO, Scarlata CJ, Sluiter AD, Templeton DW. Compositional analysis of lignocellulosic feedstocks. 1. Review and description of methods. J Agric Food Chem. 2010;58(16):9043–53.
24. Torget R, Himmel ME, Grohmann K. Dilute sulfuric acid pretreatment of hardwood bark. Bioresource Technol. 1991;35(3):239–46.
25. Overend RP, Chornet E, Gascoigne JA. Fractionation of lignocellulosics by steam-aqueous pretreatments [and discussion]. Philos Trans R Soc. 1987;321(1561):523–36.
26. Hoyer K, Galbe M, Zacchi G. Production of fuel ethanol from softwood by simultaneous saccharification and fermentation at high dry matter content. J Chem Technol Biotechnol. 2009;84:570–7.
27. Alfani F, Gallifuoco A, Saporosi A, Spera A, Cantarella M. Comparison of SHF and SSF processes for the bioconversion of steam-exploded wheat straw. J Ind Microbiol Biot. 2000;25(4):184–92.
28. Tomás-Pejó E, Oliva JM, Ballesteros M, Olsson L. Comparison of SHF and SSF processes from steam-exploded wheat straw for ethanol production by xylose-fermenting and robust glucose-fermenting *Saccharomyces cerevisiae* strains. Biotechnol Bioeng. 2008;100(6):1122–31.
29. Tejirian A, Xu F. Inhibition of enzymatic cellulolysis by phenolic compounds. Enzyme Microb Tech. 2011;48(3):239–47.
30. Palmqvist E, Hahn-Hägerdal B. Fermentation of lignocellulosic hydrolysates. II: inhibitors and mechanisms of inhibition. Bioresour Technol. 2000;74(1):25–33.
31. Olofsson K, Bertilsson M, Liden G. A short review on SSF - an interesting process option for ethanol production from lignocellulosic feedstocks. Biotechnol Biofuels. 2008;1:7.
32. Rudolf A, Alkasrawi M, Zacchi G, Lidén G. A comparison between batch and fed-batch simultaneous saccharification and fermentation of steam pretreated spruce. Enzyme Microb Tech. 2005;37(2):195–204.
33. Hoyer K, Galbe M, Zacchi G. Effects of enzyme feeding strategy on ethanol yield in fed-batch simultaneous saccharification and fermentation of spruce at high dry matter. Biotechnol Biofuels. 2010;3(1):14.
34. Zhu JY, Pan XJ. Woody biomass pretreatment for cellulosic ethanol production: technology and energy consumption evaluation. Bioresource Technol. 2010;101(13):4992–5002.
35. Palmqvist E, Hahn-Hägerdal B, Galbe M, Larsson M, Stenberg K, Szengyel Z, et al. Design and operation of a bench-scale process development unit for the production of ethanol from lignocellulosics. Bioresource Technol. 1996;58(2):171–9.
36. Taherzadeh MJ, Liden G, Gustafsson L, Niklasson C. The effects of pantothenate deficiency and acetate addition on anaerobic batch fermentation of glucose by *Saccharomyces cerevisiae*. Appl Microbiol Biotechnol. 1996;46(2):176–82.
37. Sluiter A, Hames B, Hyman D, Payne C, Ruiz R, Scarlata C, et al. Determination of total solids in biomass and total dissolved solids in liquid process samples. In: Laboratory analytical procedure. Golden, Colorado: National Renewable Energy Laboratory; 2008.
38. Sluiter A, Hames B, Ruiz R, Scarlata C, Sluiter J, Templeton D, et al. Determination of structural carbohydrates and lignin in biomass. In: Laboratory analytical procedure. Golden, Colorado: National Renewable Energy Laboratory; 2008.
39. Sluiter A, Hames B, Ruiz R, Scarlata C, Sluiter J, Templeton D, et al. Determination of sugars, byproducts, and degradation products in liquid fraction process samples. In: Laboratory analytical procedure. Golden, Colorado: National Renewable Energy Laboratory; 2006.
40. Sluiter A, Hames B, Ruiz R, Scarlata C, Sluiter J, Templeton D. Determination of ash in biomass. National Renewable Energy Laboratory: Golden, Colorado; 2008.
41. Sluiter A, Ruiz R, Scarlata C, Sluiter J, Templeton D. Determination of extractives in biomass. In: Laboratory analytical procedure. Golden, Colorado: National Renewable Energy Laboratory; 2005. p. 1617.

5

2G ethanol from the whole sugarcane lignocellulosic biomass

Sandra Cerqueira Pereira[1], Larissa Maehara[1,2], Cristina Maria Monteiro Machado[3] and Cristiane Sanchez Farinas[1,2*]

Abstract

Background: In the sugarcane industry, large amounts of lignocellulosic residues are generated, which includes bagasse, straw, and tops. The use of the whole sugarcane lignocellulosic biomass for the production of second-generation (2G) ethanol can be a potential alternative to contribute to the economic viability of this process. Here, we conducted a systematic comparative study of the use of the lignocellulosic residues from the whole sugarcane lignocellulosic biomass (bagasse, straw, and tops) from commercial sugarcane varieties for the production of 2G ethanol. In addition, the feasibility of using a mixture of these residues from a selected variety was also investigated.

Results: The materials were pretreated with dilute acid and hydrolyzed with a commercial enzymatic preparation, after which the hydrolysates were fermented using an industrial strain of *Saccharomyces cerevisiae*. The susceptibility to enzymatic saccharification was higher for the tops, followed by straw and bagasse. Interestingly, the fermentability of the hydrolysates showed a different profile, with straw achieving the highest ethanol yields, followed by tops and bagasse. Using a mixture of the different sugarcane parts (bagasse-straw-tops, 1:1:1, in a dry-weight basis), it was possible to achieve a 55% higher enzymatic conversion and a 25% higher ethanol yield, compared to use of the bagasse alone. For the four commercial sugarcane varieties evaluated using the same experimental set of conditions, it was found that the variety of sugarcane was not a significant factor in the 2G ethanol production process.

Conclusions: Assessment of use of the whole lignocellulosic sugarcane biomass clearly showed that 2G ethanol production could be significantly improved by the combined use of bagasse, straw, and tops, when compared to the use of bagasse alone. The lower susceptibility to saccharification of sugarcane bagasse, as well as the lower fermentability of its hydrolysates, can be compensated by using it in combination with straw and tops (sugarcane trash). Furthermore, given that the variety was not a significant factor for the 2G ethanol production process within the four commercial sugarcane varieties evaluated here, agronomic features such as higher productivity and tolerance of soil and climate variations can be used as the criteria for variety selection.

Keywords: Sugarcane bagasse, Sugarcane trash, Second-generation ethanol, Whole sugarcane lignocellulosic biomass, Cellulosic ethanol, Bioethanol

Background

The growing concern over the shortage of oil reserves, together with the need to preserve the environment, is the main drivers of the search for viable alternative renewable sources for the production of sustainable fuels. Furthermore, since the transportation sector accounts for about 16% of global greenhouse gas (GHG) emissions [1] and about 27% of total US GHG emissions [2], the replacement of oil-derived fuels by biofuels could contribute to the reduction of environmental impacts as well as provide socio-economic benefits [3,4]. There is already a consensus that a shift of the global energy scenario towards increasing the share of renewable sources is necessary and that ethanol will undoubtedly be an important biofuel [5]. The need to produce inexpensive renewable fuels to substitute fossil fuels is revealed in the political agendas of several countries [6]. Therefore, the future energy economy is likely to be based on a wide range of alternative

* Correspondence: cristiane.farinas@embrapa.br
[1]Embrapa Instrumentation, Rua XV de Novembro 1452, 13560-970 São Carlos, SP, Brazil
[2]Graduate Program of Chemical Engineering, Federal University of São Carlos, Rodovia Washington Luiz, km 235, 13565-905 São Carlos, SP, Brazil
Full list of author information is available at the end of the article

energy platforms. Among the potential bioenergy sources, the lignocellulosic biomass has been identified as an important feedstock for the production of biofuels and other value-added products, thus contributing to the global energy supply. In this context, Brazil is notable in that the country has made considerable progress towards replacing fossil fuels by renewable ethanol from sugarcane. Moreover, Brazil is the world's largest producer of sugarcane and together with US leads the global production of bioethanol [7-9]. In the 2014/15 harvest, it is estimated that more than 640 million tons of sugarcane will be processed by the Brazilian sugar-alcohol mills, which will result in an ethanol production of approximately 28 billion liters and a sugar production of about 36 million tons [7].

In the sugarcane industry, large amounts of lignocellulosic residues (bagasse and trash) are generated during the production of ethanol and sugar. After the sugarcane harvest, the trash remaining consists of three main components, namely, dry leaves, green leaves, and tops [10]. The tops are the segments of the sugarcane plant between the upper end and the last stalk node with attached green leaves, while the dry leaves are known as straw. The milling of the sugarcane to extract the juice generates the bagasse. In this process, every ton of sugarcane processed generates 140 kg of bagasse and 140 kg of trash, on a dry basis (db) [11]. In contemporary facilities, the bagasse is currently used as fuel, that is, it is burned in boilers to meet the energy demands of the industrial plant, with the surplus electric energy being exported to the grid [12-14]. The sugarcane trash was previously burned prior to the harvest in order to facilitate the harvest procedure, but is now mostly left in the field for agricultural purposes as fertilizer and for pest control [14]. There is currently an ongoing progressive shift in the sugarcane harvesting method, from manual harvesting of burned sugarcane to mechanical harvesting of unburned sugarcane, with the trash remaining on the ground [14-16]. This increasing amount of lignocellulosic material being left on the ground could be partially recovered and used for energy generation in the mills, thus improving the overall energy balance [14].

The second-generation (2G) ethanol produced from lignocellulosic biomass has been considered to be the biofuel with the greatest potential to replace oil-based fuels [17,18]. However, 2G technology is not as mature as that of conventional first-generation (1G) ethanol, and the process is still less economically feasible [12,19]. In energy terms, it is known that the energy content (MJ/kg) of sugarcane trash is similar to that of the bagasse, with the two components accounting for approximately two thirds of the total energy content of sugarcane [11,15]. Despite the fact that the trash is currently considered the main unexploited energy reserve in the sugarcane agro-industry, its most likely use in the short to medium term will be as boiler fuel [11,15]. Nonetheless, considering the amounts of residues generated and that all the cellulose present in these materials could be converted into ethanol, the use of the whole sugarcane lignocellulosic biomass (including trash and bagasse) could greatly increase ethanol production without the need for expansion of cultivated areas. It is worth mentioning that there is still no consensus about the optimal amount of trash that should be left on the ground. It has been observed that 65% of the residue from sugarcane harvesting can be removed from the fields as dry leaves (straw), without causing negative agricultural effects [20].

There have been several studies of 2G ethanol production processes using sugarcane bagasse [21-29]. In one study, evaluation was made of the bagasse from 115 varieties of sugarcane developed by classical and precision breeding technologies, in terms of fiber composition and fermentable sugar yields after dilute acid pretreatment and enzymatic hydrolysis [21]. It was suggested that variety selection could contribute to the development of a cost-effective pretreatment and saccharification process. After the screening reported previously [21], the authors also described the optimization of pretreatment of selected varieties, used in the cellulosic ethanol production process in order to fully demonstrate the benefits of variety selection in terms of the yields of fermentable sugar and ethanol from the bagasse [23,24].

On the other hand, there have been few reports concerning the individual assessment of other parts of the sugarcane biomass (straw and tops) for the development of 2G ethanol production processes [30-32] or the combined use of lignocellulosic sugarcane residues [33,34]. In fact, the combined use of sugarcane residues has been investigated only regarding the enzymatic hydrolysis step, employing either hydrothermal [33] or dilute acid [34] pretreated materials. Nevertheless, to the best of our knowledge, there have not been any systematic studies of 2G ethanol production using the whole lignocellulosic biomass from different commercial sugarcane varieties with known agronomic advantages in terms of productivity as well as tolerance of soil and climate variability.

The aim of the present study was to contribute to the development of 2G ethanol production processes by evaluating a biotechnological approach employing the whole sugarcane lignocellulosic biomass. The lignocellulosic residues (bagasse, straw, and tops) from four commercial sugarcane varieties (SP79-1011, RB867515, SP81-3250, and RB92579) were assessed for 2G ethanol production, with determination of the individual responses obtained for different process stages (dilute sulfuric acid pretreatment, enzymatic hydrolysis, and alcoholic fermentation). The parameters considered were chemical composition, susceptibility to saccharification, and fermentability. Investigation was also carried out on the combined use of a mixture of bagasse, straw, and tops, under the same conditions.

Results and discussion

Chemical composition of the raw material

Three lignocellulosic residue parts (bagasse, straw, and tops) from the processing of four commercial sugarcane varieties (SP79-1011, RB867515, SP81-3250, and RB92579, represented here by K, M, Q, and X, respectively) were evaluated according to their chemical composition in terms of hemicellulose, lignin, cellulose, and ash contents (*w/w*, dry weight basis) (Table 1). As can be seen, the lignocellulosic residue parts differed considerably in their chemical composition. Overall, cellulose and hemicellulose were the main components of the biomass, accounting for 34.1 to 42.1% and 28.5 to 38.8% of the dry material, respectively. Lignin accounted for 5.6 to 13.8%, and the ash content of the raw material was in the range 0.8 to 3.8%.

For all the varieties, the highest cellulose content was found for the straw, followed by the bagasse and tops (Table 1). On the other hand, all the varieties showed the highest levels of hemicellulose in the tops, followed by the straw and bagasse. For all the varieties, the lignin content was highest in the bagasse, followed by the straw and tops, while the ash content was highest in the tops, followed by the straw and bagasse. It can therefore be seen that determination of chemical composition is vital because the conversion of vegetal biomass into biofuel (by means of the sequential stages of pretreatment, enzymatic hydrolysis, and alcoholic fermentation) is closely connected to the inherent features of these lignocellulosic materials. Overall, the raw materials showed similar chemical composition profiles, irrespective of the variety.

The overall average cellulose, hemicellulose, lignin, and ash contents of the untreated straw from the four varieties of sugarcane were 41.1 ± 0.9, 36.2 ± 0.9, 11.4 ± 0.4, and 2.2 ± 0.5%, respectively (Table 1). These values differ from those described in the literature, mainly because the majority of studies were carried out with sugarcane trash as the raw material and not with straw alone. For example, in one study, the contents of cellulose, hemicellulose, lignin, and ash were 33.6, 28.9, 31.8, and 5.7%, respectively [35], while other work reported values of 39.8, 28.6, 22.5, and 2.4%, respectively [30]. Moreover, the average contents of cellulose, hemicellulose, lignin, and ash found for the untreated bagasse from the four varieties of sugarcane were 37.6 ± 1.4, 29.4 ± 1.1, 13.2 ± 0.5, and 0.9 ± 0.2%, respectively, and differed only slightly from the values reported in the literature [21,25-27,34,36]. Few studies have evaluated the chemical composition of the tops separately. For the four varieties, the average contents of cellulose, hemicellulose, lignin, and ash in the tops were 35.2 ± 1.3, 37.7 ± 1.4, 8.4 ± 0.7, and 3.6 ± 0.2%, respectively. In the case of cellulose and hemicellulose (but not lignin and ash), the contents found here for the tops were comparable to those reported in the literature [32,37].

These differences in chemical composition for the same type of biomass were expected, due to the influence of numerous factors including plant variety, growth environment, and processing conditions, as well as the procedures employed for the compositional analysis [38]. Furthermore, it is difficult to compare the compositions of samples from different origins when the analyses are performed by laboratories that do not use the same methods [10]. Nevertheless, in terms of the contents of cellulose, hemicellulose, lignin, and ash, the sugarcane characterization performed here resulted in a range of

Table 1 Chemical composition of the three lignocellulosic residue parts (straw, tops, and bagasse)

Variety	Symbol	Residue	Cellulose	Hemicellulose	Lignin	Ash
SP79-1011	K	Straw	41.09	35.07	11.13	2.25
		Tops	34.68	35.92	8.00	3.63
		Bagasse	36.60	28.45	13.81	0.80
RB867515	M	Straw	42.12	35.93	12.04	2.81
		Tops	35.02	37.27	8.95	3.65
		Bagasse	38.47	28.86	13.45	0.79
SP81-3250	Q	Straw	41.27	36.53	11.33	1.72
		Tops	34.05	38.79	9.08	3.82
		Bagasse	36.13	29.19	12.83	1.14
RB92579	X	Straw	39.91	37.07	11.26	2.07
		Tops	37.08	38.76	7.59	3.30
		Bagasse	39.07	31.02	12.89	0.88

Data refer to the four varieties of sugarcane (K, M, Q, and X), in terms of cellulose, hemicellulose, lignin, and ash contents (*w/w*, dry weight basis). Data are means of three replicates.

values similar to those found for lignocellulosic feedstocks in general [39].

Many sugarcane varieties are available in Brazil, which is highly advantageous for the profitability of the sugar-alcohol sector because it enables the selection of varieties that are better adapted to adverse conditions of soil and weather, more resistant to pests and diseases, and suitable for specific harvesting systems. The commercial sugarcane varieties evaluated in this study (SP79-1011, RB867515, SP81-3250, and RB92579) were selected because they are among the most widely used in Brazilian sugarcane plantations, due to their good agronomic characteristics including low demand for water, suitability for mechanical harvesting, and good transportation efficiency, among others.

Effect of dilute sulfuric acid pretreatment on the feedstock

Table 2 shows the chemical composition of the biomass obtained after dilute acid pretreatment (H_2SO_4, 1.5% *w/w*, 1:10 solid/liquid ratio, 121°C, 30 min) of the three residue parts (bagasse, straw, and tops) obtained from processing of the four varieties of sugarcane (here denoted K, M, Q, and X). There were only slight differences in the chemical compositions of the pretreated materials. Cellulose was the main component, followed by lignin, while hemicellulose and ash accounted for much smaller proportions of the dry samples. Overall, the contents of cellulose, hemicellulose, lignin, and ash were in the ranges 47.0 to 51.9, 6.2 to 7.9, 29.6 to 32.9, and 0.5 to 1.9%, respectively.

As also found for the raw materials (Table 1), the chemical composition profiles of the pretreated residues (bagasse, straw, and tops) were similar for the four sugarcane varieties. Nevertheless, it can be seen from Table 2 that the magnitude of the variability between the three sugarcane residues in terms of the components (cellulose, hemicellulose, lignin, and ash) was smaller than for the raw materials, suggesting that pretreatment with dilute sulfuric acid increased the similarity of the materials, irrespective of sugarcane variety. The highest cellulose and ash contents, as well as the lowest lignin contents, were found for the tops, while straw presented the lowest cellulose and hemicellulose contents and the highest lignin contents. Finally, bagasse showed the highest hemicellulose contents and the lowest ash contents. In order to observe more clearly the effect of dilute acid pretreatment on the sugarcane biomass, an order of chemical composition was established, before and after the pretreatment, in terms of the average contents of hemicellulose, lignin, cellulose, and ash for the different types of residue (bagasse, straw, and tops), as shown in Table 3. It is interesting to note that only the order of ash content remained unaltered after the dilute acid pretreatment.

The process of lignocellulosic ethanol production by means of bioconversion consists of three critical steps: pretreatment of the biomass, hydrolysis of sugar polymers to fermentable sugar monomers, and fermentation of sugar monomers to ethanol [40]. The pretreatment step is essential for effective ethanol production because of the natural recalcitrance of lignocellulosic biomass, which limits the access of enzymes to the structural matrix [41-44]. Among the various types of chemical biomass pretreatment, dilute acid is fast and is one of the oldest and most widely used methods, since it satisfies the majority of the requirements of pretreatment processes [10,45,46]. Lignocellulosic biomass submitted to dilute acid pretreatment undergoes partial hemicellulose solubilization and lignin redistribution [47]. As a result, the contents of cellulose and lignin are enhanced in the pretreated material, compared to the untreated biomass [21,24,26,34,48]. In the present study, the dilute acid pretreatment (H_2SO_4, 1.5% *w/w*, 1:10 solid/

Table 2 Chemical composition of the three pretreated residue parts (straw, tops, and bagasse)

Variety	Symbol	Residue	Cellulose	Hemicellulose	Lignin	Ash
SP79-1011	K	Straw	48.06	6.37	31.82	0.89
		Tops	51.34	7.54	29.62	1.75
		Bagasse	49.37	7.69	30.37	0.51
RB867515	M	Straw	47.44	6.38	32.32	0.99
		Tops	49.18	7.73	30.15	1.84
		Bagasse	47.98	7.88	30.85	0.59
SP81-3250	Q	Straw	47.03	6.89	32.91	1.01
		Tops	48.87	7.26	31.41	1.86
		Bagasse	47.72	7.54	32.48	0.66
RB92579	X	Straw	49.21	6.18	32.94	0.92
		Tops	51.88	7.54	30.37	1.76
		Bagasse	50.50	7.77	31.99	0.52

Data refer to the four varieties of sugarcane (K, M, Q, and X), in terms of cellulose, hemicellulose, lignin, and ash contents (*w/w*, dry weight basis). Data are means of three replicates.

Table 3 Effect of the dilute sulfuric acid pretreatment on the order of the different residues

Component	Raw materials	Pretreated materials
Cellulose	Straw > bagasse > tops	Tops > bagasse > straw
Hemicellulose	Tops > straw > bagasse	Bagasse > tops > straw
Lignin	Bagasse > straw > tops	Straw > bagasse > tops
Ash	Tops > straw > bagasse	Tops > straw > bagasse

Chemical composition orders based on average data for each type of residue (bagasse, straw, and tops) obtained for the four varieties (K, M, Q, and X).

liquid ratio, 121°C, 30 min) provided effective enrichment of the biomass in terms of the cellulose and lignin contents, regardless of the residue type (bagasse, straw, or tops) or sugarcane variety (K, M, Q, or X). This was mainly achieved by removing the hemicellulosic fraction (Table 2).

The average contents of cellulose, hemicellulose, lignin, and ash found for the pretreated bagasse from the different sugarcane varieties (48.9 ± 1.3, 7.7 ± 0.1, 31.4 ± 1.0, and 0.6 ± 0.1% for varieties K, M, Q, and X, respectively) were slightly different to those reported in the literature using a dilute mixed acid pretreatment (1% *w/v* sulfuric acid and 1% *w/v* acetic acid solution, 1:10 solid/liquid ratio, 190°C, 10 min) [27] and comparable to a pretreatment employing dilute sulfuric acid (H_2SO_4, 2% *w/w*, 1/15 solid/liquid ratio, 150°C, 30 min) [47]. The results obtained here with respect to the chemical composition of sugarcane bagasse after the pretreatment with dilute acid (H_2SO4, 1.5% *w/w*, 1:10 solid/liquid ratio, 121°C, 30 min) were similar to those reported previously using a solids loading of 15% (*w/w*) and a different cultivar of sugarcane [26].

As mentioned above, there have been few studies conducted to evaluate the chemical composition of sugarcane tops alone, and only hydrothermal pretreatment has been employed [32,37]. The same applies to the sugarcane trash, since there are few studies on the evaluation of its chemical composition in comparison to sugarcane bagasse. The effect of dilute acid pretreatment on hemicellulose solubilization was investigated in a recent study of the enzymatic hydrolysis of pretreated sugarcane trash [34]. Other pretreatments of sugarcane trash that have been studied are steam explosion [30,31] and hydrothermal [33] processes. Dilute acid, hydrothermal, and steam explosion pretreatments have all been extensively studied and used for cellulosic ethanol production, where they provide effective removal of the hemicellulosic fraction, opening up or improving the access of enzymes to the cellulosic fraction in the subsequent enzymatic hydrolysis step [49,50].

In earlier work, the efficiency of enzymatic hydrolysis of hydrothermally pretreated sugarcane trash (which includes dry leaves, green leaves, and tops) was investigated (1:10 solid/liquid ratio, 190°C, 10 min) and cellulose, hemicellulose, lignin, and ash contents of 51.6, 16.8, 30.0, and 1.4%, respectively, were reported for tops. The corresponding values for dry leaves (straw) were 53.9, 12.3, 29.5, and 3.5%, respectively [32]. The results of the dilute acid pretreatment (H_2SO_4, 1.5% *w/w*, 1:10 solid/liquid ratio, 121°C, 30 min) conducted here were superior in terms of the solubilization of hemicellulose, since the average contents of hemicellulose in the pretreated tops and straw were 7.5 ± 0.2 and 6.5 ± 0.3%, respectively.

Enzymatic conversion of cellulose in the pretreated materials

Figure 1 presents the temporal profiles of glucose release during the enzymatic saccharification of bagasse, straw, and tops from the different varieties of sugarcane previously pretreated with dilute sulfuric acid (H_2SO_4, 1.5% *w/w*, 1:10 solid/liquid ratio, 121°C, 30 min). All the enzymatic hydrolysis assays were conducted under the same conditions to ensure the validity of comparisons. The enzymatic hydrolysis profiles showed similar patterns, regardless of sugarcane variety (K, M, Q, or X), as can be clearly seen in Figure 1. In this ethanol production step, the tops were found to be most susceptible to enzymatic degradation, with an average glucose concentration of 39.8 g/L reached after 24 h. The bagasse showed the lowest enzymatic digestibility, with 22.2 g/L glucose obtained in the same time. Glucose release from the tops was about 80% higher than from the bagasse. The straw provided an average of 31.0 g/L glucose, which was 40% higher compared to the bagasse and 28% lower in relation to the tops. An important finding was therefore that the qualitative order of sugar released from the sugarcane biomass was as follows: tops > straw > bagasse.

For a more precise analysis of the enzymatic hydrolysis step of the sugarcane lignocellulosic biomass pretreated with dilute acid, the data in terms of enzymatic conversion of cellulose are presented in Table 4. In an initial examination, the tops were most susceptible to enzymatic hydrolysis, followed by the straw and bagasse, with average values for cellulose conversion to glucose of 66.9, 51.8, and 35.5%, respectively. The sugarcane variety was not a significant factor in the enzymatic hydrolysis step, since the enzymatic conversion achieved for each type of residue (bagasse, straw, or tops) was statistically the same for all the sugarcane varieties (K, M, Q, and X), as revealed by Tukey's test with a significance level of $P < 0.05$ (Table 4). In addition, the responses of the different residue types to saccharification were significantly different, corroborating the order of susceptibility to enzymatic conversion mentioned above (tops > straw > bagasse).

In work concerning the optimization of dilute sulfuric acid pretreatment to maximize the sugar yield (pentose and hexose sugars) from sugarcane bagasse for ethanol production, considerable variations in sugar yields from

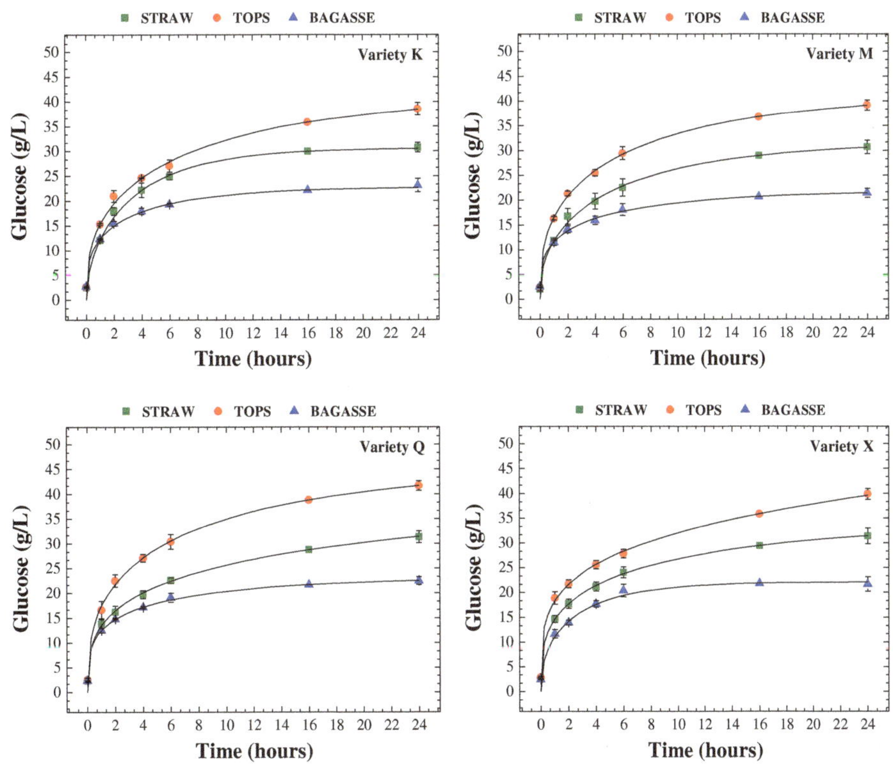

Figure 1 Temporal profiles of glucose release. Temporal profiles of glucose release during the enzymatic hydrolysis of bagasse, straw, and tops from different varieties of sugarcane (K, M, Q, and X) pretreated with dilute sulfuric acid. The lines are models fitted according to the Chrastil [61] approach.

enzymatic hydrolysis were observed for different varieties of sugarcane [24]. In a subsequent study, the same authors reported that there were significant differences in the enzymatic digestibility of different varieties of sugarcane [23]. Different responses in the enzymatic hydrolysis step have similarly been reported for other varieties of sugarcane or alternative feedstocks such as switchgrass [36,47,51]. In this regard, the earlier findings differ from the results obtained here. However, it is important to consider the set of samples analyzed. It is likely that the sugarcane varieties employed here (SP79-1011, RB867515, SP81-3250, and RB92579) have similar physical-chemical-morphological characteristics. Accordingly, additional studies of these varieties of sugarcane are currently being conducted using spectrometric and microscopic techniques such as nuclear magnetic resonance, Fourier transform infrared spectroscopy, X-ray diffraction, and scanning electron microscopy, in order to elucidate these and other important issues.

Table 4 Enzymatic conversion of cellulose (ECC, %) after 24 h of hydrolysis

Variety	Symbol	Residue	Cellulose conversion (%)
SP79-1011	K	Straw	52.53^{B}
		Tops	65.54^{A}
		Bagasse	35.82^{C}
RB867515	M	Straw	51.68^{B}
		Tops	67.45^{A}
		Bagasse	35.45^{C}
SP81-3250	Q	Straw	51.26^{B}
		Tops	70.20^{A}
		Bagasse	35.91^{C}
RB92579	X	Straw	51.76^{B}
		Tops	64.37^{A}
		Bagasse	34.98^{C}

Data refer to the hydrolysis of straw, bagasse, and tops. from four varieties of sugarcane (K, M, Q, and X) pretreated with dilute acid. Means with different subscript capital letters are statistically different (Tukey's test, $P < 0.05$).

The enzymatic hydrolysis of sugarcane bagasse and trash pretreated with dilute acid (H_2SO_4, 2.9% *w*/*v*, 1:4 solid/liquid ratio, 130°C, 30 min) was investigated by Moutta et al. [34]. The optimal conditions were established in a previous study [46]. In agreement with the present results, the sugarcane trash showed higher enzymatic digestibility than bagasse, and the authors suggested that it was due to the morphological characteristics and the distribution of structural macromolecules (cellulose, hemicellulose, and lignin) in the cell walls [34]. As result, cellulose was less accessible in the bagasse, which explained the lower yield of glucose than obtained using trash [34]. Interestingly, the aforementioned study employed loadings of enzymes per gram of biomass, while here, we used enzymatic loadings per gram of cellulose, with the aim of ensuring equal starting conditions for all the materials tested, since the enzyme/substrate ratio (E/S) has a large influence on the rate of the catalytic reaction. Another similar study from the same group assessed the effect of hydrothermal pretreatment (1:10 solid/liquid ratio, 195°C, 10 min) on the enzymatic conversion of bagasse and trash, and reported that after 24 h, the glucose concentrations from the pretreated trash and bagasse reached 17.2 and 13.5 g/L, corresponding to cellulose hydrolysis yields of 84.4 and 49.1%, respectively [33]. These findings emphasize the lower susceptibility of bagasse to enzymatic hydrolysis, compared to the trash (straw and tops), irrespective of whether the pretreatment is performed using dilute acid or a hydrothermal process.

An important point is that lignin is considered the most recalcitrant component of the plant cell wall macromolecules. It is predominantly found in the secondary cell wall and plays a key role in pathogen resistance, water regulation, and maintaining the integrity of the cell wall structure [50]. The relation between lignin content and enzymatic degradation of vegetal biomass has been extensively studied in recent years [36,52,53]. In general, it is reported that the lower the lignin content, the greater will be the bioavailability of the substrate for the production of ethanol [50]. Here, all the sugarcane biomass parts showed similar lignin contents after the pretreatment step (Table 2), with the following order for this component: straw > bagasse > tops (Table 3). Therefore, the lower susceptibility to enzymatic conversion observed here for bagasse is not only related to the lignin content, and other factors related to the physical-chemical characteristics of the biomass could also play a role, as further discussed in the following sections.

Cellulosic hydrolysate fermentability

After production of the cellulosic hydrolysates, alcoholic fermentation was performed in order to evaluate the fermentability of these media. The fermentation step followed the same conditions for all assays, in order to ensure the same basis for comparison of the individual responses for each type of sugarcane residue (bagasse, straw, or tops) from the four sugarcane varieties. Figures 2 and 3 present the temporal profiles of glucose consumption and ethanol production, respectively, throughout 8 h of alcoholic fermentation employing an industrial strain of *Saccharomyces cerevisiae*. Overall, the alcoholic fermentation profiles (glucose consumption and ethanol production) showed similar patterns, regardless of sugarcane variety (Figures 2 and 3). In this step of the cellulosic ethanol production process, the hydrolysates obtained from the straw of all varieties provided the best fermentability, consuming on average more than 95% of the available glucose (Figure 2) and reaching an average ethanol concentration of 35.5 g/L after 8 h (Figure 3). The lowest fermentability was found for the media derived from the bagasse of all varieties, with consumption of slightly more than 65% of the glucose (Figure 2) and production of 20.0 g/L ethanol (Figure 3). Ethanol production was about 78% higher for the hydrolysates from straw, compared to those from bagasse. The results obtained for the hydrolysates from the tops were intermediate between those for straw and bagasse, with consumption of around 80% of the glucose (Figure 2) and production of 26.5 g/L ethanol (Figure 3), corresponding to an ethanol production that was 33% higher compared to the hydrolysates from bagasse and 34% lower in relation to those from straw. The qualitative order of fermentability for the hydrolysates from the different sugarcane residues was therefore as follows: straw > tops > bagasse.

Determination of the ethanol yield (% of theoretical) provides essential information for evaluation of the ethanol production process, as shown in Table 5. The responses of each type of hydrolysate (from bagasse, straw, or tops) to alcoholic fermentation showed statistically significant differences, with the hydrolysates produced from straw being most fermentable, followed by those from tops and bagasse, with average ethanol yields of 74.1, 65.1, and 56.3%, respectively (Table 5). Statistical analysis indicated that the variety of sugarcane was not a significant factor in the alcoholic fermentation step, with the ethanol yields for each individual type of hydrolysate (from bagasse, straw, or tops) being statistically the same (Tukey's test, $P < 0.05$) for all the varieties (Table 5).

The influence of variety on cellulosic ethanol production has been evaluated for different feedstocks such as sugarcane, poplar, and wheat [22,23,36,47,51,54,55]. It has generally been found that different varieties of the same raw material differ in terms of overall ethanol production, suggesting that the selection step should identify varieties requiring minimal pretreatment and enzymatic hydrolysis steps, which would reduce the total cost of the process. For example, use of varieties of sugarcane bagasse with lower lignin content and highly substituted xylan resulted in higher sugar and ethanol yields, with milder pretreatment conditions and reduced enzyme dosage [23]. On the other hand, the results found here concerning the influence of sugarcane variety on the fermentability of the hydrolysates revealed no significant differences in the ethanol yield. As mentioned before, this could have been due to the similar physical-chemical-morphological characteristics of the set of sugarcane varieties evaluated here.

The commercial sugarcane varieties employed (SP79-1011, RB867515, SP81-3250, and RB92579) are among the most widely cultivated in central-south Brazil [56]

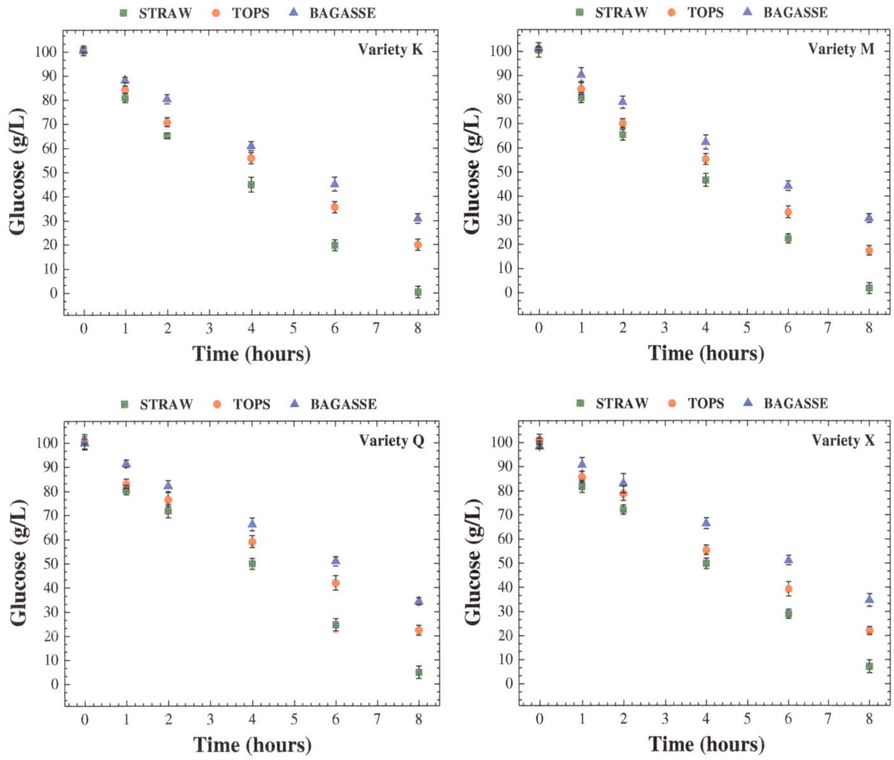

Figure 2 Temporal profiles of glucose consumption. Temporal profiles of glucose consumption throughout the alcoholic fermentation of the cellulosic hydrolysates resulting from the enzymatic saccharification of bagasse, straw, and tops from different varieties of sugarcane (K, M, Q, and X) pretreated with dilute sulfuric acid.

and show similar agronomic characteristics including low water demand, good mechanical harvesting, and efficient transportation. Differences between the varieties are the greater soil requirement of SP79-1011, later maturation of RB867515, and poorer drought tolerance of SP81-3250 and RB92579. Acknowledging that the set of varieties evaluated here was relatively small, the variety of sugarcane was not a factor that significantly influenced the cellulosic ethanol production process. Selection of variety for cultivation should therefore focus on agronomic aspects such as adaptation to regional soil and weather conditions, in order to maximize productivity and yield while minimizing management costs. Furthermore, this finding is an important indication of the possibility of cultivating different varieties of sugarcane without affecting the final production of cellulosic ethanol in the mills. It is vital that the producer has a diversity of sugarcane varieties in the field, in order to decrease the likelihood of damage due to proliferation of pests and diseases within the plantation [56].

Another important fermentation parameter is the volumetric ethanol productivity (Q_P). The highest productivities were achieved for the hydrolysates from straw (Q_P = 4.4 g/L.h), followed by those from tops (Q_P = 3.3 g/L.h) and bagasse (Q_P = 2.5 g/L.h). These values were averages for each type of hydrolysate (from bagasse, straw, or tops) after 8 h of alcoholic fermentation and were statistically different. Therefore, the volumetric ethanol productivity for bagasse hydrolysates was about 76% lower compared to straw hydrolysates and approximately 32% lower in relation to tops hydrolysates. This can be explained by the negative effect of aromatic substances (mainly phenolic compounds) on volumetric ethanol productivity. These substances can be released from residual lignin and interfere in the substrate assimilation rate, hence affecting ethanol productivity [47].

In terms of alcoholic fermentation, another essential aspect concerns the inhibition of microorganisms, especially given the known inhibition of ethanol-producing yeast by degradation products produced during pretreatment of biomass. The generation of pretreatment by-products is strongly dependent on the feedstock and pretreatment method. Dilute sulfuric acid pretreatment is recognized to produce by-products considered inhibitory to microbial fermentation [10], such as furan derivatives, weak acids, and phenolic compounds [57,58]. The main furan derivatives are furfural and 5-hydroxymethyl furfural (HMF) derived from degradation of pentoses and

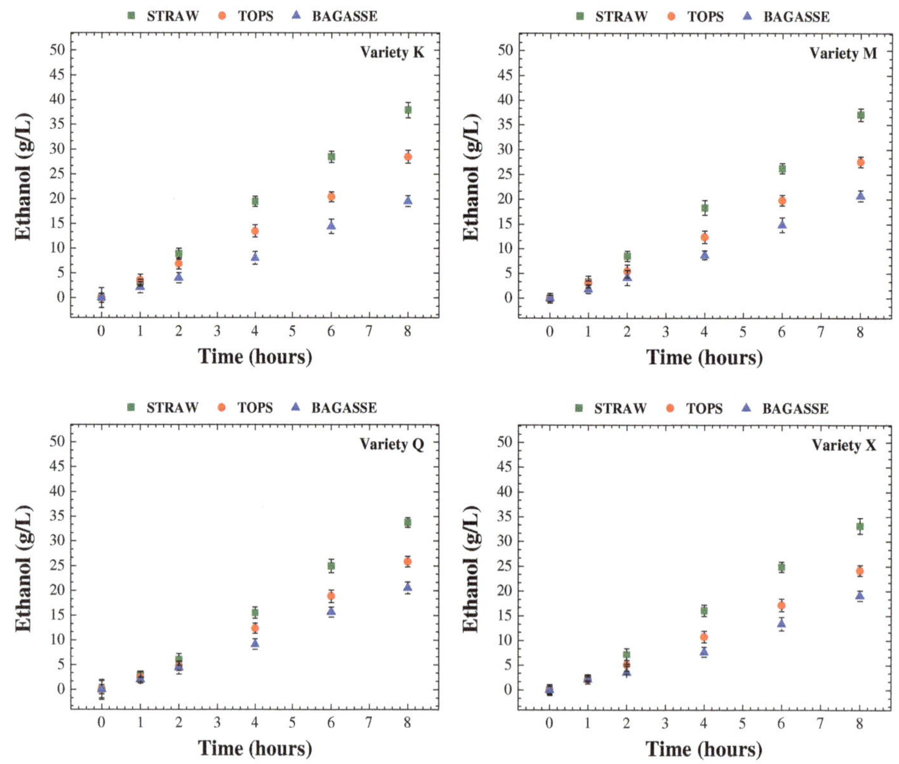

Figure 3 Temporal profiles of ethanol production. Temporal profiles of ethanol production during the course of the alcoholic fermentation of the cellulosic hydrolysates from the enzymatic hydrolysis of bagasse, straw, and tops from different varieties of sugarcane (K, M, Q, and X) pretreated with dilute sulfuric acid.

Table 5 Ethanol yield (% of theoretical) after 8 h of alcoholic fermentation

Variety	Symbol	Residue	Ethanol yield (%)
SP79-1011	K	Straw	74.70^A
		Tops	65.52^B
		Bagasse	54.69^C
RB867515	M	Straw	74.19^A
		Tops	64.47^B
		Bagasse	55.91^C
SP81-3250	Q	Straw	73.90^A
		Tops	63.85^B
		Bagasse	58.16^C
RB92579	X	Straw	73.74^A
		Tops	66.38^B
		Bagasse	56.30^C

Data are from the cellulosic hydrolysates resulting from the enzymatic hydrolysis of straw, tops, and bagasse from four varieties of sugarcane (K, M, Q, and X) pretreated with dilute acid.
Means with different subscript capital letters are significantly different (Tukey's test, $P < 0.05$).

hexoses, respectively, while weak acids such as acetic acid are formed from the acetic groups present in the hemicellulosic fraction [41].

The temporal profiles of glucose consumption (Figure 2) and ethanol production (Figure 3) during the alcoholic fermentation of the hydrolysates (from bagasse, straw, or tops) clearly showed that the presence of inhibitory substances could have interfered in the metabolism of the yeast strain used in this study. Since there were no significant differences among the varieties of sugarcane evaluated here in terms of the overall process of ethanol production, measurements of inhibitors (acetic acid, HMF, and furfural) were only made during alcoholic fermentation of the hydrolysates (from bagasse, straw, and tops) obtained using the variety K. No detectable levels of furfural were found in any of the hydrolysates after 8 h of alcoholic fermentation. On the other hand, very low levels of acetic acid and HMF were detected, with the highest amounts in the hydrolysate from bagasse (481.8 and 12.2 mg/L, respectively) and the lowest concentrations in the hydrolysate from straw (154.7 and 10.2 mg/L, respectively). The hydrolysate from tops presented intermediate levels of both inhibitors (318.5 mg/L of acetic acid and 11.5 mg/L of HMF). These findings were in correlation with the order of fermentability described above (straw > tops > bagasse).

It has been found previously that after milling and pretreatment of sugarcane bagasse and trash, the bagasse hydrolysates contained higher levels of acetic acid, compared to the trash hydrolysates, resulting in an ethanol yield that was somewhat higher when trash hydrolysates were used [35], supporting the results found here. The present findings obtained from the assessment of separate sugarcane residues therefore clearly showed that ethanol production from sugarcane biomass could be improved by the combined use of bagasse, straw, and tops. This approach is an interesting alternative because use of the whole sugarcane lignocellulosic biomass could greatly improve ethanol productivity per hectare, without any need to expand the areas under cultivation. To the best of our knowledge, there have been no previous studies of this strategy including all the steps of cellulosic ethanol production.

Combined use of the whole sugarcane lignocellulosic biomass for ethanol production

After assessing the responses for each type of residue (bagasse, straw, and tops) obtained after processing of the four sugarcane varieties (K, M, Q, and X), investigation was made of the feasibility of using a mixture of the three sugarcane residues for the conversion of cellulose into glucose and ethanol production. For this purpose, variety K was arbitrarily selected, since there were no significant differences among the varieties in terms of enzymatic conversion (Table 4). Figure 4A shows the temporal profiles of glucose release during the enzymatic hydrolysis step for the three separate residues and a mixture of them (bagasse-straw-tops, 1:1:1, dry weight basis). The chemical composition of the pretreated mixture was calculated as a weighted average of the compositions of the individual pretreated residues.

As expected, the enzymatic saccharification of the mixture showed a pattern that was intermediate to those for the individual residues (Figure 4A). Figure 4B presents the data for the enzymatic conversion of cellulose in these processes, together with a statistical analysis (Tukey's test, $P < 0.05$). The conversion value obtained for the mixture was significantly higher than that for bagasse and significantly lower than that for the tops. The conversion value for the mixture was statistically the same as that for the straw. These results are encouraging in terms of the use of whole sugarcane lignocellulosic biomass because they show that enzymatic saccharification can be performed using the residues from the processing of sugarcane together in a mixture (bagasse-straw-tops). In this biotechnological approach, the lower susceptibility to enzymatic degradation of bagasse can be balanced by the higher susceptibility of straw and tops (trash), as can be seen in Figure 4A,B. Importantly, the enzymatic conversion obtained using the mixture was 55% higher, compared to the use of bagasse alone. Therefore, the potential application of bagasse (which is already being studied) could be further extended if combined with the use of tops and straw.

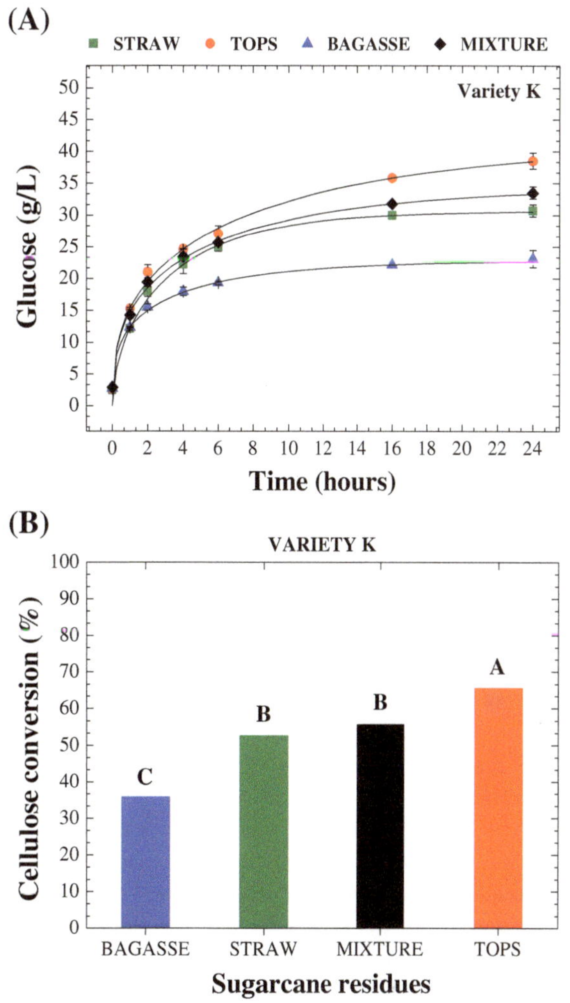

Figure 4 Temporal profiles of glucose release and statistical analysis of the data for enzymatic conversion. Comparison of the temporal profiles of glucose release **(A)** and statistical analysis of the data for enzymatic conversion of cellulose after 24 h **(B)** for the saccharification of bagasse, straw, tops, and the combination of them (bagasse-straw-tops, 1:1:1 mixture), using variety K. The lines are models fitted according to the Chrastil [61] approach. Means with different capital letters are statistically different (Tukey's test, $P < 0.05$).

In previous work, the influence of biomass chemical composition on enzymatic conversion of cellulose into glucose has been evaluated by correlating the yield of glucose with the contents of lignin, hemicellulose, and ash. No direct relation between residual lignin or hemicellulose and the enzymatic conversion of cellulose was observed for bagasse derived from different sugarcane hybrids [47]. Nevertheless, the lignin/hemicellulose (L/H) ratio showed a strong correlation with conversion yield ($R^2 = 0.90$), with maximum conversion values achieved when the L/H ratio was in the range from 3 to 4 [47]. On the other hand, several other studies have reported negative correlations between cellulose digestibility and the lignin and ash contents [21,22,24,36,55]. Nonetheless, the enzymatic conversion of biomass into glucose is not exclusively determined by the lignin and ash contents, but it also depends on the contents of structural carbohydrates as well as other physical and chemical properties of these lignocellulosic materials [55]. Moreover, even though lignin is usually assumed to be one of the most important factors limiting aspects of the saccharification, the presence of residual hemicellulose cannot be ignored, because its close association with the cellulose fibrils hinders access of the cellulases to the cellulose surface [47].

Here, no direct correlations were observed between cellulose conversion and the lignin or hemicellulose contents of the pretreated materials, and there was no relation between conversion and the L/H ratio. However, there was a strong negative correlation ($R^2 = 0.96$) between the enzymatic conversion values for the residues (bagasse, straw, and tops) from the four sugarcane varieties and the (lignin + hemicellulose)/ash ratio (Figure 5). Hence, the highest enzymatic conversion was achieved when this ratio was smallest. This correlation was able to explain the observed order of enzymatic digestibility (tops > straw > bagasse), since the three sugarcane residues obtained from all varieties were distributed in three independent groups. Inclusion of the conversion value obtained for the mixture of the three residues (bagasse-straw-tops, 1:1:1, dry weight basis) showed that the enzymatic conversion of the whole sugarcane lignocellulosic biomass into glucose could be correlated with the (lignin + hemicellulose)/ash ratio (Figure 5). Since statistical analysis showed that there was no significant

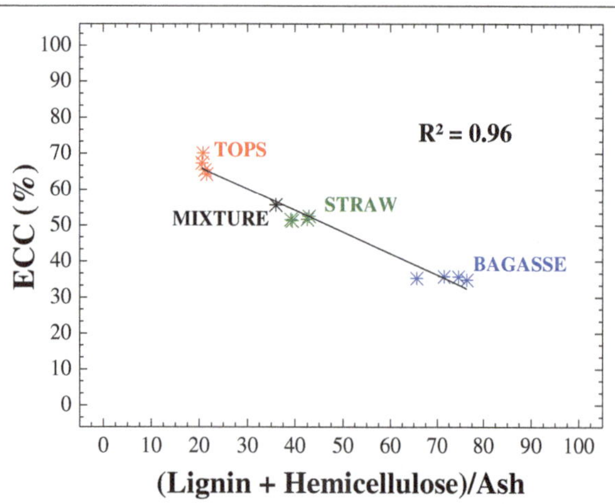

Figure 5 Negative linear correlation between the enzymatic conversion of cellulose and the (lignin + hemicellulose)/ash ratio.

difference between the straw and the mixture, in terms of the conversion of cellulose (Figure 4B), the influence of the chemical composition of the materials was not related to the individual components, but rather to the ratios between them, indicating the importance of the distribution of the different components.

After individual assessment of the different hydrolysates (from bagasse, straw, and tops) in terms of their fermentability, evaluation was made of the fermentation behavior of a hydrolysate from enzymatic hydrolysis of a mixture of the three sugarcane residues (bagasse-straw-tops, 1:1:1, dry weight basis) pretreated with dilute sulfuric acid. Figure 6 presents the temporal profiles of glucose consumption (A) and ethanol production (B). In this phase of the study, it was important to examine if the fermentation of a hydrolysate generated after enzymatic conversion of a sugarcane residues mixture would show any unexpected effects, or even be detrimental to the production of ethanol. Since there were no significant differences among the varieties in terms of alcoholic fermentation (Table 5), variety K was again arbitrarily selected for this comparison. The alcoholic fermentation of hydrolysate from the mixture (bagasse-straw-tops, 1:1:1, dry weight basis) exhibited an intermediate pattern, compared to the hydrolysates from bagasse, straw, and tops, as can be clearly seen from the data for glucose consumption (Figure 6A) and ethanol production (Figure 6B). The fermentation of hydrolysate from the mixture was monitored in terms of the contents of furfural, HMF, and acetic acid. Again, there were no detectable levels of furfural. The acetic acid and HMF levels were intermediate (267.6 and 10.8 mg/L, respectively) compared to the hydrolysates from bagasse, straw, and tops. Furthermore, statistical analysis (Tukey's test, $P < 0.05$) of the ethanol yield data for these fermentation processes showed that there were no significant differences between the hydrolysates from the mixture and from tops or straw (Figure 6C). On the other hand, the ethanol yield value for the mixture was significantly higher than for bagasse. The volumetric productivity of ethanol for the hydrolysate from the mixture was also calculated ($Q_P = 3.9$ g/L.h) and was approximately 56% higher compared to the productivity of ethanol for the hydrolysate from bagasse ($Q_P = 2.5$ g/L.h).

An important point is that the yield of ethanol found for the hydrolysate from the mixture was about 25% higher than for the hydrolysate from bagasse (Figure 6C). These results are highly encouraging for the strategy of using the whole sugarcane plant for the production of cellulosic ethanol, since the fermentability of hydrolysates obtained from the enzymatic hydrolysis of individual sugarcane residues was comparable to that for a mixture of them (bagasse-straw-tops, 1:1:1, dry weight basis), so use of the mixture had no negative effects on the overall process. Thus, the lower fermentability presented by the

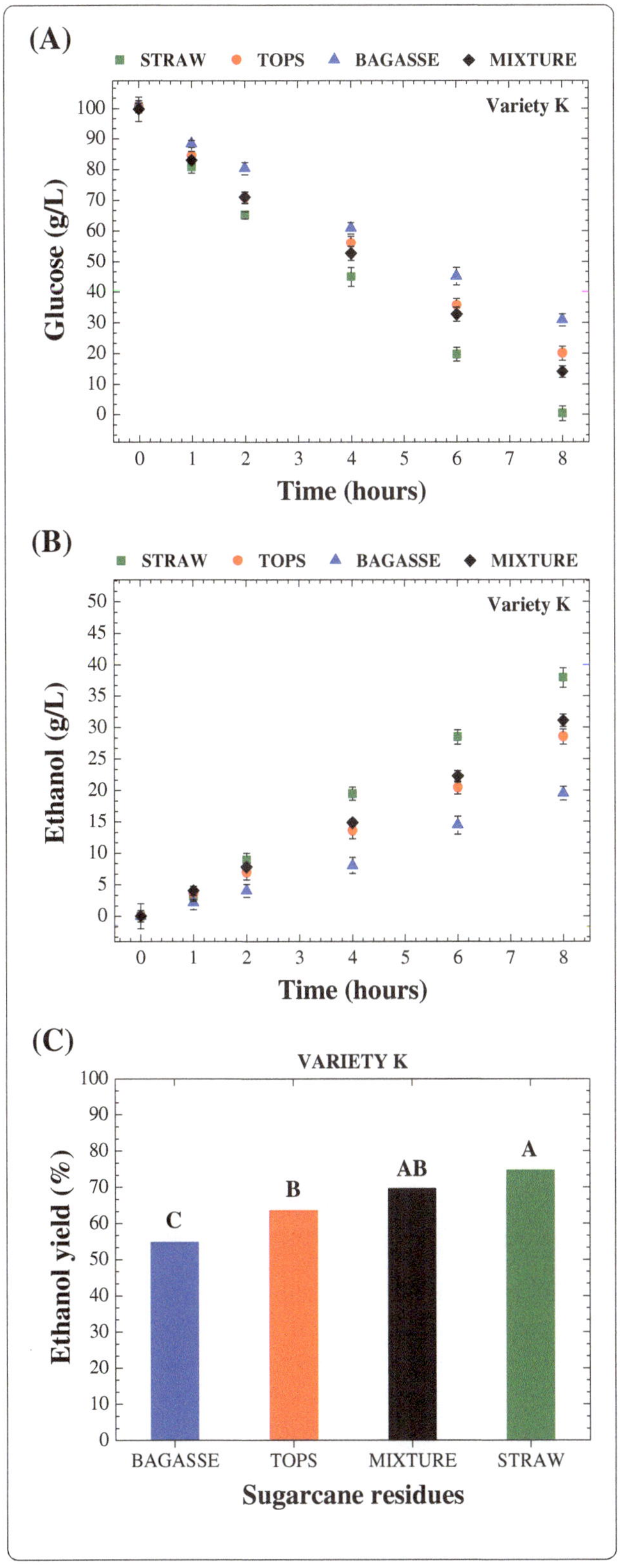

Figure 6 Temporal profiles of glucose consumption and ethanol production and the data for the ethanol yield. Comparison of the temporal profiles of glucose consumption **(A)** and ethanol production **(B)**, and statistical analysis of the data for the ethanol yield (% of the theoretical yield) after 8 h **(C)** for the alcoholic fermentation of the cellulosic hydrolysates from bagasse, straw, tops, and the combination of them (bagasse-straw-tops, 1:1:1 mixture), using variety K. Means with different capital letters are statistically different (Tukey's test, $P < 0.05$).

hydrolysate from bagasse could be balanced by higher fermentability found for the hydrolysates from straw and tops (trash), as shown in Figure 6. It can be concluded that overall cellulosic ethanol production can be increased by using a mixture of the whole sugarcane lignocellulosic biomass (bagasse, straw, and tops), instead of bagasse alone.

In order to explain how the chemical composition of biomass could affect the production of ethanol, it was reported previously that the ethanol concentration presented a strong inverse correlation ($R^2 = 0.91$ to 0.99) with lignin content [23]. In the present study, there were no direct correlations between ethanol yield (or volumetric productivity of ethanol) and the residual contents of lignin or hemicellulose from the pretreated materials or between the ethanol yield and the L/H ratio. On the other hand, it is expected that there is a positive correlation between the glucose yield in the enzymatic saccharification and the ethanol yield from the alcoholic fermentation, that is, an increase in the enzymatic conversion of cellulose into glucose will result in greater amount of fermentable sugar readily available for the ethanol production step. Here, the best cellulose conversions were obtained for the tops from the different varieties, whose hydrolysates resulted in intermediate volumetric ethanol productivity values (or ethanol yields). The highest fermentability values were obtained for the hydrolysates from straw. In order to better understand these observations, the volumetric productivity of ethanol (Q_P, g/L.h) and the enzymatic conversion of cellulose (ECC, %) were examined for each hydrolysate type (bagasse, straw, and tops) from the four sugarcane varieties. A meaningful relationship ($R^2 = 0.85$) was found between Q_P and ECC, which could be fitted using a second-order polynomial (Figure 7A). As mentioned previously, a strong inverse linear correlation was established between the conversion of cellulose (ECC, %) and a ratio involving the contents of lignin, hemicellulose, and ash (Figure 5). Investigation of the influence of the latter on ethanol productivity revealed a strong relationship ($R^2 = 0.84$), and a good fit was obtained using a second-order polynomial (Figure 7B).

The relationships illustrated in Figure 7A,B provided a qualitative order of fermentability, straw > tops > bagasse, with all the sugarcane varieties contained in three distinct

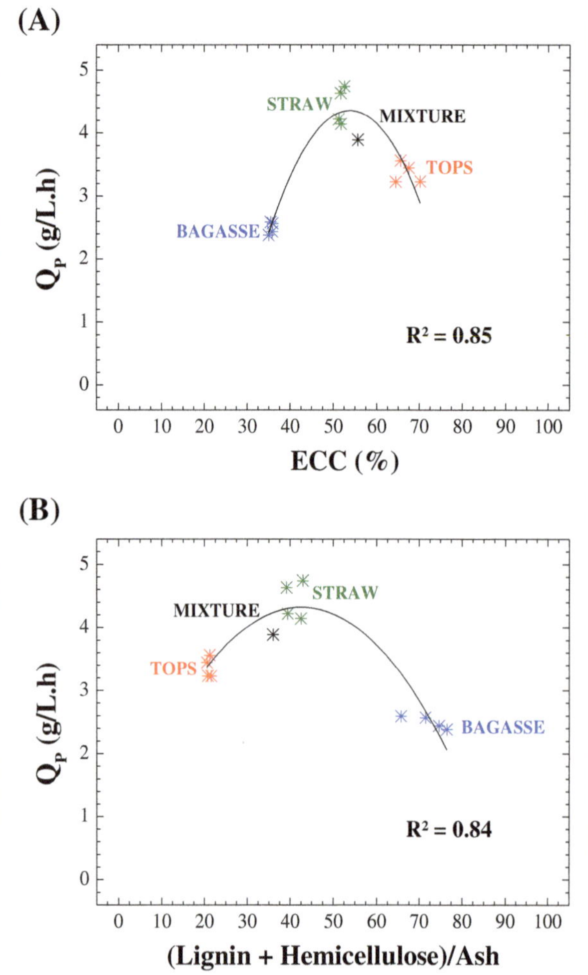

Figure 7 Ethanol productivity and cellulose conversion and chemical composition of the pretreated biomass. Relationship between volumetric ethanol productivity and enzymatic conversion of cellulose **(A)**, and relationship between the volumetric ethanol productivity and the (lignin + hemicellulose)/ash ratio **(B)**.

groups corresponding to the three residues. The value for the mixture of the three residues (bagasse-straw-tops, 1:1:1, dry weight basis) was in agreement with the polynomial fit, showing that the ethanol productivity could be related to both the cellulose conversion and the chemical composition of the pretreated biomass (Figure 7). A possible explanation for the fact that the greater susceptibility to enzymatic degradation displayed by the tops did not lead to a higher yield of ethanol for the hydrolysates from the tops could have been due to a greater propensity for generation of degradation products. This was supported by the presence of inhibitory substances (HMF and acetic acid) at higher concentrations in the hydrolysates from the tops, compared to the hydrolysates from the straw (see the previous section). The greater potential of the sugarcane tops to produce degradation products, compared to the straw, seems to be closely related to the chemical composition of these lignocellulosic materials. These factors could help to explain the interesting trends illustrated in Figure 7A,B. Notwithstanding, this promising biotechnological approach, employing the whole sugarcane lignocellulosic biomass for ethanol production, can help in overcoming the difficulties experienced in the enzymatic hydrolysis and alcoholic fermentation steps using bagasse alone because the mixture of sugarcane residues (bagasse-straw-tops) provides a means of compensating for the lower degradability and fermentability of bagasse.

Conclusions

Within the context of the current trend towards using the whole sugarcane lignocellulosic biomass, the lignocellulosic residues (bagasse, straw, and tops) from four varieties of sugarcane (SP79-1011, RB867515, SP81-3250, and RB92579) were assessed for the production of ethanol, under the same conditions. The chemical compositions of the residues pretreated with dilute sulfuric acid showed only slight differences, indicating that the pretreatment was able to produce similar contents of the main components (cellulose, hemicellulose, lignin, and ash) in the three types of pretreated residue. Susceptibility to the enzymatic saccharification step was in the order: tops > straw > bagasse. For the alcoholic fermentation step, the fermentability of the hydrolysates from the different sugarcane residues was in the order: straw > tops > bagasse. Sugarcane variety was not a significant factor in the overall ethanol production process. The proposed strategy is a promising way in which the whole sugarcane lignocellulosic biomass can be used in combination for the production of cellulosic ethanol. The use of the mixture (bagasse-straw-tops, 1:1:1, dry weight basis) resulted in an intermediate pattern, compared to the residues used individually, without any notable deleterious effects. In addition, the enzymatic conversion of cellulose obtained with the mixture was about 55% higher than for the bagasse alone. Similarly, the ethanol yield for the hydrolysate derived from the mixture was 25% higher than obtained using bagasse alone. These findings show that the lower susceptibility of sugarcane bagasse to enzymatic degradation, as well as the lower fermentability of the hydrolysates from sugarcane bagasse, can be assisted by using the bagasse in combination with the tops and straw (sugarcane trash). The use of bagasse, which is already being widely studied as a feedstock for production of 2G ethanol, can be further extended by employing the whole sugarcane lignocellulosic biomass.

Materials and methods

Raw materials and biomass preparation

Three lignocellulosic residue parts (bagasse, straw, and tops) from four varieties of sugarcane (SP79-1011,

RB867515, SP81-3250, and RB92579, here represented by K, M, Q and X, respectively) were kindly supplied by Sumaúma Mill (Marechal Deodoro, Brazil). These lignocellulosic materials were dried in an oven at 45°C until reaching moisture content below 10%. The samples were then milled in a knife mill, sieved at 2 mm, and stored at room temperature for later use. The raw materials were chemically characterized (on a dry matter basis) in terms of their contents of structural carbohydrates (cellulose and hemicellulose), lignin, and ash according to Gouveia et al. [59].

Dilute sulfuric acid pretreatment

The pretreatment of the lignocellulosic materials was carried out by a chemical pretreatment method using dilute acid. This process employed a solution of dilute sulfuric acid (1.5%, *w/w*) and a solid loading of 10%. The reactions were performed at 121°C for 30 min in an autoclave. After the pretreatment, the liquid and solid fractions (hemicellulosic hydrolysates and cellulignins, respectively) were separated by vacuum filtration. The hemicellulosic hydrolysates were discarded. The pretreated solid materials were thoroughly washed with distilled water to remove the soluble components and used directly for the next steps. To assess the combination of the residues, a mixture of bagasse, straw, and tops (1:1:1, dry weight basis) was prepared and pretreated following the same procedure described previously. The pretreated materials were also chemically characterized according to Gouveia et al. [59].

Enzymatic hydrolysis step

A commercial enzyme preparation (Cellic®Ctec2), kindly provided by Novozymes A/S (Bagsværd, Denmark), was used for the enzymatic hydrolysis assays of the pretreated materials. The enzymatic activity of commercial extract, in terms of filter paper units (FPU), was determined according to Ghose [60]. The enzymatic hydrolyses of the pretreated materials were performed in 500-mL Erlenmeyer flasks containing 0.1 M sodium citrate buffer (pH 5.0), at a solid/liquid ratio of 1:10 and using an enzyme loading of 30 FPU per gram of residual cellulose in the pretreated materials. The reaction mixtures were incubated for 24 h at 50°C and 200 rpm. The saccharification of the mixture (bagasse-straw-tops, 1:1:1, dry weight basis) was carried out under the same conditions described above. Samples were periodically withdrawn (0, 1, 2, 4, 6, and 24 h) for quantification of glucose release using an enzymatic kit Doles® (Goiânia, Brazil). In addition, in order to simulate the sampling after 16 h, the Chrastil approach for modeling enzymatic reactions was used to fit the experimental data for glucose release [61].

Alcoholic fermentation step

An industrial strain of *S. cerevisiae* CAT-1 acquired from the Jalles Machado Mill (Goianésia, Brazil) was used for the alcoholic fermentation assays of the cellulosic hydrolysates, carried out in 250 mL Erlenmeyer flasks. After the saccharification step, the residual solid was separated by centrifugation. The levels of glucose in the cellulosic hydrolysates were measured, and anhydrous glucose was added up to a concentration of 100 g/L. The media were then inoculated with 25 g of yeast cells (on dry weight basis) per liter of cellulosic hydrolysate. The experiments were conducted for 8 h using a shaker at 31°C and 100 rpm. For the cellulosic hydrolysates from the enzymatic hydrolysis of the mixture (bagasse-straw-tops, 1:1:1, dry weight basis), the fermentation step was conducted under the same conditions described above. Throughout the alcoholic fermentation, the consumption of glucose was followed using an enzymatic kit Doles® (Goiânia, Brazil) and ethanol was monitored by HPLC (see next section).

Analytical methods

The concentrations of monomeric sugars (cellobiose, glucose, xylose, and arabinose), organic acids (formic and acetic), furans (furfural and hydroxymethylfurfural), and ethanol were determined by high-performance liquid chromatography (HPLC) according to Gouveia et al. [59]. The determination of monomeric sugars, organic acids, and ethanol used a refractive index detector and an Aminex HPX-87H column. The chromatographic separation was conducted using isocratic elution with 5 mM sulfuric acid as the mobile phase, at a flow rate of 0.6 mL/min, and oven and detector cell temperatures of 50°C. Quantification of the furans employed a UV–vis detector (274 nm) and an Agilent Sorbex C18 column (250 × 2.5 mm, particle size 5 μm). The chromatographic separation was conducted using isocratic elution, with a solution of acetonitrile/water (1:8) containing 1% (*v/v*) acetic acid as the mobile phase, at a flow rate of 0.8 mL/min, and an oven temperature of 25°C.

Data and statistical analysis

All the experimental steps (chemical characterization, dilute acid pretreatment, enzymatic hydrolysis, and alcoholic fermentation) were carried out in triplicate, and the results were calculated as means ± standard deviations. In all cases, the standard deviations were less than 5%. Thus, only the mean values are presented in the Tables. The error bars shown in the figures represent the standard deviations for triplicate assays, and where error bars are not visible, they are smaller than the symbol. The mean values were analyzed statistically using Tukey's test with a confidence level of 95% (Origin 8.0 software).

Calculations

For the enzymatic saccharification step, the enzymatic conversion of cellulose (ECC, %) was calculated according to Equation 1:

$$\mathrm{ECC} = \left[\frac{\left(m_{\mathrm{glucose}}^{24\mathrm{h}} - m_{\mathrm{glucose}}^{0\mathrm{h}}\right)}{\left(m_{\mathrm{cellulose}}^{0\mathrm{h}}\right) * 1.11}\right] * 100 \quad (1)$$

For the alcoholic fermentation step, three parameters were calculated: the ethanol yield factor ($Y_{P/S}$), described by Equation 2, which was used to calculate the ethanol yield (% of the theoretical yield), according to Equation 3, and the volumetric ethanol productivity (Q_P, g/L.h), described by Equation 4.

$$Y_{P/S} = \frac{\left(C_{\mathrm{ethanol}}^{8\mathrm{h}} - C_{\mathrm{ethanol}}^{0\mathrm{h}}\right)}{\left(C_{\mathrm{glucose}}^{0\mathrm{h}} - C_{\mathrm{glucose}}^{8\mathrm{h}}\right)} \quad (2)$$

$$\mathrm{Ethanolyield} = \left(\frac{Y_{P/S}}{0.511}\right) * 100 \quad (3)$$

$$Q_P = \frac{\left(C_{\mathrm{ethanol}}^{8\mathrm{h}} - C_{\mathrm{ethanol}}^{0\mathrm{h}}\right)}{8\mathrm{h}\text{-}0\mathrm{h}} \quad (4)$$

In the above equations, m is the mass and C is the concentration of the compounds. The value 1.11 is the theoretical yield factor for the enzymatic conversion of cellulose to glucose (Equation 1). Likewise, 0.511 is the theoretical yield factor for the alcoholic fermentation process. In some studies, the ethanol yield (Equation 3) is sometimes called the fermentation efficiency.

Abbreviations

2G: second generation; 1G: first generation; db: dry basis; ECC: enzymatic conversion of cellulose; FPU: filter paper units; GHG: global greenhouse gas; HMF: 5-hydroxymethyl furfural; HPLC: high-performance liquid chromatography; Q_P: ethanol productivity; $Y_{P/S}$: ethanol yield factor.

Competing interests

The authors declare they have no competing interests.

Authors' contributions

SCP participated in the design and planning of this study, carried out the laboratory work, interpreted the data, and drafted the manuscript. LM participated in the planning and execution of the laboratory work and editing of the manuscript. CMMM participated in the designing of this study and editing of the manuscript. CSF conceived the study, coordinated the design, supervised all other tasks, and edited the manuscript. All authors read and approved the final manuscript.

Acknowledgements

The authors would like to thank the Brazilian research funding agencies CAPES, CNPq, and FAPESP, and the Brazilian Agricultural Research Corporation (Embrapa).

Author details

[1]Embrapa Instrumentation, Rua XV de Novembro 1452, 13560-970 São Carlos, SP, Brazil. [2]Graduate Program of Chemical Engineering, Federal University of São Carlos, Rodovia Washington Luiz, km 235, 13565-905 São Carlos, SP, Brazil. [3]Embrapa Agroenergy, Parque Estação Biológica s/n°, 70770-901 Brasília, DF, Brazil.

References

1. Fuglestvedt J, Berntsen T, Myhre G, Rypdal K, Skeie RB. Climate forcing from the transport sectors. Proc Natl Acad Sci U S A. 2008;105:454–8.
2. EPA (United States Environmental Protection Agency). Trasnportation sector greenhouse gas emissions. http://www.epa.gov/otaq/climate/documents/420f13033a.pdf. Accessed 15 Jan 2014.
3. IPCC (Intergovernmental Panel on Climate Change). Renewable energy sources and climate change mitigation. http://www.ipcc.ch/pdf/special-reports/srren/SRREN_Full_Report.pdf. Accessed 15 Jan 2014.
4. Walter A, Dolzan P, Quilodran O, de Oliveira JG, da Silva C, Piacente F, et al. Sustainability assessment of bio-ethanol production in Brazil considering land use change, GHG emissions and socio-economic aspects. Energy Policy. 2011;39:5703–16.
5. Goldemberg J. Ethanol for a sustainable energy future. Science. 2007;315:808–10.
6. Sorda G, Banse M, Kemfert C. An overview of biofuel policies across the world. Energy Policy. 2010;38:6977–88.
7. CONAB (Companhia Nacional de Abastecimento). Acompanhamento da safra brasileira de cana-de-açúcar. http://www.conab.gov.br/OlalaCMS/uploads/arquivos/14_12_19_09_02_49_boletim_cana_portugues_-_3o_lev_-_2014-15.pdf. Accessed 15 Jan 2014.
8. IEA (International Energy Agency). Sustainable production of second-generation biofuels. http://www.iea.org/publications/freepublications/publication/second_generation_biofuels.pdf. Accessed 15 Jan 2014.
9. RFA (Renewable Fuels Association). World fuel ethanol production. http://ethanolrfa.org/pages/World-Fuel-Ethanol-Production. Accessed 15 Jan 2014.
10. Canilha L, Chandel AK, dos Santos Milessi TS, Fernandes Antunes FA, da Costa Freitas WL, Almeida Felipe MG, et al. Bioconversion of sugarcane biomass into ethanol: an overview about composition, pretreatment methods, detoxification of hydrolysates, enzymatic saccharification, and ethanol fermentation. J Biomed Biotechnol. 2012;2012:1–15.
11. Alonso Pippo W, Luengo CA, Alonsoamador Morales Alberteris L, Garzone P, Cornacchia G. Energy recovery from sugarcane-trash in the light of 2nd generation biofuels. Part 1: current situation and environmental aspects. Waste Biomass Valoriz. 2011;2:1–16.
12. Furlan FF, Tonon Filho R, Pinto FHPB, Costa CBB, Cruz AJG, Giordano RLC, et al. Bioelectricity versus bioethanol from sugarcane bagasse: is it worth being flexible? Biotechnol Biofuels. 2013;6:142–53.
13. Chandel AK, da Silva SS, Carvalho W, Singh OV. Sugarcane bagasse and leaves: foreseeable biomass of biofuel and bio-products. J Chem Technol Biotechnol. 2012;87:11–20.
14. Leal MRLV, Galdos MV, Scarpare FV, Seabra JEA, Walter A, Oliveira COF. Sugarcane straw availability, quality, recovery and energy use: a literature review. Biomass Bioenergy. 2013;53:11–9.
15. Alonso Pippo W, Luengo CA, Alonsoamador Morales Alberteris L, Garzone P, Cornacchia G. Energy recovery from sugarcane-trash in the light of 2nd generation biofuel. Part 2: socio-economic aspects and techno-economic analysis. Waste Biomass Valoriz. 2011;2:257–66.
16. Galdos M, Cavalett O, Seabra JEA, Horta Nogueira LA, Bonomi A. Trends in global warming and human health impacts related to Brazilian sugarcane ethanol production considering black carbon emissions. Appl Energy. 2013;104:576–82.
17. Seabra JEA, Tao L, Chum HL, Macedo IC. A techno-economic evaluation of the effects of centralized cellulosic ethanol and co-products refinery options with sugarcane mill clustering. Biomass Bioenergy. 2010;34:1065–78.
18. Macrelli S, Mogensen J, Zacchi G. Techno-economic evaluation of 2nd generation bioethanol production from sugar cane bagasse and leaves integrated with the sugar-based ethanol process. Biotechnol Biofuels. 2012;5:22–39.
19. Macrelli S, Galbe M, Wallberg O. Effects of production and market factors on ethanol profitability for an integrated first and second generation ethanol plant using the whole sugarcane as feedstock. Biotechnol Biofuels. 2014;7:26–41.
20. Franco HCJ, Magalhães PSG, Cavalett O, Cardoso TF, Braunbeck OA, Bonomi A, et al. How much trash to removal from sugarcane field to produce bioenergy. 1st Brazilian BioEnergy Science and Technology Conference: Campos do Jordão; 2011.

21. Benjamin Y, Cheng H, Goergens JF. Evaluation of bagasse from different varieties of sugarcane by dilute acid pretreatment and enzymatic hydrolysis. Ind Crop Prod. 2013;51:7–18.
22. Benjamin Y, Goergens JF, Joshi SV. Comparison of chemical composition and calculated ethanol yields of sugarcane varieties harvested for two growing seasons. Ind Crop Prod. 2014;58:133–41.
23. Benjamin Y, Garcia-Aparicio MP, Goergens JF. Impact of cultivar selection and process optimization on ethanol yield from different varieties of sugarcane. Biotechnol Biofuels. 2014;7:60–76.
24. Benjamin Y, Cheng H, Goergens JF. Optimization of dilute sulfuric acid pretreatment to maximize combined sugar yield from sugarcane bagasse for ethanol production. Appl Biochem Biotechnol. 2014;172:610–30.
25. Sasaki M, Adschiri T, Arai K. Fractionation of sugarcane bagasse by hydrothermal treatment. Bioresour Technol. 2003;86:301–4.
26. Canilha L, Santos VTO, Rocha GJM, Almeida e Silva JB, Giulietti M, Silva SS, et al. A study on the pretreatment of a sugarcane bagasse sample with dilute sulfuric acid. J Ind Microbiol Biotechnol. 2011;38:1467–75.
27. de Moraes Rocha GJ, Martin C, Soares IB, Souto Maior AM, Baudel HM, Moraes de Abreu CA. Dilute mixed-acid pretreatment of sugarcane bagasse for ethanol production. Biomass Bioenergy. 2011;35:663–70.
28. Aguiar RS, Luciano Silveira MH, Pitarelo AP, Corazza ML, Ramos LP. Kinetics of enzyme-catalyzed hydrolysis of steam-exploded sugarcane bagasse. Bioresour Technol. 2013;147:416–23.
29. de Albuquerque Wanderley MC, Martin C, de Moraes Rocha GJ, Gouveia ER. Increase in ethanol production from sugarcane bagasse based on combined pretreatments and fed-batch enzymatic hydrolysis. Bioresour Technol. 2013;128:448–53.
30. Oliveira FMV, Pinheiro IO, Souto-Maior AM, Martin C, Goncalves AR, Rocha GJM. Industrial-scale steam explosion pretreatment of sugarcane straw for enzymatic hydrolysis of cellulose for production of second generation ethanol and value-added products. Bioresour Technol. 2013;130:168–73.
31. Oliveira LRM, Nascimento VM, Goncalves AR, Rocha GJM. Combined process system for the production of bioethanol from sugarcane straw. Ind Crop Prod. 2014;58:1–7.
32. Junqueira Franco HC, Borges Pimenta MT, Nunes Carvalho JL, Graziano Magalhaes PS, Vaz Rossell CE, Braunbeck OA, et al. Assessment of sugarcane trash for agronomic and energy purposes in Brazil. Sci Agric. 2013;70:305–12.
33. Moutta RO, Silva MC, Corrales RCNR, Cerulho MAS, Ferreira-Leitão VS, Bon EPS. Comparative response and structural characterization of sugarcane bagasse, straw and bagasse-straw 1:1 mixtures subjected to hydrothermal pretreatment and enzymatic conversion. J Microb Biochem Tech. 2013;S12:1–8.
34. Moutta RDO, Ferreira-Leitao VS, Da Silva Bon EP. Enzymatic hydrolysis of sugarcane bagasse and straw mixtures pretreated with diluted acid. Biocatalysis Biotransformation. 2014;32:93–100.
35. da Silva ASA, Inoue H, Endo T, Yano S, Bon EPS. Milling pretreatment of sugarcane bagasse and straw for enzymatic hydrolysis and ethanol fermentation. Bioresour Technol. 2010;101:7402–9.
36. Masarin F, Gurpilhares DB, Baffa DCF, Barbosa MHP, Carvalho W, Ferraz A, et al. Chemical composition and enzymatic digestibility of sugarcane clones selected for varied lignin content. Biotechnology Biofuels. 2011;4:55–64.
37. Jeon YJ, Xun Z, Rogers PL. Comparative evaluations of cellulosic raw materials for second generation bioethanol production. Lett Appl Microbiol. 2010;51:518–24.
38. Hatfield R, Fukushima RS. Can lignin be accurately measured? Crop Sci. 2005;45:832–9.
39. Chandel AK, da Silva SS, Singh OV. Detoxification of lignocellulose hydrolysates: biochemical and metabolic engineering toward white biotechnology. Bioenergy Res. 2013;6:388–401.
40. Kumar D, Murthy GS. Impact of pretreatment and downstream processing technologies on economics and energy in cellulosic ethanol production. Biotechnol Biofuels. 2011;4:27–45.
41. Alvira P, Tomas-Pejo E, Ballesteros M, Negro MJ. Pretreatment technologies for an efficient bioethanol production process based on enzymatic hydrolysis: a review. Bioresour Technol. 2010;101:4851–61.
42. Cardona CA, Quintero JA, Paz IC. Production of bioethanol from sugarcane bagasse: status and perspectives. Bioresour Technol. 2010;101:4754–66.
43. Meng X, Ragauskas AJ. Recent advances in understanding the role of cellulose accessibility in enzymatic hydrolysis of lignocellulosic substrates. Curr Opin Biotechnol. 2014;27:150–8.
44. Li J, Lin J, Zhou P, Wu K, Liu H, Xiong C, et al. One-pot simultaneous saccharification and fermentation: a preliminary study of a novel configuration for cellulosic ethanol production. Bioresour Technol. 2014;161:171–8.
45. Sun Y, Cheng J. Hydrolysis of lignocellulosic materials for ethanol production: a review. Bioresour Technol. 2002;83:1–11.
46. Moutta RO, Chandel AK, Rodrigues RCLB, Silva MB, Rocha GJM, Silva SS. Statistical optimization of sugarcane leaves hydrolysis into simple sugars by dilute sulfuric acid catalyzed process. Sugar Tech. 2012;14:53–60.
47. Cortez DV, Roberto IC, Barbosa MHP, Milagres AMF. Evaluation of cellulosic and hemicellulosic hydrolysate fermentability from sugarcane bagasse hybrids with different compositions. Biomass Conversion Biorefinery. 2014;4:351–6.
48. Martin C, Alriksson B, Sjode A, Nilvebrant N-O, Jonsson LJ. Dilute sulfuric acid pretreatment of agricultural and agro-industrial residues for ethanol production. Appl Biochem Biotechnol. 2007;137:339–52.
49. Agbor VB, Cicek N, Sparling R, Berlin A, Levin DB. Biomass pretreatment: fundamentals toward application. Biotechnol Adv. 2011;29:675–85.
50. Pu Y, Hu F, Huang F, Davison BH, Ragauskas AJ. Assessing the molecular structure basis for biomass recalcitrance during dilute acid and hydrothermal pretreatments. Biotechnol Biofuels. 2013;6:15–27.
51. Kim Y, Mosier NS, Ladisch MR, Pallapolu VR, Lee YY, Garlock R, et al. Comparative study on enzymatic digestibility of switchgrass varieties and harvests processed by leading pretreatment technologies. Bioresour Technol. 2011;102:11089–96.
52. Costa THF, Masarin F, Bonifacio TO, Milagres AMF, Ferraz A. The enzymatic recalcitrance of internodes of sugar cane hybrids with contrasting lignin contents. Ind Crop Prod. 2013;51:202–11.
53. Rahikainen J, Mikander S, Marjamaa K, Tamminen T, Lappas A, Viikari L, et al. Inhibition of enzymatic hydrolysis by residual lignins from softwood-study of enzyme binding and inactivation on lignin-rich surface. Biotechnol Bioeng. 2011;108:2823–34.
54. Duan X, Zhang C, Ju X, Li Q, Chen S, Wang J, et al. Effect of lignocellulosic composition and structure on the bioethanol production from different poplar lines. Bioresour Technol. 2013;140:363–7.
55. Lindedam J, Andersen SB, DeMartini J, Bruun S, Jorgensen H, Felby C, et al. Cultivar variation and selection potential relevant to the production of cellulosic ethanol from wheat straw. Biomass Bioenergy. 2012;37:221–8.
56. AGEITEC (Agência Embrapa de Informação Tecnológica). http://www.agencia.cnptia.embrapa.br/gestor/cana-de-acucar/arvore/CONTAG01_42_1110200717570.html. Accessed 15 Jan 2014.
57. Jonsson LJ, Alriksson B, Nilvebrant N-O. Bioconversion of lignocellulose: inhibitors and detoxification. Biotechnol Biofuels. 2013;6:16–25.
58. Klinke HB, Thomsen AB, Ahring BK. Inhibition of ethanol-producing yeast and bacteria by degradation products produced during pre-treatment of biomass. Appl Microbiol Biotechnol. 2004;66:10–26.
59. Gouveia ER, do Nascimento RT, Souto-Maior AM, de Moraes Rocha GJ. Validation of methodology for the chemical characterization of sugar cane bagasse. Quim Nova. 2009;32:1500–3.
60. Ghose TK. Measurement of cellulase activities. Pure Appl Chem. 1987;59:257–68.
61. Chrastil J. Enzymic product formation curves with the normal or diffusion limited reaction-mechanism and in the presence of substrate receptors. Int J Biochem. 1988;20:683–93.

Transcriptomic analysis revealed the mechanism of oil dynamic accumulation during developing Siberian apricot (*Prunus sibirica* L.) seed kernels for the development of woody biodiesel

Jun Niu[1†], Jiyong An[1†], Libing Wang[2], Chengliang Fang[3†], Denglong Ha[3], Chengyu Fu[4], Lin Qiu[3], Haiyan Yu[2], Haiyan Zhao[3], Xinyu Hou[1], Zheng Xiang[1], Sufan Zhou[1], Zhixiang Zhang[1], Xinyi Feng[1] and Shanzhi Lin[1*]

Abstract

Background: Siberian apricot (*Prunus sibirica* L.) has emerged as a novel potential source of biodiesel in China, but the molecular regulatory mechanism of oil accumulation in Siberian apricot seed kernels (SASK) is still unknown at present. To better develop SASK oil as woody biodiesel, it is essential to profile transcriptome and to identify the full repertoire of potential unigenes involved in the formation and accumulation of oil SASK during the different developing stages.

Results: We firstly detected the temporal patterns for oil content and fatty acid (FA) compositions of SASK in 7 different developing stages. The best time for obtaining the high quality and quantity of SASK oil was characterized at 60 days after flowering (DAF), and the representative periods (10, 30, 50, 60, and 70 DAF) were selected for transcriptomic analysis. By Illumina/Solexa sequencings, approximately 65 million short reads (average length = 96 bp) were obtained, and then assembled into 124,070 unigenes by Trinity strategy (mean size = 829.62 bp). A total of 3,000, 2,781, 2,620, and 2,675 differentially expressed unigenes were identified at 30, 50, 60, and 70 DAF (10 DAF as the control) by DESeq method, respectively. The relationship between the unigene transcriptional profiles and the oil dynamic patterns in developing SASK was comparatively analyzed, and the specific unigenes encoding some known enzymes and transcription factors involved in acetyl-coenzyme A (acetyl-CoA) formation and oil accumulation were determined. Additionally, 5 key metabolic genes implicated in SASK oil accumulation were experimentally validated by quantitative real-time PCR (qRT-PCR). Our findings could help to construction of oil accumulated pathway and to elucidate the molecular regulatory mechanism of increased oil production in developing SASK.

Conclusions: This is the first study of oil temporal patterns, transcriptome sequencings, and differential profiles in developing SASK. All our results will serve as the important foundation to further deeply explore the regulatory mechanism of SASK high-quality oil accumulation, and may also provide some reference for researching the woody biodiesel plants.

Keywords: Siberian apricot, Woody biodiesel, Transcriptome sequencing, Differential expression, Oil accumulation mechanism

* Correspondence: szlin@bjfu.edu.cn
†Equal contributors
[1]College of Biological Sciences and Biotechnology, College of Nature Conservation, National Engineering Laboratory for Tree Breeding, Key Laboratory of Genetics and Breeding in Forest Trees and Ornamental Plants, Ministry of Education, Beijing Forestry University, Beijing 10083, China
Full list of author information is available at the end of the article

Background

With the increasing energy demand and environmental pollution problems caused by the fossil fuels, the renewable and clean of energy sources have become an inevitable choice for sustainable development of society and economy. Biodiesel, an alternative diesel fuel, was produced from biological sources such as plant oils or animal fats and has become a popular and environment-friendly fuel with renewability, low-emissions, security, and biodegradability [1]. However, the high cost of biodiesel has been the major barrier to its commercialization, and the application of edible oils to biodiesel production would lead to the problem of food versus fuel [2,3]. Recently, woody seed oils with a notable advantage over conventional feedstocks have been used as potential raw materials for biodiesel production [2,3]. Thus, it is important to develop woody resources for biodiesel. In our previous study, Siberian apricot has been characterized as a novel potential raw material for biodiesel from the ten candidate oleiferous tree species [3].

Siberian apricot (*Prunus sibirica* L.), a member of the family Rosaceae and the genus *Prunus*, is widely distributed in the temperate and mountainous districts of northern and northeastern China, eastern and southeastern regions of Mongolia, eastern Siberia regions, and maritime territory of Russia [4]. In China, Siberian apricot is one of the most economically and ecologically important and intensively studied tree species owing to its very plentiful resource, superior adaptability, and ecological benefits [5,6]. Moreover, the annual production of Siberian apricot seeds is above 192,500 tons and the average yield of seed oils is approximately 0.8 ton/ha in China [7]. Recently, the oils of SASK have been used in many fields, such as edible oils, lubricants, cosmetics, and surfactants, especially in biodiesel feedstocks [5,6,8]. Based on our evaluations of 17 selected Siberian apricot germplasms (oil content ranged from 44.73% to 57.83%) for biodiesel properties, the best Siberian apricot germplasm for biodiesel production has been identified with a high percentage of oleic (65.23% ± 4.97%) and linoleic acid (28.92% ± 4.62%), showing the potential for using SASK oil as a new source of biodiesel feedstocks in China [9-11]. Although considerable progress of Siberian apricot have been made in oil contents, FA compositions, biodiesel traits, ecological uses, afforestation, and medicinal utilizations [5,6,8-11], little is known about the molecular regulatory mechanism of oil accumulation in SASK. Thus, the transcriptomic analysis of developing SASK has become an imperative for the development of woody biodiesel.

In recent years, advances in low cost next-generation sequencing technology have made RNA sequencing (RNA-Seq) to become an effective choice for gene expression studies [12]. Although RNA-Seq has been performed on the transcriptomic analysis of many oilseed plants such as soybean, peanut, rapeseed, *Ricinus communis*, *Euonymus alatus*, *Landoltia punctata*, and *Tropaeolum majus*, and some known enzymes and genes were characterized to be involved in FA and triacylglycerol (TAG) biosynthesis [13-18], up to now reports about this situation only in oil palm, olive, and coconut for woody plants [19,20].

Recently, transcriptomic analysis for different tissues of Siberian apricot has been performed by 454 sequencing, and some candidate unigenes were annotated in oil biosynthesis [21], but the obtained results are not still suitable for us to deeply explore the biochemical and molecular mechanism of oil dynamic accumulation in developing SASK. In this work, the first kinetic patterns of oil contents and FA compositions were detected at different developing stages (10, 20, 30 40, 50, 60, and 70 days after flowering (DAF)), and the optimal periods for high quality and quantity of SASK oils and for comparative deep transcriptomic analysis were determined. The transcriptomes of developing SASK were sequenced by using Illumina technology, and then the assembled unigenes were functionally annotated. Moreover, the differentially expressed genes for the enzymes and transcription factors involved in oil accumulation of developing SASK were screened by DESeq method, and experimental validation of some key genes was performed by qRT-PCR. Overall, the temporal accumulated patterns of oils and FA compositions, and the transcriptional profiles of transcriptional regulatory factors and metabolic enzymes associated with the biosynthesis of pyruvate, acetyl-coenzyme A (acetyl-CoA), FA, TAG, and oil body were systematically analyzed in developing SASK, which will contribute to elucidate the molecular and metabolic mechanisms leading to increased oil biosynthesis and accumulation in SASK.

Results and discussion

Dynamic changes of oil content and FA compositions in developing SASK

In general, high oil content is considered to be the vital metric for an energy-efficient plant. The study in different Siberian apricot germplasms has shown that the oil content of the fully ripened SASK, ranged from 44.73% to 57.83% [7,9,10], was higher than that of traditional oil plants such as *Jatropha curcas* (38.09%) and *Sapium sebiferum* (12% to 29%) [22], indicating a high commercial value for SASK oils. To explore the dynamic accumulated patterns of oils in developing SASK, we evaluated the SASK oil contents at different developing stages (Figure 1). There was a gradual increase in SASK oil content from 10 DAF (4.00% ± 0.39%) to 60 DAF (50.68% ± 4.18%), followed by approximately 2% decline at 70 DAF (fully ripe), indicating that the optimal

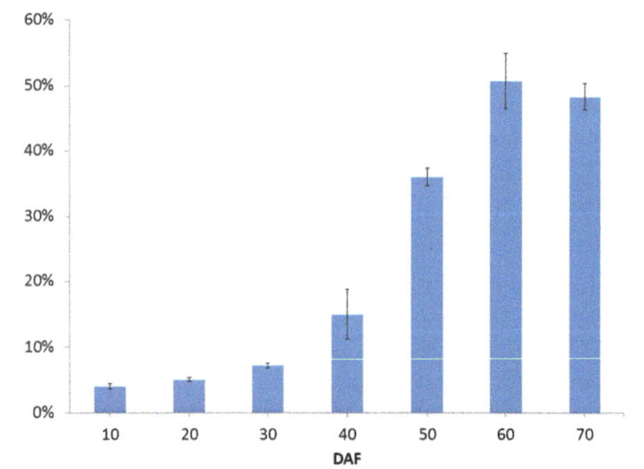

Figure 1 The SASK oil content at different development period.

harvest time for obtaining the maximum SASK oil content was at 60 DAF. It is important to note that the relative proportion of saturated, monounsaturated, and polyunsaturated FAs is the critical factor affecting biodiesel fuel properties [3]. In this study, the eight kinds of FAs were firstly detected in SASK oils at different developing stages, and the temporal patterns of their relative proportions were analyzed (Figure 2). We found that the C18:1 (oleic acid) relative proportion gradually increased from 10 DAF (34.37% ± 1.57%) to 70 DAF (67.41% ± 2.35%) with a remarkable elevated degree at 40 to 50 DAF, and the percentage of C18:2 (linoleic acid) exhibited a peak value (50.07% ± 2.87%) only at 30 DAF, but almost no significant alteration (23.29% to 38.45%) for the other different developing periods. Additionally, the other saturated or long chain FAs with a lower relative proportion showed a slight modification in developing SASK. Impressively, the total relative proportion of C18:1 and C18:2 in SASK oils ranged from 60.66% to 91.74% at 10 to 70 DAF (Figure 2), and therefore a higher ratio of oleic and linoleic acid revealed SASK oils with a high quality as a novel potential woody biodiesel feedstock in China, which corresponded to the previous investigations on SASK oils [7,10].

Taken together, our findings on the highest oil content (50.68% ± 4.18%) and the near-maximal total percentage of C18:1 and C18:2 (91.64%) in SASK at 60 DAF indicated that the very best harvest time for SASK oils with high quality and quantity is at 60 DAF. Moreover, according to the temporal patterns of oil content and the total relative proportion of C18:1 and C18:2 in developing SASK, the samples from five crucial periods (10, 30, 50, 60, and 70 DAF) were selected as the experimental materials for comparative deep transcriptomic analysis to better explore the molecular and metabolic regulatory mechanism of SASK oil increased-accumulation.

Illumina sequencing and *de novo* assembly of developmental SASK

To clarify a global overview of the gene expressing profiles in developing SASK, a total of five cDNA libraries were constructed from different developing SASK RNA samples and were respectively sequenced by the Illumina transcriptome sequence. More than 5 GB raw data were respectively generated from 5 RNA-Seq samples, and then an average (55,994,330) of short reads were obtained after removing low-quality reads and adaptor sequences (Table 1). The numbers of raw data and short reads were larger in developing SASK than in longan

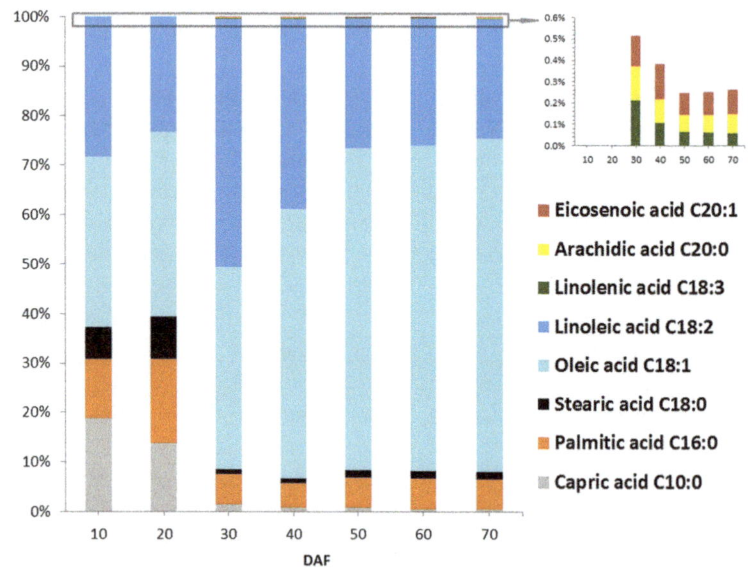

Figure 2 Changes in the fatty acid composition during SASK development. The data represent the mean of three independent measurements.

[23] and peanut [15], suggesting that SASK Illumina sequencing could construct a complete read database. All high-quality reads were deposited in the NCBI Short Read Archive (SRA) database under accession numbers SRR1564517, SRR1568273, SRR1568275, SRR1568789, and SRR1568805.

At present, there are no standard criteria to evaluate the quality of transcriptome assemblies [24]. To ensure accuracy and reliability of our SASK transcriptome assembly, 500,000 reads randomly selected from all reads were compared with NCBI database of nucleotide for pollution monitoring. From the Figure 3, the first *Prunus dulcis*, belonging to the same family Rosaceae with our experimental material (*P. sibirica* L.), was shown to have over three times as many reads as the second *Vitis vinifera*, indicating the reliability of our Illumina transcriptome sequencing data. Additionally, all these reads obtained from SASK of five developing stages were assembled by the Trinity software [25], and then low-complexity and low-quality reads were filtered out, resulting in 226,682 transcripts with GC content of 49.09%. By clustering through the Trinity program, we obtained 124,070 unigenes (N50: 1,603 bp) with the mean length of 829.62 bp (Additional file 1: Table S1) that is longer than those assembled from *Centella asiatica* (474 bp) [26], Yellow Horn (462 bp) [27], and peanut (751 bp) [15]. We also statistically found that there were 16,698 unigenes (13.46%) in the length range of 1,000 to 2,000 bp and 13,507 unigenes (10.88%) with length more than 2,000 bp (Additional file 2: Figure S1). Thus, a total of 30,205 unigenes (≥1,000 bp) in developing SASK, nearly 3 times more than our previous transcriptome analysis for different tissues of Siberian apricot [21], could massively fill and enrich the dataset of Siberian apricot.

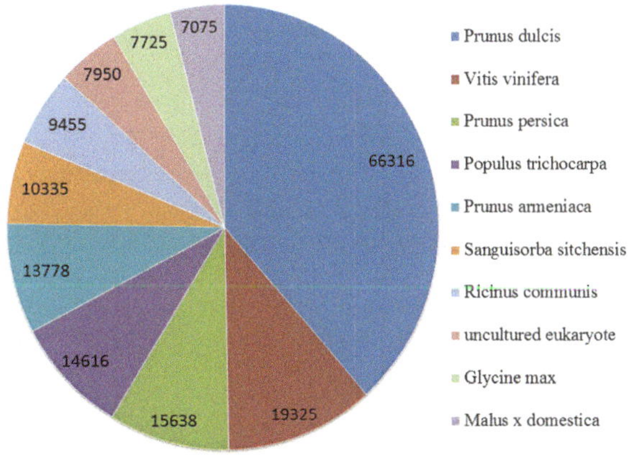

Figure 3 Species distribution of the top BLAST hits for reads in Nr database.

Functional annotation and classification of SASK unigenes

In order to understand specific functions of all the assembled unigenes in developing SASK, they were predicted via Basic Local Alignment Search Tool (BLAST) with a cut-off E-value of 10 to 5 in NCBI non-redundant protein (Nr) database, Swiss-Prot protein database (SwissProt), Arabidopsis proteome (AP) database, Clusters of Orthologous Groups (COG) database, Gene Ontology (GO), and Kyoto Encyclopedia of Genes and Genomes (KEGG). The BLAST results showed that 47,406 (38.21%), 30,511 (24.59%), 46,949 (37.84%), 17,870 (14.40%), 36,553 (29.46%), and 18,807 (15.10%) unigenes of developing SASK were highly similar with known proteins in Nr, SwissProt, AP, COG, GO, and KEGG databases, respectively (Additional file 3: Table S2 and Additional file 4: Table S3). Of all SASK unigenes, 49,904 (40.22%) homologous unigenes showed indeed that a considerable functional contig size was obtained. However, those unmatched SASK unigenes (74,166, 59.78%) might be as the putative tissue-specific novel genes or probably resulted from the shorter sequences with a lack of characterized protein domain to get BLAST hits [27].

The Gene Ontology (GO) annotation for the assembled unigenes was used to categorize the functions of the predicted SASK unigenes [28], and a total of 36,553 unigenes were assigned to the 3 main GO categories and 65 subcategories. Additional file 5: Figure S2 shows that 'metabolic process,' 'cellular process,' 'binding,' and 'catalytic activity' are the most dominant categories involving more than 20,000 unigenes, but only a few of genes were associated with terms such as 'viral reproduction,' 'cell junction,' and 'channel regulator activity'. To further evaluate the completeness of our transcriptomic libraries and the effectiveness of the

Table 1 Raw data and valid data statistics

Sample (DAF)	Raw data		Valid data			Valid ratio (reads) (%)
	Read	Base	Read	Base	Average length	
10	60,940,126	6,094,012,600	53,830,008	5,182,918,614	96.28	88.33
30	65,980,058	6,598,005,800	58,413,976	5,636,764,333	96.50	88.53
50	64,572,540	6,457,254,000	56,984,496	5,479,436,388	96.16	88.25
60	72,078,976	7,207,897,600	57,262,494	5,368,833,289	93.76	79.44
70	62,898,984	6,289,898,400	53,480,678	5,112,244,068	95.59	85.03

annotation processes, all SASK unigenes were aligned to the COG database to predict and classify by possible function [29]. The resulting 17,870 (14.4%) unigenes were assigned to the 25 COG classifications (Additional file 6: Figure S3), among which the cluster for 'Post-translational modification, protein turnover, chaperones' represented the largest group (3,875, 21.68%) followed by 'Signal transduction mechanisms' (3,379, 18.91%) and 'General function prediction only' (2,662, 14.90%). Subsequently, to identify the biological functions and interactions of genes in developing SASK, the KEGG was applied to annotate the reference canonical pathways [30]. In total, 18,807 sequences were assigned to 313 KEGG pathways. Most of them were 'Ribosome' (1,254, 6.67%), 'Protein processing in endoplasmic reticulum' (604, 3.21%), 'Oxidative phosphorylation' (534, 2.84%), 'Glycolysis/gluconeogenesis' (494, 2.63%), and 'Plant-pathogen interaction' (490, 2.61%). Moreover, we also annotated several signaling networks of reproductive biology, such as 'Non-homologous end-joining' and 'HIF-1 signaling pathway'.

Analysis of differential gene expression for developing SASK

To fully explore the differential expressions of the specific genes that are actually implicated in developing SASK, the clean reads from each library were mapped to the transcriptomic database for profiling the expressions of unigenes. The utilization rate of reads and the expression rate of unigenes showed that 10 DAF samples had the most unigenes (91,563, 74.28%) mapped to our constructed unigene database, whereas 50 DAF sample had the least unigenes (75,667, 60.9%) (Table 2 and Additional file 7: Figure S4). Also, the RPKM values were statistically calculated to select differential unigenes from developing SASK by using the DESeq method [31,32] (Additional file 8: Figure S5 and Additional file 9: Table S4). Our results revealed that 10:30 DAF exhibited the most up-regulated (1,513) and the least down-regulated unigenes (1,487), whereas 10:70 DAF had the least up-regulated (726) and the most down-regulated unigenes (1,949) (Figure 4A). By Venn diagram analysis, 838, 198, 74, and 90 up-regulated unigenes and 602, 354, 370, and 295 down-regulated unigenes were identified to be specific expression at 30, 50, 60, and 70 DAF (Figure 4B,C), respectively. Thus, combining with the temporal patterns of SASK oil at developing stages (Figure 1), we suggested that the candidate unigenes associated with oil accumulation pathway may be expressed at early stage of developing SASK.

Transcript patterns for enzymes and transporters involved in generating pyruvate for SASK FA synthesis

Sucrose is known as the most source of carbon, and its major metabolic flux through glycolytic pathway is to provide the large amounts of pyruvate required for *de novo* FA biosynthesis in oilseed plants [19,33,34]. In general, plant glycolysis occurs independently in two subcellular compartments of the cytosol and plastid. ATP-dependent phosphofructokinase (PFK) and pyruvate kinase (PK) are considered as two regulatory enzymes of glycolytic pathway in both cytosol and plastid, while pyrophosphate-dependent phosphofructokinase (PFP) only in cytosol has been characterized as a key enzyme to reversibly catalyze the conversion of fructose-6-phosphate to fructose-1,6-bisphosphate [16,19,35]. Moreover, the interchange of glycolytic intermediates between cytosol and plastid in plants has been implicated in the highly selective plastid transporters, including glycolipid transporter (GLT1), triose phosphate transporter (TPT), glucose-6-phosphate transporter (GPT2), and phosphoenolpyruvate transporter (PPT1) [19,33,36,37]. It is well known that the glycolytic substrates of hexose phosphates, generated from sucrose in the cytosol, are generally transported by GLT1 or GPT2 for plastidial glycolysis [34]. Recently, the transcripts of the plastid-localized enzymes (PFK and PK) and the transporters (GLT1 and PPT1) exhibited an obvious up-regulation, but PFP remained unchanged during ripening, indicating that plastid glycolysis plays a major role in providing pyruvate for high rates of FA synthesis in oil palm [34,35]. However, our transcriptional profiles revealed the stable-high expressions of cytosolic PFK (AT4G26270) and PK (AT3G52990) and the lower expressions of plastidial PFK (AT2G22480) and cytosolic PFP (AT1G12000) in developing SASK (Figure 5), suggesting a greater proportion of hexose to pyruvate flux in cytosol. Also consistent with this conclusion, the decreasing expressions in the translocators of GLT1 (AT2G33470), TPT (AT5G46110), and GPT2 (AT1G61800) were observed on developing SASK. It is, therefore, considered that there is a

Table 2 Statistics of reads and the utilization of unigenes

DAF	Valid data	Map data	Data (%)	All unigene	Exp unigene	Unigene (%)
10	53,830,008	49,899,224	92.70	124,070	92,163	74.28
30	58,413,976	54,258,107	92.89	124,070	84,912	68.44
50	56,984,496	53,171,111	93.31	124,070	76,668	61.79
60	57,262,494	51,322,910	89.63	124,070	75,667	60.99
70	53,480,678	47,986,764	89.73	124,070	84,030	67.73

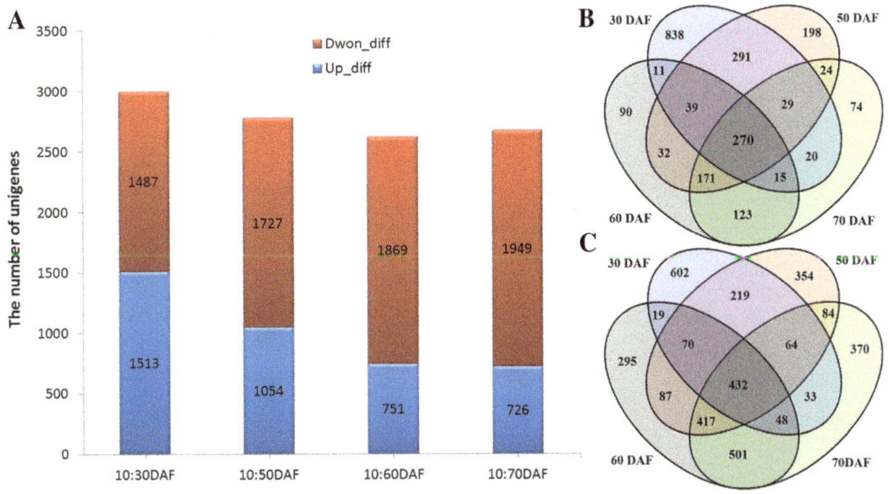

Figure 4 Number and distribution of up-regulated and down-regulated genes. (A) The number of up-regulated or down-regulated unigenes during developing SASK. **(B)** The distribution of up-regulated genes in differently developmental periods. **(C)** The distribution of down-regulated genes in differently developmental periods. The detailed sequences were showed in Additional file 9: Table S4.

lower capacity to provide glycolytic substrates (hexose and hexose phosphate) and intermediate (triose phosphate) from cytosol to plastid in developing SASK. Interestingly, an up-regulated expression of the plastidial PK (AT5G52920) and its corresponding PPT1 (AT5G33320) were noted in SASK only at 30 DAF (Figure 5), the period that was important for initial FA biosynthesis (Figure 1). This implies a transportation of PEP from cytosol to plastid and the production of pyruvate required for SASK FA biosynthesis. Together, our results revealed that the cytosolic carbon supply via glycolytic pathway is crucial for *de novo* FA biosynthesis of developing SASK.

Transcriptional patterns for the enzymes of acetyl-CoA formation for SASK *de novo* FA biosynthesis

Acetyl-CoA is an important precursor for *de novo* FA biosynthesis in plastids, various phytochemical biosynthesis and FA elongation in cytosol, and TCA cycle in mitochondria [38]. However, acetyl-CoA must be synthesized by specific synthetase in each subcellular compartment for a variety of important biosynthesis owing to the impermeability of biological membranes [39], indicating the multiple mechanisms of generating acetyl-CoA for different acetyl-CoA-requiring metabolisms. To our knowledge, acetyl-CoA synthetase (ACS) and pyruvate dehydrogenase complex (PDC) are thought to be implicated in the conversion of pyruvate to acetyl-CoA for *de novo* FA biosynthesis in plastids of plants, and four subunits (E1-α, E1-β, E2, and E3) of PDC have been characterized in Arabidopsis [16,39]. In an attempt to determine the importance of the enzymes (ACS and PDC) involved in acetyl-CoA formation for SASK *de novo* FA biosynthesis, their transcriptional patterns in developing SASK were analyzed in this study. The SASK homologies of Arabidopsis plastidial PDC E1-α (At1g01090), E1-β (At2g34590), E2 (At1g34430), and E3 (At4g16155) were identified as a significantly higher expression at 30 DAF, but the homologous gene of ACS (AT5G36880) displayed a lower expression (Figure 5), indicating that plastidial PDC may play an important role in the supply of acetyl-CoA for *de novo* FA synthesis in developing SASK. Attentively, PDC has also been characterized to locate in mitochondria of plants [40]. Indeed, our annotated four subunits of PDC in developing SASK exhibited high similarity with Arabidopsis mitochondrial E1-α (At1g59900), E1-β (At5g50850), E2 (At3g52200), and E3 (At1g48030), but their transcriptional patterns were shown no significant change (Figure 5), which could help to distinguish the PDC distribution between plastids and mitochondria in SASK at transcriptional level. Overall, the generation of acetyl-CoA might be tightly regulated in response of SASK to different developing stages.

Identification of the *de novo* FA biosynthetic pathway-specific unigenes in developing SASK

De novo FA synthesis from acetyl-CoA precursors, as the first part of the lipogenesis process, is important in controlling the content and chain length (up to 18 carbons) of the saturated FAs in seed oils. Indeed, it has been reported that a multi-subunit acetyl-CoA carboxylase (ACC, EC: 6.4.1.2), regulated by phosphorylation and redox status as well as PII interactions, is the first committed key enzyme in the *de novo* FA biosynthesis pathway, and its physiological function is to provide the substrate malonyl-CoA for the biosynthetic saturated FAs [41,42]. Recently, it has been proposed that the intracellular accumulation of free C18:1, 18:1-CoA, and

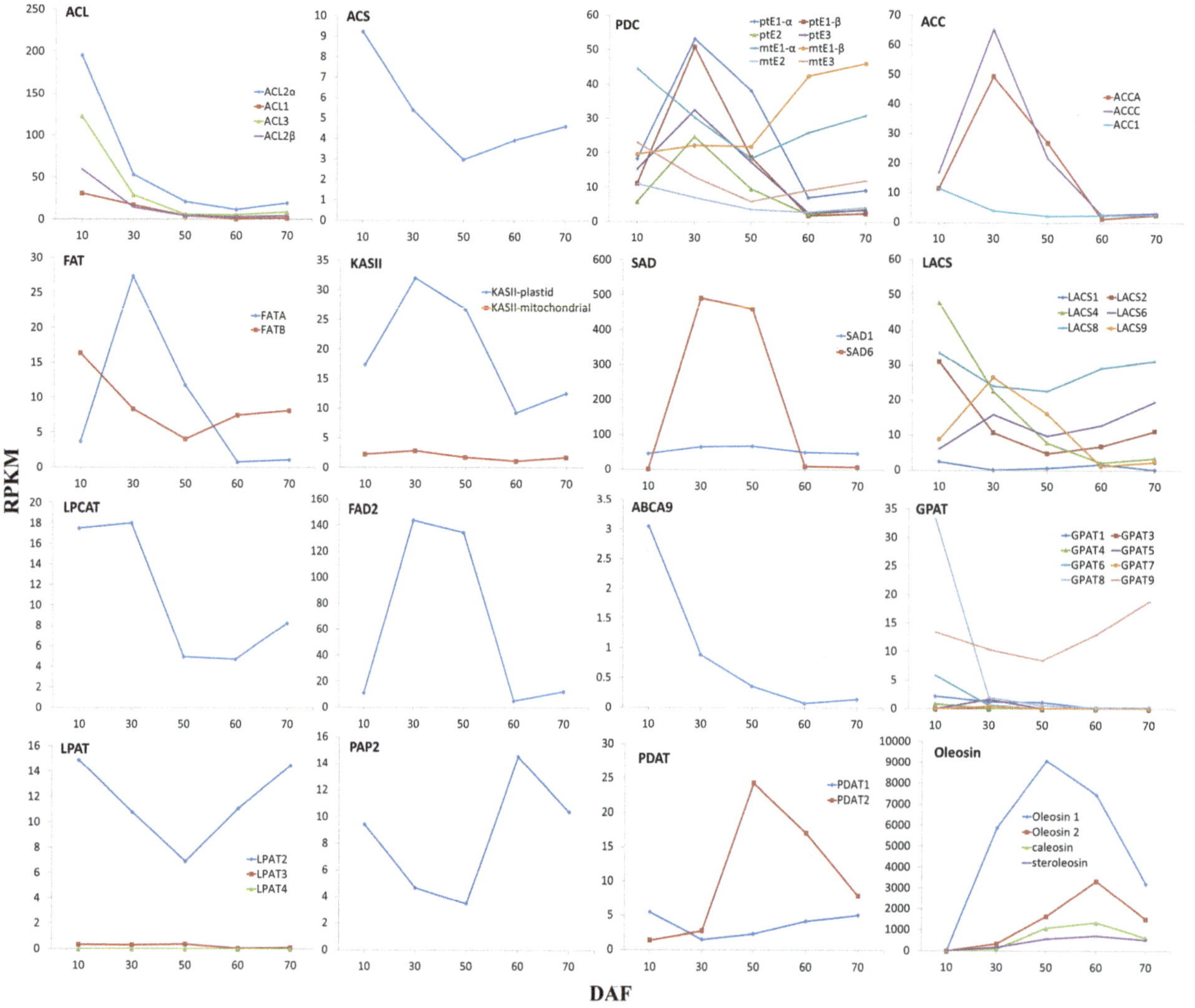

Figure 5 Temporal profile of RPKM for genes encoding enzymes involved in the conversion of sucrose to TAG. Abbreviations: PFK, phosphofructokinase; PFP, pyrophosphate-dependent phosphofructokinase; PK, pyruvate kinase; GLT, glycolipid transporter; TPT, triose phosphate transporter; GPT2, glucose-6-phosphate transporter; PPT1, phosphoenolpyruvate transporter; ACS, acetyl-CoA synthetase; PDC, pyruvate dehydrogenase complex; ACC, acetyl-CoA carboxylase; FAT, acyl-ACP thioesterase; KASII, ketoacylACP synthase II; SAD, stearoyl-ACP desaturase; LACS, long chain acyl-CoA synthetase; acyl-CoA: LPCAT, lysophosphatidylcholine acyltransferase; FAD2, fatty acid desaturase 2); acyl-CoA: glycerol-3-phosphate acyltransferase; LPAT, acyl-CoA: lyso-phosphatidic acid acyltransferase; PAP, phosphatidic acid phosphatase; PDCT, PC: DAG cholinephosphotransferase; and PDAT, phospholipid:DAG acyltransferase.

18:1-acyl-carrier protein (ACP) in *Brassica napus* cell-suspension culture by feeding oleic acid (18:1) Tween esters could result in feedback inhibition of plastidial ACC activity, leading to a reduction of FA synthesis [43]. To characterize whether the transcriptional expressions of plastidial ACCs were linked to the FA biosynthesis in developing SASK, we performed a concurrent analysis of functional annotations and differential profiles for all the obtained unigenes. Here, a total of 10 SASK unigenes were characterized to be homologous with ACC subunits of other species, including 6 Arabidopsis ACC carboxyl transferase subunit alpha (ACCA), 2 Arabidopsis ACC1, and 2 *Populus trichocarpa* ACC biotin carboxylase subunit (ACCC), but only 2 homologies of *accA* and *accC* exhibited the highest expressing levels at 30 DAF followed by a dramatic decrease at 50 to 60 DAF (Figure 5). Combined with our observations on an obvious increase of FAs (C18:1 is one major component) at 50 DAF and the decreasing rate of FA biosynthesis after 50 DAF (Figures 1 and 2), we considered that a lower expressions of *acc* genes might be related with a high intracellular accumulation of FAs in developing SASK. Thus, our results prompt us to believe that an inhibitory expression of *acc* genes may be at 50 DAF, giving rise to a declined capacity of ACC to generate malonyl-CoA for *de novo* FA biosynthesis. However, further research is needed in transcriptional regulatory mechanism for plastidial *acc* as the target gene in *de novo* FA synthesis of developing SASK.

Before entering the synthesis pathway of FAs, malonyl-CoA from Acetyl-CoA is firstly transferred to malonyl-ACP by malonyl-CoA-ACP transferase (MAT, EC: 2.3.1.39) to provide two-carbon unit at each step of elongation. There are four reactions of condensation, reduction, and dehydration, and reduction occurred in each synthesized cycle, which is catalyzed by a series of regulatory enzymes. Our results of functional annotations revealed that one MAT, three 3-ketoacyl ACP synthase II (KASII, EC: 2.3.1.179), one 3-ketoacyl ACP synthase III (KASIII, EC: 2.3.1.180), two 3-ketoacyl ACP reductase (KAR, EC: 1.1.1.100), and one enoyl-ACP reductase (EAR, EC: 1.3.1.9) involved in FA preliminary synthesis in developing SASK. After seven elongated cycles, the production of saturated C16:0-ACP can either be converted to free C16:0 by the fatty acyl-ACP thioesterase B (FATB, EC:3.1.2.14 3.1.2.-) or further elongated by KASII to 18:0-ACP. Subsequently, 18:0-ACP can be hydrolyzed to free C18:0 by FATB or be desaturated to 18:1-ACP by 18:0-ACP desaturase (SAD, EC:1.14.19.2), and then 18:1-ACP is hydrolyzed to free C18:1 by the fatty acyl-ACP thioesterase A (FATA, EC:3.1.2.14 3.1.2.-) [44]. In plants, the relative proportion of FAs was remarkably altered by manipulation of any of these four enzymes (FATA, FATB, SAD, and KASII), but FATA and FATB have shown to be not directly involved in FAs biosynthesis [45,46]. In this study, the *fatB* genes with a stable expression were observed in developing SASK, while the *fatA* genes were notably up-regulated only at 30 DAF (Figure 5), suggesting that the resulting free C18:1 rather than C16:0 and C18:0 may be as one main product of plastid FA synthesis in developing SASK, which was coincide with our investigations on FA compositions (Figure 2). Additionally, a total of six *kasII* unigenes were found in our experiments, where 1 mitochondria *kasII* unigene with expressive stability and 5 plastid *kasII* unigenes with a high expression were identified at 30 DAF (Figure 5). It is interesting to note that 2 *sad6* genes were sharply up-regulated over 200 folds at 30 and 50 DAF, whereas 2 highly-expressed *sad1* genes remained stable in developing SASK (Figure 5). Thus, we considered that the increasing trend of C18:1 (oleic acid) content paralleled to the temporally transcriptional patterns for *sad6* and *kasII* in developing SASK (Figures 2 and 5) could explain SAD6 and KASII as the key enzymes for oleic acid formation in developing SASK. This also was strongly supported by the fact that a reduction in the activities of KASII and SAD caused by antisense gene silencing could lead to a decrease of 18:1 accumulation in Brassica and Arabidopsis seeds [47,48]. In conclusion, 18:1-ACP, as a major metabolite of FA biosynthesis, may be responsible for transportation of FA from plastid to cytosol in developing SASK.

Determination of the unigenes-regulated polyunsaturated FA and TAG biosynthesis in developing SASK

The synthesized free FAs by the FATA/B thioesterases are exported from plastid to cytosol and then converted to the fatty acyl-CoAs by long-chain acyl CoA synthetase (LACS), which can provide the substrate for esterification in endoplasmic reticulum (ER) [49]. In this study, we observed a variety of kinetic transcriptional profiles for *lacs* in developing SASK. For example, *lacs4* and *lacs2* were up-regulated at 10 and 30 DAF, and two *lacs6* genes displayed high expression at 60 and 70 DAF, but *lacs8* showed stable expression during the whole developing periods, (Figure 5). It is concluded that multigene *lacs* genes may compensate for the lack of function each other in developing SASK, which could be evidenced by the fact that double (*lacs1* and *lacs9*) or triple (*lacs1*, *lacs8*, and *lacs9*) *lacs* mutant notably affect the FA content of Arabidopsis seeds [50].

For polyunsaturated FAs synthesis, C18:1 must be esterified by acyl-CoA: lysophosphatidylcholine acyltransferase (LPCAT) from cytosol to phosphatidylcholine (PC) pool in ER and then desaturated by omega-6 FA desaturase (FAD2) in PC pool [51]. In this work, four highly expressed *lpcat* genes at 10 to 30 DAF and two drastically up-regulated *fad2* genes at 30 to 50 DAF were detected in developing SASK (Figure 5), indicating that the polyunsaturated FAs may be as one of main components in SASK oil, which was also experimentally demonstrated on an accumulation of linolenic acid as the secondary main component in developing SASK (Figure 2). It is generally accepted that acyl-CoAs must be transported from the cytosol to ER for glycerolipid synthesis through diffusion by soluble carriers or by some more efficient inter-membrane transporters [44]. Recently, an ABC transporter (AtABCA9) has been identified to be involved in the delivery of FA to the ER in Arabidopsis [52]. However, SASK transcriptional profiles revealed that five unigenes encoding ABC transporter displayed a lower expression in developing SASK, of which two unigenes (comp63523_c1 and comp67480_c0) showed highly homologous with AtABCA9 transporter protein (Figure 5). Therefore, how this ABC transport system influence FAs transport in developing SASK remains to be further investigated.

The *de novo* formation of TAG occurs in ER via the Kennedy pathway, in which glycerol-3-phosphate (G3P) is initially acylated to produce TAG by the *sn*-G3P acyltransferase (GPAT, EC: 2.3.1.15) [53]. The previous studies of Arabidopsis GPATs showed that GPAT1 and GPAT4-8 with *sn*-2 regiospecificity were not required for TAG synthesis, and GPAT9 with *sn*-1 regiospecificity might be involved in TAG synthesis, but the function of GPAT2-3 with undetermined regioselectivity remained unknown [54,55], suggesting a functional divergence of the members in GPAT family. In this study, 2 genes of *gpat9* with a high and stable expression and 11 genes of *gpat1* and *gpat3-7* with an ultralow expression were observed at different developing stages, but 2 genes of *gpat8* with specific up-regulation were only found at 10 DAF (Figure 5), indicating the difference of transcriptional expressions for the members of *gpat* multigene family in developing SASK. This investigation allowed us to speculate that the expressions of *gpat* multigene family may be tightly regulated in response to developing SASK. Moreover, together with the dynamic accumulated patterns of oils in developing SASK (Figure 1), we concluded that the GPAT9 may function as a key initial acyltransferase for SASK TAG assembly.

In the second step of *de novo* TAG assembly, lysophosphatidyl acyltransferase (LPAT, EC: 2.3.1.51) and phosphatidic acid phosphohydrolases (PAP, EC: 3.1.3.4) play the essential roles in the formation of phosphatidic acid (PA) and in the removal of the phosphate group from PA to form diacylglycerol (DAG), respectively. Also, several isozymes of LPAT and PAP have been demonstrated in higher plants [16,19,56-58]. Overexpression of rapeseed microsomal LPAT isozymic gene (*BAT1.13* or *BAT1.5*) could result in an increase in oil content of Arabidopsis seeds [56], but oil biosynthesis was not significantly affected by double knockout of two PAP genes (*PAH1* and *PAH2*) [57]. In order to assess the contribution of LPAT and PAP to DAG biosynthesis in developing SASK, the transcriptional patterns for the homologies of LPATs and PAPs were comparatively analyzed. Here, one LPAT homologous gene *lpat2* (AT3G57650) and four PAP homologous genes of *pah1* (At3g09560), *pah2* (AT5G42870), *lppδ* (AT3G58490), and *lpp3* (At3g02600) were characterized with an obvious increasing expression at 50 to 60 DAF (Figure 5), and therefore they are likely to be responsible for DAG biosynthesis in developing SASK. Recently, genetic analysis revealed that mutation of Arabidopsis *rod1* (encoding PC:DAG cholinephosphotransferase, PDCT) resulted in a marked decrease in polyunsaturated FAs (18:2 and 18:3) and a concomitant increase in 18:1 in seed, suggesting that PDCT may play an important role in the transfer of 18:1 into PC pool for desaturation and also in the reverse transfer of 18:2 and 18:3 into the TAG synthesis pathway in developing seeds [59]. In this work, the SASK homologies for PDCT in Arabidopsis (At3g15820) showed a much lower expression at 50 to 60 DAF (Figure 6). Thus, the contribution of PDCT to the accumulation of polyunsaturated FAs was very limited in developing SASK, which was in accordance with our detective results of FA compositions (Figure 2).

For the last step of the Kennedy pathway, multiple acyl-CoA:DAG acyltransferase (DGAT) enzymes have been demonstrated for the formation of TAG from DAG and Acyl-CoA in plants. For example, DGAT1, DGAT2, and DGAT3 are responsible for TAG production in Arabidopsis, tung tree, and peanut [60-62]. We annotated all the unigenes of developing SASK in this work. Surprisingly, only two *dgat1* genes were incrementally expressed in developing SASK, suggesting that DGAT1 might be one major enzyme specific to TAG biosynthesis of SASK. It was also noted that plant TAG synthesis could be catalyzed by phospholipid: diacylglycerol acyltransferase (PDAT) using the FAs from the PC pool [63]. We found two *pdat1* genes with a stable expression during the developing stages and two *pdat2* genes with a high expression at 50 to 60 DAF (Figure 5). Interestingly, the transcriptional level of *pdat2* was temporally paralleled to oil accumulation in developing SASK (Figure 1), implying that PDAT2 may play a major role in TAG synthesis of developing SASK.

Taken together, our investigations could help to characterize which isoforms of these large gene families are mainly responsible for SASK TAG biosynthesis, and also attested to the fact that Siberian apricot possesses many specific enzymes in TAG assembly (Figure 6), providing the important references for the study of lipogenesis process in woody oil trees.

The transcriptionally temporal comparisons of oil body genes in developing SASK

TAGs can be stored as a form of oil bodies surrounded by a phospholipid monolayer and abundant amphipathic proteins such as oleosin, caleosin, and steroleosin in mature seeds [16,19,64]. By the transcriptionally temporal comparisons, we characterized the homologous oleosin-1 and oleosin-2 of Arabidopsis (AT4G25140 and AT5G40420), caleosin of *Ficus pumila var* (ABV72237.1), and steroleosin of *Sesamum indicumm* (AAM46847.1) with an increasing or stable-high RPKM values in developing SASK (Figure 5). However, the RPKM for oleosin-1 was far higher than that for oleosin-2, steroleosin, or caleosin (Figure 5), which was very similar to the temporal patterns in developing oilseeds of *Brassica napus* [16].

Detection of SASK transcription factors involved in oil synthesis

In previous studies, the transcription factors (TFs) such as ABI3, LEC1, LEC2, FUS3, WRI1, PKL, FIE, or GL2 play the key roles in seed oil synthesis and deposition [65]. To gain an insight into transcriptional regulatory mechanism of SASK oil biosynthesis by TFs, we analyzed all the unigenes of developing SASK by BLASTX against Arabidopsis TFs. The resulting 25 unigenes at developing SASK were identified as 9 homologies of ABI3 (AT3G24650), WRI1 (AT3G54320), LEC2 (AT1G28300), FUS3 (AT3G26790), ADOF1 (AT1G51700), PKL (AT2G25170), AP2 (AT4G36920), GL2 (AT1G79840), and HSI2 (AT2G30470) in Arabidopsis, but none of these

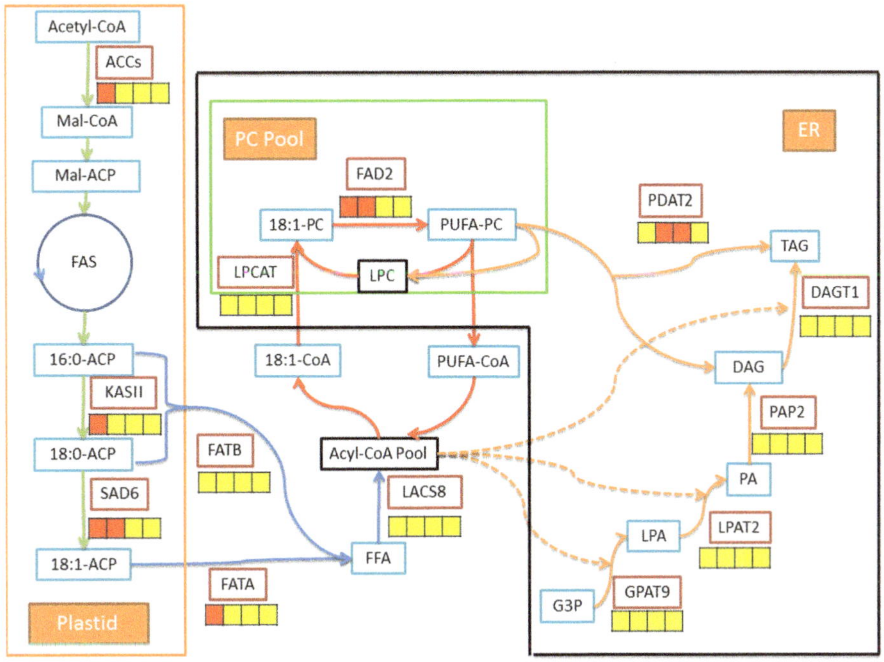

Figure 6 The temporal patterns for enzymes involved in oil synthesis. The icons below each enzyme show the result of DESeq analysis, from left to right: 10:30DAF, 10:50DAF, 10:60DAF, and 10:70DAF; red, up-regulation; yellow, no significant difference; green, down-regulation. Substrate abbreviations: ACP, acyl carrier protein; DAG, diacylglycerol; FFA, free fatty acid; G3P, glycerol-3-phosphate; LPA, lyso-phosphatidic acid; PA, phosphatidic acid; PC, phosphatidylcholine; PUFA, polyunsaturated fatty acids; TAG, triacylglycerol. Enzymatic abbreviations: ACC, acetyl-CoA carboxylase; DAG cholinephosphotransferase; DGAT1, acyl-CoA:DAG acyltransferase 1; FAD2, fatty acid desaturase 2; FATA/B, acyl-ACP thioesterase A/B; GPAT9, acyl-CoA:G3P acyltransferase 9; KASII, ketoacyl-ACP synthase II; LACS8, long chain acyl-CoA synthetase 8; LPAT2, acyl-CoA:LPA acyltransferase 2; LPCAT, acyl-CoA: lysophosphatidylcholine acyltransferase; PAP, PA phosphatase; PDAT2, phospholipid:DAG acyltransferase 2; PDCT: PC:DAG cholinephosphotransferase; SAD6, Stearoyl-ACP desaturase 6.

unigenes showed homology to LEC1, FIE, EMF2, or SWN (Additional file 10: Table S5). Moreover, our transcriptional profiling revealed that ABI3, WRI1, and LEC2 participated in positive regulation of the genes for oil synthesis in developing SASK, while AP2 and GL2 acted as negative regulatory factors. It is noteworthy that WRI1, the AP2-domain-containing transcription factor, is known to be up-regulated by LEC1, LEC2, ABI3, and FUS3 in Arabidopsis [66]. In our experiments, both FUS3 and WRI1 were shown with a high similar transcriptional profile in developing SASK (Figure 7), suggesting that the WRI1 gene might be a direct target of FUS3 in developing SASK. Additionally, the fact that plastidial *pk*, *ppt1*, *pdc*, and *acc* (*accA and accC*) displayed a coordinated transcriptional profile with *wri1* in developing SASK (Figures 5 and 7) indicated an involvement of WRI1 in the transcriptional regulation of target genes, as with Arabidopsis seeds [66]. Thus, SASK-WRI1 may play a contextual role in regulatory network of oil accumulation in developing SASK.

Experimental validation and analysis of key genes involved in SASK oil accumulation

To validate RPKM data, the expression profiles for *accC*, *fatB*, *fatA*, *dgat1*, and *pdat2* related to oil biosynthesis in developing SASK were analyzed by using qRT-PCR (Table 3). Our results showed that the relative expression levels of these genes were mostly consistent with the RPKM comparative ratios (10 DAF as the control). However, only *dgat1* gene, not obtained further by DESeq method (Figure 6), was detected with significantly up-regulated expression in developing SASK by qRT-PCR analysis (Figure 8). Thus, differential expression analysis of SASK unigenes by DESeq method was reliable in this study.

It was hypothesized that the temporal transcriptional patterns for the genes involved in the biosynthesis of FAs and TAGs are very different in developing seed [16], which has been confirmed in our qRT-PCR results that *accC* and *fatA* for FA synthesis, and *dgat1* and *pdat2* for TAG formation in SASK were identified with a high expression at 10 to 30 DAF and 50 to 70 DAF, respectively (Figure 8). This appears to indicate the beginning synthesis of the FAs and TAGs before 30 DAF and after 50 DAF, respectively. It was reported that *dgat* or *pdat* mutation had minimal impact on oil content in Arabidopsis seeds, showing that the overlapping functions of PDAT and DGAT are essential for TAG biosynthesis [62]. Notably, our investigations on the maximum expressed

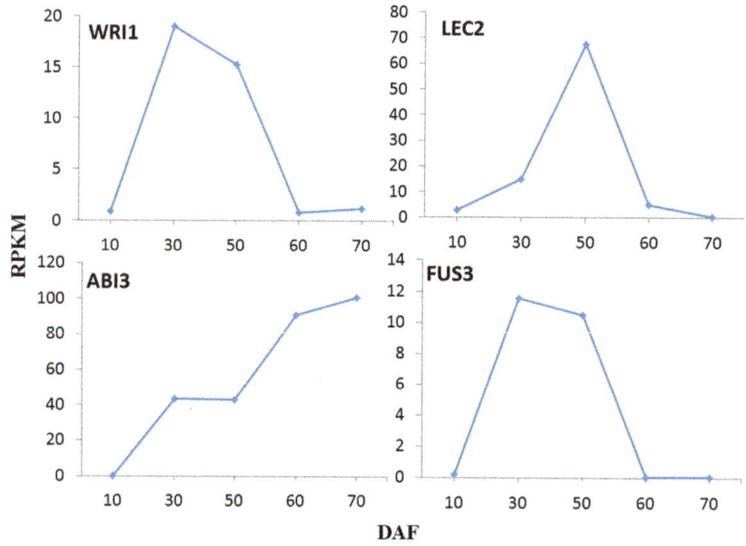

Figure 7 Temporal pattern of RPKM for SASK transcription factors.

value of *pdat2* and *dgat1* at 50 and 60 DAF respectively revealed an alteration of TAG accumulation in developing SASK by PDAT2 though PC pool or by DGAT1 via Kennedy pathway. All the above results evidenced that FAs and TAGs biosynthesis in developing SASK possessed complex dynamic regulation mechanisms, which deserve further research.

Conclusions

The Siberian apricot, a potential woody oil species for biodiesel, has a high quality and quantity of oil and excellent adaptability to different growing conditions. To fully understand temporally accumulated patterns of oil content and FA compositions, seven samples from the whole developmental SASK were firstly detected and analyzed. As a result, 60 DAF were characterized as the best time to obtain the high quality and quantity of SASK oil, and 5 experimental samples from the representative periods (10, 30, 50, 60, and 70 DAF) were selected for comparative deep transcriptome analysis. Using the Illumina sequencing technology and Trinity program, 124,070 unigenes (mean length = 829.62 bp) were obtained from clean reads assembly and annotated in a public database, which could massively fill and enrich the dataset of Siberian apricot. There were 3,000, 2,781, 2,620, and 2,675 differentially expressed unigenes identified at 30, 50, 60, and 70 DAF using DESeq method (10 DAF as control), respectively. To reveal the accumulated mechanism of oil in developing SASK, the transcriptional profiles for the crucial transcription factors and metabolic enzymes involving in the biosynthesis of acetyl-CoA, FA, TAG, and oil body were systematically analyzed. We characterized that two transcription factors (WRI1 and FUS3) and some regulatory enzymes (PPT1, ptPK, ptPDC, ACC, KASII, SAD6, FATA, LACS8, FAD2, GPAT9, LPAT2, DGAT1, and PDAT2) play an important role in oil accumulation of developing SASK. Moreover, we also validated the temporal expression levels of five key genes (*accC, fatB, fatA, dgat1, and pdat2*) by qRT-PCR in developing SASK. All of these findings will be conducive to provide a substantial molecular biology background for researches on breeding, genetic diversity, and gene excavation of Siberian apricot and will promote the significant development for woody

Table 3 Genes and primer sequences used for real-time PCR

Abbreviation	Gene name	Accession number	Primer forward	Primer reverse	Amplicon size (bp)
accC	Acetyl-CoA carboxylase. biotin carboxylase subunit	KM364594	CCAGGAAGAATAACCGCCTAC	GGGAGTCATAGTTTGGAGGAAC	100
fatB	Fatty acyl-ACP thioesterase B	KM364595	GGCCCAAATATGCCAACAATC	GACGATCCATCCAGCCTTAAA	97
fatA	Fatty acyl-ACP thioesterase A	KM364596	TGATGTTCGGGAAGAGCATTTA	GGAGCAGGATCTTCCAGTTTAG	111
dgat1	Diacylglycerol acyltransferase 1	KM364597	CAGCCTATGTGTCTGCTTCTAT	CTCGATCCACACCTTGAGATT	98
pdat2	Phospholipid:diacylglycerol acyltransferase 2	KM364598	TGATGACAGTGTGCCTGTATTG	CTTGTGCTGGTACTCCCTTATG	115

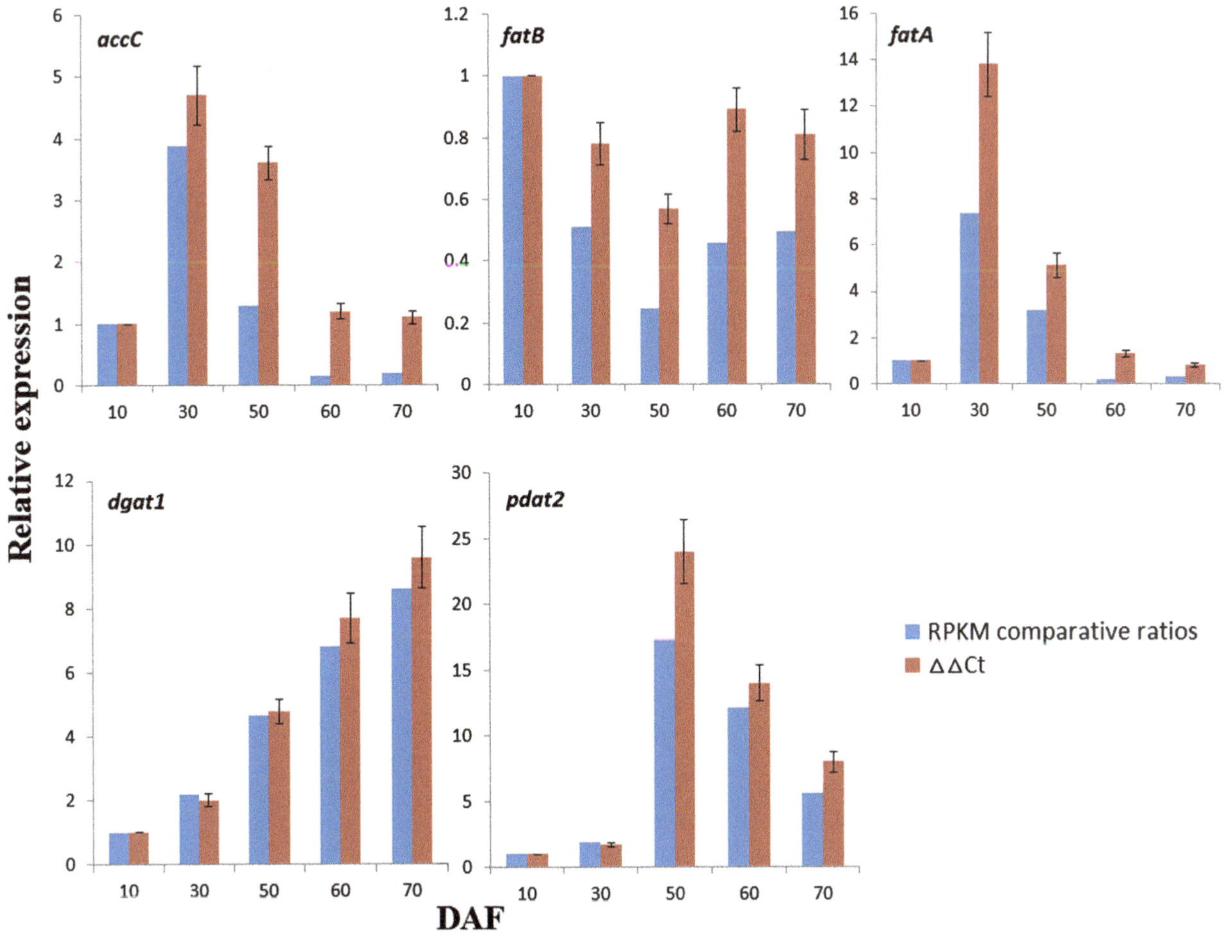

Figure 8 Gene expression data obtained by qRT-PCR analysis.

biodiesel. Certainly, we only analyzed a small subset of the mass data during SASK development and anticipate that other researchers would find valuable information from the SASK datasets available on request.

Materials and methods

Plant materials and extraction of SASK Oil

Siberian apricot is widely distributed in China, so it has not been listed as an endangered or protected species and does not require approval. The different developmental stages of SASK were obtained from the same tree located at the Beijing Forestry University experimental station, Beijing, China. The developmental processes of SASK from flowering to seed maturity were observed from May to July 2012. Flowers with the same anthesis were marked, and then seeds were harvested at 10 DAF (immature stage), 20 DAF, 30 DAF, 40 DAF, 50 DAF, 60 DAF, and 70 DAF (fully matured stage), respectively. After removing the sarcocarp, the seeds were immediately frozen in liquid nitrogen and stored at −80°C until use. The oil content of SASK and FA composition was determined using the previously described method [7].

cDNA Library preparation and sequence data analysis and assembly

According to the detective results of SASK oil content and FA composition, the experimental SASK from the five crucial periods (10, 30, 50, 60, and 70 DAF) were selected as materials for comparative deep transcriptome and differential gene expression analysis. The equal weight of three biological samples from every developmental stage was mixed, and then total RNA was extracted from the mixture using RNeasy Plant Mini Kits (Qiagen, Inc., Valencia, CA, USA) according to the manufacturer's protocol. Extracted RNA was qualified and quantified using a Nanodrop ND-1000 Spectrophotometer (Nanodrop Technologies, Wilmington, DE, USA), and all the samples showed a 260/280 nm ratio from 1.9 to 2.1; no sign of degradation was found. The samples for transcriptome analysis were prepared using Illumina's kit following manufacturer's protocol (Illumina, San Diego, CA, USA). Briefly, the poly-(A) mRNA was purified from the total RNA by Oligotex mRNA Mini Kit (Qiagen, Inc., Valencia, CA, USA) following the manufacturer's protocol. cDNA library construction and normalization were performed using protocols described previously [67].

The cDNA library (two independent biological replicates) was sequenced on the Illumina sequencing platform (HiSeqTM 2000). After removal of the adapter sequences, the low-quality sequences (reads with ambiguous bases 'N') and reads with more than 10% Q < 20 bases, the quality of the clean reads were assessed by comparing the randomly selected 500,000 reads from all data to NT database with an E-value cut-off of 10^{-10} and coverage more than 80%. Subsequently, the clean reads were assembled into unigenes with the Trinity program [25].

Sequence annotation

To understand their functions, the Siberian apricot unigenes were annotated using BLASTX alignment with an E-value cut-off of 10^{-5} against the following protein databases: NCBI nonredundant, Arabidopsis proteome, SwissProt, and eukaryotic orthologous groups. GO functional classifications and KEGG pathway assignments were performed, as was described previously and a detailed description of the workflow and settings used in the data analysis is given in [27].

Differential expression analysis of unigenes

The unigenes expression levels were calculated using reads per kilobase transcriptome per million mapped reads (RPKM), which could eliminated the influence of gene length and sequencing level on the calculation of gene expression. The levels of unigenes expression in the different samples were compared using the DESeq method described in [31].

qRT-PCR validation

Total RNA was extracted as the description for the cDNA library preparation and was reverse transcribed using the Reverse transcription System (Promega). The amplification primers were designed using PrimerQuest (http://www.idtdna.com/PrimerQuest/Home/Index) software with melting temperatures at 62°C, and the absence of secondary structures was verified by the UNAFold program (http://eu.idtdna.com/UNAFold) according to D'haene et al. [68]. According to our previous studies, cyclophilin gene and ubiquitin-conjugating enzyme gene were used as an internal control [69]. The qRT-PCR was performed using the SYBR Premix *Ex Taq* Kit (TaKaRa) according to the manufacturer's protocol. Negative controls consisting of nuclease-free water instead of template, and reverse transcriptase controls prepared by substituting reverse transcriptase for nuclease-free water in the cDNA synthesis step were included in all analyses for each primer pair. Three technical repetitions were performed for qRT-PCR.

Additional files

Additional file 1: Table S1. Summary of SASK sequencing and assembly.

Additional file 2: Figure S1. Length distribution of the unigenes.

Additional file 3: Table S2. All unigenes' individual annotations, nt length, and levels of transcripts in RPKM.

Additional file 4: Table S3. Detailed information of the annotated genes based on *Arabidopsis* proteome.

Additional file 5: Figure S2. Histogram presentation of Gene Ontology classification. The results are summarized in three main categories: biological process, cellular component, and molecular function. The y-axis on the top indicates the number of genes, and the y-axis on the blow means the percent of genes in a category.

Additional file 6: Figure S3. COG classification. A total of 17,870 unigenes were assigned to 25 classifications. The capital letters in x-axis indicates the COG categories as listed on the right of the histogram, and the y-axis indicates the number of unigenes.

Additional file 7: Figure S4. RPKM distribution of unigene.

Additional file 8: Figure S5. Comparative unigene expression profiles of five SASK developmental periods.

Additional file 9: Table S4. All up- or down-regulated genes and function annotations in distinct developing SASK by DESeq analysis.

Additional file 10: Table S5. The prediction of transcription factors involving in oil biosynthesis based on *Arabidopsis* protein.

Abbreviations

acetyl-CoA: acetyl-coenzyme A; ACC: acetyl-CoA carboxylase; ACCA: ACC carboxyl transferase subunit alpha; ACCC: ACC biotin carboxylase subunit; ACP: acyl-carrier protein; ACS: acetyl-CoA synthetase; AP: Arabidopsis proteome; BLAST: Basic Local Alignment Search Tool; COG: clusters of orthologous groups; DAF: days after flowering; DAG: diacylglycerol; DGAT: acyl-CoA:DAG acyltransferase; EAR: enoyl-ACP reductase; ER: endoplasmic reticulum; FA: fatty acid; FAD2: omega-6 FA desaturase; FATA: fatty acyl-ACP thioesterase A; FATB: fatty acyl-ACP thioesterase B; GLT1: glycolipid transporter; GO: gene ontology; G3P: glycerol-3-phosphate; GPAT: *sn-1* G3P acyltransferase; GPT2: glucose-6-phosphate transporter; KAR: 3-ketoacyl ACP reductase; KASII: 3-ketoacyl ACP synthase II; KASIII: 3-ketoacyl ACP synthase II; KEGG: Kyoto Encyclopedia of Genes and Genomes; LPAT: lysophosphatidyl acyltransferase; LACS: long-chain acyl CoA synthetase; LPCAT: acyl-CoA:lysophosphatidylcholine acyltransferase; MAT: malonyl-CoA-ACP transacylase; Nr: NCBI non-redundant protein; PA: phosphatidic acid; PAP: PA phosphohydrolases; PC: phosphatidylcholine; PDC: pyruvate dehydrogenase complex; PDAT: phospholipid: diacylglycerol acyltransferase; PDCT: PC:DAG cholinephosphotransferase; PFK: ATP-dependent phosphofructokinase; PFP: pyrophosphate-dependent phosphofructokinase; PK: pyruvate kinase; PPT1: phosphoenolpyruvate transporter; qRT-PCR: quantitative real-time PCR; RNA-Seq: RNA sequencing; SAD: 18:0-ACP desaturase; SASK: Siberian apricot seed kernels; SRA: Short Read Archive; SwissProt: Swiss-Prot protein database; TAG: triacylglycerol; TCA cycle: tricarboxylic acid cycle; TPT: triose phosphate transporter.

Competing interests

The authors declare that they have no competing interests.

Authors' contributions

JN and JA organized and planned the research, and drafted the manuscript. SZ, CIF and HY performed the sequence analysis. LW, XF, ZX, and XH determined the oil content and FA composition. LQ, CyF, DH, and HZ prepared the materials and cDNA samples. ZZ and SL provided funding, computational guidance, and were substantially involved in research design, data analysis, and manuscript preparation. All authors have read and approved the manuscript.

Acknowledgements
This research was supported by the National Natural Sciences Foundation of China (No. J1103516), and the CentralPublic-Interest Scientific Institution Basal Research Fund (RIF2013-02) (No. 201004001).

Author details
[1]College of Biological Sciences and Biotechnology, College of Nature Conservation, National Engineering Laboratory for Tree Breeding, Key Laboratory of Genetics and Breeding in Forest Trees and Ornamental Plants, Ministry of Education, Beijing Forestry University, Beijing 10083, China. [2]Research Institute of Forestry, Chinese Academy of Forestry, Beijing 10091, China. [3]Jigongshan National Nature Reserve, Xingyang 464133, China. [4]Liaocheng Food and Drug Administration, Liaocheng 252000, Shandong, China.

References

1. Ghadge SV, Raheman H. Biodiesel production from mahua (*Madhuca indica*) oil having high free fatty acids. Biomass Bioenergy. 2005;28:601–5.
2. Lin SF, Shen Y. Thinking about the wood-oil biodiesel industry. Issues of Forestry Econ. 2010;30:131–5.
3. Wang LB, Yu HY, He XH, Liu RY. Influence of fatty acid composition of woody biodiesel plants on the fuel properties. J Fuel Chem Technol. 2012,40.397–404.
4. Wang LB. Resource investigation and distribute regular of three Armeniaca species. Forest Resource Manage. 2011;5:65–70.
5. Zhang JY, Wei ZZ, Zhao HL. Determination of ability of several arbor and shrub species to endure and survive extreme aridity with limited-areas methods under field conditions in Horqin Sandy Land. Acta Ecol Sin. 2006;26:467–74.
6. Wang LB. Geographic distribution and botanical characters of 3 Armeniaca plant in China. For Res. 2010;3:435–9.
7. Wang LB. Properties of Manchurian apricot (*Prunus mandshurica* Skv.) and Siberian apricot (*Prunus sibirica* L.) seed kernel oils and evaluation as biodiesel feedstocks. Ind Crop Prod. 2013,50:838–43.
8. Zhang JY. Chinese Fruit Species. Volume apricot. Beijng: China Forestry Publishing House; 2003. p. 627.
9. Wang LB. Evaluation of Siberian Apricot (*Prunus sibirica* L.) germplasm variability for biodiesel properties. J Am Oil Chem Soc. 2012;89:1743–7.
10. Wang LB, Yu HY. Biodiesel from Siberian apricot (*Prunus sibirica* L.) seed kernel oil. Bioresour Technol. 2012;112:355–8.
11. Wang LB, Chu J. Optimization of biodiesel production from Siberian Apricot (*Prunus sibirica* L.) oil using response surface methodology. Asian J Chem. 2013;25:2577–82.
12. Dillies MA, Rau A, Aubert J, Hennequet-Antier C, Jeanmougin M, Servant N, et al. A comprehensive evaluation of normalization methods for Illumina high-throughput RNA sequencing data analysis. Brief Bioinform. 2012;14:671–83.
13. Huang JY, Jie ZJ, Wang LJ, Yan XH, Wei WH. Analysis of the differential expression of the genes related to *Brassica napus* seed development. Mol Biol Rep. 2011;38:1055–61.
14. Chen H, Wang FW, Dong YY, Wang N, Sun YP, Li XY, et al. Sequence mining and transcript profiling to explore differentially expressed genes associated with lipid biosynthesis during soybean seed development. BMC Plant Biol. 2012;12:122.
15. Yin DM, Wang Y, Zhang XG, Li HM, Lu X, Zhang JS, et al. *De novo* assembly of the peanut (*Arachis hypogaea* L.) seed transcriptome revealed candidate unigenes for oil accumulation pathways. PLoS ONE. 2013;8:e73767.
16. Troncoso-Ponce MA, Kilaru A, Cao X, Durrett TP, Fan J, Jensen JK, et al. Comparative deep transcriptional profiling of four developing oilseeds. Plant J. 2011;68:1014–27.
17. Kroon J, Wei W, Simon WJ, Slabas AR. Identification and functional expression of a type 2 acyl-CoA: diacylglycerol acyltransferase (DGAT2) in developing castor bean seeds which has high homology to the major triglyceride biosynthetic enzyme of fungi and animals. Phytochemistry. 2006;67:2541–9.
18. Tao X, Fang Y, Xiao Y, Jin YL, Ma XR, Zhao Y, et al. Comparative transcriptome analysis to investigate the high starch accumulation of duckweed (*Landoltia punctata*) under nutrient starvation. Biotechnol Biofuels. 2013;6:72.
19. Bourgis F, Kilaru A, Cao X, Ngando-Ebongue G-F, Drira N, Ohlrogge JB, et al. Comparative transcriptome and metabolite analysis of oil palm and date palm mesocarp that differ dramatically in carbon partitioning. Proc Natl Acad Sci U S A. 2011;108:12527–32.
20. Fan HK, Xiao Y, Yang YD, Xia W, Mason AS, Xia ZH, et al. RNA-Seq analysis of *Cocos nucifera*: transcriptome sequencing and *de novo* assembly for subsequent functional genomics approaches. PLoS ONE. 2013;8:e59997.
21. Dong SB, Liu YL, Niu J, Ning Y, Lin SZ, Zhang ZX. *De novo* transcriptome analysis of the Siberian apricot (*Prunus sibirica* L.) and search for potential SSR markers by 454 pyrosequencing. Gene. 2014;544:220–7.
22. Wang R, Hanna MA, Zhou WW, Bhadury PS, Chen Q, Song BA, et al. Production and selected fuel properties of biodiesel from promising non-edible oils: *Euphorbia lathyris* L., *Sapium sebiferum* L. and *Jatropha curcas* L. Bioresour Technol. 2011;102:1194–9.
23. Lai ZX, Lin YL. Analysis of the global transcriptome of longan (*Dimocarpus longan* Lour.) embryogenic callus using Illumina paired-end sequencing. BMC Genomics. 2013;14:561.
24. Martin J, Bruno VM, Fang Z, Meng X, Blow M, Zhang T, et al. Rnnotator: an automated *de novo* transcriptome assembly pipeline from stranded RNA-Seq reads. BMC Genomics. 2010;11:663.
25. Grabherr MG, Haas BJ, Yassour M, Levin JZ, Thompson DA, Amit I, et al. Full-length transcriptome assembly from RNA-Seq data without a reference genome. Nat Biotechnol. 2011;29:644–52.
26. Sangwan RS, Tripathi S, Singh J, Narnoliya LK, Sangwan NS. *De novo* sequencing and assembly of *Centella asiatica* leaf transcriptome for mapping of structural, functional and regulatory genes with special reference to secondary metabolism. Gene. 2013;525:58–76.
27. Liu YL, Huang ZD, Ao Y, Li W, Zhang ZX. Transcriptome analysis of Yellow Horn (*Xanthoceras sorbifolia* Bunge.): a potential oil-rich seed tree for biodiesel in China. PLoS ONE. 2013;8:e74441.
28. Consortium GO. The Gene Ontology (GO) database and informatics resource. Nucleic Acids Res. 2004;32:D258–61.
29. Tatusov RL, Fedorova ND, Jackson JD, Jacobs AR, Kiryutin B, Koonin EV, et al. The COG database: an updated version includes eukaryotes. BMC Bioinformatics. 2003;4:41.
30. Kanehisa M, Goto S. KEGG: kyoto encyclopedia of genes and genomes. Nucleic Acids Res. 2000;28:27–30.
31. Anders S, Huber W. Differential expression of RNA-Seq data at the gene level–the DESeq package. Heidelberg, Germany: European Molecular Biology Laboratory (EMBL); 2012. Available at http://watson.nci.nih.gov/bioc_mirror/packages/2.11/bioc/vignettes/DESeq/inst/doc/DESeq.pdf.
32. Mortazavi A, Williams BA, McCue K, Schaeffer L, Wold B. Mapping and quantifying mammalian transcriptomes by RNA-Seq. Nat Methods. 2008;5:621–8.
33. Rawsthorne S. Carbon flux and fatty acid synthesis in plants. Prog Lipid Res. 2002;41:182–96.
34. Andriotis VM, Kruger NJ, Pike MJ, Smith AM. Plastidial glycolysis in developing Arabidopsis embryos. New Phytol. 2010;185:649–62.
35. Plaxton WC. The organization and regulation of plant glycolysis. Annu Rev Plant Biol. 1996;47:185–214.
36. Weber AP, Schwacke R, Flügge U-I. Solute transporters of the plastid envelope membrane. Annu Rev Plant Biol. 2005;56:133–64.
37. Knappe S, Löttgert T, Schneider A, Voll L, Flügge UI, Fischer K. Characterization of two functional phosphoenolpyruvate/phosphate translocator (PPT) genes in Arabidopsis-*AtPPT1* may be involved in the provision of signals for correct mesophyll development. Plant J. 2003;36:411–20.
38. Fatland BL, Ke J, Anderson MD, Mentzen WI, Cui LW, Allred CC, et al. Molecular characterization of a heteromeric ATP-citrate lyase that generates cytosolic acetyl-coenzyme A in Arabidopsis. Plant Physiol. 2002;130:740–56.
39. Ke J, Behal RH, Back SL, Nikolau BJ, Wurtele ES, Oliver DJ. The role of pyruvate dehydrogenase and acetyl-coenzyme A synthetase in fatty acid synthesis in developing Arabidopsis seeds. Plant Physiol. 2000;123:497–508.
40. Randall DD, Miernyk JA, Fang TK, Budde RJ, Schuller KA. Regulation of the pyruvate dehydrogenase complexes in plants. Ann N Y Acad Sci. 1989;573:192–205.
41. Nikolau BJ, Ohlrogge JB, Wurtele ES. Plant biotin-containing carboxylases. Arch Biochem Biophys. 2003;414:211–22.

42. Bourrellier ABF, Valot B, Guillot A, Ambard-Bretteville F, Vidal J, Hodges M. Chloroplast acetyl-CoA carboxylase activity is 2-oxoglutarate-regulated by interaction of PII with the biotin carboxyl carrier subunit. Proc Natl Acad Sci U S A. 2010;107:502–7.
43. Andre C, Haslam RP, Shanklin J. Feedback regulation of plastidic acetyl-CoA carboxylase by 18: 1-acyl carrier protein in *Brassica napus*. Proc Natl Acad Sci U S A. 2012;109:10107–12.
44. Bates PD, Stymne S, Ohlrogge J. Biochemical pathways in seed oil synthesis. Curr Opin Plant Biol. 2013;16:358–64.
45. Cahoon EB, Clemente TE, Damude HG, Kinney AJ. Oil Crops Handbook of Plant Breeding (Modifying Vegetable Oils for Food and Non-Food Purposes). Volume 4. New York: Springer; 2010. p. 31–56.
46. Voelker TA, Hayes TR, Cranmer AM, Turner JC, Davies HM. Genetic engineering of a quantitative trait: metabolic and genetic parameters influencing the accumulation of laurate in rapeseed. Plant J. 1996;9:229–41.
47. Knutzon DS, Thompson GA, Radke SE, Johnson WB, Knauf VC, Kridl JC. Modification of Brassica seed oil by antisense expression of a stearoyl-acyl carrier protein desaturase gene. Proc Natl Acad Sci U S A. 1992;89:2624–8.
48. Pidkowich MS, Nguyen HT, Heilmann I, Ischebeck T, Shanklin J. Modulating seed β-ketoacyl-acyl carrier protein synthase II level converts the composition of a temperate seed oil to that of a palm-like tropical oil. Proc Natl Acad Sci U S A. 2007;104:4742–7.
49. Koo AJ, Ohlrogge JB, Pollard M. On the export of fatty acids from the chloroplast. J Biol Chem. 2004;279:16101–10.
50. Zhao LF, Katavic V, Li FL, Haughn GW, Kunst L. Insertional mutant analysis reveals that long-chain acyl-CoA synthetase 1 (LACS1), but not LACS8, functionally overlaps with LACS9 in Arabidopsis seed oil biosynthesis. Plant J. 2010;64:1048–58.
51. Behrouzian B, Buist PH. Fatty acid desaturation: variations on an oxidative theme. Curr Opin Chem Biol. 2002;6:577–82.
52. Kim S, Yamaoka Y, Ono H, Kim H, Shim D, Maeshima M, et al. AtABCA9 transporter supplies fatty acids for lipid synthesis to the endoplasmic reticulum. Proc Natl Acad Sci U S A. 2013;110:773–8.
53. Barron EJ, Stumpf P. Fat metabolism in higher plants XIX. The biosynthesis of triglycerides by avocado-mesocarp enzymes. Biochim Biophys Acta. 1962;60:329–37.
54. Yang W, Simpson JP, Li-Beisson Y, Beisson F, Pollard M, Ohlrogge JB. A land-plant-specific glycerol-3-phosphate acyltransferase family in Arabidopsis: substrate specificity, sn-2 preference, and evolution. Plant Physiol. 2012;160:638–52.
55. Gidda SK, Shockey JM, Rothstein SJ, Dyer JM, Mullen RT. *Arabidopsis thaliana* GPAT8 and GPAT9 are localized to the ER and possess distinct ER retrieval signals: Functional divergence of the dilysine ER retrieval motif in plant cells. Plant Physiol Biochem. 2009;47:867–79.
56. Maisonneuve S, Bessoule J-J, Lessire R, Delseny M, Roscoe TJ. Expression of rapeseed microsomal lysophosphatidic acid acyltransferase isozymes enhances seed oil content in Arabidopsis. Plant Physiol. 2010;152:670–84.
57. Eastmond PJ, Quettier A-L, Kroon JT, Craddock C, Adams N, Slabas AR. PHOSPHATIDIC ACID PHOSPHOHYDROLASE1 and 2 regulate phospholipid synthesis at the endoplasmic reticulum in Arabidopsis. Plant Cell. 2010;22:2796–811.
58. Kim HU, Li Y, Huang AH. Ubiquitous and endoplasmic reticulum-located lysophosphatidyl acyltransferase, LPAT2, is essential for female but not male gametophyte development in Arabidopsis. Plant Cell. 2005;17:1073–89.
59. Lu C, Xin Z, Ren Z, Miquel M. An enzyme regulating triacylglycerol composition is encoded by the ROD1 gene of Arabidopsis. Proc Natl Acad Sci U S A. 2009;106:18837–42.
60. Saha S, Enugutti B, Rajakumari S, Rajasekharan R. Cytosolic triacylglycerol biosynthetic pathway in oilseeds. Molecular cloning and expression of peanut cytosolic diacylglycerol acyltransferase. Plant Physiol. 2006;141:1533–43.
61. Shockey JM, Gidda SK, Chapital DC, Kuan J-C, Dhanoa PK, Bland JM, et al. Tung tree DGAT1 and DGAT2 have nonredundant functions in triacylglycerol biosynthesis and are localized to different subdomains of the endoplasmic reticulum. Plant Cell. 2006;18:2294–313.
62. Zhang M, Fan J, Taylor DC, Ohlrogge JB. DGAT1 and PDAT1 acyltransferases have overlapping functions in Arabidopsis triacylglycerol biosynthesis and are essential for normal pollen and seed development. Plant Cell. 2009;21:3885–901.
63. Mhaske V, Beldjilali K, Ohlrogge J, Pollard M. Isolation and characterization of an *Arabidopsis thaliana* knockout line for phospholipid: diacylglycerol transacylase gene (At5g13640). Plant Physiol Biochem. 2005;43:413–7.
64. Jolivet P, Roux E, d'Andrea S, Davanture M, Negroni L, Zivy M, et al. Protein composition of oil bodies in *Arabidopsis thaliana* ecotype WS. Plant Physiol Biochem. 2004;42:501–9.
65. Palaniswamy SK, James S, Sun H, Lamb RS, Davuluri RV, Grotewold E. AGRIS and AtRegNet. a platform to link *cis*-regulatory elements and transcription factors into regulatory networks. Plant Physiol. 2006;140:818–29.
66. Baud S, Lepiniec L. Physiological and developmental regulation of seed oil production. Prog Lipid Res. 2010;49:235–49.
67. Xu Q, Zhao WM, Chen Y, Tong YY, Rong GH, Huang ZY, et al. Transcriptome profiling of the goose (*Anser cygnoides*) ovaries identify laying and broodiness phenotypes. PLoS ONE. 2013;8:e55496.
68. Løvdal T, Saha A. Reference gene selection in *Carnobacterium maltaromaticum*, *Lactobacillus curvatus* and *Listeria innocua* subjected to temperature and salt stress. Mol Biotechnol. 2014;56:210–22.
69. Niu J, Zhu BQ, Cai J, Li PX, Wang LB, Dai HT, et al. Selection of reference genes for gene expression studies in Siberian Apricot (*Prunus sibirica* L.) germplasm using Quantitative Real-Time PCR. PLoS ONE. 2014;9:e103900.

A transcriptomic analysis of *Neurospora crassa* using five major crop residues and the novel role of the sporulation regulator *rca-1* in lignocellulase production

Bang Wang[1,2], Pengli Cai[1,2], Wenliang Sun[1], Jingen Li[1,2], Chaoguang Tian[1*] and Yanhe Ma[1]

Abstract

Background: Crop residue is an abundant, low-cost plant biomass material available worldwide for use in the microbial production of enzymes, biofuels, and valuable chemicals. However, the diverse chemical composition and complex structure of crop residues are more challenging for efficient degradation by microbes than are homogeneous polysaccharides. In this study, the transcriptional responses of *Neurospora crassa* to various plant straws were analyzed using RNA-Seq, and novel beneficial factors for biomass-induced enzyme production were evaluated.

Results: Comparative transcriptional profiling of *N. crassa* grown on five major crop straws of China (barley, corn, rice, soybean, and wheat straws) revealed a highly overlapping group of 430 genes, the biomass commonly induced core set (BICS). A large proportion of induced carbohydrate-active enzyme (CAZy) genes (82 out of 113) were also conserved across the five plant straws. Excluding 178 genes within the BICS that were also upregulated under no-carbon conditions, the remaining 252 genes were defined as the biomass regulon (BR). Interestingly, 88 genes were only induced by plant biomass and not by three individual polysaccharides (Avicel, xylan, and pectin); these were denoted as the biomass unique set (BUS). Deletion of one BUS gene, the transcriptional regulator *rca-1*, significantly improved lignocellulase production using plant biomass as the sole carbon source, possibly functioning via de-repression of the regulator *clr-2*. Thus, this result suggests that *rca-1* is a potential engineering target for biorefineries, especially for plant biomass direct microbial conversion processes.

Conclusions: Transcriptional profiling revealed a large core response to different sources of plant biomass in *N. crassa*. The sporulation regulator *rca-1* was identified as beneficial for biomass-based enzyme production.

Keywords: *Neurospora crassa*, Crop residues, Transcriptional profiling, Biomass regulon, *rca-1*

Background

Renewable plant biomass is a potential low-cost feedstock for the microbial production of lignocellulolytic enzymes, biofuels, or value-added chemicals. This process represents a promising strategy to supply the world's energy demand and reduce greenhouse gas emissions derived from burning fossil fuels [1,2]. Crop residues are ideal raw materials for bioconversion because they are abundant worldwide, low-cost, compatible with food security, and >50% (*w*/*w*) cellulose/hemicellulose, which can be used for biorefinery [3,4]. However, plants have evolved rigid and complex cell walls to support upright growth, transport nutrients, and protect against microbial invasion. Composed of various polymeric building blocks (typically including cellulose, hemicellulose, pectin, lignin, and structural proteins) [5], the plant cell wall presents a greater challenge to lignocellulolytic enzyme digestion than individual polysaccharides like microcrystalline cellulose (e.g., Avicel), xylan, or pectin, which are commonly used in biorefinery studies. High complexity and heterogeneity in chemical composition and

* Correspondence: tian_cg@tib.cas.cn
[1]Key Laboratory of Systems Microbial Biotechnology, Tianjin Institute of Industrial Biotechnology, Chinese Academy of Sciences, Tianjin 300308, China
Full list of author information is available at the end of the article

organization makes plant biomass a bioprocessing feedstock with uneven chemical and physical properties [6,7]. Therefore, understanding the commonalities and differences in microbial degradation of diverse plant cell wall types will be useful for developing a case-by-case or universal strategy for plant biomass biorefinery.

Filamentous fungi are currently the main source for commercial carbohydrate hydrolytic enzyme production [1]. Investigation of fungal responses to a wide spectrum of biomass substrates is needed not only to gain more knowledge of plant cell wall degradation and utilization but also to rationally engineer filamentous fungi for biomass-based enzyme, ethanol, and chemical production. This information will be especially helpful in using plant biomass direct microbial conversion (DMC), a recently suggested, promising strategy [7,8] whereby filamentous fungi are used to generate valuable products, such as fatty acids produced by *Neurospora crassa* [9] and ethanol by *Trichoderma reesei* and *Fusarium oxysporum* [10]. Although transcriptome analysis has been performed in different fungal systems digesting various types of plant biomass (such as *Miscanthus giganteus*, barley, oat, canola, alfalfa, wheat straw, and bagasse) [11-16], the comparative analysis of genome-wide profiling on a wide spectrum of biomass substrates in a single fungus species can offer outstanding knowledge on specific and common microbe responses to different crop residues, which will be useful for engineering fungal DMC using various sources of plant biomass.

The filamentous fungus *N. crassa* has the ability to robustly degrade and utilize lignocellulosic materials [12,17]. Its available genome-wide deletion database [18,19] offers great advantages for ongoing studies of the mechanism of digesting polysaccharides (such as Avicel, xylan, and pectin) [12,20-22]. Studies using this model organism have begun to reveal a more thorough perspective on lignocellulolytic enzyme induction, regulation, and production in filamentous fungi [23]. These advances can be rationally applied to *N. crassa* as well as other filamentous fungi to further improve their production of lignocellulolytic enzymes and even target chemicals via DMC [9].

In this study, we report a large-scale comparison of *N. crassa* transcriptome induction by five crop straws (barley, corn, rice, soybean, and wheat straws; BS, CS, RS, SS, and WS, respectively) that are highly abundant in China [24] and around the world [25]. Differential gene expression profiling of *N. crassa* grown on these crop residues versus sucrose showed a large "core" and a small "shell" response to a variety of plant straws. Carbohydrate-active enzymes (CAZys) [26] were significantly induced by all five straws and displayed a highly overlapping gene pool, indicating a conserved CAZy set for plant biomass degradation in *N. crassa*. This CAZy set contained a high proportion (10 of 14) of plant cell wall-induced lytic polysaccharide monooxygenase (LPMO) genes. Finally, screening mutants of transcription factors which were induced only by complex plant cell wall samples yielded a sporulation regulator *rca-1* that was beneficial for lignocellulolytic enzyme production on plant biomass in *N. crassa*.

Results

Chemical composition of crop residues

The five crop straws contained similar proportions of cellulose (30%–40%; Figure 1 and Additional file 1: Table S1), the major component of agro-residues [3]. Four (BS, CS,

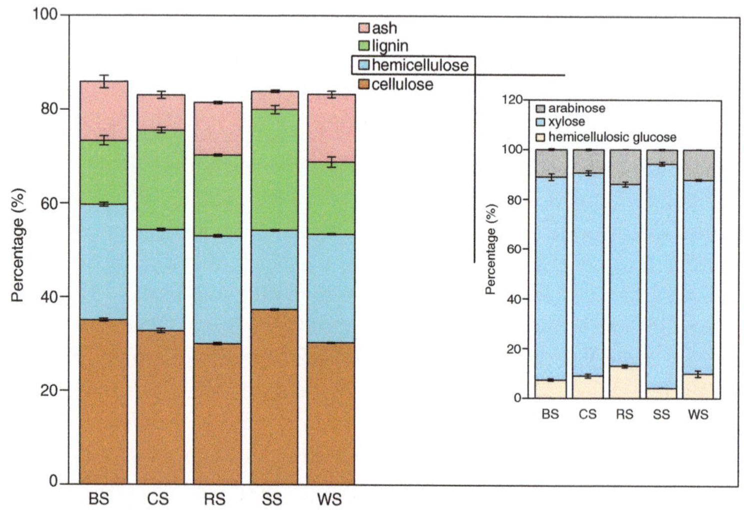

Figure 1 Chemical composition (left panel) and hemicellulosic component distribution in hemicellulose (right panel) of five crop residues. BS, barley straw; CS, corn straw; RS, rice straw; SS, soybean straw; WS, wheat straw. Error bars represent the respective standard deviation of three replicates.

RS, and WS) comprised more than 20% hemicelluloses, while soybean straw contained less hemicellulose but the highest lignin content among five crop residues (Figure 1). Moreover, soybean straw had the lowest proportion of arabinose in hemicellulose (5.6%) compared with the other four straws (9.4%–13.8%; Figure 1 and Additional file 1: Table S1). Generally, lignin and ash compositions in the five crop straws were more varied than that of cellulose/hemicellulose (Figure 1). The variety in chemical composition combined with complex organization account for the high heterogeneity in chemical and physical properties of plant biomass.

Differential expression analysis of biomass-induced profiling data revealed common and unique responses to crop residues

Cellulase and xylanase activities and secreted proteins of *N. crassa* (FGSC 2489) cultured on ground corn straw were determined over time. Culture supernatant rarely accumulated protein until day 2 (Figure 2), but weak cellulase and xylanase activities could be detected at 30 h (Figure 2). Based on a previous study [12], lignocellulolytic enzymes can be transcriptionally highly induced by plant biomass before 40 h in *N. crassa*. By checking the 30-h cultures grown on each crop residue medium, the mycelia were phenotypically (hyphal length and branching) comparable to the 16-h culture on sucrose (Additional file 2: Figure S1), which were further used for transcriptome profiling (RNA-Seq). RNA-Seq data showed good accordance with the results of quantitative real-time reverse transcription polymerase chain reaction (qRT-PCR)

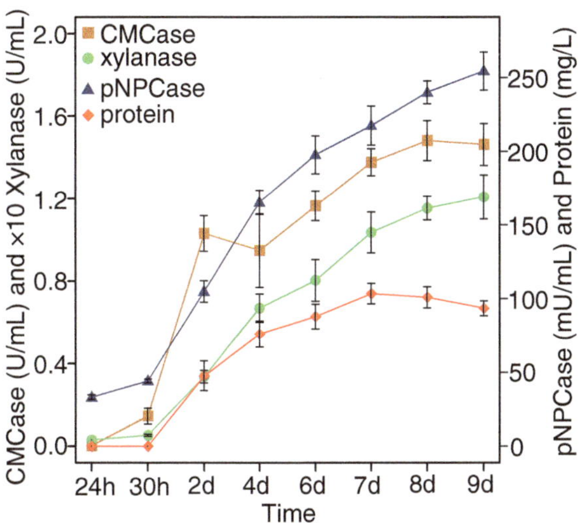

Figure 2 Growth curve of *Neurospora crassa* on 2% (*w/v*) ground corn straw as the sole carbon source. The cellulose hydrolysis activity (CMCase), exo-glucanase activity (pNPCase), xylanase activity, and total protein in supernatants were determined between 24 h and 9 days. Values represent the means of at least three replicates; error bars show standard deviation.

using the Pearson correlation analysis (Figure 3). All selected cellulases (NCU07340, *cbh-1*; NCU09680, *gh6-2*; and NCU00762, *gh5-1*) and hemicellulases (NCU08189, *gh10-2*; NCU05924, *gh10-1*; and NCU02855, *gh11-1*) were highly induced after 30 h of growth on crop residues (Figure 3).

For further analysis, differential gene expression analysis was performed by generalized fold change (GFOLD) [27] and DEGseq [28] using the 16-h sucrose-grown sample as a control (Additional file 3: Table S2; Methods). Totals of 520–890 genes were significantly elevated, with greater than threefold changes (GFOLD ≥ 1.6, *P* value < 1e-4) in the five substrates compared with the sucrose condition (Table 1). Of these genes, 430 genes were commonly induced on all five straws (Figure 4A). Conversely, a small number of genes (76–253 genes) showed decreased expression based on the same criteria (GFOLD ≤ −1.6, *P* value < 1e-4) (Table 1). For those upregulated genes, approximately 60% (300–560 genes) also exhibited threefold upregulation when mycelia pre-grown on sucrose for 16 h were transferred to minimal medium ("Methods") with no-carbon source for 4 h (Table 1), indicating a starvation response that was synergic with plant cell wall degradation and utilization in *N. crassa* [20]. However, these expression changes under the no-carbon condition were not of the same order of magnitude as those under the lignocellulosic conditions (Additional file 3: Table S2).

Except the 430 biomass commonly induced core set (BICS) genes, many genes showed different induction patterns under the five straws. For example, 148 genes were specifically induced by corn straw, accounting for a greater percentage of all induced genes than in other crop straws (Figure 4B). Most "CS-unique" genes (85 of 148) were also induced by the no-carbon condition (Additional file 4: Table S3, sheet 1), suggesting that corn straw may elicit broader carbon de-repression than other crops. Conversely, soybean straw induced the fewest genes (Table 1), as well as "shared" and "unique" genes (Figure 4A, B). Furthermore, a second large set of shared genes in Figure 4A (128 genes induced in BS, CS, RS, and WS) was not upregulated on soybean straw. This feature was also verified by principle component analysis (PCA) based on transcriptome expression profiling, which showed that soybean straw (a dicot crop) was separated a little from the other four monocot straws (Figure 4C). These observations might reflect the transcriptional responses of *N. crassa* to the chemical composition distinctions displayed in Figure 1.

A detailed functional analysis (FunCat [29]) showed that polysaccharide or carbohydrate metabolism was also enriched in all five biomass "shared" and "unique" sets, which was not the case in the "common" set or starvation condition (no carbon) (Additional file 4:

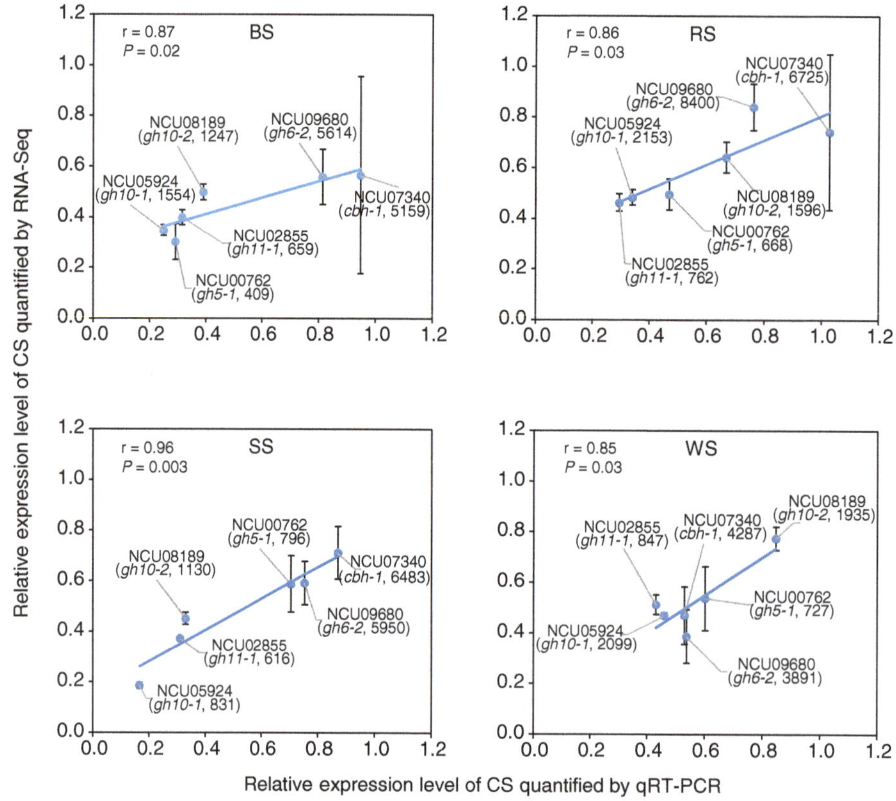

Figure 3 Correlation comparison between qRT-PCR results and RNA-Seq data. BS, barley straw; CS, corn straw; RS, rice straw; SS, soybean straw; WS, wheat straw. Relative expression levels of selected CAZy genes were normalized by the expression under CS. Values of the corresponding gene expression levels by RNA-Seq (RPKM) are shown in brackets. Pearson correlation coefficients and *P* values are shown on the top left of each figure. For qRT-PCR (horizontal coordinate), values represent the means of triplicates; error bars show standard deviation.

Table S3, sheet 1). Although these 23 polysaccharide/carbohydrate metabolism genes (Additional file 4: Table S3, sheet 1, after removing redundant ones) in the “shared” and “unique” sets induced by each straw were not the same, most (17 out 23) were induced by at least two crop straws. For example, all six CAZy genes (NCU09904, *gh16*; NCU09175, *gh17*; NCU09042, *gh2*; NCU07351, *gh67*; NCU04526, *ce3*; and NCU01059, *gh47*) were upregulated under at least two straw conditions (Additional file 4: Table S3, sheet 1). However, corn straw uniquely induced many genes unrelated to C-compound metabolism but associated with secondary metabolism and detoxification (Additional file 4: Table S3, sheet 1). For example, NCU05780, a glutathione S-transferase gene, was highly induced only on corn straw (RPKM (reads per kilobase per million mapped reads) = 224.6, Additional file 3: Table S2), indicating a reduction reaction was required for detoxifying oxides when *N. crassa* was grown on corn residue. Intriguingly, rice straw induced some cation-metabolism genes and soybean straw induced nucleoside-metabolism genes, reflecting diverse impurities, such as iron contamination [15], in different crop straws. Functional analysis of all downregulated genes in the five straws revealed that protein-folding, structural protein, inorganic element metabolism, and disease-defense genes were mainly enriched in this gene set (Additional file 4: Table S3, sheet 2).

Table 1 Differential gene expression profile on five crop residues

	BS	CS	RS	SS	WS
Upregulated genes	656	888	738	524	742
“Common” rate[a]	65.5%	48.4%	58.3%	82.1%	58.0%
“No-carbon” rate[b]	60.7%	63.6%	62.6%	57.6%	62.1%
Downregulated genes	76	253	117	141	101
“Common” rate	72.4%	21.7%	47.0%	39.0%	54.5%
“No-carbon” rate	60.5%	78.3%	63.2%	46.1%	63.4%

BS barley straw, *CS* corn straw, *RS* rice straw, *SS* soybean straw, *WS* wheat straw.

[a]Genes commonly regulated by all five straws versus sucrose as a sole carbon source; “no-carbon” data were derived from a reference [20].

[b]Genes regulated in the no-carbon condition versus sucrose as a sole carbon source.

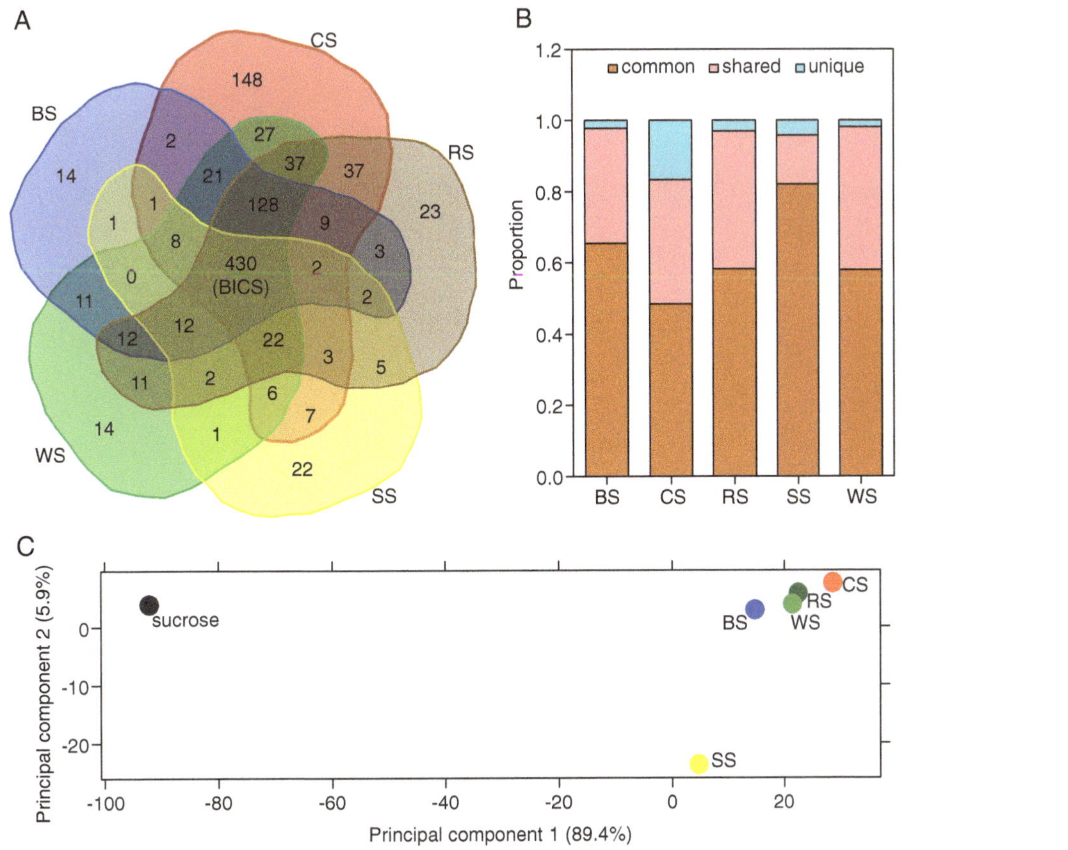

Figure 4 Overview of *Neurospora crassa* transcriptome induced by five crop residues. BICS, biomass commonly induced core set; BS, barley straw; CS, corn straw; RS, rice straw; SS, soybean straw; WS, wheat straw. **(A)** Venn diagram of genes upregulated by each crop straw. **(B)** Proportions of genes upregulated by each crop straw. "Common" represents genes induced by all five straws; "shared" represents genes induced by at least two but not all straws; "unique" represents genes uniquely induced by one crop residue. **(C)** Principal component analysis of transcription profiling data performed by DESeq. The RNA-Seq reads of *N. crassa* grown on sucrose, BS, CS, RS, SS, and WS were used for the principal component plot. Each point on the graph represents one biological sample. Proportions of the principal components 1 and 2 are shown in brackets.

Comparative analysis of biomass regulon revealed common and specific features with polysaccharides regulons

The 430 genes commonly induced by all five crop residues (the BICS genes) were primarily assigned into three clusters using hierarchical clustering with comparison to the no-carbon condition [30] (Figure 5A; Additional file 5: Table S4, sheet 1). Clusters 1 (C1) and 2 (C2) genes were mostly upregulated by carbon starvation. These two clusters were mainly related to nutrient metabolism and transport and enzyme cofactors (Figure 5B, C). This result might indicate reliance on cellular organic metabolism, transformation, and energy homeostasis in carbon starvation conditions. The biggest cluster (C3) harbored 252 genes that were specifically induced by biomass (Figure 5A). This 252-gene set was upregulated on all five crop residues compared with both sucrose (carbon catabolite repression) and no-carbon (starvation) conditions and was denoted as the "biomass regulon" (BR) [20,22]. FunCat results showed that C-compound or extracellular carbon metabolism and polysaccharide binding were prevalent in the BR group (Figure 5D). Two recently characterized, conserved cellulose degradation regulators *clr-1* (NCU07705) and *clr-2* (NCU08042) [20] were also included in this set (Additional file 6: Table S5, sheet 1).

The regulons of three main plant cell wall polysaccharides, cellulose (Avicel), hemicellulose (xylan), and pectin, have been published and include 212, 117, and 189 genes, respectively [20,22]. Venn diagram comparison of these three datasets showed that 164 BR genes were shared with the polysaccharide regulons, donated as the "biomass shared set" (BSS), but another 88 genes were not shared and comprised the "biomass unique set" (BUS) (Figure 6A). As shown in Figure 6A, the 252 BR genes observed in the present study, combined with the polysaccharide regulons identified previously (212 for Avicel [20], 117 for xylan [22], and 189 for pectin [22]), a total of 461 genes were obtained after removing the redundant ones. These 461 genes were then manually

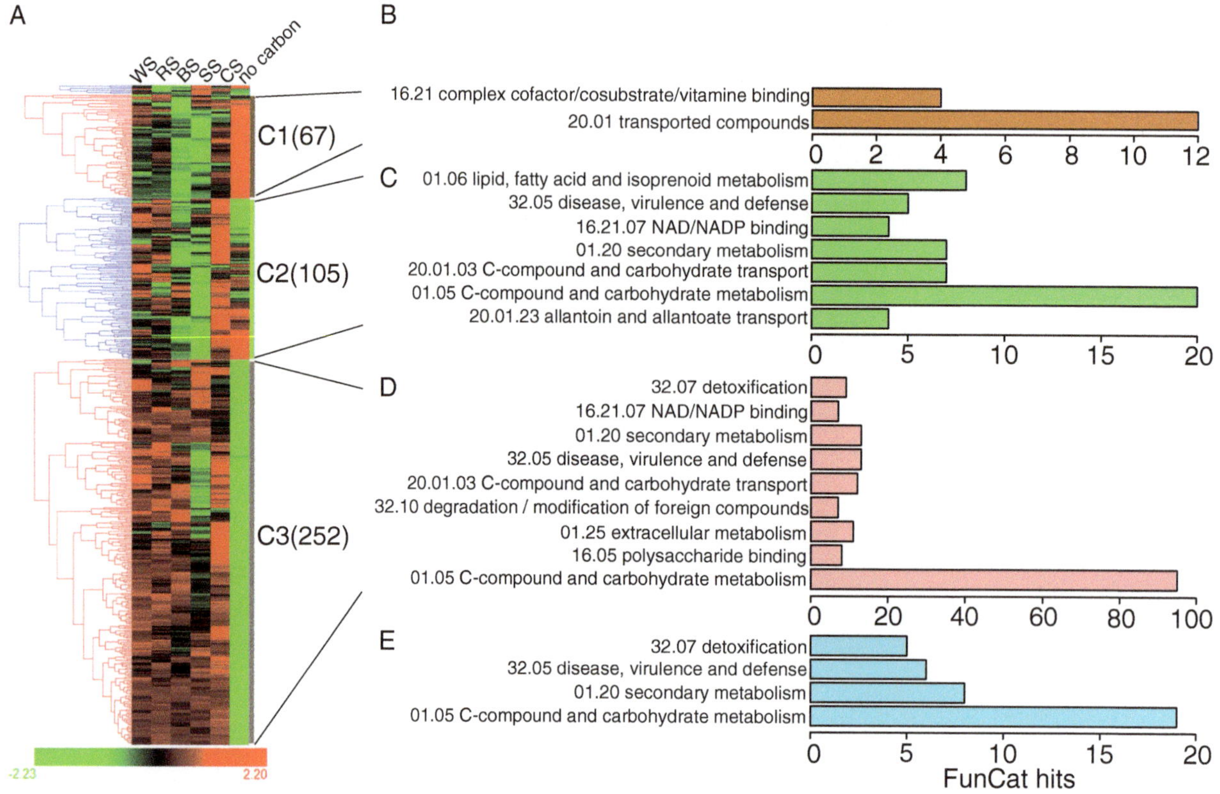

Figure 5 Hierarchical clustering and functional category analysis of 430-gene biomass commonly induced core set (BICS). BS, barley straw; C1, cluster 1; C2, cluster 2; C3, cluster 3; CS, corn straw; RS, rice straw; SS, soybean straw; WS, wheat straw. **(A)** Hierarchical clustering analysis of $\log_2$ fold changes in each crop residue versus the sucrose condition. The gene numbers for each cluster are shown in brackets. "No carbon" data were derived from a reference [20]. **(B–E)** Functional category analysis [29] of C1, C2, C3 (biomass regulon, BR), and biomass unique set (BUS, including 88 genes) from C3 compared with three polysaccharide regulons (see Figure 6A), respectively. Categories with FunCat hits more than three and P value < 0.05 are shown.

assigned to several functional categories based on FunCat [29] and CAZy [26,31] (Additional file 6: Table S5, sheet 2). Hierarchical clustering analysis of functional category showed that the BR was the closest to the Avicel regulon (Figure 6B). The 164-gene BSS was especially enriched for auxiliary activities family (mainly belonging to AA9, previously GH61), hemicellulase, and pectinase genes. The BUS consisted of some unclassified CAZy genes, carbon metabolism genes, and 38 other unclassified genes (Figure 5E; Additional file 6: Table S5, sheet 2). These genes might be of importance for plant biomass degradation and utilization [17]. For example, by checking the phenotypes of transcription regulator mutants from BUS, a limited characterized transcription factor *rca-1* (NCU01312) [32] was found to affect the expression and production of cellulase and xylanase (see mutant screening, below). Not surprisingly, the xylan and pectin regulons were enriched for hemicellulose and pectin hydrolytic genes but showed limited induction for AA genes (Figure 6B).

The CAZome was induced on a large scale and functioned on various lignocellulosic materials as a conserved hydrolytic enzyme pool

Two hundred twenty-five genes in the *N. crassa* genome are predicted to encode glycosyl hydrolases based on the latest version of CAZy [31], called the CAZome (Additional file 7: Table S6, sheet 1). Of the 225 CAZy genes, approximately 100 were significantly upregulated threefold on most of the tested crop residues (Figure 7A and Additional file 7: Table S6, sheet 2). This upregulated gene set was slightly larger than those on any one of the three polysaccharides (Avicel, xylan, and pectin), especially xylan (Figure 7A). Moreover, the CAZy groups induced by the five crop residues displayed similar distributions of six CAZy functional categories; all five crop residues elicited an especially high number (11–13 of 14) of AA9 genes (Figure 7A). A Venn diagram of these five crop residue-induced CAZy groups showed that they shared the majority (82 genes) of their upregulated genes (Figure 7B).

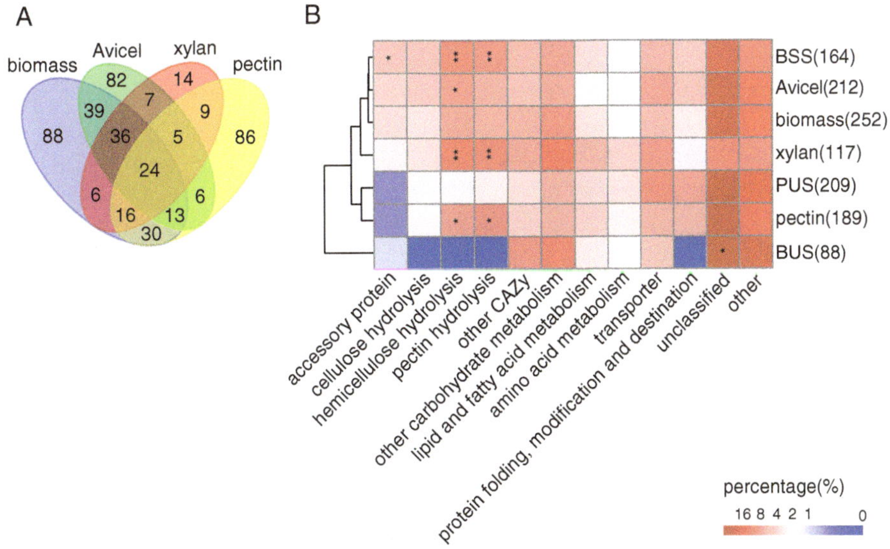

Figure 6 Comparative analysis of 252 BR genes versus three polysaccharide regulons (Avicel, xylan, and pectin) in *Neurospora crassa* [20,22]. BSS, biomass shared set; BUS, biomass unique set; PUS, polysaccharide unique set. **(A)** Venn diagram of the BR compared with three polysaccharide regulons. **(B)** Functional enrichment of BR versus polysaccharide regulons. Gene numbers for each group are shown in brackets. The percentages of functional categories were displayed using a heat map. Enrichment analysis was performed by one-tailed Fisher's exact test with the background of all lignocellulolytic regulon genes (BR plus three polysaccharide regulons with redundant genes removed). $^{*}P < 0.05$; $^{**}P < 0.01$.

The transcriptional regulator *rca-1* is involved in lignocellulolytic enzyme synthesis for *N. crassa* grown on plant biomass

Multiple studies have already shown that several transcriptional regulators and a set of hydrolytic enzymes are tightly associated with polysaccharide degradation in *N. crassa* [20,21,33]. These genes were mainly identified under homo-polysaccharide conditions. This study showed that many genes were uniquely induced by various crop residues (Figure 6A). Screening the single gene deletion strain database from Fungal Genetics Stock Center (FGSC) [19] is an efficient way to identify critical components for plant cell wall degradation in *N. crassa* [20,22]. Thus, knockout strains of transcriptional regulators from the BUS (Figure 6A and Additional file 6: Table S5, sheet 2) were screened using Vogel's minimal medium with 2% ground corn straw as the sole carbon source ("Methods").

A sporulation regulator (NCU01312) deletion mutant Δ*rca-1* secreted significantly more proteins than wild type (WT) when grown on corn straw (Figure 8A). The mycelia dry weight of Δ*rca-1* was slightly higher than that of WT, but not significantly so by Student's *t*-test (Figure 8B). However, the Δ*rca-1* mutant showed deficient growth on glucose and on sucrose (Figure 8B). This deficient phenotype was also observed on Avicel but not on xylan and pectin (Additional file 8: Figure S2), suggesting that the higher enzyme secretion in Δ*rca-1* was related to the plant biomass and not pure cellulose induction. Protein sodium dodecyl sulfate polyacrylamide gel electrophoresis (SDS-PAGE) of Δ*rca-1* showed similar banding patterns but remarkably increased protein abundances as compared with WT grown on corn straw (Figure 8C). These bands were probably NCU07340 (CBH-1, ~70 kD)/NCU09680 (GH6-2, ~70 kD), NCU07190 (GH6-3, ~40 kD), and NCU07326 (GH43-6, ~37 kD) based on a comparison with the SDS-PAGE of the *N. crassa* secretome from a previous report [12]. The CMCase, pNPCase, and xylanase activities confirmed this observation (Figure 8D), especially that of xylanase, which was approximately twofold higher xylanase activity than that of WT. The transcript abundance of all selected main cellulases (NCU09680, NCU07340, NCU04952, and NCU00762) and xylanases (NCU07326 and NCU05924) in mutant Δ*rca-1* grown for 16 h on sucrose and then transferred to corn straw for 4 h caused at least a twofold higher induction than in WT (Figure 8E). This finding was in accord with the greater extracellular protein levels and enzyme activities in Δ*rca-1* (Figure 8A, D), which were also observed when *N. crassa* was grown on BS, RS, and WS as the sole carbon sources (Additional file 9: Figure S3). Although the culture supernatant of SS showed low secreted protein levels (data not shown), 4 h induction by SS when transferred from 16 h of pre-growth on sucrose showed that the transcript abundances of cellulase and xylanase were elevated in the *rca-1* mutant (Additional file 9: Figure S3C).

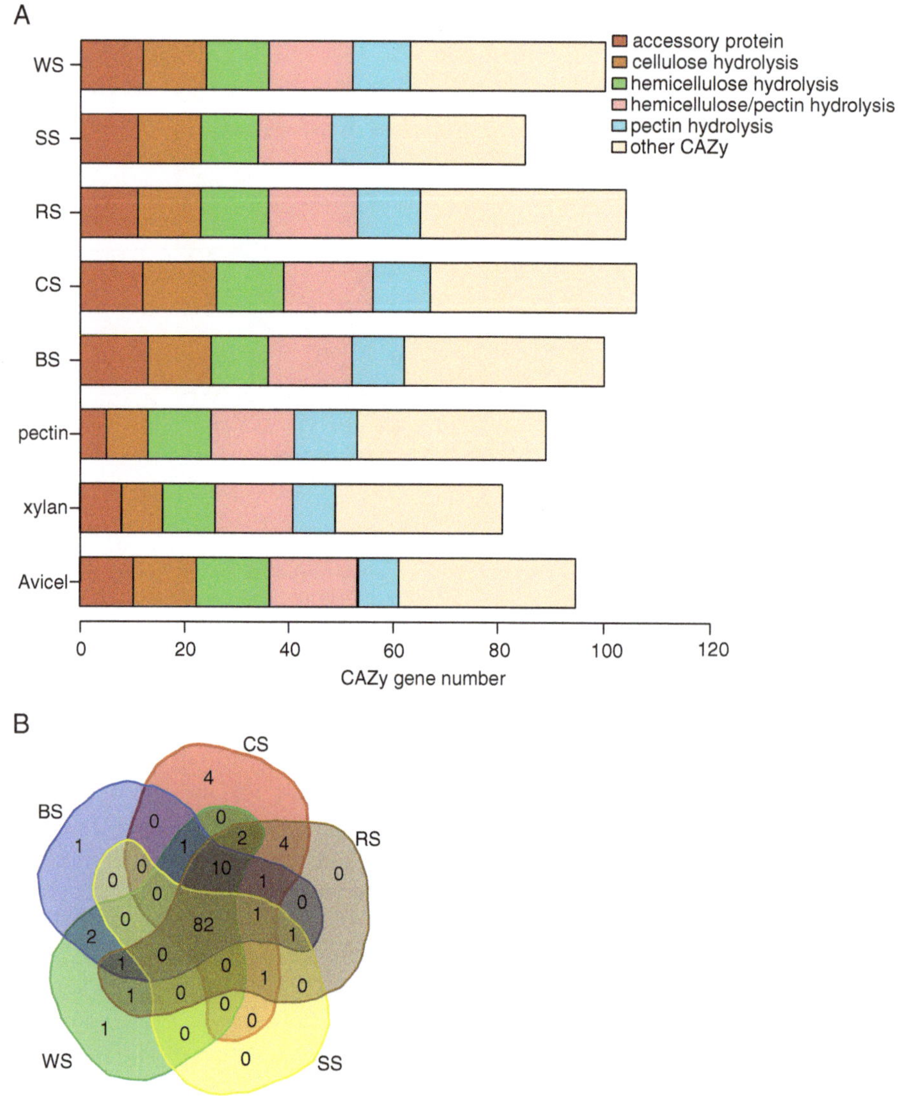

Figure 7 Comparative analysis of *Neurospora crassa* CAZome induced by five crop residues and three polysaccharides (Avicel, xylan, and pectin) [20,22]. BS, barley straw; CS, corn straw; RS, rice straw; SS, soybean straw; WS, wheat straw. **(A)** Composition distribution of the CAZome of *N. crassa* induced by various lignocellulosic substrates. **(B)** Venn comparison of the CAZome induced by five crop residues.

N. crassa rca-1 is a homolog of the sporulation regulator *flbD* in *Aspergillus nidulans* [34]. It encodes a protein with a Myb-like DNA-binding domain that can complement the *A. nidulans flbD* mutant but has nearly unidentifiable effects on *N. crassa* conidiation [32]. However, loss of this gene significantly increased lignocellulolytic enzyme production in *N. crassa* when exposed to plant cell wall (Figure 8A, C and Additional file 9: Figure S3A). To further determine the regulatory differences in cellulase expression between Δ*rca-1* and WT, the transcript expression levels of major lignocellulase regulators in *N. crassa*, including *clr-1/2* (NCU07705/NCU08042) [20], *xlr-1* (NCU06971) [21], and *cre-1* (NCU08807) [33], were assessed by qRT-PCR. *clr-2* consistently displayed a two folds higher transcript level in Δ*rca-1* versus WT when mycelia pre-grown for 16 h in minimal medium with 2% sucrose were transferred to corn straw for 4–6 h (Figure 8F), indicating that *rca-1* may have a negative effect on *clr-2* expression when mycelia are exposed to plant straws. *clr-2* also showed remarkably higher expression level in the Δ*rca-1* mutant when mycelia were grown on soybean straw (Additional file 9: Figure S3C). Conversely, the transcript level of *rca-1* did not show any remarkable changes in Δ*clr-1/2*, Δ*xlr-1*, or Δ*cre-1* mutants versus WT [20,21,33]. All these observations suggest that *rca-1* affects the expression of plant polysaccharide-degrading enzyme genes by upstream regulation/modulation of the cellulose regulator *clr-2* when grown on plant biomass. Additionally, the double mutant F1-Δ*rca-1*; Δ*cre-1* (Additional file 10: Figure S4) had approximately two folds higher lignocellulolytic enzyme

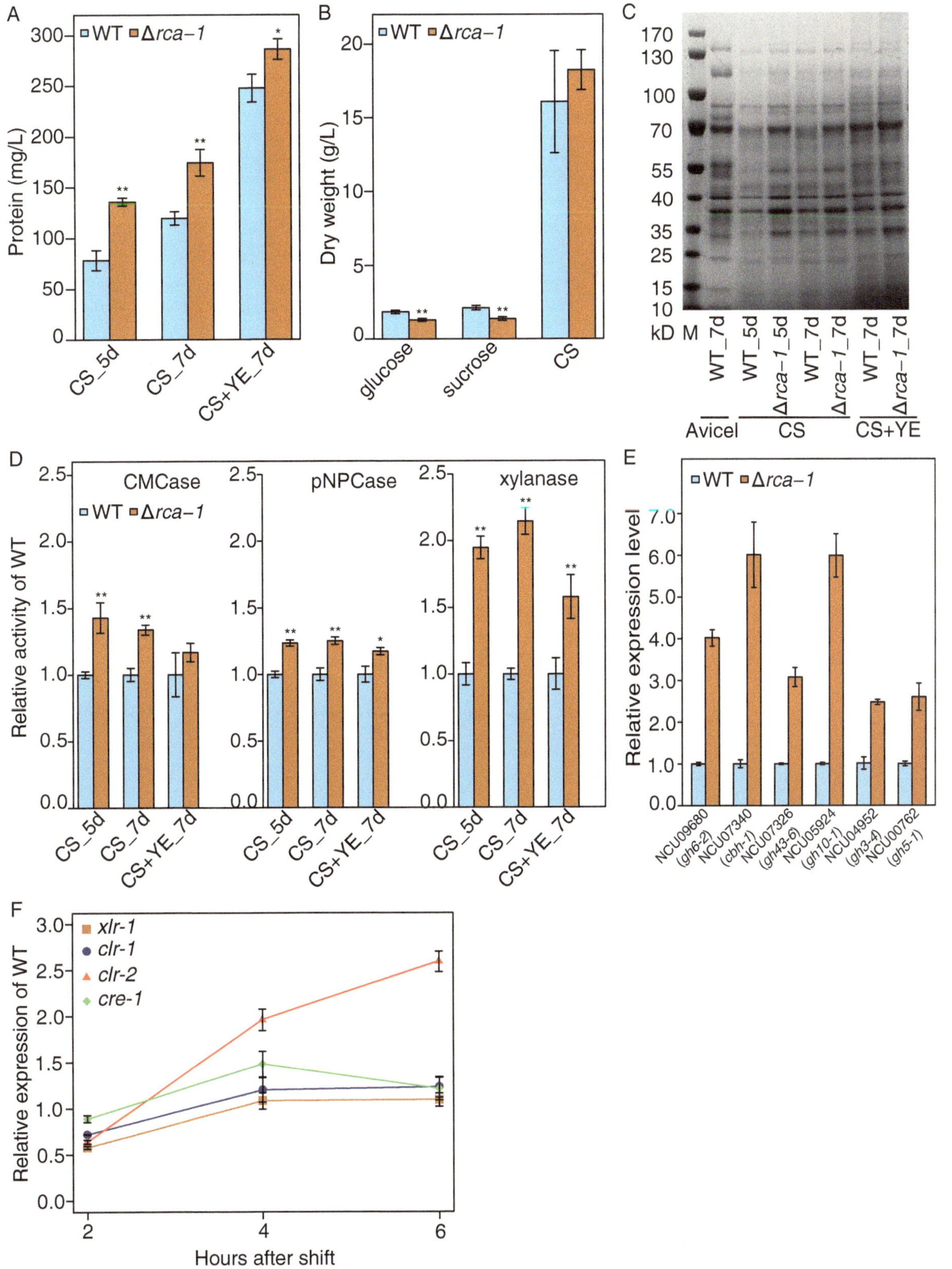

Figure 8 (See legend on next page.)

(See figure on previous page.)
Figure 8 Phenotype of mutant Δ*rca-1* versus wild type (WT) grown on corn straw. CS, corn straw; M, protein marker; YE, yeast extract. **(A-D)** Secreted proteins, mycelial dry weight (mycelia grown on 2% glucose, or sucrose for 16 h, and 2% corn straw for 7 days), SDS-PAGE, and enzymatic activities, including cellulose hydrolysis activity (CMCase), exo-glucanase activity (pNPCase), and xylanase activity, of Δ*rca-1* versus WT when grown for the indicated times. **(E)** Induction of selected CAZy genes in WT and Δ*rca-1* mutant after transfer 16-h sucrose grown mycelia to corn straw for 4 h. **(F)** Time course of relative expression levels of plant cell wall-degrading associated regulators in Δ*rca-1* mutant versus WT after transfer from 16-h sucrose culture to corn straw for the indicated times. Values represent the means of at least three biological replicates; error bars show standard deviation. Statistical significance was performed using a two-tailed Student's *t*-test. $^*P < 0.05$; $^{**}P < 0.01$.

production than that of Δ*rca-1* (Figure 9) and significantly higher pNPCase activity than that of Δ*cre-1* (Figure 9A), suggesting a possible novel strategy of strain improvement for lignocellulase production by co-disruption of *rca-1* and *cre-1*.

Discussion

Despite some differences in chemical composition and organization among five major crop residues, especially the dicotyledonous soybean straw, which was slightly more distinct than the other four monocot straws in composition as well as transcriptional response in *N. crassa*, transcriptome profiling showed a large conserved core response to these straws in the model cellulolytic fungus *N. crassa*, including a conserved CAZome. Of 430 core genes responding to growth on plant biomass, a 252-gene set was identified as the BR. Many genes (65%) in this set overlapped with either the Avicel, xylan, or pectin regulons, including cellulase genes, hemicellulase genes, pectinase genes, and LPMO genes (Additional file 7: Table S6, sheet 2), indicating that the polysaccharide block organization of plant straws extensively induced the *N. crassa* CAZome to degrade complex lignocellulosic substrates. Previous work [22] has identified a 29-gene set commonly induced by all three polysaccharides and suggested that this gene pool could be used for carbon scouting under any carbon-inducing condition. Twenty-four (82%) of these genes were also induced by crop straws in the present study (Figure 6A), supporting the hypothesis that a conserved gene set is involved in carbon scouting in *N. crassa*. A conserved response to crop straws increases the feasibility of universal plant biomass-based biorefinery by fungi, as in DMC. Further investigation of this core response during the plant biomass DMC process may offer potential targets for engineering

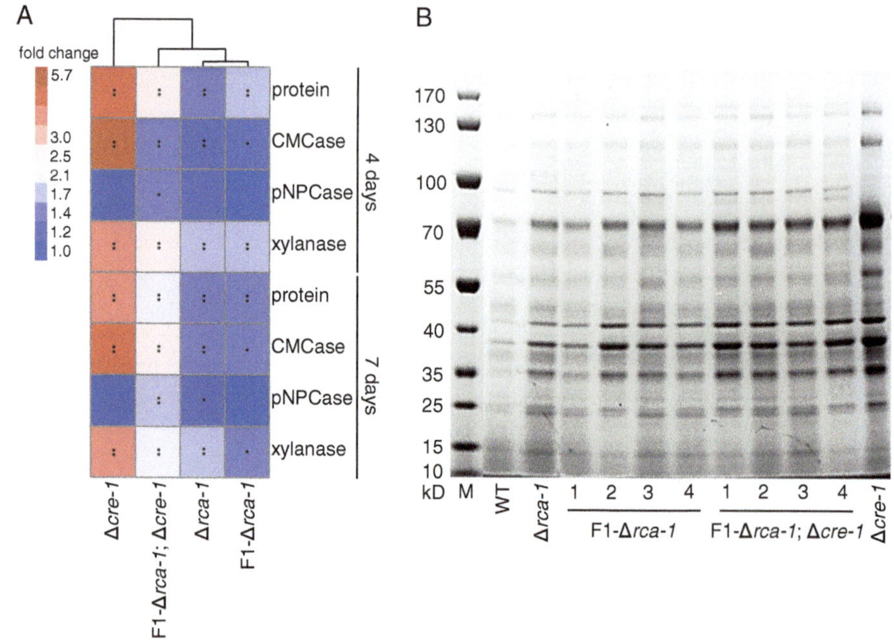

Figure 9 Phenotype of Δ*rca-1* × Δ*cre-1* progeny. F1-Δ*rca-1*, *rca-1* deletion progeny; F1-Δ*rca-1*; Δ*cre-1*, *rca-1* and *cre-1* double deletion progeny; M, protein marker. **(A)** Heat map of secreted proteins, cellulose hydrolysis activity (CMCase), exo-glucanase activity (pNPCase), and xylanase activity of progenies, Δ*rca-1*, and Δ*cre-1* grown on corn straw for 4 and 7 days compared with wild type (WT). Relative activities versus WT are displayed by color codes. Statistical significance was performed using a two-tailed Student's *t*-test. $^*P < 0.05$; $^{**}P < 0.01$. **(B)** SDS-PAGE of secreted proteins of progenies, Δ*rca-1*, Δ*cre-1*, and WT when grown on corn straw for 4 days. Values represent the means of at least three biological replicates.

strains to increase production of cell wall-degrading enzymes and the end-product conversion rate.

Eighty-eight genes were upregulated only on plant biomass, rather than on Avicel, xylan, or pectin [20,22]. Athough most genes in this set encoded unknown or unclassified proteins, they could be a resource for identifying novel components involved in lignocellulose deconstruction [17]. For example, the sporulation regulator *rca-1* (NCU01312) screened from this gene set dramatically increased lignocellulase production in *N. crassa* when grown on plant biomass but not on Avicel. Previous studies show that *rca-1* can complement *flbD* deletion in *A. nidulans*, but deletion had no detectable sporulation defect in *N. crassa* [32]. Intriguingly, loss of the *flbD* downstream conidiation regulator *brlA* in *Penicillium decumbens* not only blocked conidiation but also significantly improved cellulase production when grown on cellulose supplemented with wheat bran [35]. A bioinformatics search predicted that many sequences upstream of CAZy genes harbor *BrlA*-binding sites in *P. decumbens* [35], but evidence for a direct binding interaction has not yet been revealed. The RCA-1 regulator is a MYB family member that contains two imperfect repeats [32] for binding the *cis*-element 5′-AACTGNCW-3′ [36]. Searching the upstream 1-kb promoters of 225 CAZy genes by PATSER in RSAT [37] yielded 51 promoter regions containing candidate Myb-binding sites, but most harbored only one or two sites, with the exception of NCU07035 (Additional file 11: Table S7). The gene NCU07035, a member of glycosyl hydrolase family 18 (mainly representing chitinase), showed elevated expression levels when grown on all five tested crop residues (Additional file 3: Table S2) and also on Avicel when grown for 30 h [12]. Although the transcript of NCU07035 was not highly expressed, the role of its family in *N. crassa* grown on lignocellulose needs to be further investigated.

Examination of the expression levels of lignocellulolytic regulators *xlr-1*, *clr-1/2*, and *cre-1* showed that *clr-2* was upregulated in the *rca-1* deletion mutant compared with WT, indicating that the mutant might enhance the expression of biomass-degrading genes via de-repression of *clr-2*. Cellulolytic transcription factors *clr-1/2* were necessary to fully induce all major cellulase and some major hemicellulase genes in *N. crassa* [20]. Notably, induction of *clr-2* is dependent on *clr-1* in *N. crassa*. Therefore, RCA-1 might repress *clr-2* expression by directly regulating and (or) interacting with CLR-1 to modulate its induction, although *clr-1* was not suppressed by RCA-1 at the mRNA level (Figure 8F). Additionally, deletion of *rca-1* led to slower growth on glucose or sucrose than in the WT (Figure 8B) and also observed that *rca-1* was de-repressed when extracellular glucose levels increased (unpublished data), suggesting that this sporulation regulator plays a role in carbon or glucose metabolism/signaling. However, how this sporulation regulator in *N. crassa* affects cell growth in carbon-depleted or -plentiful environment requires further investigation of the correlation between colony survival and sporulation as well as biochemical identification of the targets of conidiation regulators in *N. crassa*.

Given that *rca-1* influences production of plant biomass-degrading enzymes on a wide spectrum of biomass substrates, *rca-1* might be an advantageous target for engineering the biorefinery of plant biomass through DMC. Functional characterization of BUS genes is a potential strategy to gain a more complete view of *N. crassa* growth on plant biomass. Further engineering of growth properties on plant biomass in conjunction with DMC product yield based on BUS might facilitate the fungal bioconversion process.

Conclusions

Transcriptional profiling of *N. crassa* grown on different plant biomass revealed a conserved core gene set for plant cell wall utilization, including a conserved pool of CAZy genes. Eighty-eight genes (the biomass unique set, BUS) were specifically induced by plant biomass but not by Avicel, xylan, or pectin. Deletion of one of these genes, the sporulation transcription factor *rca-1*, was identified as beneficial for the production of lignocellulases on plant straw and might be a potential target to engineer strains for biomass-based biorefinery through direct microbial conversion technology.

Methods

Strains, media, and culture conditions

The *N. crassa* WT strain (FGSC 2489) and the *rca-1* deletion strain (NCU01312, FGSC 11209) were obtained from FGSC [38]. Mutant Δ*cre-1* was a gift from the laboratory of Professor N. Louise Glass. For RNA-Seq preparation, conidia of *N. crassa* WT were inoculated at 10^6 conidia/mL into 100 mL 1× Vogel's salts (minimal medium) with 2% (*w*/*v*) sucrose for 16 h or with 2% (*w*/*v*) of the indicated ground crop-straw carbon sources for 30 h and grown at 25°C with constant light shaking at 200 rpm. The 50× Vogel's salts were prepared as previously described [39]. The straws included BS, CS, RS, SS, and WS were harvested from Anhui Province, China, and mechanically ground. Mycelia grown for 30 h on various straws were photographed under a microscope (OLYMPUS, Japan).

For protein assays, 2 mL of culture supernatants were collected at each time point, centrifuged at 15,294 × *g* for 8 min to remove mycelia and stored at 4°C for analysis within 1 day or at −20°C for SDS-PAGE. For media shift experiments, *N. crassa* cultures were first grown on 100 mL 1× Vogel's salts supplemented with 2% (*w*/*v*) sucrose for 16 h, and mycelia were filtered through six layers of gauze and immediately washed with sterilized water at least five times. Then, mycelia were transferred into new media containing 100 mL 1× Vogel's salts with 2% (*w*/*v*) ground straws for the indicated time (2, 4, and 6 h) at 25°C, 200 rpm.

For transcription factor mutant screening, conidia were inoculated at 10^6 conidia/mL into a 250-mL flask with 100 mL 1× Vogel's minimal medium with 2% ground corn straw as the sole carbon source then cultured for 7 days at 25°C, with shaking at 200 rpm under constant light.

RNA extraction, sequencing, and data analysis

Cultured mycelia were harvested after the indicated growth time via filtration and immediately frozen in liquid nitrogen. Total RNA from frozen samples was isolated with TRIzol reagent (Invitrogen, Carlsbad, CA, USA) as previously described [12] and further treated with DNase I (RNeasy Mini Kit, QIAGEN, Hilden, Germany). RNA concentration was measured with a Nanodrop 2000c (Thermo Scientific, Waltham, MA, USA), and RNA integrity was checked by agarose gel electrophoresis. The total RNA of two biological replicate samples was extracted separately, mixed together after measuring the quality of each sample, and used for high-throughput RNA sequencing.

The qualified RNA was prepared using the standard protocol from Shenzhen BGI (Shenzhen, China) and sequenced on the Illumina HiSeq™ 2000 platform (San Diego, CA, USA). Prior to reads mapping, adaptors and low quality reads were removed based on the BGI standard process (www.genomics.cn). Total clean reads were mapped against predicted transcripts from the *N. crassa* OR74A genome (version 12) [40] using TopHat (version 2.0.8b) [41] with at most two mismatches. Raw counts of reads mapped to unique exons were calculated by HTSeq (version 0.6.0) [42] and used for both normalizing transcript abundance (RPKM, reads per kilobase per million mapped reads [43]) and differential gene expression analysis using GFOLD (version 1.1.0) [27] and DEGseq [28]. Genes with GFOLD ≥ 1.6 ($\log_2$ fold change ≥ 1.6) and DEGseq *P* value < 1e-4 were considered significantly differentially expressed between growth conditions. RNA-Seq raw data are available at the Gene Expression Omnibus under accession number GSE60986. The GFOLD values of genes commonly upregulated by the five crop residues were hierarchically clustered with HCE software (version 3.5) [30]. The complete linkage method with Euclidean distance as the similarity metric was used for cluster generation.

For PCA, the RNA-Seq reads of *N. crassa* grown on sucrose, BS, CS, RS, SS, and WS were the input data of DESeq [44], which calculated the eigenvectors for each sample by eigenvalue decomposition of the corresponding sample covariance matrix. The first two principle components of each eigenvector were plotted by it [44], with each point representing a biological sample. This type of analysis is useful for determining of sample-to-sample distances and batch effects [44]. Although PCA represents the total variance, the PCA plot displayed only the most important component of each sample.

Quantitative real-time PCR

qRT-PCR was performed using the iScript cDNA Synthesis Kit and IQ SYBR Green Supermix (Bio-Rad, Hercules, CA, USA) or SYBR Green Realtime PCR Master Mix (TOYOBO, Osaka, Japan) according to the manufacturers' instructions on a CFX96 real-time PCR detection system (Bio-Rad). Each reaction was done in triplicate. The actin gene (NCU04173) was used as an endogenous control for all experiments. All primers used in this study are listed in Additional file 12: Table S8. Relative expression level of each gene was calculated using the Livak method ($2^{-\Delta\Delta Ct}$).

Measurement of crop residue composition

Agricultural crop residue composition was measured according to the analytical procedure of the National Renewable Energy Laboratory (NREL; http://www.nrel.gov/). Briefly, 300 ± 1 mg biomass powder was transferred to a capped vessel with the addition of 3 mL 72% (*w*/*w*) sulfuric acid and incubated at 30°C in a water bath for 1 h. During this time, the vessel was vortexed at a maximum speed of 5 s every 10 min. After incubation, 84 mL deionized water was added, and the mixture was carefully transferred to a capped bottle. The mixture was then autoclaved for 1 h at 121°C. The hemicellulosic glucose analysis was previously described [45]. Briefly, 50 ± 1 mg biomass was shortly mixed with 14.5 mL of 4% (*w*/*w*) sulfuric acid before autoclaving. Finally, the concentration of monosaccharides was determined by high-performance liquid chromatography (HPLC; Waters e2695, Manchester, UK) using a Waters 2414 refractive index detector and Aminex HPX-87H Column (Bio-Rad). Elution was performed at 63°C with 5 mM sulfuric acid at a flow rate of 0.6 mL/min. The content of cellulose and hemicellulose in the crop residues was calculated by the following formulas: %cellulose $(w/w) = [(C_{\text{total glucose}} - C_{\text{hemicellulosic glucose}}) \times 86.73 \times 0.9 \times 100]/300$, %hemicellulose $(w/w) = [(C_{\text{xylose}} + C_{\text{arabinose}}) \times 86.73 \times 0.88 \times 100 + C_{\text{hemicellulosic glucose}} \times 86.73 \times 0.9 \times 100]/300$, where, $C_{\text{monosaccharide}}$ is the concentration (mg/mL) of a sugar as determined by HPLC.

For acid insoluble lignin, the precipitate of the autoclaved mixture above was dried at 75°C until constant weight then ashed using a muffle furnace with a ramping program at 575°C for 3 h. The sample was weighed again after cooling. The content of acid insoluble lignin was calculated by the formula: %acid insoluble lignin $(w/w) = (W_{precipitation} - W_{acid\text{-}insoluble\ ash}) \times 100/300$. The absorbance at 320 nm of the hydrolysis supernatant was measured with a UV-1800 spectrophotometer (Shimadzu, Kyoto, Japan) and used to calculate acid soluble lignin by the formula: %acid soluble lignin $(w/w) = OD_{320} \times 86.73 \times 100/30/300$.

For ash determination, 600 ± 1 mg dried biomass powder was ashed using a muffle furnace at 575°C for 3 h. The content was calculated by the formula: %ash $(w/w) = W_{ash} \times 100/600$.

Enzyme activity and dry weight assays

The total extracellular protein content was determined using the Bradford (Bio-Rad) method with BSA as a standard. The endo-glucanase and endo-xylanase activities were measured using the azo-CMC/xylan kit (Megazyme, Wicklow, Ireland) according to the manufacturer's instructions. The exo-glucanase activity was assessed using *p*-nitrophenol-D-cellobioside (pNPC; Sigma-Aldrich) as the substrate. Briefly, a 250-μL culture supernatant that was diluted with 50 mM sodium citrate (pH 4.8) was added to 250 μL of 1 mg/mL pNPC, then incubated immediately for 10 min at 50°C. The reaction mixture was terminated by adding 500 μL of 1 M sodium carbonate and determined at 420 nm by a UV-1800 spectrophotometer (Shimadzu). One unit (U) of exo-glucanase activity was defined as the amount of enzyme that liberates 1 μmol of pNP per min.

Mycelia grown on sucrose or glucose for 16 h were harvested, dried, and weighed. Biomass dry weight of corn straw cultures was indirectly measured as previously described [46]. Briefly, a 5-mL culture broth was centrifuged at 3,220 × *g* for 5 min, the supernatant was discarded, and 3 mL acetic acid (80%, *v*/*v*):nitrate (10:1, *v*/*v*) reagent was added to solubilize fungal biomass by boiling in water for 2 h. The reaction mixture was then centrifuged, dried, and weighed. Mycelial dry weight was defined as the dry weight of the original 5-mL culture minus that of the reaction mixture.

Statistical significance tests and data plotting

Statistical significance tests between two conditions were performed using a two-tailed Student's *t*-test. Enrichment analysis involved the identification of one functional category or gene group that was overrepresented in a whole gene collection. Significant enrichment was analyzed by a one-tailed Fisher's exact test (http://www.langsrud.com/stat/fisher.htm) unless otherwise indicated. For all tests, $^*P < 0.05$, $^{**}P < 0.01$. Data were plotted using the R program platform (http://www.r-project.org/). Venn diagrams were constructed with a web tool from Ghent University (http://bioinformatics.psb.ugent.be/webtools/Venn/).

Additional files

Additional file 1: Table S1. Chemical composition of five crop residues. BS, barley straw; CS, corn straw; RS, rice straw; SS, soybean straw; WS, wheat straw. Values represent the mean ± standard deviation of at least three replicates. [a]The percentages of three monosaccharides in each residual hemicellulose are shown in brackets in the order hemicellulosic glucose, xylose, and arabinose.

Additional file 2: Figure S1. Phenotype of *Neurospora crassa* grown on five crop residues (2%, *w/v*) for 30 h and 2% sucrose for 16 h. Scale bar = 100 um.

Additional file 3: Table S2. Gene expression profiles and differential expression analysis of RNA-Seq data. For all genes RPKM, $\log_2$ fold change, and *P* value on each crop residue versus sucrose were calculated.

Additional file 4: Table S3. Functional category analysis (FunCat) of up/downregulated genes. Sheet 1: functional category analysis (FunCat) of upregulated genes excluding the 430-gene biomass commonly induced core set; sheet 2: functional category analysis (FunCat) of downregulated genes.

Additional file 5: Table S4. Hierarchical clustering and functional category analysis of 430 biomass commonly induced core set genes. Sheet 1: clustering analysis of 430-gene biomass commonly induced core set (BICS); sheet 2: functional category analysis (FunCat) of 430-gene BICS.

Additional file 6: Table S5. Functional classification of biomass and polysaccharide regulons. Sheet 1: regulons induced by various lignocellulosic substrates in *N. crassa*; sheet 2: functional classification of lignocellulolytic regulons.

Additional file 7: Table S6. CAZome elicited by five crop residues. Sheet 1: the CAZome of *N. crassa* genome; sheet 2: upregulated CAZy genes of *N. crassa* induced by various lignocellulosic substrates.

Additional file 8: Figure S2. Phenotype of mutant Δ*rca-1* versus wild type (WT) grown on three polysaccharides, Avicel for 7 days, xylan for 4 days, and pectin for 4 days. (A)–(D) Relative levels of secreted proteins, cellulose hydrolysis activity (CMCase), exo-glucanase activity (pNPCase), and xylanase activity of Δ*rca-1* versus WT grown on the indicated carbon source. Values represent the means of three replicates; error bars show standard deviation.

Additional file 9: Figure S3. Phenotype of mutant Δ*rca-1* versus WT grown on four non-corn crop residues. (A) Relative levels of secreted proteins and xylanase activities of Δ*rca-1* versus WT grown on indicated crop straws for 4 days. (B) SDS-PAGE of secreted proteins of WT and Δ*rca-1*. (C) Induction of tested CAZy genes and lignocellulolytic regulators in WT and Δ*rca-1* mutant after transfer to soybean straw for 4 h from 16-h sucrose grown culture. Values represent the means of three replicates; error bars show standard deviation.

Additional file 10: Figure S4. Genotypes of mutant Δ*rca-1* × Δ*cre-1* progeny identified by PCR and 1.5% (*w/v*) agarose gel electrophoresis. F1-Δ*rca-1*, the Δ*rca-1* progeny; F1-Δ*rca-1*; Δ*cre-1*, the *rca-1* and *cre-1* double deletion progeny; *hph*, hygromycin phosphotransferase gene; M, oligo-nucleotide marker. For each isolate, from left to right, PCR products present the genes *cre-1*, *hph* in *cre-1* locus, *rca-1*, and *hph* in *rca-1* locus. If one isolate had DNA bands only in the lanes "cre-1" and "rca-1", it was wild type. Conversely, if one isolate had DNA bands in the lanes "cre-1-hph" and "rca-1-hph" and no bands in the "cre-1" or "rca-1" lanes, it was an F1-Δ*rca-1*; Δ*cre-1* progeny. If it only showed bands in "cre-1" and "rca-1-hph" lanes, it was an F1-Δ*rca-1* progeny.

Additional file 11: Table S7. Myb-binding site prediction in the upstream 1-kb promoter region of the CAZome of *N. crassa*. "F" represents forward strand; "C" represents complement strand.

Additional file 12: Table S8. Primers used in this study.

Abbreviations

AA: Auxiliary activity family; BICS: Biomass commonly induced core set; BR: Biomass regulon; BS: Barley straw; BSS: Biomass shared set; BUS: Biomass unique set; CAZome: Carbohydrate-active enzyme proteome; CAZy: Carbohydrate-active enzyme; CE: Carbohydrate esterase; CMCase: Carboxy-methyl-cellulose enzyme; CS: Corn straw; DMC: Direct microbial conversion; FGSC: Fungal Genetics Stock Center; GEO: Gene expression omnibus; GH: Glycoside hydrolase; HPLC: High-performance liquid chromatography; LPMO: Lytic polysaccharide monooxygenase; PCA: Principle component analysis; pNPC: *p*-nitrophenol-D-cellobioside; pNPCase: *p*-nitrophenol-D-cellobioside enzyme; PUS: Polysaccharides unique set; qRT-PCR: Quantitative real-time reverse transcription polymerase chain reaction; RPKM: Reads per kilobase per million mapped reads; RS: Rice straw; SDS-PAGE: Sodium dodecyl sulfate polyacrylamide gel electrophoresis; SS: Soybean straw; WT: Wild type; YE: Yeast extract.

Competing interests

The authors declare that they have no competing interests.

Authors' contributions

BW, PC, CT, and YM conceived and designed the project. BW, PC, and WS performed the experiments. BW and CT analyzed the data. BW, PC, WS, and CT wrote the manuscript. JL participated in the bioinformatics analysis. All authors have read and approved the final manuscript.

Acknowledgements

We acknowledge Professor N. Louise Glass for the generous gift of the Δ*cre-1* deletion strain. We also thank Dr. Liangcai Lin for the critical discussions and Dr. Le Gao and Mr. Yong Chen for the technical assistance. This work was financially supported by the 973 Program of China (2011CB707403), the 863 Program of China (SS2014AA021301), the National Natural Science Foundation of China (31171207), and Tianjin Zhuanxiang (11ZCZDSY08400, 13ZCDZSY05000).

Author details

[1]Key Laboratory of Systems Microbial Biotechnology, Tianjin Institute of Industrial Biotechnology, Chinese Academy of Sciences, Tianjin 300308, China. [2]University of Chinese Academy of Sciences, Beijing 100049, China.

References

1. Kubicek CP. Fungi and lignocellulosic biomass. New York: John Wiley & Sons; 2012.
2. Carroll A, Somerville C. Cellulosic biofuels. Annu Rev Plant Biol. 2009;60:165–82.
3. Sarkar N, Ghosh SK, Bannerjee S, Aikat K. Bioethanol production from agricultural wastes: an overview. Renew Energ. 2012;37:19–27.
4. Menon V, Rao M. Trends in bioconversion of lignocellulose: biofuels, platform chemicals & biorefinery concept. Prog Energ Combust. 2012;38:522–50.
5. Popper ZA, Michel G, Hervé C, Domozych DS, Willats WGT, Tuohy MG, et al. Evolution and diversity of plant cell walls: from algae to flowering plants. Annu Rev Plant Biol. 2011;62:567–90.
6. Alvira P, Tomás-Pejó E, Ballesteros M, Negro MJ. Pretreatment technologies for an efficient bioethanol production process based on enzymatic hydrolysis: a review. Bioresour Technol. 2010;101:4851–61.
7. Parisutham V, Kim TH, Lee SK. Feasibilities of consolidated bioprocessing microbes: from pretreatment to biofuel production. Bioresour Technol. 2014;161:431–40.
8. Lynd LR, Zyl WH, McBride JE, Laser M. Consolidated bioprocessing of cellulosic biomass: an update. Curr Opin Biotechnol. 2005;16:577–83.
9. Roche CM, Glass NL, Blanch HW, Clark DS. Engineering the filamentous fungus *Neurospora crassa* for lipid production from lignocellulosic biomass. Biotechnol Bioeng. 2014;111:1097–107.
10. Xu Q, Singh A, Himmel ME. Perspectives and new directions for the production of bioethanol using consolidated bioprocessing of lignocellulose. Curr Opin Biotechnol. 2009;20:364–71.
11. Kolbusz MA, Di Falco M, Ishmael N, Marqueteau S, Moisan M-C, Baptista CS, et al. Transcriptome and exoproteome analysis of utilization of plant-derived biomass by Myceliophthora thermophila. Fungal Genet Biol. 2014;72:10–20.
12. Tian C, Beeson WT, Iavarone AT, Sun J, Marletta MA, Cate JHD, et al. Systems analysis of plant cell wall degradation by the model filamentous fungus *Neurospora crassa*. Proc Natl Acad Sci U S A. 2009;106:22157–62.
13. Delmas S, Pullan ST, Gaddipati S, Kokolski M, Malla S, Blythe MJ, et al. Uncovering the genome-wide transcriptional responses of the filamentous fungus *Aspergillus niger* to lignocellulose using RNA sequencing. PLoS Genet. 2012;8:e1002875.
14. Berka RM, Grigoriev IV, Otillar R, Salamov A, Grimwood J, Reid I, et al. Comparative genomic analysis of the thermophilic biomass-degrading fungi *Myceliophthora thermophila* and *Thielavia terrestris*. Nat Biotech. 2011;29:922–7.
15. Bischof R, Fourtis L, Limbeck A, Gamauf C, Seiboth B, Kubicek CP. Comparative analysis of the *Trichoderma reesei* transcriptome during growth on the cellulase inducing substrates wheat straw and lactose. Biotechnol Biofuels. 2013;6:127.
16. Ries L, Pullan S, Delmas S, Malla S, Blythe M, Archer D. Genome-wide transcriptional response of Trichoderma reesei to lignocellulose using RNA sequencing and comparison with *Aspergillus niger*. BMC Genomics. 2013;14:541.
17. Znameroski EA, Glass NL. Using a model filamentous fungus to unravel mechanisms of lignocellulose deconstruction. Biotechnol Biofuels. 2013;6:6.
18. Colot HV. A high-throughput gene knockout procedure for *Neurospora* reveals functions for multiple transcription factors. Proc Natl Acad Sci U S A. 2006;103:10352–7.
19. Dunlap JC, Borkovich KA, Henn MR, Turner GE, Sachs MS, Glass NL, et al. Enabling a community to dissect an organism: overview of the *Neurospora* functional genomics project. Adv Genet. 2007;57:49–96.
20. Coradetti ST, Craig JP, Xiong Y, Shock T, Tian C, Glass NL. Conserved and essential transcription factors for cellulase gene expression in ascomycete fungi. Proc Natl Acad Sci U S A. 2012;109:7397–402.
21. Sun J, Tian C, Diamond S, Glass NL. Deciphering transcriptional regulatory mechanisms associated with hemicellulose degradation in *Neurospora crassa*. Eukaryot Cell. 2012;11:482–93.
22. Benz JP, Chau BH, Zheng D, Bauer S, Glass NL, Somerville CR. A comparative systems analysis of polysaccharide-elicited responses in *Neurospora crassa* reveals carbon source-specific cellular adaptations. Mol Microbiol. 2014;91:275–99.
23. Glass NL, Schmoll M, Cate JH, Coradetti S. Plant cell wall deconstruction by ascomycete fungi. Annu Rev Microbiol. 2013;67:477–98.
24. Gang L, Lei S. Quantitative appraisal of biomass energy and its geographical distribution in China. J Nat Resour. 2007;22:9–19.
25. Kim S, Dale BE. Global potential bioethanol production from wasted crops and crop residues. Biomass Bioenerg. 2004;26:361–75.
26. Cantarel BL, Coutinho PM, Rancurel C, Bernard T, Lombard V, Henrissat B. The Carbohydrate-Active EnZymes database (CAZy): an expert resource for glycogenomics. Nucleic Acids Res. 2009;37:D233–8.
27. Feng J, Meyer CA, Wang Q, Liu JS, Liu XS, Zhang Y. GFOLD: a generalized fold change for ranking differentially expressed genes from RNA-seq data. Bioinformatics. 2012;28:2782–8.
28. Wang L, Feng Z, Wang X, Wang X, Zhang X. DEGseq: an R package for identifying differentially expressed genes from RNA-seq data. Bioinformatics. 2010;26:136–8.
29. Ruepp A, Zollner A, Maier D, Albermann K, Hani J, Mokrejs M, et al. The FunCat, a functional annotation scheme for systematic classification of proteins from whole genomes. Nucleic Acids Res. 2004;32:5539–45.
30. Seo J, Bakay M, Chen Y-W, Hilmer S, Shneiderman B, Hoffman EP. Interactively optimizing signal-to-noise ratios in expression profiling: project-specific algorithm selection and detection p-value weighting in affymetrix microarrays. Bioinformatics. 2004;20:2534–44.
31. Levasseur A, Drula E, Lombard V, Coutinho PM, Henrissat B. Expansion of the enzymatic repertoire of the CAZy database to integrate auxiliary redox enzymes. Biotechnol Biofuels. 2013;6:41.
32. Shen W-C, Wieser J, Adams TH, Ebbole DJ. The *Neurospora rca-1* gene complements an *aspergillus flbD* sporulation mutant but has no identifiable role in *Neurospora* sporulation. Genetics. 1998;148:1031–41.
33. Sun J, Glass NL. Identification of the CRE-1 Cellulolytic regulon in Neurospora crassa. PLoS ONE. 2011;6:e25654.
34. Wieser J, Adams TH. *flbD* encodes a Myb-like DNA-binding protein that coordinates initiation of *Aspergillus nidulans* conidiophore development. Genes Dev. 1995;9:491–502.
35. Qin Y, Bao L, Gao M, Chen M, Lei Y, Liu G, et al. *Penicillium decumbens* BrlA extensively regulates secondary metabolism and functionally associates with

the expression of cellulase genes. Appl Microbiol Biotechnol. 2013;97:10453–67.
36. Mathelier A, Zhao X, Zhang AW, Parcy F, Worsley-Hunt R, Arenillas DJ, et al. JASPAR 2014: an extensively expanded and updated open-access database of transcription factor binding profiles. Nucleic Acids Res. 2014;42:D142–7.
37. Thomas-Chollier M, Defrance M, Medina-Rivera A, Sand O, Herrmann C, Thieffry D, et al. RSAT 2011: regulatory sequence analysis tools. Nucleic Acids Res. 2011;39:W86–91.
38. McCluskey K. The Fungal Genetics Stock Center: from molds to molecules. Adv Appl Microbiol. 2003;52:245–62.
39. Vogel HJ. A convenient growth medium for *Neurospora*. Microb Genet Bull. 1956;13:42–6.
40. Galagan JE, Calvo SE, Borkovich KA, Selker EU, Read ND, Jaffe D, et al. The genome sequence of the filamentous fungus *Neurospora crassa*. Nature. 2003;422:859–68.
41. Trapnell C, Roberts A, Goff L, Pertea G, Kim D, Kelley DR, et al. Differential gene and transcript expression analysis of RNA-seq experiments with TopHat and Cufflinks. Nat Protoc. 2012;7:562–78.
42. Anders S, Pyl PT, Huber W. HTSeq - a Python framework to work with high-throughput sequencing data. Bioinformatics. 2015;31:166–9.
43. Mortazavi A, Williams BA, McCue K, Schaeffer L, Wold B. Mapping and quantifying mammalian transcriptomes by RNA-Seq. Nat Methods. 2008;5:621–8.
44. Anders S, Huber W. Differential expression analysis for sequence count data. Genome Biol. 2010;11:R106.
45. Bauer S, Ibáñez AB. Rapid determination of cellulose. Biotechnol Bioeng. 2014;111:2355–7.
46. Updegraff DM. Semimicro determination of cellulose in biological materials. Anal Biochem. 1969;32:420–4.

8

Impact of two ionic liquids, 1-ethyl-3-methylimidazolium acetate and 1-ethyl-3-methylimidazolium methylphosphonate, on *Saccharomyces cerevisiae*: metabolic, physiologic, and morphological investigations

Nasir Mehmood[1], Eric Husson[1], Cédric Jacquard[2], Sandra Wewetzer[3], Jochen Büchs[3], Catherine Sarazin[1] and Isabelle Gosselin[1*]

Abstract

Background: Ionic liquids (ILs) are considered as suitable candidates for lignocellulosic biomass pretreatment prior enzymatic saccharification and, obviously, for second-generation bioethanol production. However, several reports showed toxic or inhibitory effects of residual ILs on microorganisms, plants, and animal cells which could affect a subsequent enzymatic saccharification and fermentation process.

Results: In this context, the impact of two hydrophilic imidazolium-based ILs already used in lignocellulosic biomass pretreatment was investigated: 1-ethyl-3-methylimidazolium acetate [Emim][OAc] and 1-ethyl-3-methylimidazolium methylphosphonate [Emim][MeO(H)PO_2]. Their effects were assessed on the model yeast for ethanolic fermentation, *Saccharomyces cerevisiae*, grown in a culture medium containing glucose as carbon source and various IL concentrations. Classical fermentation parameters were followed: growth, glucose consumption and ethanol production, and two original factors: the respiratory status with the oxygen transfer rate (OTR) and carbon dioxide transfer rate (CTR) of yeasts which were monitored online by respiratory activity monitoring systems (RAMOS). In addition, yeast morphology was characterized by environmental scanning electron microscope (ESEM).

The addition of ILs to the growth medium inhibited the OTR and switched the metabolism from respiration (conversion of glucose into biomass) to fermentation (conversion of glucose to ethanol). This behavior could be observed at low IL concentrations ($\leq$5% IL) while above there is no significant growth or ethanol production. The presence of IL in the growth medium also induced changes of yeast morphology, which exhibited wrinkled, softened, and holed shapes. Both ILs showed the same effects, but [Emim][MeO(H)PO_2] was more biocompatible than [Emim][OAc] and could be better tolerated by *S. cerevisiae*.

Conclusions: These two imidazolium-derived ILs were appropriate candidates for useful pretreatment of lignocellulosic biomass in the context of second-generation bioethanol production. This fundamental study provides additional information about the toxic effects of ILs. Indeed, the investigations highlighted the better tolerance by *S. cerevisiae* of [Emim][MeO(H)PO_2] than [Emim][OAc].

Keywords: *Saccharomyces cerevisiae*, Ionic liquid, Second-generation bioethanol, Respirofermentative metabolism, Environmental scanning electronic microscopy

* Correspondence: isabelle.gosselin@u-picardie.fr
[1]Unité Génie Enzymatique et Cellulaire, FRE-CNRS 3580, Université de Picardie Jules Verne, 33 rue Saint-Leu, 80039 Amiens Cedex, France
Full list of author information is available at the end of the article

Background

Depletion of fossil fuels, excess of greenhouse gas, and global planet warming require the exploration of renewable energies that are sustainable from economical, ecological, and environmental points of view [1]. Lignocellulosic plant biomass is generally considered as the most promising renewable feedstock for bioproduction of transportation fuels and commodity chemicals, without alimentary competition [2,3]. Lignocellulosic biomass is composed of three main polymers: cellulose, hemicellulose, and lignin which are tightly linked and organized [1]. For that reason, the bioethanol production from lignocellulosic biomass begins with a pretreatment step to increase cellulose accessibility and digestibility by the cellulolytic enzymes into glucose monomers (improving the saccharification step), followed by an ethanolic fermentation and ends by a distillation step of bioethanol [4].

Several types of pretreatment exist, classified into four categories: chemical, physical, physicochemical, and biological. Chemical pretreatments are the most commonly used for lignocellulosic biomass and include alkali or acid pretreatments, ozonolysis, and organosolv process [1,4].

From the last decade, ionic liquids (ILs) have paid attention in the pretreatment of lignocellulosic biomass under mild conditions with improved yields of reducing sugars [4-7]. ILs are salts typically composed of organic cations and organic or inorganic anions, which exist as liquids at temperatures below 100°C, often at room temperature. They are non-volatile, non-flammable, and present high chemical and thermal stabilities [5-7]. ILs are emerging solvents of interest for lignocellulosic biomass pretreatment, but the complete removal of ILs after pretreatment is technically unrealizable, especially for processes at large scale [8], and residual ILs are still present in the pretreated substrate even after extensive washing [4,5]. Thus, it is essential to investigate their compatibility with all the steps of the bioethanol production process, particularly with the fermentation phase.

Concerning the impact of ILs on microorganisms, conflicting reports are available. Ouellet et al. [9] observed that residual 1-ethyl-3-methylimidazolium acetate [Emim][OAc] at a low concentration of 0.1% (v/v) which is remaining after pretreatment of corn stover and switchgrass had significant inhibitory effects on the growth of the yeast *Saccharomyces cerevisiae* and ultimately ethanol production. However Li et al. [10] described that corn cob pretreatment with 1-methyl-3-methylimidazolium dimethylphosphite [Mmim][DMP] had no notable effect on enzymatic saccharification, cell growth, and accumulation of lipid of the bacteria *Rhodococcus opacus*. In the same manner, Nakashima et al. [11] showed that 1-ethyl-3-methylimidazolium diethylphosphate [Emim][DEP], 1-ethyl-3-methylimidazolium chloride [Emim][Cl], and [Emim][OAc] used for cellulose pretreatments had no negative impact on *S. cerevisiae* growth. Lee et al. [12] and Ganske and Bornscheuer [13] observed toxic effects of 1-butyl-3-methylimidazolium hexafluorophosphate [BMIM][PF6] at a concentration of 1% on *Escherichia coli* growth, while no harmful action was found by Pfruender et al. [14] at a concentration of 20% on the same bacteria. In another study, Sendovski et al. [15] found that [BMIM][PF6] was toxic for *S. cerevisiae*, while no negative repercussion was observed by Pfruender et al. [14] with the same microorganism at 20%. Hence, understanding the origin of IL toxicity on cells is gaining interest.

In this work, the impact of two hydrophilic imidazolium-based ILs was investigated on *S. cerevisiae* grown in a culture medium containing glucose as carbon source. The first IL, [Emim][OAc], is commonly used for various lignocellulosic substrate pretreatments, and the second one, 1-ethyl-3-methylimidazolium methylphosphonate [Emim][MeO(H)PO_2], was recently demonstrated as an efficient alternative to [Emim][OAc] in lignocellulose pretreatment [5,16-19]. The effects of both ILs were observed on the model yeast physiology by following growth, glucose consumption, and ethanol production. Moreover, two novel approaches were used here. The first one is the respiratory activity (oxygen transfer rate OTR and carbon dioxide transfer rate CTR) of yeasts which was followed online by respiratory activity monitoring systems (RAMOS). The second one is the yeast morphology that was observed by the environmental scanning electron microscope (ESEM). To our knowledge, this is the first report of a complete investigation of IL impacts on yeast cells, including physiologic, metabolic, and morphologic parameters.

Results and discussion

S. cerevisiae growth in the presence of [Emim][OAc] or [Emim][MeO(H)PO_2] without previous adaptation

The impact of [Emim][OAc] and [Emim][MeO(H)PO_2] ILs on yeast growth was first assessed by biomass measurements, expressed as cell dry weight (CDW), in YMD culture medium (Figure 1). Without IL, *S. cerevisiae* type II from Sigma-Aldrich grew in YMD medium after an 8-h lag phase to a maximal biomass value of 5.8 g CDW/L ($OD_{600} \approx 13$). The [Emim][OAc] IL reduced drastically the growth (Figure 1A): when 0.5% (v/v) was added, the maximum biomass was 2.3 g CDW/L after the same lag phase, and no growth could be observed when 1% or more [Emim][OAc] was added to YMD medium. When [Emim][MeO(H)PO_2] was added to the YMD medium (Figure 1B), the stationary growth phase was obtained with a biomass value of 3.5 g CDW/L after an 8-h lag phase at 0.5% IL and 2.8 g CDW/L after a 24-h lag phase with 1% IL. Growth was no longer observed with an addition of 2% [Emim][MeO(H)PO_2] or more (data not shown). The decrease of yeast growth was in agreement with the IL toxic effects already reported in the literature [6,20].

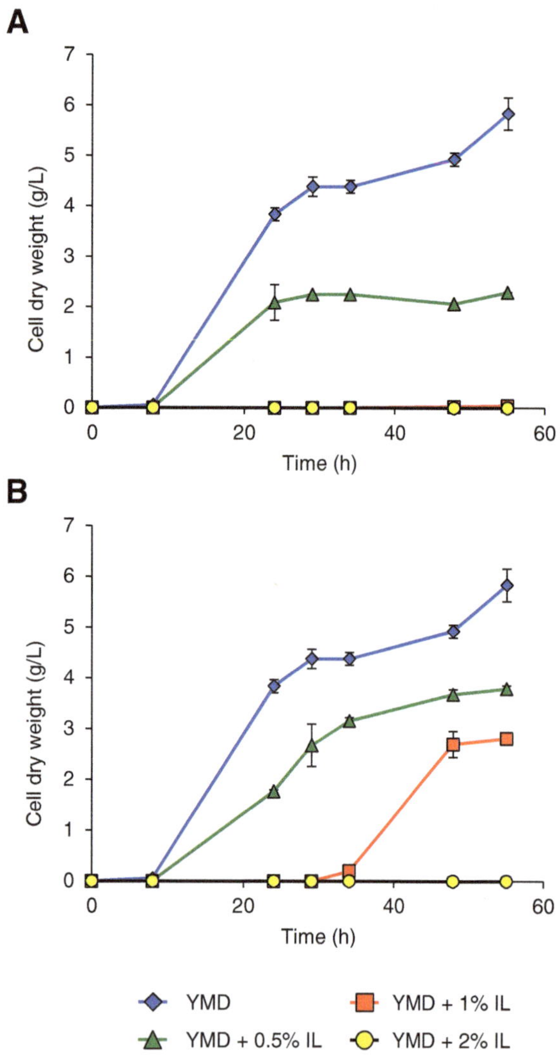

Figure 1 Growth of non-adapted *S. cerevisiae* cells in the presence of various concentrations of ILs in YMD culture medium (A [Emim][OAc]; B [Emim][MeO(H)PO_2]). The results are mean of two experiments, and error bars represent standard deviations from mean value.

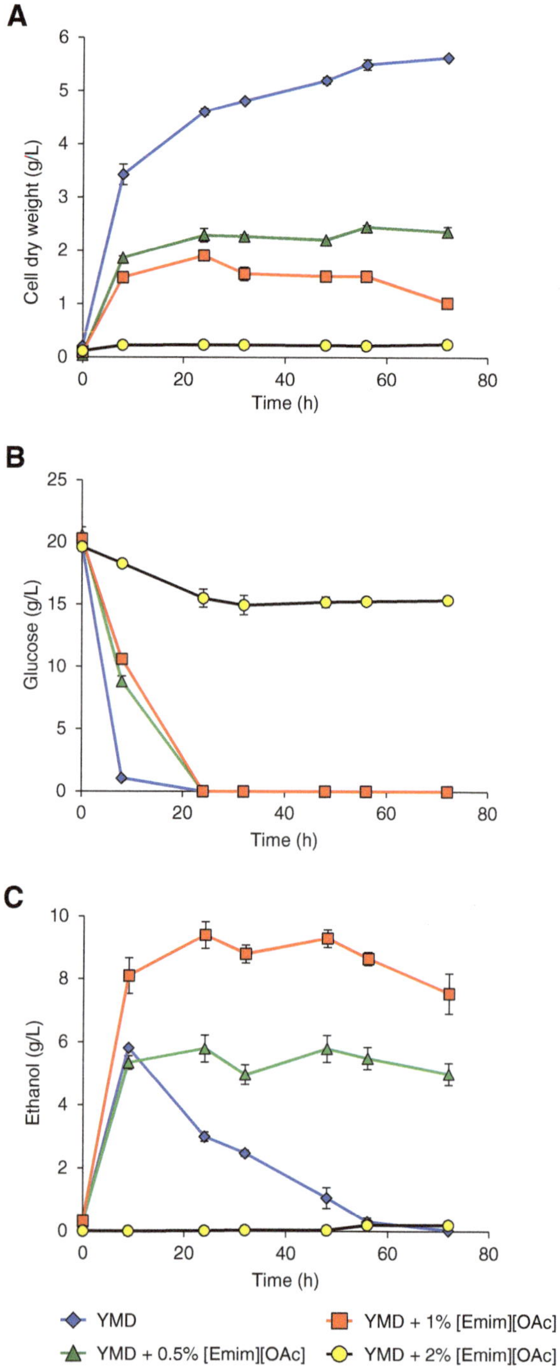

Figure 2 Growth of IL-adapted *S. cerevisiae* in the presence of various concentrations of [Emim][OAc] in YMD culture medium (A cell dry weight (g/L); B glucose concentration; C ethanol concentration). The results are mean of two experiments, and error bars represent standard deviations from mean value.

S. cerevisiae adaptation to ILs

In order to increase the yeast tolerance to both ILs, successive cultures of *S. cerevisiae* were realized in YMD supplemented with increasing IL %, separated by a spread on YMD plates (see the "Methods" part). The aim of this protocol was the selection of yeasts with better IL tolerance. This IL-adapted *S. cerevisiae* strain was used for all the further results presented in this study.

S. cerevisiae growth, glucose consumption, and ethanol production in the presence of [Emim][OAc] IL

Figure 2 shows the results obtained with [Emim][OAc] addition to the YMD medium. The maximal biomass was 2.4 g CDW/L with 0.5% IL and 1.5 g CDW/L with 1% IL, both without lag phase. Growth was no longer observed with 2% [Emim][OAc] addition (Figure 2A).

The glucose consumption was measured with an initial concentration of 20 g/L in the YMD medium (Figure 2B) and was proportional to the growth results: glucose was fully consumed for IL addition of 0%, 0.5%, and 1% but poorly decreased with 2% [Emim][OAc].

The ethanol production was also quantified (Figure 2C). Without [Emim][OAc], the measured maximum ethanol yield (5.8 g/L) was obtained at 8 h when glucose was completely metabolized and then ethanol decreased to 0 at 72 h. After glucose depletion, no increase in ethanol concentration was observed which indicated that glucose was the only substrate for ethanol production in YMD medium. The other components of the culture medium, i.e. yeast extract, malt extract, and peptone, could not serve as substrates for ethanol production but however may contribute to the growth of *S. cerevisiae* because the biomass value was 3.4 g CDW/L at 8 h while 5.6 g CDW/L at 72 h.

The addition of [Emim][OAc] modified the ethanol profiles: with 0.5% IL, the observed maximal ethanol concentration was similar to the control without IL (5.8 g/L at 24 h when glucose was totally consumed) but remained constant till the end of the culture (5.0 g/L at 72 h). With 1% [Emim][OAc], the measured maximal ethanol concentration was 9.4 g/L at 24 h when glucose was depleted and showed few variations till 72 h (7.6 g/L).

The constancy of ethanol concentration implied that the ethanol decrease observed for the control without [Emim][OAc] was due to a consumption by the yeasts when glucose was exhausted.

It is worth to notice that the addition of 1% [Emim][OAc] to the fermentation medium allowed to increase the maximal ethanol concentration by a factor 1.6 whereas the yeast growth was reduced by a factor 2.9 (Figure 2A).

S. cerevisiae growth, glucose consumption, and ethanol production in the presence of [Emim][MeO(H)PO_2] IL

Then, the impact of the second IL, [Emim][MeO(H)PO_2], was investigated on *S. cerevisiae* growth, glucose consumption, and ethanol production. The addition of [Emim][MeO(H)PO_2] was better tolerated by *S. cerevisiae* than [Emim][OAc] and growth could be observed until a supplementation of 6% IL (v/v) (Figure 3A), indicating that *S. cerevisiae* adaptation to [Emim][MeO(H)PO_2] was more efficient than to [Emim][OAc]. The maximal biomass decreased progressively with the [Emim][MeO(H)PO_2] addition from 3.8 g CDW/L (OD_{600} = 8.5) with 0.5% IL (instead of 5.6 g CDW/L for control without IL) till 1.0 g CDW/L with 6% IL. No lag phase was observed except for 6% [Emim][MeO(H)PO_2]. The totality of glucose contained in the YMD medium was consumed for all the condition tested with a consumption rate inversely proportional to the growth (Figure 3B). The ethanol production was measured (Figure 3C) with a maximum at 5.8 g/L at 9 h for the control without IL followed by a decrease to 0 at 72 h. The [Emim][MeO(H)PO_2] addition until 6% (v/v) led to a measured maximum ethanol concentration higher than the control with an optimum at 8.0 g/L for 4% [Emim][MeO(H)PO_2]. Moreover, the decrease in ethanol concentration observed without IL after glucose depletion was less pronounced with [Emim][MeO(H)PO_2] addition and the ethanol was reduced from 47% at 1% IL, 23% at 2% IL, and 17% at 4% IL between the glucose depletion and the end of the culture at 72 h.

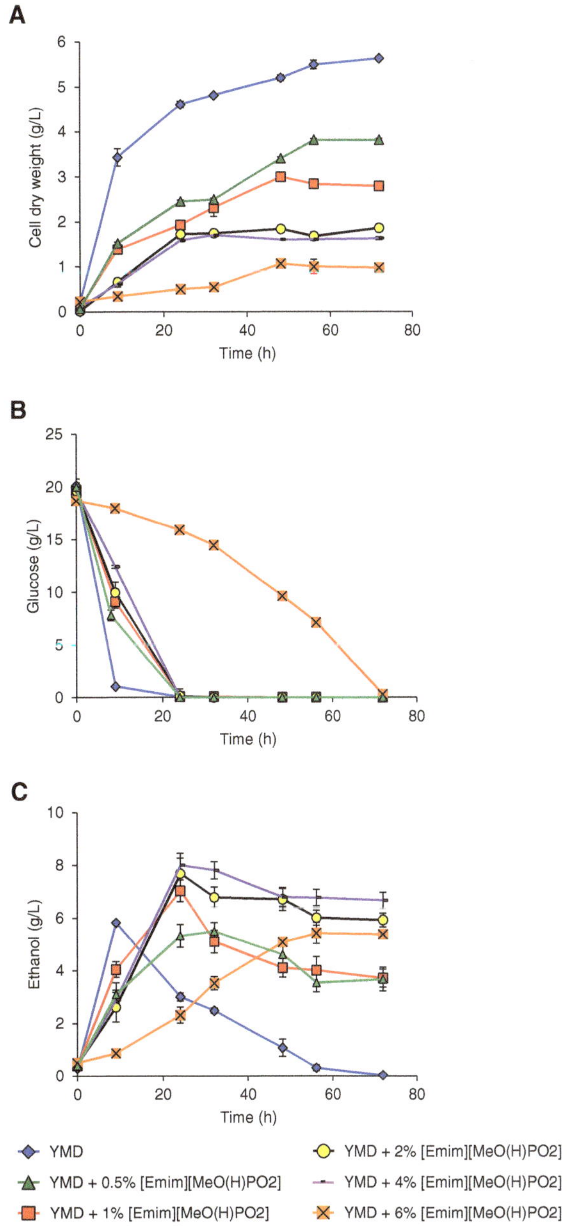

Figure 3 Growth of IL-adapted *S. cerevisiae* in the presence of various concentrations of [Emim][MeO(H)PO_2] in YMD culture medium (A cell dry weight (g/L); B glucose concentration; C ethanol concentration). The results are mean of two experiments, and error bars represent standard deviations from mean value.

These results showed that both imidazolium-based ILs, [Emim][OAc] and [Emim][MeO(H)PO_2], increased significantly the measured maximum ethanol yield with 9.4 g/L for an addition of 1% [Emim][OAc] and 8.0 g/L for 4% [Emim][MeO(H)PO_2] instead of 5.8 g/L ethanol for the control without IL. These values with ILs were closer than the control to the theoretical maximum ethanol yield that could be obtained from glucose at 20 g/L contained in the YMD medium, i.e. 10.2 g/L ethanol [21]. Furthermore, the ethanol consumption by the yeasts after glucose depletion was not observed with ILs addition and instead of a null yield at 72 h for the condition without IL, the ethanol concentrations measured were 7.6 g/L for 1% [Emim][OAc] and 6.7 g/L for 4% [Emim][MeO(H)PO_2] at 72 h.

The nature of the anion associated to the imidazolium cation seemed to play a deleterious role since the *S. cerevisiae* growth was more drastically affected by the acetate anion (growth until 1%) than by the methylphosphonate one (growth until 6%). But the yeast growth was not correlated to the ethanol production, and the better ethanol yields were obtained with addition of 1% [Emim][OAc] (9.4 g/L ethanol) and 4% [Emim][MeO(H)PO_2] (8 g/L ethanol) instead of 5.8 g/L ethanol for control without IL.

As *S. cerevisiae* is a facultative aero-anaerobic yeast, these first results led to investigate the respirofermentative status of the cells which is directly linked to the ethanol biosynthesis [22-25]. For that purpose, the oxygen transfer rate (OTR) was followed during growth in absence or presence of ILs. Secondly, the previous results showed that ethanol was consumed by the yeasts after glucose depletion, underlying an ethanol oxidation into carbon dioxide and water. This hypothesis was verified by measuring the carbon dioxide transfer rate (CTR). Both OTR and CTR were simultaneously followed online during the cultivation time by a RAMOS device proposed by Anderlei and Büchs [26] and Anderlei et al. [27]. To our knowledge, this is the first report of OTR and CTR measurements with a RAMOS device applied to yeast grown in presence of ILs. These results are presented in Figure 4 for [Emim][OAc] IL and Figure 5 for [Emim][MeO(H)PO_2] IL.

OTR and CTR profiles of *S. cerevisiae* culture in the presence of [Emim][OAc] IL

For *S. cerevisiae* in YMD medium without IL, the OTR increased exponentially at the beginning of the culture from 0 to 7.6 mmol/L/h at 14 h (Figure 4A) and slowly raised to 8.1 mmol/L/h at 20 h. The OTR remained constant until 47 h then suddenly dropped to nearly 0 at 60 h. The OTR increase at the beginning of the culture was correlated with exponential growth phase and was consistent with the carbon substrate consumption [26,28]. The OTR plateau between 20 and 47 h represented the maximum oxygen transfer capacity of the system [27,29] and implied that *S. cerevisiae* fermentation was oxygen limited during this phase. The sharp drop in the OTR at 60 h indicated that the carbon source was completely consumed [26,27].

In the presence of [Emim][OAc], the OTR profiles were different. At 0.5% LI, a small increase was observed at the beginning of the culture from 0 to 1.1 mmol/L/h at 2 h, followed by a slight steady decrease to 0.2 mmol/L/h at 52 h. At 1% LI, the OTR weakly raised to 0.8 mmol/L/h at 2 h, became null at 12 h, and remained constant till the end of the culture. At 2% LI, the OTR value stayed zero during all the experimental time. OTR profiles in the presence of IL concentrations as low as 0.5 or 1% (v/v) showed that [Emim][OAc] greatly modified the oxygen status of the culture and the yeast metabolic behavior. In the absence of [Emim][OAc], glucose was aerobically consumed and converted in ethanol thanks to the short-term Crabtree effect in *S. cerevisiae* which inhibits respiration in the presence of high glucose concentrations and allows aerobic alcoholic fermentation [30-33], thus generating a high OTR maximal value (8.1 mmol/L/h). In the presence of 0.5% and 1% [Emim][OAc], glucose was consumed anaerobically, leading to low OTR maximal values around 1 mmol/L/h [34].

The CTR evolution (Figure 4B) showed for the condition without IL a peak at 32.5 mmol/L/h at 8 h, then a decrease followed by a plateau at 4.8 mmol/L/h from 13 to 47 h, ended by a null value at 50 till 60 h. The first phase of the CTR profile could be correlated with the consumption of glucose converted in carbon dioxide and ethanol until the sugar was depleted at 8 h (Figure 2B, C). Then, the steady CTR at a value well above null between 13 and 47 h implied that another carbon source was consumed by *S. cerevisiae* during that time which was different from the glucose and generating a lower CO_2 liberation rate. This is consistent with the diminution of ethanol concentration pointed in Figure 2C from 8 to 72 h and showed that *S. cerevisiae* consumed the ethanol produced from glucose during the first phase of the culture [27,34]. When ethanol was totally consumed, the CTR value fell down to 0 at 60 h, underlying that the amino acid nutrients contained in the YMD medium (i.e. yeast extract, malt extract, peptone) could not be used solely by the yeast and that all the carbon sources were exhausted [27].

When 0.5% (v/v) [Emim][OAc] was added to the growth medium (Figure 4B), a CTR peak reaching 19.6 mmol/L/h was observed at 11 h, corresponding to the glucose depletion (Figure 2B); but contradictory to the control condition without IL, the CTR value directly decreased to zero without a plateau around 5 mmol/L/h and stayed null till 52 h. This meant that ethanol was not consumed by the yeast after glucose depletion in

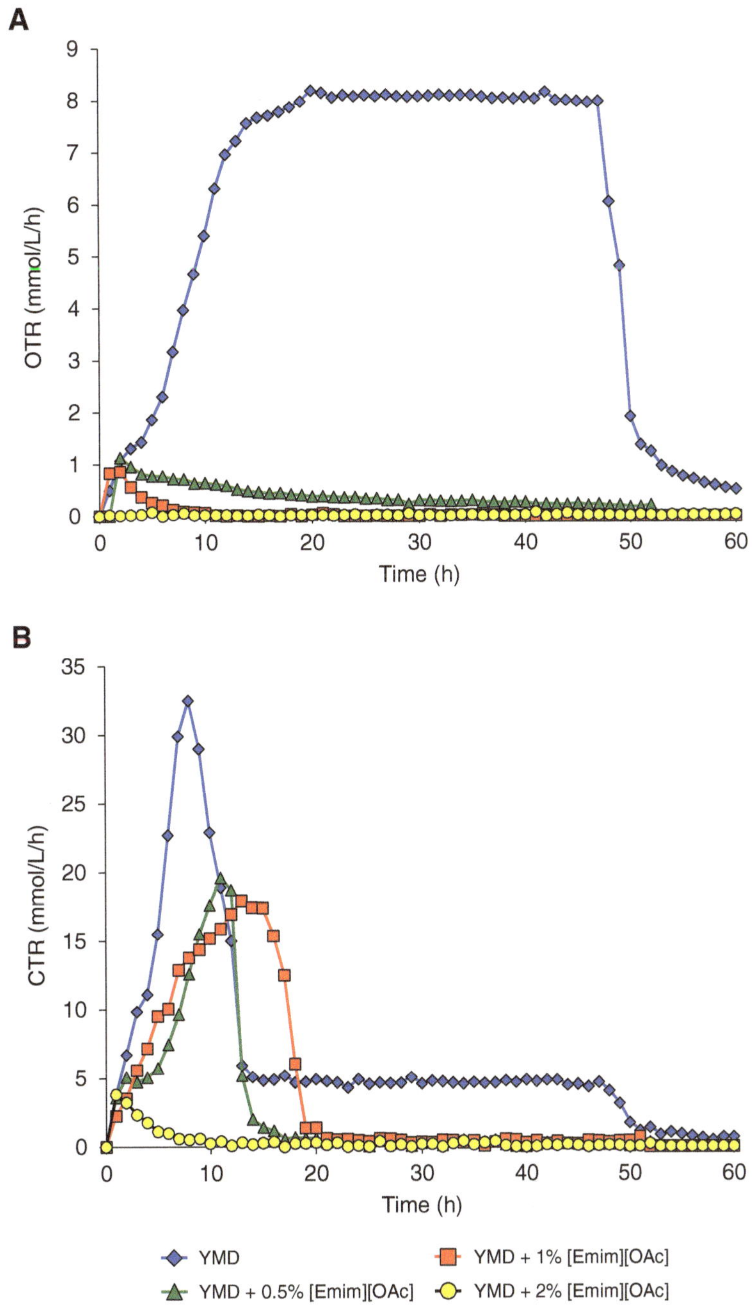

Figure 4 Oxygen transfer rate OTR (A) and carbon dioxide transfer rate CTR (B) of *S. cerevisiae* in the presence of various concentrations of [Emim][OAc] in YMD culture medium. The results are mean of two experiments. Error bars are not represented to avoid overloading the figure.

that condition and was consistent with the ethanol concentration profile in Figure 2C indicating that ethanol remained constant till the end of the fermentation. At 1% (v/v) [Emim][OAc], the *S. cerevisiae* CTR profile was similar and raised to 17.9 mmol/L/h at 13 h before becoming null at 20 h until the end of the fermentation. With 2% (v/v) [Emim][OAc], the CTR showed a slight peak at 3.8 mmol/L/h at 1 h followed by a decrease to 0 at 10 h and remained constant till the end of the culture at 52 h, which is consistent with the very low yeast growth and glucose consumption observed in Figure 2A, B.

Thus, the maximal CTR value decreased when [Emim][OAc] was added to the medium culture: 32.5 mmol/L/h without IL, 19.6 mmol/L/h with 0.5% IL, and 17.9 mmol/L/h with 1% IL. This indicated that the yeast metabolism had switched from aerobic to anaerobic consumption of glucose with the addition of [Emim][OAc], leading to a lower CO_2 liberation rate. It is known that anaerobic

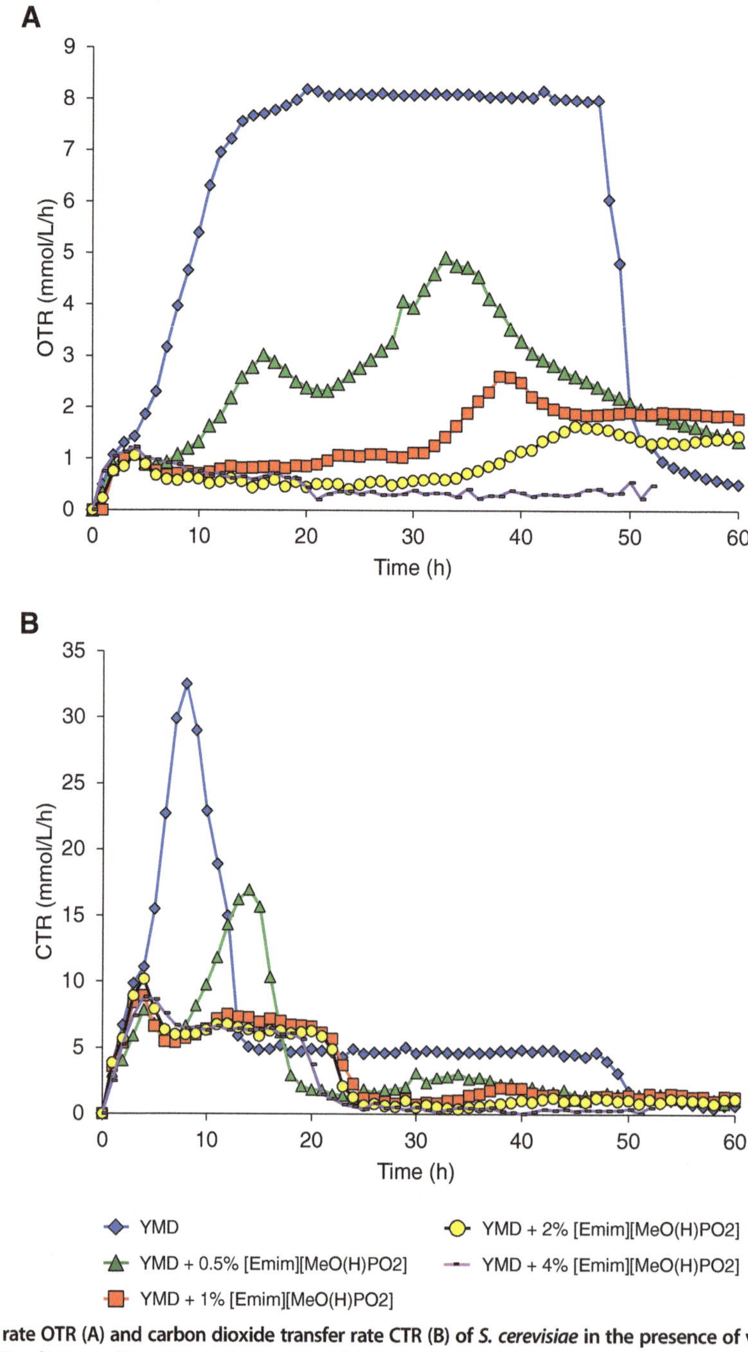

Figure 5 Oxygen transfer rate OTR (A) and carbon dioxide transfer rate CTR (B) of *S. cerevisiae* in the presence of various concentrations of [Emim][MeO(H)PO_2] in YMD culture medium. The results are mean of two experiments. Error bars are not represented to avoid overloading the figure.

conditions stimulate ethanol biosynthesis at the expense of biomass production [31,34,35] and could explain the highest ethanol concentration observed in Figure 2C whereas the growth was low (Figure 2A).

OTR and CTR profiles of *S. cerevisiae* culture in the presence of [Emim][MeO(H)PO_2] IL

The OTR evolution in presence of the other IL, [Emim][MeO(H)PO_2], is presented in Figure 5A. The control condition without IL was already described above. When [Emim][MeO(H)PO_2] was added to the culture medium, the OTR decreased systematically with increasing percentages in IL. With 0.5% IL, the OTR increased from 0 to 1 mmol/L/h at 4 h; then, two OTR peaks were observed, the first at 16 h with 3.0 mmol/L/h and the second at 33 h with 4.9 mmol/L/h. After then, the OTR decreased and reached 1.5 mmol/L/h at 60 h. For conditions with 1% and 2% [Emim][MeO(H)PO_2], the OTR also increased from 0 to almost 1 mmol/L/h at 4 h and remained constant until 30 h then slightly increased to a maximal OTR value of 2.6 mmol/L/h at 38 h for 1% IL and 1.6 mmol/L/h at 45 h for 2% IL. Then, the OTR decreased and stayed constant to 1.8 mmol/L/h for 1% IL and 1.6 mmol/L/h for 2% IL until 60 h. With addition of 4% [Emim][MeO(H)PO_2], the OTR increased to 1 mmol/L/h at 4 h and slowly decreased to 0.4 mmol/L/h at the end of the culture.

The effect of [Emim][MeO(H)PO_2] addition to the culture medium produced the same effect as [Emim][OAc] and induced a transition of the glucose consumption from an aerobic catabolism without IL to an anaerobic behavior in presence of low IL concentrations. But the effect of [Emim][MeO(H)PO_2] seemed to be less drastic than [Emim][OAc] on *S. cerevisiae*, and the profile obtained with 0.5% [Emim][OAc] was identical to those with 4% [Emim][MeO(H)PO_2]. This allowed an interesting observation of the transition between aerobic to anaerobic glucose consumption which could be illustrated with the 0.5% [Emim][MeO(H)PO_2] condition. The OTR profile showed clearly a diauxic growth with two peaks at 16 and 33 h. These two points corresponded to the glucose depletion for the first one (Figure 3B) and to the ethanol consumption for the second one, explaining the decrease in ethanol concentration observed after 32 h in Figure 3C. The diauxic growth with 0.5% [Emim][MeO(H)PO_2] is also observed in the Figure 3A with a small growth deceleration at 32 h, followed by a second exponential phase until 56 h. When [Emim][MeO(H)PO_2] percentages increased, the glucose catabolism became more and more anaerobic generating lower yeast growths (Figure 3A) and higher ethanol concentrations (Figure 3C).

The CTR was measured during growth in presence of [Emim][MeO(H)PO_2] (Figure 5B) and confirmed the previous results. When 0.5% [Emim][MeO(H)PO_2] was added to the growth medium, the CTR indicated the glucose depletion at 14 h with a peak at 17.0 mmol/L/h and the ethanol consumption at 33 h with an increase to 3.0 mmol/L/h. The ethanol consumption in the second part of the fermentation was less and less observed as the [Emim][MeO(H)PO_2] percentage raised, for being almost absent in the condition with 4% IL (Figure 5B), exhibiting an ethanol concentration nearly maximal at 72 h (Figure 3C). With 4% IL, the maximal CTR value was 8.8 mmol/L/h observed at 4 h, which is almost four times lower than the control without IL (32.5 mmol/L/h) confirming higher fermentation rates than respiration.

As previously discussed, the two imidazolium-based ILs [Emim][OAc] and [Emim][MeO(H)PO_2] could decrease the growth, slow down the glucose consumption, and promote ethanol production by inhibiting yeast respiration and avoiding further consumption of ethanol. Then, the impact of both ILs was investigated on *S. cerevisiae* shape and morphology by environmental scanning electron microscopy (ESEM). Here again, to our knowledge, this is the first report of the direct visualization of IL effects on the yeast morphology by electronic microscopy.

Impact of ILs on *S. cerevisiae* morphology

Figure 6A presents *S. cerevisiae* in the control condition, i.e. YMD medium without IL. The cells were oval, smooth, and swollen, with the typical appearance of ovoid yeasts. Scars from previous buddings could also been observed at the surface of some cells. When [Emim][MeO(H)PO_2] was added to the culture medium between 1% and 3% (Figure 6B, D), the cell appearance changed: they presented holes clearly visible in the cell wall, some yeasts were wrinkled and exhibited a softer and irregular surface. With higher [Emim][MeO(H)PO_2] concentrations (4% and 5% in Figure 6E, F, respectively), the yeasts seemed to dissolve and take a gel-like appearance. The cell walls were not clearly distinguishable anymore and cells agglutinated to each other in a sticky and slimy mass.

As already observed, the other imidazolium derived IL, [Emim][OAc], had the same impact on *S. cerevisiae* morphology (Figure 7), but the deleterious effect was more pronounced and the same damages could be observed with lower IL concentrations. The Figure 7A showed an ESEM micrography of *S. cerevisiae* in the presence of 0.5% [Emim][OAc]. The yeasts presented wrinkled surfaces and exhibited holes in their cell wall, more numerous than with 3% [Emim][MeO(H)PO_2]. At 1% [Emim][OAc] (Figure 7B), the gelified and smooth aspect of the culture was retrieved. A defect in cell division could also be observed leading to longer cell shapes [36], corresponding to around three or four times the normal yeast length.

These ESEM results showed for the first time the deleterious effects of [Emim][OAc] and [Emim][MeO(H)PO_2] on *S. cerevisiae* morphology. An explanation could rely on the structure of the yeast cell wall, representing 30% of the cell dry weight and composed of 85% polysaccharides. All the components are linked to each other as a polysaccharide-mannoprotein complex, composed of chains of β-1,3-linked glucose residues branched by β-1,6 linkages, forming a fibrillar glucan serving as a backbone to which is linked chitin, β-1,4-linked N-acetylglucosamine polysaccharide, and some mannoproteins [36,37]. Both imidazolium-based ILs [Emim][OAc] and [Emim][MeO(H)PO_2] are known and used for their ability to destructure the rigid supramolecular architecture of cellulose, which is a β-1,4-linked glucose polysaccharide [16,17]. It is worth to consider that ILs interacted in a similar manner on the polysaccharides composing the yeast cell wall, inducing holed, softened, and gelified structures as observed in Figure 7. In that context, some authors described ILs as efficient

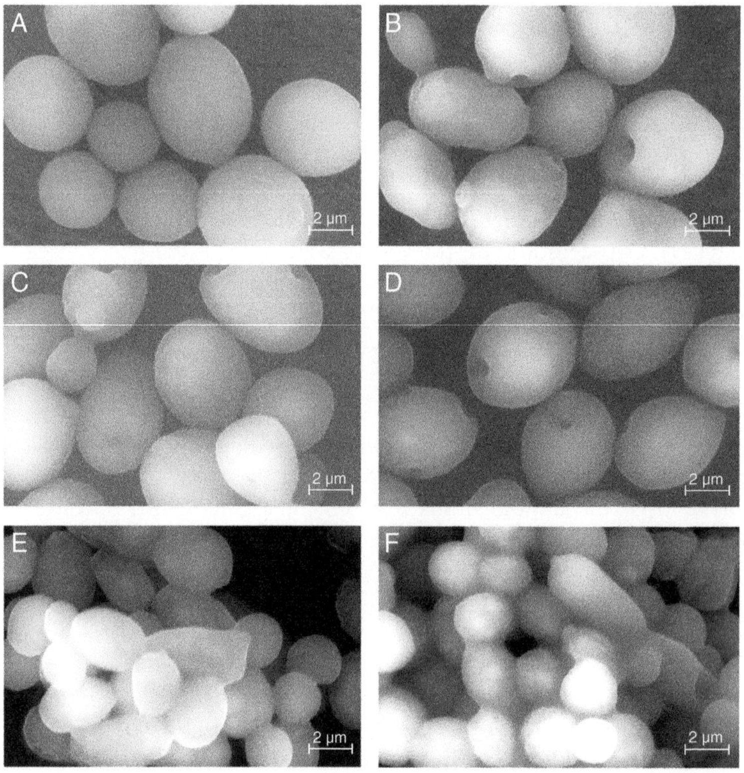

Figure 6 ESEM micrographs of *S. cerevisiae* cells in the presence of various concentrations of [Emim][MeO(H)PO_2] in YMD culture medium: (A) YMD; (B) YMD + 1% [Emim][MeO(H)PO_2]; (C) YMD + 2% [Emim][MeO(H)PO_2]; (D) YMD + 3% [Emim][MeO(H)PO_2]; (E) YMD + 4% [Emim][MeO(H)PO_2]; (F) YMD + 5% [Emim][MeO(H)PO_2].

lysis reagents, allowing protein extraction in yeast cells [38]. The IL anionic moiety is describe to be responsible for the disruption of the hydrogen bonds network between the glucosidic monomers in the cellulosic matrix [5,7]. It is probable that the methylphosphonate anion has a lower destructuring ability than the acetate on the yeast cell wall polysaccharides, which could explain the milder deleterious effect of [Emim][MeO(H)PO_2].

Impact of ILs on *S. cerevisiae* viability

After visualizing the damages in yeast cell structures due to low concentrations of [Emim][OAc] and [Emim][MeO(H)PO_2], the cell viability in ILs was quantified by measuring the colony-forming units (CFU) after 24 h of growth in YMD medium supplemented with increasing ILs concentrations. These mixtures were inoculated by 10^7 cells (Figure 8). In the absence of IL, the *S. cerevisiae* culture contained 6.0×10^8 CFU/mL after 24 h. The addition of 0.5% and 1% [Emim][OAc] was fungistatic and the yeasts grew but to a lesser extent than the control without IL. Cell counts of 1.8×10^8 and 6.1×10^7 CFU/mL, respectively, were reached after 24 h. The [Emim][OAc] became fungicide from 2% (v/v) (7.0×10^5 CFU/mL) and no viable cell could be detected with a 3% addition. The [Emim][MeO(H)PO_2] was less toxic for *S. cerevisiae* cells: the number of CFU obtained

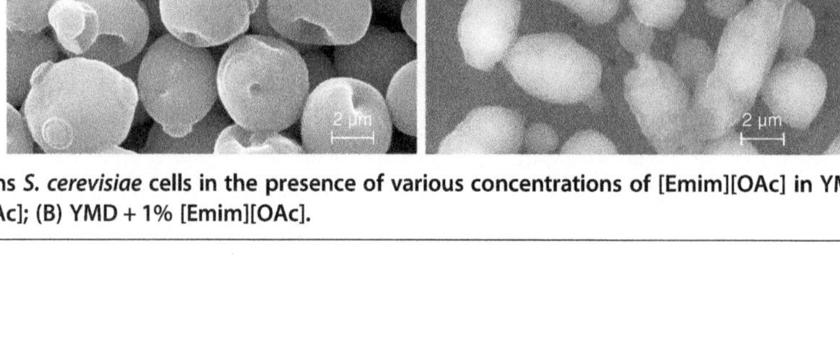

Figure 7 ESEM micrographs *S. cerevisiae* cells in the presence of various concentrations of [Emim][OAc] in YMD culture medium: (A) YMD + 0.5% [Emim][OAc]; (B) YMD + 1% [Emim][OAc].

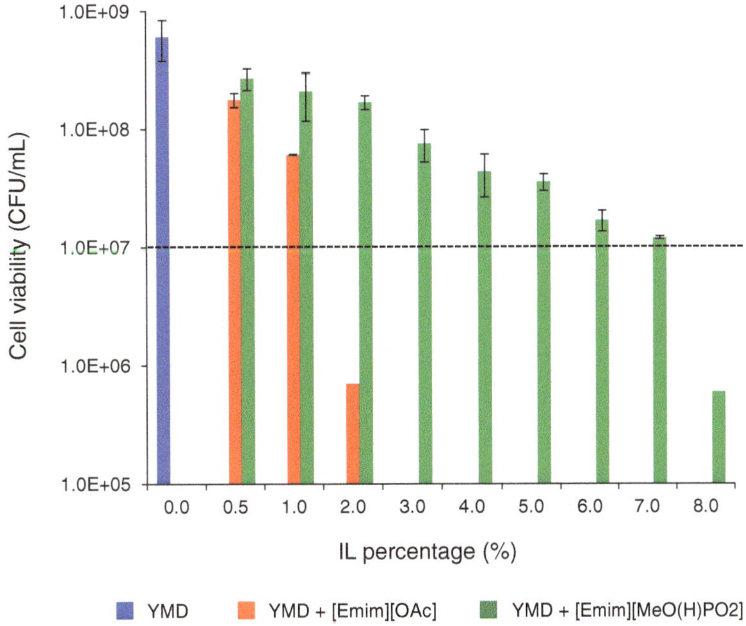

Figure 8 Viability of *S. cerevisiae* in the presence of various concentrations of [Emim][OAc] and [Emim][MeO(H)PO$_2$] in YMD culture medium. Inoculum was constituted by 10^7 cells (horizontal dashed line). The colony-forming units were numbered after 24 h of growth.

after 24 h of growth was superior to the inoculum until 7% (v/v), indicating fungistatic concentrations. At 8%, [Emim][MeO(H)PO$_2$] became fungicide (6.0×10^5 CFU/mL) and no viable cell could be observed from 9% IL (data not shown).

The viability results were consistent with all the previous parts and indicated that [Emim][MeO(H)PO$_2$] was more biocompatible with *S. cerevisiae* cells than [Emim][OAc]. It also meant that yeasts retained metabolic activity and reproductive ability, even if they appear damaged with holes in their softened cell walls.

Conclusions

Altogether, our results showed that both imidazolium-based ILs [Emim][OAc] and [Emim][MeO(H)PO$_2$] promoted bioethanol production by *S. cerevisiae*. Addition of low IL concentrations to the growth medium generated higher ethanol yields than the control without IL, i.e. +58.2% with 1% [Emim][OAc] and +37.9% with 4% [Emim][MeO(H)PO$_2$]. These increased yields were due to inhibition of the respiration activity of the yeasts, indicated by measured OTR in the culture medium, thereby generating a transition from a respiratory metabolism to a fermentative consumption of glucose with lower biomass production but higher ethanol yields. Moreover, the ethanol consumption by the yeasts observed for the control without IL after glucose depletion was reduced or absent in certain IL conditions. Deleterious effects of ILs were also observed on *S. cerevisiae* cell wall.

The number of ILs is virtually infinite because of unlimited capabilities to exchange constitutive cations and anions. Hence, it is highly probable that new ILs could be designed at will to reduce toxicity and help yeasts to tolerate higher IL concentrations, while retaining the properties of improved ethanolic yields and destructuring the lignocellulosic biomass. But this implies firstly to deepen the knowledge on IL cytotoxicity and how they interact with the cells: some authors pointed the toxic role of the cationic moiety [20,39], others stated for an anionic origin [40], while others evidenced deleterious effects for both anionic and cationic parts [9,41]. Differences between all these reports may originate from the provenance of the ILs, and several works deal with impurities found in commercial ILs that could interact with cells and could lead to distorted toxicity results [42-44].

The future of this fundamental study is to gather the three steps of the second-generation bioethanol production in a one-pot process, regrouping IL pretreatment of the lignocellulosic biomass, enzymatic saccharification, and ethanolic fermentation. A previous study on cellulose pretreatment with [Emim][MeO(H)PO$_2$] combined to a simultaneous enzymatic saccharification with *Trichoderma reesei* cellulases showed the best glucose yield with a concentration of 10% [Emim][MeO(H)PO$_2$] [17]. We evidenced in our study that *S. cerevisiae* was not viable anymore at that IL concentration and there is more work to do to increase yeast tolerance to ILs and achieve a one-pot process from lignocellulosic biomass to bioethanol, in order to optimize the industrial process and assure a cost-effective ethanolic production [45-47]. This also implies to focus on IL reuse which is still a technological lock even if several works are done in that purpose but only in a separate phase of IL pretreatment [17,18,48]. This will be more difficult to achieve when the ILs are used in diluted media containing enzymes for saccharification and cells for fermentation.

Methods

Chemicals

Yeast extract, peptone, and malt extract were obtained from Merck (Darmstadt, Germany) while glucose, osmium tetraoxide (OsO_4), glutaraldehyde, and agar were purchased from Sigma-Aldrich (Steinheim, Germany). Both ionic liquids, 1-ethyl-3-methylimidazolium acetate $[Emim]^+[CH_3COO]^-$, and 1-ethyl-3-methylimidazolium methylphosphonate $[Emim]^+[MeO(H)PO_2]^-$ were acquired from Solvionic SA (Veniole, France), with purity higher than 98%.

Strain and culture conditions

The yeast strain used in this study was *S. cerevisiae* Type II from Sigma-Aldrich (Steinheim, Germany).

The growth medium was the yeast-malt-dextrose (YMD) medium: glucose 20 g/L, yeast extract 3 g/L, malt extract 3 g/L, peptone 3 g/L, pH 4.8, eventually solidified by agar 20 g/L, and sterilized at 121°C for 20 min. The culture conditions were 30°C at 150 rpm with a shaking diameter of 2.5 cm.

The *S. cerevisiae* strain was conserved at +4°C onto YMD plate. For each experiment, one yeast colony was used to inoculate a 50 mL YMD preculture. At mid-log phase, preculture was used to inoculate the culture at an initial optical density at 600 nm (OD_{600nm}) of 0.25 into a 250 mL non-baffled flask containing 50 mL of YMD or YMD + IL from 0.5% to 7% (v/v) added after medium sterilization. The yeast growth was routinely followed by measuring OD_{600nm} values. The biomass was determined as cell dry weight (g/L) after filtration through dried and pre-weighed membrane filter with 0.22-µm pore size (Millipore, Ireland) with subsequent washing in physiological water. The samples were freeze dried and weighted. The CDW was obtained by the difference between the masses of the filter alone and the filter with the yeast cells. All the experiments were realized at least twice.

The yeast viability was measured as number of colony-forming unit (CFU/mL) after sampling a 24-h growth culture, serial dilutions up to 10^{-8} in sterile physiological water, spread of 100 µL of each dilution on YMD agar plates followed by incubation 48 h at 30°C, and counting colonies on plates containing between 30 and 300 colonies.

In order to increase the tolerance of *S. cerevisiae* to both ILs, a series of precultures and cultures was realized in YMD supplemented with increasing IL concentrations (0.5% to 7% v/v). Each run of preculture/culture was separated by a plating on YMD medium, and a colony of this plating was used to inoculate the next run of preculture/culture in increasing IL%. The hypothesis for this adaptation was that only those yeast cells which are resistant to culture with IL% will be able to grow on YMD agar plates. No growth could be observed in culture supplemented with 2% [Emim][OAc] and 8% $[Emim][MeO(H)PO_2]$, respectively, or higher IL concentrations, even after yeast adaptation. The yeast cells adapted to 1% [Emim][OAc] and 7% $[Emim][MeO(H)PO_2]$, respectively, were transferred to YMD agar plates. All the further experiments in this study were realized from a colony of these plates.

Glucose and ethanol quantification

The concentration of glucose was monitored by HPAEC-PAD (high pressure anionic exchange chromatography coupled with pulsed amperometric detection) using an analytical CarboPac PA-20 column kept at 25°C. Elution was carried out at a flow rate of 0.5 mL/min with a gradient method described previously [16,17]. The retention time of glucose was 11.96 min (±0.56%). Quantification was based on calibration curves established using standard glucose.

The ethanol concentration was measured by enzymatic kit from BioLabo SA (Maizy, France).

Oxygen transfer rate (OTR) and carbon dioxide transfer rate (CTR) measurements

Online measurements of the OTR and CTR in shaken flasks were carried out in a self-made respiratory activity monitoring system (RAMOS) described by Anderlei and Büchs [26] and Anderlei et al. [27]. Commercial versions are available from Kühner (Birsfelden, Switzerland) or HiTec Zang (Herzogenrath, Germany). Specially designed flasks were used for cultures in the RAMOS device, but hydrodynamics and gas phase concentrations were similar to those in standard Erlenmeyer flasks with cotton plugs. The measuring cycle was separated into a measuring and a rinsing phase. During the rinsing phase, air with a calculated flow rate flowed through the measuring flask. To protect the measuring flask against contamination, two sterile filters were installed at the inlet and the outlet. The partial pressure of oxygen in the headspace of the measuring flask was detected by an oxygen gas sensor. At the beginning of the measuring phase, the inlet and outlet valves of the measuring flask were closed. The continuing respiration activities of the microorganisms subsequently led to a decrease in the partial pressure of oxygen and to an increase in the carbon dioxide partial pressure in the headspace of the measuring flask. The OTR and CTR of the particular measuring flask were automatically calculated from the change in the partial pressure over time. After the measuring phase, the valves were reopened and "fresh" air could flow through the flask; new measuring cycle started. The measuring cycles were completed approximately in 30 min. A newly developed calibration strategy ensured the compensation of the drift of the oxygen sensors [27]. Due to low solubility of oxygen in culture medium (approximately 8 mg/L at 25°C), it can be assumed that OTR is equal to oxygen uptake rate (OUR). OUR can also be described as the cellular oxygen demand or consumption [49].

Morphology observations

Microscopic preparations were performed according to a protocol adapted by Gognies et al. [50]. The culture samples were immersed in a fixative solution composed of 1% glutaraldehyde in 0.1 M potassium buffer (pH 6.8), at room temperature for 16 h. Samples were subsequently washed with 0.1 M potassium buffer (pH 6.8), post-fixed for 4 h in 2% osmium tetroxide, then washed again with buffer and dehydrated in successive ethanol baths. The morphology of yeasts was observed on a Philips ESEM-FEG XL30 scanning electron microscope in an environmental scanning electron microscope (ESEM) mode.

Abbreviations

CDW: cell dry weight; CFU: colony forming unit; CTR: carbon dioxide transfer rate; [Emim][OAc]: 1-ethyl-3-methylimidazolium acetate; [Emim][MeO(H)PO_2]: 1-ethyl-3-methylimidazolium methylphosphonate; ESEM: environmental scanning electron microscope; HPAEC-PAD: high pressure anionic exchange chromatography coupled with pulsed amperometric detection; IL: ionic liquid; OD_{600}: optical density at 600 nm; OTR: oxygen transfer rate; RAMOS: respiratory activity monitoring system.

Competing interests

The authors declare that they have no competing interests.

Authors' contributions

NM designed and executed all the experiments. EH participated in the design of the project and helped for manipulation of ILs. CJ contributed to the preparation of cell samples for electronic microscopy experiments. SW and JB led and managed the RAMOS experiments. CS initiated the collaboration and revised the manuscript. IG coordinated the overall project and wrote the manuscript. All authors provided critical input to the manuscript and read and approved the final manuscript.

Acknowledgements

We are grateful to the Conseil Regional de Picardie (CRP) and European Union for funding this project (SISAF). Europe is engaged in Picardy with the European Fund of Regional Development (FEDER). We acknowledge the support of the Competitiveness Cluster Industries & Agro-Resources. Nasir Mehmood was funded by CRP. We are also very thankful to Frederic Nadaud from the Plateform of Electronic Microscopy of the University of Technology Compiègne, France, for ESEM analysis.

Author details

[1]Unité Génie Enzymatique et Cellulaire, FRE-CNRS 3580, Université de Picardie Jules Verne, 33 rue Saint-Leu, 80039 Amiens Cedex, France. [2]Unité de Recherche Vignes et Vins de Champagne—UPRES-EA 4707, Université de Reims Champagne-Ardenne, BP103951687 Reims Cedex 2, France. [3]AVT—Biochemical Engineering, RWTH Aachen University, Aachen, Germany.

References

1. Balat M. Production of bioethanol from lignocellulosic materials via the biochemical pathway: a review. Energ Convers Manag. 2011;52:858.
2. Naik SN, Goud VV, Rout PK, Dalai AK. Production of first and second generation biofuels: a comprehensive review. Renew Sust Energ Rev. 2010;14:578.
3. Sims REH, Mabee W, Saddler JN, Taylor M. An overview of second generation biofuel technologies. Bioresour Technol. 2010;101:1570.
4. Alvira P, Tomás-Pejó E, Ballesteros M, Negro MJ. Pretreatment technologies for an efficient bioethanol production process based on enzymatic hydrolysis: a review. Bioresour Technol. 2010;101:4851.
5. Olivier-Bourbigou H, Magna L, Morvan D. Ionic liquids and catalysis: recent progress from knowledge to applications. Appl Catal A. 2010;373:1.
6. Quijano G, Couvert A, Amrane A. Ionic liquids: applications and future trends in bioreactor technology. Bioresour Technol. 2010;101:8923.
7. Mora-Pale M, Meli L, Doherty TV, Linhardt RJ, Dordick JS. Room temperature ionic liquids as emerging solvents for the pretreatment of lignocellulosic biomass. Biotechnol Bioeng. 2011;108:1229.
8. Zhao H, Jones CL, Baker GA, Xia S, Olubajo O, Person VN. Regenerating cellulose from ionic liquids for an accelerated enzymatic hydrolysis. J Biotechnol. 2009;139:47.
9. Ouellet M, Datta S, Dibble DC, Tamrakar PR, Benke PI, Li C, et al. Impact of ionic liquid pretreated plant biomass on *Saccharomyces cerevisiae* growth and biofuel production. Green Chem. 2011;13:2743.
10. Li Q, Jiang X, He Y, Li L, Xian M, Yang J. Evaluation of the biocompatible ionic liquid 1-methyl-3-methylimidazolium dimethylphosphite pretreatment of corn cob for improved saccharification. Appl Microbiol Biotechnol. 2010;87:117.
11. Nakashima K, Yamaguchi K, Taniguchi N, Arai S, Yamada R, Katahira S, et al. Direct bioethanol production from cellulose by the combination of cellulase-displaying yeast and ionic liquid pretreatment. Green Chem. 2011;13:2948.
12. Lee SM, Chang WJ, Choi AR, Koo YM. Influence of ionic liquids on the growth of *Escherichia coli*. Korean J Chem Eng. 2005;22:687.
13. Ganske F, Bornscheuer U. Growth of *Escherichia coli*, *Pichia pastoris* and *Bacillus cereus* in the presence of the ionic liquids [BMIM][BF4] and [BMIM][PF6] and organic solvents. Biotechnol Lett. 2006;28:465.
14. Pfruender H, Jones R, Weuster-Botz D. Water immiscible ionic liquids as solvents for whole cell biocatalysis. J Biotechnol. 2006;124:182.
15. Sendovski M, Nir N, Fishman A. Bioproduction of 2-phenylethanol in a biphasic ionic liquid aqueous system. J Agr Food Chem. 2010;58:2260.
16. Husson E, Buchoux S, Avondo C, Cailleu D, Djellab K, Gosselin I, et al. Enzymatic hydrolysis of ionic liquid-pretreated celluloses: contribution of CP-MAS ^{13}C NMR and SEM. Bioresour Technol. 2011;102:7335.
17. Auxenfans T, Buchoux S, Djellab K, Avondo C, Husson E, Sarazin C. Mild pretreatment and enzymatic saccharification of cellulose with recycled ionic liquids towards one-batch process. Carbohydr Polym. 2012;90:805.
18. Auxenfans T, Buchoux S, Larcher D, Husson G, Husson E, Sarazin C. Enzymatic saccharification and structural properties of industrial wood sawdust: recycled ionic liquids pretreatments. Energ Convers Manag. 2014;88:1094.
19. Chen Z, Feng Y, Wang J, Wang J, Guan W, Zhang H. Effects of [C_2mim][OAc] (1-Ethyl-3-methyl-imidazolium acetate) on the growth of wheat seedlings under Cd^{2+} stress. Bull Environ Contam Toxicol. 2014;92:714.
20. Biczak R, Pawłowska B, Bałczewskia P, Rychter P. The role of the anion in the toxicity of imidazolium ionic liquids. J Hazard Mater. 2014;274:181.
21. Krishnan MS, Ho NWY, Tsao GT. Fermentation kinetics of ethanol production from glucose and xylose by recombinant *Saccharomyces* 1400(pLNH33). Appl Biochem Biotechnol. 1999;77–79:373.
22. Franzén CJ. Metabolic flux analysis of RQ-controlled microaerobic ethanol production by *Saccharomyces cerevisiae*. Yeast. 2003;20:117.
23. Alfenore S, Cameleyre X, Benbadis L, Bideaux C, Uribelarrea JL, Goma G, et al. Aeration strategy: a need for very high ethanol performance in *Saccharomyces cerevisiae* fed-batch process. Appl Microbiol Biotechnol. 2004;63:537.
24. Ishtar Snoek IS, de Steensma HY. Factors involved in anaerobic growth of *Saccharomyces cerevisiae*. Yeast. 2007;24:1.
25. Rintala E, Toivari M, Pitkänen JP, Wiebe MG, Ruohonen L, Penttilä M. Low oxygen levels as a trigger for enhancement of respiratory metabolism in *Saccharomyces cerevisiae*. BMC Genomics. 2009;10:461.
26. Anderlei T, Büchs J. Device for sterile online measurement of the oxygen transfer rate in shaking flasks. Biochem Eng J. 2001;7:157.

27. Anderlei T, Zang W, Papaspyrou M, Büchs J. Online respiration activity measurement (OTR, CTR, RQ) in shake flasks. Biochem Eng J. 2004;17:187.
28. Stöckmann C, Losen M, Dahlems U, Knocke C, Gellissen G, Büchs J. Effect of oxygen supply on passaging, stabilising and screening of recombinant *Hansenula polymorpha* production strains in test tube cultures. FEMS Yeast Res. 2003;4:195.
29. Maier U, Losen M, Büchs J. Advances in understanding and modeling the gas–liquid mass transfer in shake flasks. Biochem Eng J. 2004;17:155.
30. Crabtree HG. Observations on the carbohydrate metabolism of tumours. Biochem J. 1929;23:536.
31. Baumann K, Dato L, Graf AB, Frascotti G, Dragosits M, Porro D, et al. The impact of oxygen on the transcriptome of recombinant S cerevisiae and P pastoris—a comparative analysis. BMC Genomics. 2011;12:218.
32. Christen S, Sauer U. Intracellular characterization of aerobic glucose metabolism in seven yeast species by ^{13}C flux analysis and metabolomics. FEMS Yeast Res. 2011;11:263.
33. Hagman A, Säll T, Piškur J. Analysis on yeast short-term Crabtree effect and its origin. FEBS J. 2014;281:4805.
34. Barnett JA, Entian K-D. A history of research on yeasts 9: regulation of sugar metabolism. Yeast. 2005;22:835.
35. Frick O, Wittman C. Characterization of the metabolic shift between oxidative and fermentative growth in *Saccharomyces cerevisiae* by comparative ^{13}C flux analysis. Microb Cell Fact. 2005;4:30.
36. Cabib E, Arroyo J. How carbohydrates sculpt cells: chemical control of morphogenesis in the yeast cell wall. Nat Rev Microbiol. 2013;11:648.
37. Lesage G, Bussey H. Cell wall assembly in *Saccharomyces cerevisiae*. Microbiol Mol Biol Rev. 2006;70:317.
38. Ge L, Wang XT, Ngin Tana S, Hang Tsaic H, Yonga JWH, Huac L. A novel method of protein extraction from yeast using ionic liquid solution. Talanta. 2010;81:1861.
39. Thuy Pham TP, Cho C-W, Yun Y-S. Environmental fate and toxicity of ionic liquids: a review. Water Res. 2010;44:352.
40. Frade RF, Afonso CA. Impact of ionic liquids in environment and humans: an overview. Hum Exp Toxicol. 2010;29:1038.
41. Gouveia W, Jorge TF, Martins S, Meireles M, Carolino M, Cruz C, et al. Toxicity of ionic liquids prepared from biomaterials. Chemosphere. 2014;104:51.
42. Seddon KR, Stark A, Torres MJ. Influence of chloride, water, and organic solvents on the physical properties of ionic liquids. Pure Appl Chem. 2000;72:2275.
43. Nockmann P, Binnemans K, Driesen K. Purification of imidazolium ionic liquids for spectroscopic application. Chem Phys Lett. 2005;415:131.
44. Urbanek M, Varenne A, Gebauer P, Krivánková L, Gareil P. Determination of trace cationic impurities in butylmethylimidazolium-based ionic liquids: from transient to comprehensive single-capillary counterflow isotachophoresis-zone electrophoresis. Electrophoresis. 2006;27:4859.
45. Vicari KJ, Tallam SS, Shatova T, Joo KK, Scarlata CJ, Humbird D, et al. Uncertainty in techno-economic estimates of cellulosic ethanol production due to experimental measurement uncertainty. Biotechnol Biofuels. 2012;5:23.
46. Konda NVSNM, Shi J, Singh S, Blanch HW, Simmons BA, Klein-Marcuschamer D. Understanding cost drivers and economic potential of two variants of ionic liquid pretreatment for cellulosic biofuel production. Biotechnol Biofuels. 2014;7:86.
47. Sarks C, Jin M, Sato TK, Balan V, Dale BE. Studying the rapid bioconversion of lignocellulosic sugars into ethanol using high cell density fermentations with cell recycle. Biotechnol Biofuels. 2014;7:73.
48. da Costa Lopes AM, João KG, Rubik DF, Bogel-Łukasik E, Duarte LC, Andreaus J, et al. Pre-treatment of lignocellulosic biomass using ionic liquids: wheat straw fractionation. Bioresour Technol. 2013;142:198.
49. Mehmood N, Olmos E, Goergen JL, Blanchard F, Ullisch D, Klöckner W, et al. Oxygen supply controls the onset of pristinamycins production by *Streptomyces pristinaespiralis* in shaking flasks. Biotechnol Bioeng. 2011;108:2151.
50. Gognies S, Belarbi A, Ait BE. *Saccharomyces cerevisiae*, a potential pathogen towards grapevine, *Vitis vinifera*. FEMS Microbiol Ecol. 2001;37:143.

Identification of a laccase Glac15 from *Ganoderma lucidum* 77002 and its application in bioethanol production

Zemin Fang[1,2], Xiaoman Liu[1,2], Liyuan Chen[3], Yu Shen[3], Xuecheng Zhang[1,2], Wei Fang[1,2], Xiaotang Wang[4], Xiaoming Bao[3*] and Yazhong Xiao[1,2*]

Abstract

Background: Laccases have potential applications in detoxification of lignocellulosic biomass after thermochemical pretreatment and production of value-added products or biofuels from renewable biomass. However, their application in large-scale industrial and environmental processes has been severely thwarted by the high cost of commercial laccases. Therefore, it is necessary to identify new laccases with lower cost but higher activity to detoxify lignocellulosic hydrolysates and better efficiency to produce biofuels such as bioethanol. Laccases from *Ganoderma lucidum* represent proper candidates in processing of lignocellulosic biomass.

Results: *G. lucidum* 77002 produces three laccase isoenzymes with a total laccase activity of 141.1 U/mL within 6 days when using wheat bran and peanut powder as energy sources in liquid culture medium. A new isoenzyme named Glac15 was identified, purified, and characterized. Glac15 possesses an optimum pH of 4.5 to 5.0 and a temperature range of 45°C to 55°C for the substrates tested. It was stable at pH values ranging from 5.0 to 7.0 and temperatures lower than 55°C, with more than 80% activity retained after incubation for 2 h. When used in bioethanol production process, 0.05 U/mL Glac15 removed 84% of the phenolic compounds in prehydrolysate, and the yeast biomass reached 11.81 (optimal density at 600 nm (OD_{600})), compared to no growth in the untreated one. Addition of Glac15 before cellulase hydrolysis had no significant effect on glucose recovery. However, ethanol yield were improved in samples treated with laccases compared to that in control samples. The final ethanol concentration of 9.74, 10.05, 10.11, and 10.81 g/L were obtained from samples containing only solid content, solid content treated with Glac15, solid content containing 50% prehydrolysate, and solid content containing 50% prehydrolysate treated with Glac15, respectively.

Conclusions: The *G. lucidum* laccase Glac15 has potentials in bioethanol production industry.

Keywords: Laccase, *Ganoderma lucidum*, Identification, Characterization, Bioethanol, Detoxification

Background

Bioethanol production from renewable biomass such as lignocellulosic materials has been the focus of numerous contemporary investigations with the decreasing supply of fossil fuels and the increasing environmental concerns associated with them [1]. Lignocellulosic feed stocks such as sugar cane bagasse and corn stover are cheap, renewable, and abundant in nature. They do not compete with food production and contribute to reduce the use of fossil fuels, thus alleviating carbon dioxide emission and global warming [1]. Lignocellulosic feed stocks mainly consist of two polysaccharides, cellulose and hemicellulose, which can be hydrolyzed to provide monosaccharides used by microbial biocatalysts in fermentation processes [2,3]. Unfortunately, cellulose and hemicellulose are closely linked by lignin, a polymer that acts as cementing agent between cellulose fibers [2]. Lignocellulosic materials are thus recalcitrant to hydrolysis

* Correspondence: bxm@sdu.edu.cn; yzxiao@ahu.edu.cn
[3]The State Key Laboratory of Microbial Technology, School of Life Sciences, Shandong University, 27 Shanda Nanlu, Jinan, Shandong 250100, China
[1]School of Life Sciences, Anhui University, 111 Jiulong Road, Hefei, Anhui 230601, China
Full list of author information is available at the end of the article

and require pretreatment before being converted to bioethanol [3,4].

The combination of steam explosion and acidic catalyst such as H_2SO_4 or SO_2 is one of the most commonly used methods for lignocellulose pretreatment, since it breaks and/or removes a minor part of lignin, depolymerizes cellulose and hemicellulose, and makes the biomass more accessible to hydrolytic enzymes with relatively lower cost [5]. However, this pretreatment generates some soluble inhibitory compounds from partial degradation of sugars and lignin, which can affect enzymatic hydrolysis as well as fermentation steps, and will reduce the ethanol productivity of the microorganisms and the final ethanol yield [2,3,5].

The inhibitory compounds generated during the pretreatment process include weak acids, furan derivatives, and phenolic and other compounds [2]. Several procedures for the removal of these compounds including biological, physical, and chemical methods have been assayed. Among which, enzymatic treatment using laccases has been suggested as one of the most promising approach in lignocellulosic biomass detoxification [3,5]. Laccases (benzenediol:oxygen oxidoreductases, EC1.10.3.2) are a family of blue multicopper oxidases that are capable of oxidizing a wide range of phenolic and aromatic compounds, with concomitant reduction of molecular oxygen to water [6]. When detoxifying the lignocellulosic hydrolysate, laccase was suggested to be selective and can virtually remove all phenolic monomers. Compared with other approaches, advantages of using laccase are that enzyme preparations have higher catalytic efficiencies, fewer toxic subproducts, shorter treatment time, and lower energy costs [3,5,7]. The disadvantage of using laccase is the high enzyme production costs, as the commercial laccases are still expensive despite strides made to reduce enzyme cost through modern biotechnology, thus restricting the applications of laccases [3]. Screening for laccases with higher lignocellulosic hydrolysate-detoxifying ability may improve the present situation.

Laccases are widespread in white-rot fungi. Among which, *Ganoderma lucidum* is a representative that is able to grow on withered wood and other agro-lignocellulosic biomass materials and can decompose lignin by secreting enzymes such as laccases [8]. For instance, *G. lucidum* is one of the richest set of wood degradation enzymes among all of the sequenced basidiomycetes [9]. Based on this, and the fact that *G. lucidum* can produce high amount of laccase in liquid cultures [10-12], laccases from this fungus may promote degradation/modification of lignin or lignin-derived components [13] and therefore represent proper candidates in processing lignocellulosic biomass. To address this issue, a fungal strain *G. lucidum* 77002 was investigated for laccase production by using agricultural by-products as media, and a laccase identified as Glac15 was purified and characterized for the first time; its application potential in lignocellulosic biomass detoxification after thermochemical pretreatment was also evaluated. Our results indicated that Glac15 is a promising candidate in bioethanol production.

Results and discussion

Laccase production in liquid culture

Wheat and peanut are two of the most common agricultural crops in China. To obtain laccases with lower cost, along with the fact that wheat bran can induce high amount of lignocellulose-degrading enzymes including laccase in liquid culture [14], wheat bran (3% dry weight, *w*/*v*) and peanut powder (3% dry weight, *w*/*v*) were used as energy sources to induce *G. lucidum* 77002 to produce laccase. Laccase activity reached 141.1 ± 0.2 U/mL within 6 days when assayed using 2,2′-azino-bis-(3-ethylbenzothiazoline-6-sulfonic acid) (ABTS) as a substrate (Figure 1), indicating that these agriculture by-products could effectively induce laccase production. This is consistent with previous reports that other agriculture and forest by-products such as pine and poplar ground woods [10] and wheat bran [14] can induce laccase production in *G. lucidum*, which were 0.93 and 97.34 U/mL, respectively. Besides, tamarind shell with ethanol (3%, *v*/*v*), $CuSO_4$ (0.4 mM), and gallic acid (1 mM)-containing medium can also induce the *G. lucidum* laccase production, yielding about 74.84 U/mL laccase activity after 15 days, which was 416 times higher than the control [15]. However, a different laccase zymography was shown according to culture conditions. As indicated by native polyacrylamide gel electrophoresis (PAGE), *G. lucidum* 77002 produced three laccase isoforms in this study (Figure 2a), whereas a *G. lucidum* strain produced only two laccase isoenzymes in high-nitrogen culture and in cultures containing both poplar and pine [10]. In contrast to these

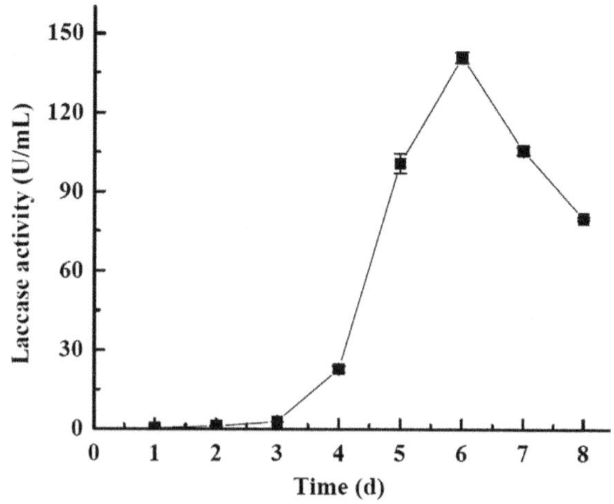

Figure 1 Laccase activity of the *G. lucidum* using wheat bran and peanut powder in liquid culture.

results, when induced by the mixture of tamarind shell, ethanol, $CuSO_4$, and gallic acid, another *G. lucidum* strain produced four laccase isoenzymes [15]. Responsive elements that are found in various locations along the laccase promoter sequences may contribute to the variation in the laccase isoforms produced among *G. lucidum* strains. Several reports have highlighted their presence and postulated that these responsive elements, including metal-responsive element, xenobiotic response element, heat shock-responsive element, and antioxidant response element, may regulate laccase gene transcription, and their locations and orientations suggest a complex picture of laccase expression regulation [16]. On the other hand, the differential expression of laccase isoenzyme genes might also arise from different ecological origins of these mushrooms [6,8].

Laccase purification and characterization

Laccase was purified from crude culture filtrate according to the methods described in the 'Methods' section. The summary of purification steps is presented in Table 1. One laccase was 14.19-fold purified from its initial culture broth with a final yield of 5.5%. The purified laccase exhibited a single band both on native PAGE and sodium dodecyl sulfate (SDS)-PAGE (Figure 2), suggesting a homogenous form. The specific activity of the laccase was 186 U/mg when estimated with ABTS as the substrate.

G. lucidum contains 16 laccase genes in its genome [17]. Based on data-dependent liquid chromatography-electrospray ionization-tandem mass spectrometry (LC-ESI-MS/MS), the purified laccase matched the laccases Glac6 and Glac15 (Additional file 1: Table S1 and Table S2) in the genome of *G. lucidum* Xiangnong No.1. Furthermore, its N-terminal sequence was GIGPTTDLTISNADI, completely matching the amino acidic sequence starting at position 22 of Glac15. However, *G. lucidum* 77002 is not the same strain compared to *G. lucidum* Xiangnong No.1, indicating that the purified protein may be Glac15, and suggested that a 21-amino acid signal peptide exists in the zymogen. As the protein obtained in this study showed the highest sequence identity with Glac15 from *G. lucidum* Xiangnong No.1, it was also designated as Glac15. To the best of our knowledge, this is the first report of Glac15 from *G. lucidum* [10-15,17-22]. Data bank homology searches (NCBI) further confirms the novelty of the purified Glac15 (Table 2) and returned a laccase from *G. lucidum* (GenBank: AHA83584) as the best hit (93% identity), followed by another two laccases from *G. lucidum* (GenBank: ABK59822 and ABK59823) with 87% identity. These three proteins were identified by whole genome sequencing and have not been characterized. When compared with the characterized laccases from *G. lucidum*, the N-terminal region of Glac15 shared the highest sequence homology of 93% with the laccase of KULac1, KULac3, KULac4, and KULac5 from *Ganoderma* sp. KU-Alk4 [22] (Table 2).

Effect of pH on Glac15 activity

The pH profiles for fungal laccases toward phenolic substrates are usually narrow bell-shaped, with optimal pH around 5, and little activity can be detected at neutral pH or above [23,24]. Similar to most fungal laccases, Glac15 could oxidize laccase-specific substrate syringaldazine at a pH range of 4.0 to 7.0, with a pH optimum of 4.5 (Figure 3a). The pH optimum for Glac15 on oxidizing other phenol substrates such as 2,6-dimethoxyphenol (2,6-DMP), guaiacol, catechol, and L-dopamine

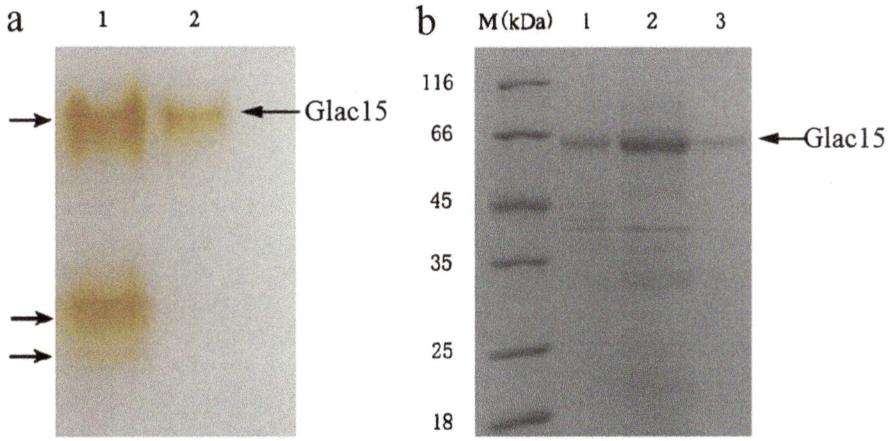

Figure 2 Native PAGE and SDS-PAGE of laccase purification. (a) Zymography of crude and purified laccase. Laccase activity was performed using 1 mM guaiacol as substrate. Lanes: 1: crude laccase; 2: purified laccase. **(b)** SDS-PAGE of crude and purified laccase. Lanes: 1: crude protein extract produced by filtering the fermentation supernatant through six layers of sterile gauze; 2: crude protein extract after ultrafiltration; 3: purified Glac15 after DEAE-Sepharose FF chromatography.

Table 1 Summary of the purification steps of extracellular laccase Glac15 from *G. lucidum* 77002

Purification step	Total volume (ml)	Total activity (U)	Total Protein (mg)	Specific activity (U/mg)	Percent recovery (%)	Purification fold
Crude culture filtrate	1,250	47,600	3,630	13.1	100.00	1.00
Ultrafiltration	100	28,000	1,100	25.5	58.8	1.94
DEAE-Sepharose FF	30	2,600	14	186	5.5	14.19

were 4.5, 5.0, 5.0, and 4.5, respectively (Table 3). In contrast, for non-phenolic substrates, ABTS and $K_4Fe(CN)_6$, Glac15 has been observed to present monotonic activity profiles, in which the rate decreased as the pH increased from 2.5 to 7.0, with the optimum pH being 2.5 (Table 3), like most fungal laccases [23]. The monotonic pH-dependent decrease of the activity has been interpreted as the sole effect of hydroxyl ion inhibition [24]. On the other hand, Glac15 was highly stable at pH 5.0 to 7.0. After incubation at 50°C for 120 min, Glac15 retained more than 70% of its original activities. In contrast, the enzyme was unstable at low pH, with approximately 20% activity left after incubation at pH 4.0 (Figure 3b) for 120 min at 50°C. This is similar to most fungal laccases, which are functional at acidic and near neutral pH but lose their activities under alkaline conditions [22,25].

Effect of temperature on Glac15 activity

Glac15 showed its maximal activity at 45°C toward syringaldazine and displayed more than 80% of the maximal activity at temperatures ranging from 35°C to 65°C for syringaldazine (Figure 3c). However, laccase from *G. lucidum* with the optimal temperature below 35°C has been described, for example, Ko *et al.* [18] reported a *G. lucidum* laccase showed the highest activity at 25°C. The thermostability of fungal laccases varies considerably [23]. In contrast to a *G. lucidum* laccase, which was inactivated at 60°C [18], Glac15 was rather stable at temperatures lower than 55°C, with about 90% activity retained after incubation at pH 5.0 and 55°C for 2 h (Figure 3d). When incubated at 60°C, Glac15 retained 40% activity after 120 min, with a half-life of 82 min (Figure 3d). Compared with other fungal laccases with high thermostability, Glac15 may be more suitable for several specific applications that run in moderate temperature, such as food and drink industries including ethanol production [23,26].

Table 2 Overview of N-terminal sequence of Glac15 to other laccases from *G. lucidum*

Protein	Sequence	Reference
Lac15	GIGPTTDLTISNADI	This study
GLac 1, 2, 3	GIGPT	[18]
-	GQNGDAVP	[19]
-	GIGPK	[20]
GLlac1	GIGPK	[21]
KULac 2	GIGPVADLTVRGGDI	[22]
KULac 1; 3; 4; 5	GIGPVTDLTISNADI	[22]
Lac1	GIGPTTDLTISNANI	AHA83584
Lac4	GIGPKTDLTISNADV	ABK59822/ABK59823

Substrate specificity

Diverse substrates were oxidized by Glac15, and the kinetic parameters of the enzyme were calculated for each of them. As summarized in Table 3, ABTS and syringaldazine were the most reactive substrates for Glac15 in terms of affinity (K_m) and catalytic efficiency (k_{cat}/K_m). The K_m values for ABTS and syringaldazine were in micromolar range. As most fungal laccases, these results confirm the specificity of ABTS and syringaldazine as non-phenolic and phenolic substrates, respectively, in laccase enzyme reaction. Glac15's substrates range was ranked as follows: ABTS > syringaldazine > 2,6-DMP > catechol > guaiacol > L-dopamine > $K_4Fe(CN)_6$. This is quite different from Tplac from *Trametes pubescens*, which showed a rank of ABTS > 2,6-DMP > L-dopa > guaiacol > syringaldazine > catechol [27]. Meanwhile, the K_m and k_{cat}/K_m values for ABTS were 19 μM and 10.5 M^{-1}/s^{-1}, respectively, higher than those of certain fungal laccases [28,29].

Effects of organic solvents, metal ions, and inhibitors on Glac15 activity

Although there are several studies on characterization of laccases from *G. lucidum*, few of them reported the effects of organic solvents on the enzyme activity [10-15,17-22]. In order to obtain more information about the laccase from *G. lucidum*, ethanol, dichloromethane, ethyl ether, ethyl acetate, dimethyl sulfoxide (DMSO), acetaldehyde, and acetone were used to test the effects of organic solvents on Glac15 activity. All of the organic solvents tested inhibited protein activity, with about 80% of the original laccase activity was retained in the presence of 5% organic solvent (*v*/*v*). This inhibitory effect of organic solvents increases as their concentration was increased (Figure 4). This is consistent with previous studies demonstrating that fungal laccases from *Trametes versicolor* [30] and *Pleurotus ostreatus* [31] were inactivated over 10% of the organic solvents. The instability of laccase in organic solvents may be attributed to the denaturation of the enzyme

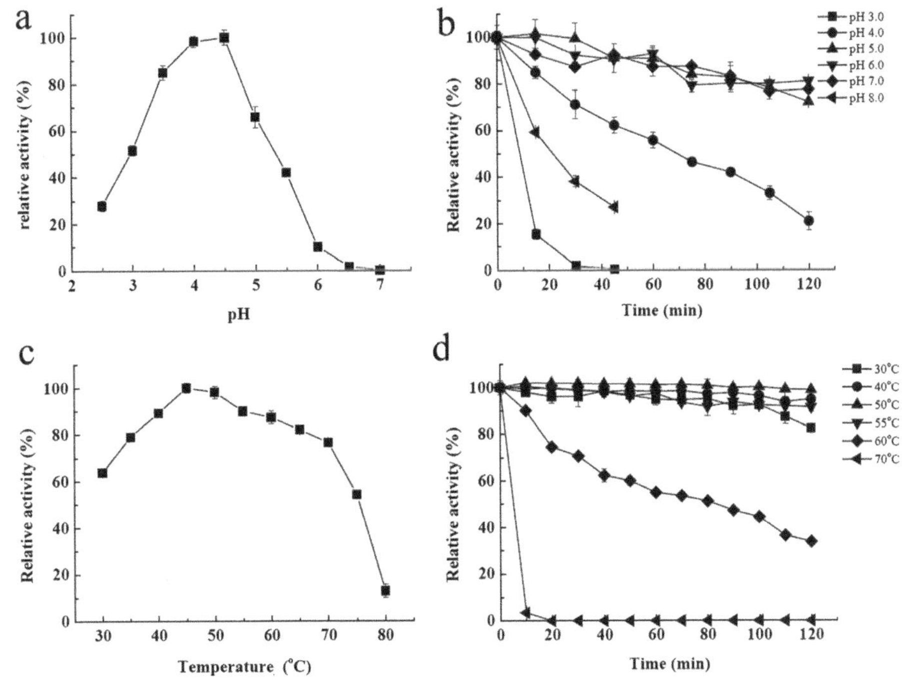

Figure 3 Effects of pH and temperature on the activity and stability of Glac15. (a) pH optimum, **(b)** pH stability, **(c)** temperature optimum, and **(d)** thermostability. Laccase activity was determined in 50 mM citrate-Na_2HPO_4 buffer.

structure caused by interfering hydrophobic effects or the substitution of water molecules by the organic solvents at the active sites [30]. However, Glac15 retained about 60% of its original activity in the presence of 15% ethanol (Figure 4). Enzymatic reactions in compatible organic solvents such as ethanol allow access of laccase to some water-insoluble substrates, which may be helpful in the detoxification of many recalcitrant organic pollutants [32].

Cu^{2+} had negative effect on Glac15 activity (Figure 4). This is apparently different from bacterial laccases, as no activity was detected for the purified bacterial laccase but resumed by addition of Cu^{2+} [33]. Similar to Cu^{2+}, Ca^{2+} also inhibited Glac15 activity. Two theories are proposed to explain the metal effects regarding laccase activity. One is that the binding of several metal ions, that is, Cu^{2+} or Ca^{2+}, induces conformational modification of the enzyme and stimulates decomposition of the trimer complex containing substrate, enzyme, and metal ion, as evidenced by non-competitive inhibition model. The other is that the metal ion binds near the TI site of laccase and acts as a competitive inhibitor of electron donors by blocking the access of substrates to the TI site or inhibiting the electron transfer at the TI active site, thereby leading to inhibition of laccase activity [27,33]. Other cations such as Mn^{2+}, Li^{+}, Zn^{2+}, and K^{+} at a final concentration of 10 mM did not have noticeable effects on Glac15 activity.

SDS is one of the anionic detergents known for its deleterious effect on the activity of most enzymes. In this study, SDS inhibited the activity, with complete inactivation (about 100%) in 5 mM SDS (Figure 4). According to Moore and Flurkey, the inhibition effect of SDS may be explained by the binding of small amounts of SDS on Glac15 which alters both its enzymatic and physical

Table 3 Substrate specificity of Glac15

Substrate	Optimal pH	Optimal temp. (°C)	K_m (M)	k_{cat} (s^{-1})	k_{cat}/K_m (M^{-1} s^{-1})
Syringaldazine	4.5	45	6.2×10^{-5}	3.87×10^{3}	6.24×10^{7}
2,6-DMP	4.5	45	1.2×10^{-4}	7.44×10^{2}	6.20×10^{6}
Guaiacol	5.0	55	7.9×10^{-4}	1.70×10^{3}	2.15×10^{6}
Catechol	5.0	45	7.8×10^{-4}	5.48×10^{3}	7.03×10^{6}
L-dopamine	4.5	50	8.9×10^{-4}	2.95×10^{3}	3.31×10^{6}
ABTS	2.5	50	1.9×10^{-5}	2.00×10^{2}	1.05×10^{7}
$K_4Fe(CN)_6$	2.5	45	4.6×10^{-3}	2.71×10^{4}	5.89×10^{6}

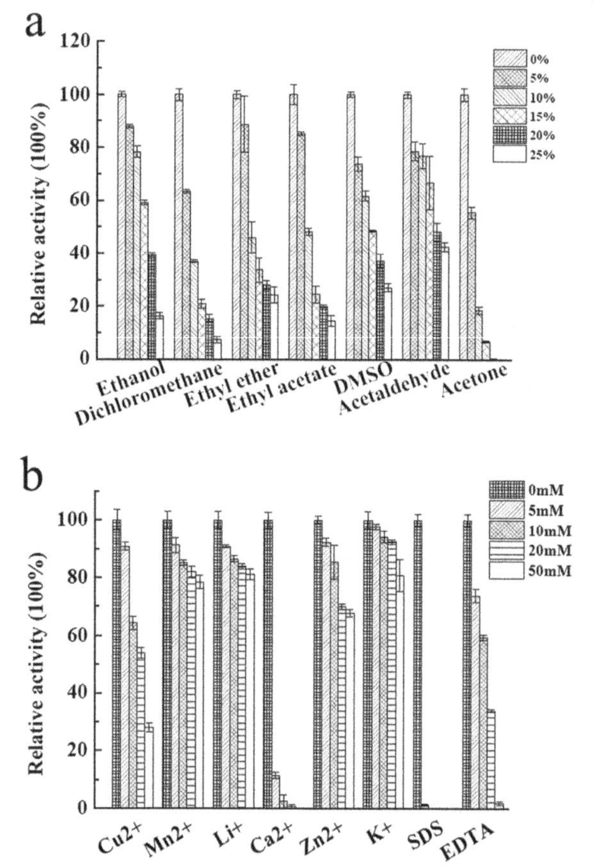

Figure 4 Effects of organic solvents, metal ions, and inhibitors on Glac15 activity. **(a)** Organic solvents and **(b)** metal ions and inhibitors.

characteristics [34]. However, only partial inhibition (about 74%) was observed for Glac15 in the presence of 5 mM ethylenediaminetetraacetic acid (EDTA) (Figure 4), compared with complete inactivation observed for laccases from *Thermus thermophilus* [35] and *Aspergillus ochraceus* [36] in the presence of 1 mM EDTA. Cabana *et al.* [37] have reported that the aggregation of laccase as cross-linked enzyme aggregates (CLEAs) improved its stability against EDTA by hindering access of EDTA to the Cu(II) ions of the protein. Thus, the resistance of Glac15 toward EDTA might be ascribed to the low accessibility of EDTA to the structural copper atoms of the protein.

Laccase treatment of corn stover prehydrolysate

After acidic steam pretreatment, the solid content and the prehydrolysate were separated. The pretreated solid content contained a total dry matter of 28.7% (*w*/*w*). Corn stover usually contains 33% to 40% cellulose, 13% to 31% hemicellulose, and 5% to 26% lignin [4]. As a consequence of hemicellulose solubilization, high xylose (11.39 g/L), arabinose (1.37 g/L), and galactose (1.35 g/L) were detected, and different degradation products were identified and quantified as soluble compounds in prehydrolysate. Among which, lignin, furfural, 5-hydroxymethylfurfural (5-HMF), and acetic acid were formed at highest concentrations (Table 4). Furfural and 5-HMF are derived from pentoses especially xylose and hexose degradation, respectively, and acetic acid is formed by the hydrolysis of acetyl groups in hemicellulose [38]. In addition, phenolic compounds at a final concentration of 28.6 mM were detected in the prehydrolysate. These compounds, derived from guaiacyl propane units and syringylpropane units present in lignin [39], are considered to be inhibitors and can affect biochemical pathways in the fermenting microorganisms and interact with cellulolytic enzymes, leading to reduced final ethanol titers and volumetric productivities [2-4].

Laccases have been applied in a few cases to remove specifically phenolic compounds in steam-exploded biomass [7]. However, little information of laccases from *G. lucidum* on removing phenolic compounds from steam-exploded biomass has been observed [7,13]. Our results showed that Glac15 can function on the phenolic compounds derived from steam-exploded corn stover. When prehydrolysate pH was adjusted from 1.4 to 3.0, 4.0, and 5.0 in control samples, total phenolic compound contents decreased from 28.61 to 25.21, 24.39, and 23.38 mM, respectively, in the first 10 min and were not significantly changed even after 24 h incubation, suggesting that phenolic compounds during steam explosion are solubilized and stable at pH 1.4 [5]. However, when pH was increased to 3.0 to 5.0, acetic acid was transformed to acetate. This might lead to the polymerization of phenolic compounds. Addition of Glac15 removed about 60%, 78%, and 84% of the phenolic compounds at pH 3.0, pH 4.0, and pH 5.0, respectively, after 24 h treatment (Figure 5). As lignin derivates generated during pretreatment are phenolic compounds similar to guaiacol and syringaldazine, the increased reduction ability of Glac15 with increment of pH could be explained by the higher laccase activity observed with phenolic substrates at pH 4.5 to 5.0 (Table 3). On the other hand, higher laccase doses (0.05, 0.1, and 0.2 U/mL) did not result in significantly higher detoxification (Figure 5). Increasing enzyme specificity for removing phenols allows decreasing enzymes doses. Therefore, low Glac15 concentrations are suitable for detoxification of steam-exploded corn stover, which is an important aspect for the application of enzymes in industrial applications [5].

Glac15-treated prehydrolysate was employed to test the effect of laccase treatment on yeast growth. When *Saccharomyces cerevisiae* CICC31014 strains were cultured in yeast nitrogen base (YNB) medium in the presence of 25%, 50%, and 75% prehydrolysate (the final

Table 4 Composition of pretreated corn stover and prehydrolysate

Corn stover		Prehydrolysate			
Component	**% weight**	**Monosaccharides**	**Concentration (g/L)**	**Inhibitors**	**Concentration**
Cellulose	23.15	Glucose	5.74	Furfural	2.28 g/L
Hemicellulose	13.44	Xylose	11.39	5-HMF	0.33 g/L
Lignin	10.92	Arabinose	1.37	Acetic acid	0.309 g/L
Extracts	34.50	Galactose	1.35	Formic acid	0.045 g/L
Unknown compounds	17.99			Lignin	4.9 g/L
				Total phenol	28.6 mM

glucose concentration was 21.44, 22.87, and 24.31 g/L, respectively), strains showed no growth in all of the untreated samples (Figure 6). In contrast, after treatment with laccase, yeast cell growth improved considerably. The biomass reached 5.55 ± 0.27, 9.33 ± 0.32, and 11.81 ± 0.33 (optimal density at 600 nm (OD_{600})) in the samples containing 25%, 50%, and 75% prehydrolysates, respectively. In accordance with Moreno *et al.* [40], who reported the treatment of slurry with laccase increased the yeast cell viability, our results also showed the increased cell viability, evidenced the detoxication effect of Glac15 on prehydrolysate. The higher cell density could be attributed to the higher concentration of glucose added to the medium with the addition of prehydrolysates. On the other hand, the higher the concentration of prehydrolysate, the higher the concentrations of other inhibitors such as furan derivates and weak acids. As a result, yeast growth in the former two samples had a shorter lag period (12 h) than that in the last sample (37 h) (Figure 6).

Effects of laccase treatment on enzymatic hydrolysis and ethanol fermentation

The effect of laccase treatment on enzymatic hydrolysis was evaluated. After treating the solid fraction with Glac15, the glucose recovery was not apparently changed when checked by a statistical analysis (Figure 7, Table S2). This is different from the previous studies reported by Tabka *et al.* [41] and Jurado *et al.* [5], who described a lower glucose recovery after enzymatic hydrolysis of steam-exploded wheat straw treated with laccase. The lower recovery was explained by the release of certain phenolic compounds by laccases inhibiting cellulolytic enzymes [5,41,42]. Based on our results, cellobiose was increased from 2.41 ± 0.17 to 3.25 ± 0.21 g/L after treating with Glac15, indicating that cellobiose was accumulated during the process. On the contrary, improved glucose recovery by laccase treatment of steam-exploded softwood has been reported, and a decrease in the unproductive binding of cellulases to lignin after laccase treatment has been suggested [43]. With the addition of prehydrolysate, cellulase activities may have been inhibited, as there was no increase in glucose concentration after addition of prehydrolysate. Statistical analysis showed that the glucose and cellulase recoveries were not significantly changed even after treating with Glac15 ($P > 0.05$), indicating that Glac15 had no effect on saccharification in the presence of prehydrolysate.

From an economical and environmental point of view, the use of prehydrolysate can increase the fermentable sugar concentration, decrease both operational costs and wastewater, and obtain a higher ethanol production [42,44,45]. As a result, four groups containing only solid content (5%, *m*/*v*), solid content (5%, *m*/*v*) and treated

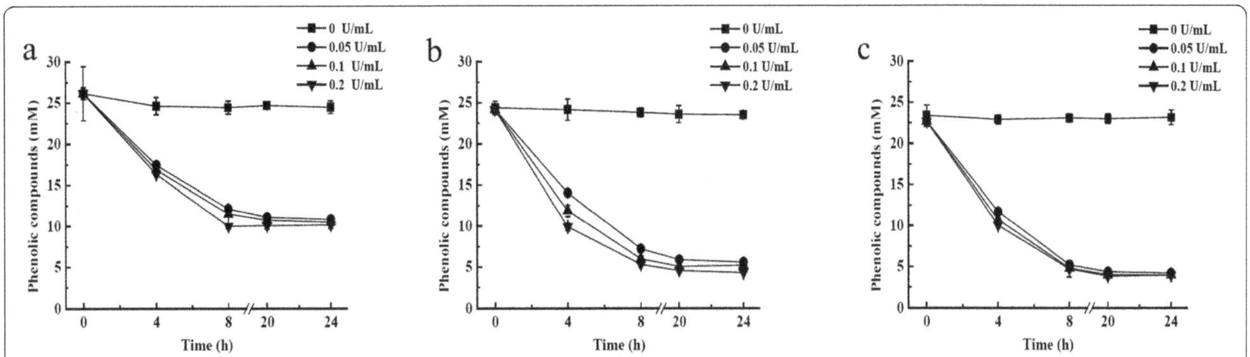

Figure 5 Time course of phenolic content of prehydrolysate during pretreatment with Glac15 at different pHs. (a) pH 3.0, **(b)** pH 4.0, and **(c)** pH 5.0.

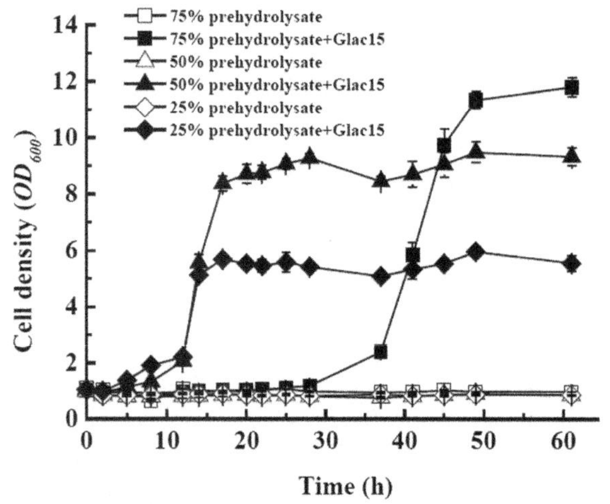

Figure 6 Time course of yeast growth in the presence of different concentrations of prehydrolysate treated or not with Glac15.

with Glac15, solid content (5%, *m*/*v*) containing 50% prehydrolysate, and solid content (5%, *m*/*v*) containing 50% prehydrolysate treated with Glac15 were employed to evaluate the effect of laccase treatment on ethanol fermentation. Fermentation was improved after detoxification with Glac15. After 60 h fermentation with *S. cerevisiae*, results showed that sugar consumption and ethanol yield were higher in samples treated with laccases than in control samples (Figure 7). The final ethanol concentrations of three groups were 9.74 ± 0.18, 10.05 ± 0.15, 10.11 ± 0.21, and 10.81 ± 0.23 g/L, respectively (Figure 7c). Laccase treatment to increase ethanol production has been reported by others. For example, Jonsson *et al.* [46] reported that ethanol productivity was increased by two to three times when treated a willow hemicellulose hydrolysate from willow with laccase and lignin peroxidase from the white-rot basidiomycete fungus *T. versicolor*. When laccase was used to detoxify water-impregnated wheat straw or acid-impregnated wheat straw, ethanol yield was two times higher than the control that has not been treated with laccase [5]. Compared to these results, an increasement of about 10% ethanol was obtained when treated with Glac15. To maximize the improvement of ethanol production caused by laccase treatment, further researches on decreasing the quantity of cellulases enzymes in enzymatic hydrolysis assays to lower activities, or increasing solid fraction consistency, are to be performed.

Conclusions

In this study, *G. lucidum* 77002 produces 141.1 ± 0.2 U/mL laccase activity within 6 days when using wheat bran and peanut powder as energy sources in liquid culture medium. A new isoenzyme named Glac15 was identified, purified, and characterized. It showed high activity toward phenolic and non-phenolic substrates syringaldazine and ABTS. When used in bioethanol production process, low amount Glac15 removed 84% of the phenolic compounds in prehydrolysate at pH 5 and considerably improved yeast growth. Addition of Glac15 before cellulase hydrolysis improved 10% ethanol yield. Our results showed that the *G. lucidum* laccase Glac15 has potentials in bioethanol production industry.

Methods

Chemicals, strains, media, and culture conditions

Filamentous fungus *G. lucidum* 77002 was preserved at culture collection of the School of Life Sciences, Anhui University, China (China Center for Type Culture Collection No. AF2013025). *S. cerevisiae* was from China Center of Industrial Culture Collection, Beijing, China (No. CICC31014). Syringaldazine, 2,6-DMP, guaiacol, catechol, L-dopamine, ABTS, and $K_4Fe(CN)_6$ were purchased from Sigma-Aldrich (St. Louis, MO, USA). Wheat bran was from Zhangyuan (Bozhou, China), and peanut powder was purchased from Huiyou (Hefei, China).

G. lucidum 77002 was maintained on potato dextrose agar (PDA, contained 20.0 g glucose, 1.0 g KH_2PO_4, 1.5 g $MgSO_4 \cdot 7H_2O$, 50.0 μg vitamin B_1, 15.0 g agar powder, and 200.0 g potato extract liquid/L) slants at 4°C. YMG

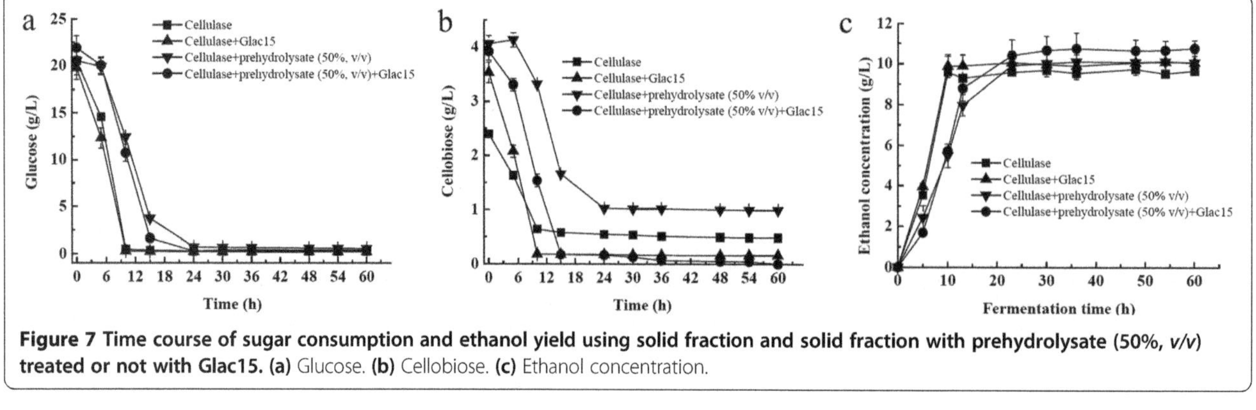

Figure 7 Time course of sugar consumption and ethanol yield using solid fraction and solid fraction with prehydrolysate (50%, *v*/*v*) treated or not with Glac15. (a) Glucose. **(b)** Cellobiose. **(c)** Ethanol concentration.

liquid medium (contained 10.0 g malt extract, 4.0 g yeast extract, and 10.0 g glucose/L) was used to prepare *G. lucidum* 77002 inoculation. The liquid fermentation medium used for producing laccase contained 30.0 g peanut powder, 30.0 g wheat bran, and 1.5 g KH_2PO_4/L. YNB medium (contained 1.7 g yeast nitrogen base, 5.0 g ammonia sulfate, and 20.0 g glucose/L) was used for culture of *S. cerevisiae.*

Laccase production in liquid culture

When producing laccases, four to five plugs (diameter in 5 mm) of *G. lucidum* 77002 actively grown on PDA plates were used to inoculate 250-mL Erlenmeyer flasks containing 100 mL YMG medium. The culture was incubated at 28°C on a rotary shaker at 120 rpm. A 4-day-old liquid culture was homogenized with a sterile blender at 3,000 rpm for 10 s, and a volume of 5 mL was inoculated into the new 400 mL (in 1-L Erlenmeyer flasks) liquid fermentation medium and was cultivated under the same conditions for 8 days.

Laccase purification

Unless otherwise stated, all procedures were performed at 4°C. The culture fluid was first filtered through six layers of sterile gauze and washed with three volumes of citrate-phosphate buffer (50 mM, pH 5.0). The aqueous solution (1.25 L) was centrifuged at 10,000 × *g* for 10 min and then concentrated to 100 mL in a Minitan Ultrafiltration System (Millipore, Bedford, MA, USA) with a low-binding regenerated cellulose membrane (Millipore, Bedford, MA, USA). The concentrate was centrifuged at 12,000 × *g* for 20 min to discard the sediment generated during the ultrafiltration, and the supernatant was then dialyzed against buffer A (20 mM citrate-phosphate, pH 7.5) overnight followed by centrifugation as before. The supernatant was applied to a DEAE-Sepharose FF column (10 × 200 mm, Amersham Pharmacia, Uppsala, Sweden) preequilibrated with buffer A. The column was then washed with approximately 100 mL buffer A to remove melanin and polysaccharide and eluted with a linear gradient of $(NH_4)_2SO_4$ (0 to 1 M in buffer A; flow rate 0.8 mL/min). The fractions containing laccase were pooled and dialyzed against 50 mM citrate-Na_2HPO_4 buffer (pH 7.5) and stored at 4°C.

Laccase identification

The purified laccase was identified according to Rühl *et al.* [47] with slight modifications. Briefly, the purified protein was recovered for SDS-PAGE [48]. Coomassie brilliant blue R-250-stained protein bands were cut with a razor blade. The gel pieces were digested with trypsin and analyzed by LC-ESI-MS/MS (LTQ, Thermo Fisher Scientific, Shanghai, China). Proteins were then identified by searching the data against a database of the *G. lucidum* Xiangnong No.1 genome [17]. To further confirm the identity of protein, purified laccase was electro-blotted directly from the SDS-PAGE gel onto a polyvinylidene difluoride membrane (Sequi-Blot PVDF, Bio-Rad, Hercules, CA, USA) and located by Coomassie blue R-250 staining. The band on the PVDF membrane corresponding to the Glac15 was excised and subjected to Edman degradation. N-terminal sequences of the proteins were determined using a PPSQ-33A Edman degradation unit (Shimadzu, Kyoto, Japan) and the data was processed by PPSQ-30 Data Processing software.

Enzyme assay

The assay mixture consisted of 10 μL of appropriately diluted laccase stock and 990 μL of 50 mM citrate-phosphate buffer (pH 4.0) containing 0.5 mM ABTS (ε_{420} = 36,000 M^{-1} cm^{-1}) [49]. The reaction was initiated by adding enzyme into the solution. After incubation at 50°C for 3 min, the absorbance was immediately measured at 420 nm. Alternative substrates for the measurement of laccase activity were 100 μM syringaldazine (ε_{525} = 65,000 M^{-1} cm^{-1}) [50], 2 mM 2,6-DMP (ε_{468} = 49,600 M^{-1} cm^{-1}) [51], 2 mM guaiacol (ε_{465} = 12,000 M^{-1} cm^{-1}) [52], 2 mM catechol (ε_{388} = 1,300 M^{-1} cm^{-1}) [51], 2 mM L-dopamine (ε_{475} = 2,835 M^{-1} cm^{-1}) [53], and 2 mM $K_4Fe(CN)_6$ (ε_{405} = 900 M^{-1} cm^{-1}) [54], respectively. Reactions with heat-treated laccase were used as the controls. One activity unit (U) was defined as the amount of enzyme required for oxidizing 1 μmol of substrate/min [33].

Laccase characterization

The effect of pH on laccase activity was examined in the pH range of 2.5 to 7.0 at 50°C in 50 mM citrate-phosphate buffer. The enzyme stability against pH was determined by measuring the residual laccase activities after incubation at 50°C in the aforementioned buffers, and residual activities were determined every 15 min. The effect of temperature on the activity was measured by incubating protein at optimal pH and a temperature range of 30°C to 80°C in an interval of 5°C. Thermostability was determined by incubating protein at various temperatures (30°C to 70°C) at pH 5.0, and residual activities were determined every 10 min.

The effects of metal ions (Cu^{2+}, Mn^{2+}, Li^{+}, Ca^{2+}, Zn^{2+}, K^{+}, using sulfate as the donor, 0 to 50 mM), SDS (0 to 50 mM), EDTA (0 to 50 mM), and organic solvents (ethanol, dichloromethane, ethyl ether, ethyl acetate, DMSO, acetaldehyde, and acetone, 0 to 25%, *v*/*v*) on the enzyme activity were investigated by adding different concentrations of each effector into the reaction system when addition of substrate ABTS. The enzymatic assays were conducted under the conditions described in enzyme assay.

Enzyme kinetics

The kinetics and the specificity of Glac15 toward each of these substrates were measured at the optimum pH of the enzyme (Table 3). The Michaelis-Menten and catalytic constants (Km, k_{cat}, and k_{cat}/Km) were determined by incubating Glac15 with various concentrations of syringaldazine, 2,6-DMP, guaiacol, catechol, L-dopamine, ABTS, and $K_4Fe(CN)_6$. The values for the kinetic parameters and their corresponding errors were then calculated.

Corn stover pretreatment

Corn stover was used as raw material. It was milled using a laboratory hammer mill, in order to obtain a chip size round 2 mm, and stored at room temperature until use. Milled biomass was pretreated by steam explosion with a raw material: H_2O proportion of 1:10 (*w*/*v*) and in the presence of 0.8% H_2SO_4 (*v*/*v*) in a 10-L reactor at 160°C for 30 min. After steam explosion treatment, the whole slurry was filtered with the aim of obtaining a liquid fraction and a solid fraction.

Laccase treatment of the prehydrolysate

The effect of laccase treatment on phenolic compounds in corn stover prehydrolysate was evaluated using the liquid fraction obtained from pretreatment. Before adding Glac15, prehydrolysate was adjusted to pH 3.0 to pH 5.0 and kept for 1 h and then incubated with 0.05, 0.1, or 0.2 U/mL of purified Glac15 for 24 h at 45°C in a rotary shaker (200 rpm) [5]. Samples were periodically taken and the supernatants analyzed for total phenols as described below. To evaluate the effect of laccase treatment on yeast growth, prehydrolysates were diluted to 75%, 50%, and 25% of original concentration by 4× YNB medium (the final concentration of YNB medium was 1×) [55]. After treatment with 0.05 U/mL Glac15 at pH 5.0 and 45°C for 8 h, *S. cerevisiae* in a final concentration of 1 OD_{600} were inoculated and were incubated at 30°C in a rotary shaker (200 rpm) for 60 h. In all cases, control assays were performed under the same conditions but without addition of laccase Glac15. All the experiments were carried out in triplicate.

Laccase treatment and enzymatic hydrolysis of thermochemical pretreated corn stover

Laccase treatment on pretreated solid fraction was performed as follows: in the first set of experiments, a total volume of 40 mL containing 2.0 g (dry weight, 5% *w*/*v*) pretreated solid fraction, 2.0 g/L $(NH_4)_2SO_4$, 4.0 g/L KH_2PO_4, 1.0 g/L $MgSO_4$, and 0.2 g/L $CaCl_2$ was used; in the second set of experiments, 5% solid fraction and 50% prehydrolysate (*v*/*v*) were added into the medium described above (the total volume was also 40 mL). Depending on the previous experiments, an enzyme loading of 0.05 U/mL of Glac15 was added and then incubated at 45°C and pH 5.0 in a rotary shaker (200 rpm). Control assay was performed under the same condition without addition of laccase. All experiments were carried out in triplicate. After 8 h of treatment, another 24-h enzymatic hydrolysis at 45°C was performed following the addition of 30 FPU/g (dry weight) cellulase (Sunson Biotech., Peking, China). Then the temperatures of hydrolyzed samples were reduced to 30°C and *S. cerevisiae* in a final concentration of 1.0 OD_{600} was aseptically inoculated. The fermentation was performed at 30°C in a rotary shaker (200 rpm) for another 60 h. Samples of different time points were recovered and centrifuged at 15,000 × *g* for 5 min; the supernatants were filtered through a 0.22-μm Millipore filter (Millipore, Bedford, MA, USA) and analyzed for glucose consumption and ethanol concentration.

Analytical methods

Biomass was determined using absorbance at 600 nm. The ethanol concentration was determined using an HPLC system according to Ji *et al.* [55]. Xylose, arabinose, galactose, furfural, 5-HMF, formic acid, and acetic acid concentrations were measured according to Moreno *et al.* [42]. Total phenols were measured according to Folin-Ciocalteu method using vanillin as a standard [56]. The data presented are the average values from triplicate technical repeats of measurements.

Additional file

Additional file 1: Tables S1 and S2 Table S1: Laccase peptides detected by LC-ESI-MS/MS and database search. Table S2: Sugar and ethanol concentrations after treated in different ways.

Abbreviations

2,6-DMP: 2,6-dimethoxyphenol; 5-HMF: 5-hydroxymethylfurfural; ABTS: 2,2′-azino-bis-(3-ethylbenzothiazoline-6-sulfonic acid); DMSO: dimethyl sulfoxide; EDTA: ethylenediaminetetraacetic acid; OD_{600}: optimal density at 600 nm; PAGE: polyacrylamide gel electrophoresis; PDA: potato dextrose agar; SDS: sodium dodecyl sulfate.

Competing interests

The authors declare that they have no competing interests.

Authors' contributions

ZF, XL, XB, and YX conceived and designed the experiments. ZF, XL, LC, and YS performed the experiments. ZF, YS, XZ, WF, XW, and YX analyzed the data. ZF, XB, and YX contributed reagents/materials/analysis tools. ZF, XZ, WX, BX, and YX contributed to the writing of the manuscript. All authors have read and approved of the final version of the manuscript.

Acknowledgements

This work was supported by the Scientific Research Foundation for Returned Scholars, Ministry of Education of China (to YX), the National Science Foundation of the Department of Education of Anhui Province (KJ2013A022 to ZF), the Introduction Project of Academic and Technology Leaders in Anhui University (32030066 to ZF), and the Innovative Research Team Program of 211 Project in Anhui University.

Author details

[1]School of Life Sciences, Anhui University, 111 Jiulong Road, Hefei, Anhui 230601, China. [2]Anhui Provincial Engineering Technology Research Center of Microorganisms and Biocatalysis, 111 Jiulong Road, Hefei, Anhui 230601, China. [3]The State Key Laboratory of Microbial Technology, School of Life Sciences, Shandong University, 27 Shanda Nanlu, Jinan, Shandong 250100, China. [4]Department of Chemistry & Biochemistry, Florida International University, Miami, FL 33199, USA.

References

1. Naik SN, Goud VV, Rout PK, Dalai AK. Production of first and second generation biofuels: a comprehensive review. Renew Sust Energ Rev. 2010;14:578–97.
2. Jönsson L, Alriksson B, Nilvebrant N-O. Bioconversion of lignocellulose: inhibitors and detoxification. Biotechnol Biofuels. 2013;6:16.
3. Parawira W, Tekere M. Biotechnological strategies to overcome inhibitors in lignocellulose hydrolysates for ethanol production: review. Crit Rev Biotechnol. 2011;31:20–31.
4. Schroyen M, Vervaeren H, Van Hulle SWH, Raes K. Impact of enzymatic pretreatment on corn stover degradation and biogas production. Bioresour Technol. 2014;173:59–66.
5. Jurado M, Prieto A, Martínez-Alcalá Á, Martínez ÁT, Martínez MJ. Laccase detoxification of steam-exploded wheat straw for second generation bioethanol. Bioresour Technol. 2009;100:6378–84.
6. Hoegger PJ, Kilaru S, James TY, Thacker JR, Kües U. Phylogenetic comparison and classification of laccase and related multicopper oxidase protein sequences. FEBS J. 2006;273:2308–26.
7. Kudanga T, Roes-Hill ML. Laccase applications in biofuels production: current status and future prospects. Appl Microbiol Biotechnol. 2014;98:6525–42.
8. Zhou XW, Cong WR, Su KQ, Zhang YM. Ligninolytic enzymes from *Ganoderma* spp: current status and potential applications. Crit Rev Microbiol. 2013;39:416–26.
9. Chen S, Xu J, Liu C, Zhu Y, Nelson DR, Zhou S, et al. Genome sequence of the model medicinal mushroom *Ganoderma lucidum*. Nat Commun. 2012;3:913.
10. D'souza TM, Merritt CS, Reddy CA. Lignin-modifying enzymes of the white rot basidiomycete *Ganoderma lucidum*. Appl Environ Microbiol. 1999;65:5307–13.
11. de Souza Silva CMM, De Melo IS, De Oliveira PR. Ligninolytic enzyme production by Ganoderma spp. Enzyme Microb Technol. 2005;37:324–9.
12. Sharma KK, Shrivastava B, Sastry VRB, Sehgal N, Kuhad RC. Middle-redox potential laccase from *Ganoderma* sp.: its application in improvement of feed for monogastric animals. Sci Rep. 2013;3:1299.
13. Sitarz AK, Mikkelsen JD, Højrup P, Meyer AS. Identification of a laccase from *Ganoderma lucidum* CBS 229.93 having potential for enhancing cellulase catalyzed lignocellulose degradation. Enzyme Microb Technol. 2013;53:378–85.
14. Songulashvili G, Elisashvili V, Wasser SP, Nevo E, Hadar Y. Basidiomycetes laccase and manganese peroxidase activity in submerged fermentation of food industry wastes. Enzyme Microb Tech. 2007;41:57–61.
15. Manavalan T, Manavalan A, Thangavelu KP, Heese K. Characterization of optimized production, purification and application of laccase from *Ganoderma lucidum*. Biochem Eng J. 2013;70:106–14.
16. Piscitelli A, Giardina P, Lettera V, Pezzella C, Sannia G, Faraco V. Induction and transcriptional regulation of laccases in fungi. Curr Genomics. 2011;12:104–12.
17. Liu D, Gong J, Dai W, Kang X, Huang Z, Zhang HM, et al. The genome of *Ganoderma lucidum* provide insights into triterpenes biosynthesis and wood degradation. Plos One. 2012;7:e36146.
18. Ko EM, Leem YE, Choi HT. Purification and characterization of laccase isozymes from the white-rot basidiomycete *Ganoderma lucidum*. Appl Microbiol Biotechnol. 2001;57:98–102.
19. Wang HX, Ng TB. A laccase from the medicinal mushroom *Ganoderma lucidum*. Appl Microbiol Biotechnol. 2006;72:508–13.
20. You LF, Liu ZM, Lin JF, Guo LQ, Huang XL, Yang HX. Molecular cloning of a laccase gene from *Ganoderma lucidum* and heterologous expression in *Pichia pastoris*. J Basic Microbiol. 2013;53:1–8.
21. Joo SS, Ryu IW, Park JK, Yoo YM, Lee DH, Hwang KW, et al. Molecular cloning and expression of a laccase from *Ganoderma lucidum*, and its antioxidative properties. Mol Cells. 2008;25:7–11.
22. Teerapatsakul C, Abe N, Bucke C, Kongkathip N, Jareonkitmongkol S, Chitradon L. Novel laccases of *Ganoderma* sp. KU-Alk4, regulated by different glucose concentration in alkaline media. World J Microbiol Biotechnol. 2007;23:1559–67.
23. Baldrian P. Fungal laccases – occurrence and properties. FEMS Microbiol Rev. 2006;30:215–42.
24. Xu F. Effects of redox potential and hydroxide inhibition on the pH activity profile of fungal laccases. J Biol Chem. 1997;272:924–8.
25. Sharma P, Goel R, Capalash N. Bacterial laccases. World J Microb Biotechnol. 2007;23:823–32.
26. Rivera-Hoyos CM, Morales-Álvarez ED, Poutou-Piñales RA, Pedroza-Rodríjuez AM, Rodríguez-Vázquez R, Delgado-Boada JM. Fungal laccases. Fungal Biol Rev. 2013;27:67–82.
27. Si J, Peng F, Cui B. Purification, biochemical characterization and dye decolorization capacity of an alkali-resistant and metal-tolerant laccase from *Trametes pubescens*. Bioresour Technol. 2013;128:49–57.
28. Guo LQ, Lin SX, Zheng XB, Huang ZR, Lin JF. Production, purification and characterization of a thermostable laccase from a tropical white-rot fungus. World J Microbiol Biotechnol. 2011;27:731–5.
29. Zhu Y, Zhang H, Cao M, Wei Z, Huang F, Gao P. Production of a thermostable metal-tolerant laccase from *Trametes versicolor* and its application in dye decolorization. Biotechnol Bioprocess Eng. 2011;16:1027–35.
30. Keum YS, Li QX. Fungal laccase-catalyzed degradation of hydroxy polychlorinated biphenyls. Chemosphere. 2004;56:23–30.
31. Liu L, Lin Z, Zheng T, Lin L, Zheng C, Lin Z, et al. Fermentation optimization and characterization of the laccase from *Pleurotus ostreatus* strain 10969. Enzyme Microb Tech. 2009;44:426–33.
32. Torres E, Bustos-Jaimes I, Le Borgne S. Potential use of oxidative enzymes for the detoxification of organic pollutants. Appl Catal B Environ. 2003;46:1–15.
33. Fang ZM, Li TL, Chang F, Zhou P, Fang W, Hong YZ, et al. A new marine bacterial laccase with chloride-enhancing, alkaline-dependent activity and dye decolorization ability. Bioresour Technol. 2012;111:36–41.
34. Moore BM, Flurkey WH. SDS activation of a plant poly-phenoloxidase. J Biol Chem. 1990;265:4982–8.
35. Miyazaki K. A hyperthermophilic laccase from *Thermus thermophilus* HB27. Extremophiles. 2005;9:415–25.
36. Telke AA, Kadam AA, Jagtap SS, Jadhav JP, Govindwar SP. Biochemical characterization and potential for textile dye degradation of blue laccase from *Aspergillus ochraceus* NCIM-1146. Biotechnol Bioproc E. 2010;15:696–703.
37. Cabana H, Jones JP, Agathos SN. Preparation and characterization of cross-linked laccase aggregates and their application to the elimination of endocrine disrupting chemicals. J Biotechnol. 2007;132:23–31.
38. Oliva JM, Sáez F, Ballesteros I, González A, Negro MJ, Manzanares P, et al. Effect of lignocellulosic degradation compounds from steam explosion pretreatment on ethanol fermentation by thermotolerant yeast *Kluyveromyces marxianus*. In: Biotechnology for fuels and chemicals. 2003. p. 141–53.
39. García-Aparicio MP, Ballesteros I, González A, Oliva JM, Ballesteros M, Negro MJ. Effect of inhibitors released during steam-explosion pretreatment of barley straw on enzymatic hydrolysis. In: Twenty-Seventh Symposium on Biotechnology for Fuels and Chemicals. 2006. p. 278–88.
40. Moreno AD, Ibarra D, Ballesteros I, González A, Ballesteros M. Comparing cell viability and ethanol fermentation of the thermotolerant yeast *Kluyveromyces marxianus* and *Saccharomyces cerevisiae* on steam-exploded biomass treated with laccase. Bioresour Technol. 2013;135:239–45.
41. Tabka MG, Herpoel-Gimbert I, Monod F, Asther M, Sigoillot JC. Enzymatic saccharification of wheat straw for bioethanol production by a combined cellulase, xylanase and feruloyl esterase treatment. Enzyme Microb Technol. 2006;39:897–902.
42. Moreno AD, Ibarra D, Fernández JL, Ballesteros M. Different laccase detoxification strategies for ethanol production from lignocellulosic biomass by the thermotolerant yeast *Kluyveromyces marxianus* CECT 10875. Bioresour Technol. 2012;106:101–9.
43. Palonen H, Viikari L. Role of oxidative enzymatic treatments on enzymatic hydrolysis of softwood. Biotechnol Bioeng. 2004;86:550–7.
44. Martín C, Galbe M, Wahlbom CF, Hahn-Häerdal B, Jösson JL. Ethanol production from enzymatic hydrolysates of sugarcane bagasse using recombinant xylose-utilising *Saccharomyces cerevisiae*. Enzyme Microb Technol. 2002;31:274–82.

45. García-Aparicio MP, Ballesteros I, González A, Oliva JM, Ballesteros M, Negro MJ. Effect of inhibitors released during steam-explosion pretreatment of barley straw on enzymatic hydrolysis. Appl Biochem Biotechnol. 2006;129:278–88.
46. Jonsson LJ, Palmqvist E, Nilvebrant N-O, Hahn-Hagerdal B. Detoxification of wood hydrolysate with laccase and peroxidase from the white-rot fungus *Trametes versicolor*. Appl Microb Biotechnol. 1998;49:691–7.
47. Rühl M, Majcherczyk A, Kües U. Lcc1 and Lcc5 are the main laccases secreted in liquid cultures of *Coprinopsis cinerea* strains. Antonie Van Leeuwenhoek. 2013;103:1029–39.
48. Laemmli UK. Cleavage of structural proteins during the assembly of the head of bacteriophage T4. Nature. 1970;227:680–5.
49. Bourbonnais R, Paice MG. Oxidation of non-phenolic substrates: an expanded role for laccase in lignin biodegradation. Febs Lett. 1990;267:99–102.
50. Harkin JM, Larsen MJ, Obst JR. Use of syringaldazine for detection of laccase in sporophores of wood rotting fungi. Mycologia. 1974;66:469–76.
51. Benfield G, Bocks SM, Bromley K, Brown B. Studies of fungal and plant laccases. Phytochemistry. 1964;3:79–88.
52. Froehner S, Eriksson K. Properties of the glycoprotein laccase immobilized by two methods. Acta Chem Scand B. 1975;29:691–4.
53. Williamson PR. Biochemical and molecular characterization of the diphenol oxidase of Cryptococcus neoformans: identification as a laccase. J Bacteriol. 1994;176:656–64.
54. Koroleva O, Gavrilova V, Yavmetdinov I, Shleev S, Stepanova E. Isolation and study of some properties of laccase from the basidiomycetes Cerrena maxima. Biochem (Moscow). 2001;66:618–22.
55. Ji L, Shen Y, Xu L, Peng B, Xiao Y, Bao X. Enhanced resistance of *Saccharomyces cerevisiae* to vanillin by expression of lacA from *Trametes* sp. AH28-2. Bioresour Technol. 2011;102:8105–9.
56. Folin O, Ciocalteu V. On tyrosine and tryptophane determinations in proteins. J Biol Chem. 1927;73:627–50.

Fungi isolated from *Miscanthus* and sugarcane: biomass conversion, fungal enzymes, and hydrolysis of plant cell wall polymers

Prachand Shrestha[1], Ana B Ibáñez[2], Stefan Bauer[2], Sydney I Glassman[3], Timothy M Szaro[1], Thomas D Bruns[1] and John W Taylor[1*]

Abstract

Background: Biofuel use is one of many means of addressing global change caused by anthropogenic release of fossil fuel carbon dioxide into Earth's atmosphere. To make a meaningful reduction in fossil fuel use, bioethanol must be produced from the entire plant rather than only its starch or sugars. Enzymes produced by fungi constitute a significant percentage of the cost of bioethanol production from non-starch (i.e., lignocellulosic) components of energy crops and agricultural residues. We, and others, have reasoned that fungi that naturally deconstruct plant walls may provide the best enzymes for bioconversion of energy crops.

Results: Previously, we have reported on the isolation of 106 fungi from decaying leaves of *Miscanthus* and sugarcane (Appl Environ Microbiol 77:5490–504, 2011). Here, we thoroughly analyze 30 of these fungi including those most often found on decaying leaves and stems of these plants, as well as four fungi chosen because they are well-studied for their plant cell wall deconstructing enzymes, for wood decay, or for genetic regulation of plant cell wall deconstruction. We extend our analysis to assess not only their ability over an 8-week period to bioconvert *Miscanthus* cell walls but also their ability to secrete total protein, to secrete enzymes with the activities of xylanases, exocellulases, endocellulases, and beta-glucosidases, and to remove specific parts of *Miscanthus* cell walls, that is, glucan, xylan, arabinan, and lignin.

Conclusion: This study of fungi that bioconvert energy crops is significant because 30 fungi were studied, because the fungi were isolated from decaying energy grasses, because enzyme activity and removal of plant cell wall components were recorded in addition to biomass conversion, and because the study period was 2 months. Each of these factors make our study the most thorough to date, and we discovered fungi that are significantly superior on all counts to the most widely used, industrial bioconversion fungus, *Trichoderma reesei*. Many of the best fungi that we found are in taxonomic groups that have not been exploited for industrial bioconversion and the cultures are available from the Centraalbureau voor Schimmelcultures in Utrecht, Netherlands, for all to use.

Keywords: Bioconversion, Biofuel, Fungi, Cellulose and hemicellulose degrading enzymes, Lignocellulose

* Correspondence: jtaylor@berkeley.edu
[1]Department of Plant and Microbial Biology, University of California, Berkeley, CA 94720-3102, USA
Full list of author information is available at the end of the article

Background

To reduce the amount of carbon dioxide released into the atmosphere from fossil fuels that are used to power vehicles, biofuels must be made from entire plants and not just the sugars squeezed from their stems or the starch produced in their fruits [1]. This complete use of plant polysaccharide (especially cellulose) would maximize the amount of fuel recovered from each plant, thereby offsetting the fossil carbon required to farm the plants and minimizing the pressure to convert natural land to agriculture [2,3].

Production of these cellulosic biofuels requires a larger investment in more diverse enzymes to convert plant cell walls to sugars than is now needed to release sugar from starch [4]. Whereas enzymes account for 4.5% of the cost to make ethanol from cornstarch, they account for 17% to 20% of the cost to make ethanol from entire plants [5,6]. For cellulosic biofuel to compete with fossil fuels, it is estimated that the cost of enzymes must account for only 8% to 10% of the total cost, a twofold reduction from present costs [7].

In addition to cost, enzyme diversity is an issue because the plant cell wall, with its many polysaccharides, is far more complex than starch. These cell wall polysaccharides comprise cellulose, hemicellulosic polymers of xylose, arabinose and other sugars, and pectins, all of which are embedded in and surrounded by the polyphenolic macromolecular lignins [8]. To make complete use of plant cell walls, cocktails of enzymes capable of orchestrated digestion of these polymers will be needed. Currently, these enzymes come from industrial fermentation of the key biofuel fungus, *Trichoderma reesei* [9].

We, and others [4,10], have reasoned that fungi that naturally deconstruct the cell walls of sugarcane and *Miscanthus* might produce enzymes with the diversity and strength of activity best suited to bioconversion of these plants. Prior efforts by others at bioprospecting for mesophilic fungi have discovered five fungi isolated from sugarcane bagasse and wood with endoglucanase activities that compared favorably to enzymes from *T. reesei* [11], and 19 fungi selected from 74 species, cultivated from temperate French forests and tropical French Guiana forests, whose secretomes increase biomass conversion of maize bran when added to commercial *T. reesei* enzyme cocktails [12]. Plant pathogenic fungi have also been studied with the finding that many of these fungi bioconvert as well or better than *T. reesei* (for example, on xylans, species of *Mucor, Rhizoctonia*, and *Cylindrocarpon* were superior to *T. reesei*), and that fungi that parasitize monocots bioconvert these plants more effectively than fungi parasitizing dicots, and *vice versa* [13]. With thermophilic and thermotolerant fungi, 27 strains isolated from sugarcane bagasse provided thermostable endoglucanases and xylanases [14]. An interesting twist on bioprospecting involved inoculating sterilized switchgrass with decaying switchgrass for 10 serial repetitions, which returned 135 strains of two *Fusarium* species, *Fusarium sporotrichioides* and *Fusarium poae*, among which were producers of thermostable cellulases and xylanases [15].

In addition to bioprospecting, there has been research on discovering and analyzing enzymes from fungi other than the production strains of *T. reesei*, the latter having been subjected to strain improvement since the 1940s. For example, when 310 strains of *T. reesei* other than the industrial strain were assessed for their ability to deconstruct switchgrass, one strain was found capable of outperforming commercial enzyme preparations [16]. More commonly, researchers investigate strains of other fungal species. When the secrotome of *Fusarium verticillioides* grown on wheat straw was added to commercial *T. reesei* enzyme preparations, additional sugars were released from cellulose (glucose) and hemicelluloses (xylose, arabinose) [17]. Similarly, when *Chrysoporthe cubensis* was grown on sugarcane bagasse, a crude enzyme extract released more glucose and xylose than commercial enzyme preparations [18]. Also, *Penicillium echinulatum* grown on sugarcane bagasse [19] and *Penicillium brasiliensis* grown on sugar beet pulp [20] produced mixtures of enzymes more complex than commercial preparations and released sugars from cellulose and hemicelluloses. Other researchers have investigated thermophilic fungi, for example *Thermoascus auraticus* grown on switchgrass [21] or *Aspergillus terreus* grown on corn stover [22], finding that unimproved strains of these fungi produce enzymes that function as well as current commercial preparations and that remain active at temperatures as high as 70°C.

We have previously reported the isolation of 106 fungal species from seven *Miscanthus* fields and ten sugarcane plantations and the demonstration that eight of the fungi were, in fact, capable of deconstructing *Miscanthus* cell walls [10]. In this paper, we extend our analysis to 30 of fungi most often cultivated from decaying leaves and stems of these plants [10] (Table 1), as well as four fungi chosen because they are well-studied for their plant cell wall deconstructing enzymes (*T. reesei*), for wood decay (*Phanerochaetae chrysosporium* and *Postia placenta*), or for genetic regulation of plant cell wall deconstruction (*Neurospora crassa*). We extend our analysis to assess not only their ability over an 8-week period to bioconvert *Miscanthus* cell walls but also their ability to secrete total protein, to secrete enzymes with the activities of xylanases, exocellulases, endocellulases, and beta-glucosidases, and to remove specific parts of *Miscanthus* cell walls, that is, glucan, xylan, arabinan, and lignin.

Table 1 Fungi studied with data on source plant, geographic location, GenBank ITS sequence, and CBS accession number

Culture #	Genebank accession #	CBS #	Taxa	Isolate ID	Collection date	Location	Country	GPS data	Host plant
1	HQ631013	134065	*Aureobasidium* aff *pullulans*	sc8d50p14-8	1/22/09	Baton Rouge LA	USA	30 16′ 19″ N, 91 5′ 43″ W	*Saccharum officinarum*
2	HQ630970	136219	*Alternaria* aff *tenuissima*	MS3p_50-33	9/26/08	Urbana IL	USA	40 2′ 29″ N, 88 13′ 28″ W	*Miscanthus giganteus*
3	HQ631009	134111	*Bipolaris sp1*	sc9d100p9-2	1/22/09	Baton Rouge LA	USA	30 1′ 18″ N, 90 47′ 00″ W	*Saccharum officinarum*
4	HQ630963	134072	*Phoma* aff *herbarum*	MS5p50-9	9/26/08	Urbana IL	USA	40 2′ 31″ N, 88 13′ 28″ W	*Miscanthus giganteus*
5	HQ630972	134059	*Epicoccum* aff *nigrum*	MS7p50-17	9/26/08	Urbana IL	USA	40 2′ 34″ N, 88 14′ 17″ W	*Miscanthus giganteus*
6	HQ630999	134109	*Phoma* aff *glomerata*	sc13d50p14-6	1/22/09	Baton Rouge LA	USA	30 0′ 11″ N, 90 44′ 34″ W	*Saccharum officinarum*
7	HQ631008	135764	*Dothideomycete sp*	sc10d50p8-8	1/22/09	Baton Rouge LA	USA	30 1′ 16″ N, 90 47′ 00″ W	*Saccharum officinarum*
8	HQ630971	-o-	*Cladosporidium* aff *cladosporioides*	MS6p50-33	9/26/08	Urbana IL	USA	40 2′ 34″ N, 88 13′ 31″ W	*Miscanthus giganteus*
9	HQ631021	134015	*Aspergillus* aff *fumigatus*	BGd1p19-4	1/22/09	Baton Rouge LA	USA	29 44′ 2″ N, 90 35′ 26″ W	*Saccharum officinarum*
10	HQ631007	134110	*Penicillium* aff *minioluteum*	BGd100p3-1	1/22/09	Baton Rouge LA	USA	29 44′ 2″ N, 90 35′ 26″ W	*Saccharum officinarum*
11	HQ631027	134014	*Exophiala* aff *spinifera*	sc12d100p8-7	1/22/09	Baton Rouge LA	USA	30 4′ 1″ N, 90 41′ 42″ W	*Saccharum officinarum*
12	HQ630990	134064	*Exophiala* aff *salmonis*	MS4p_50-2	9/26/08	Urbana IL	USA	40 2′ 29″ N, 88 13′ 30″ W	*Miscanthus giganteus*
13	HQ630981	134062	*Microdochium* aff *bolleyi*	MS5p50-32	9/26/08	Urbana IL	USA	40 2′ 31″ N, 88 13′ 28″ W	*Miscanthus giganteus*
14	HQ630982	134063	*Nigrospora* aff *oryzae*	MS5p50-34	9/26/08	Urbana IL	USA	40 2′ 31″ N, 88 13′ 28″ W	*Miscanthus giganteus*
15	HQ630961	134044	*Arthrinium* aff *sacchari*	MSbale50-22	9/26/08	Urbana IL	USA	40 5′ 38.75″ N, 88 14′ 3.10″ W	*Miscanthus giganteus*
16	HQ630967	134073	*Arthrinium* aff *phaeospermum*	MS3p_50-12	9/26/08	Urbana IL	USA	40 2′ 29″ N, 88 13′ 28″ W	*Miscanthus giganteus*
17	HQ630973	-o-	*Cephalosporium* aff *gramineum*	MS5p50-12	9/26/08	Urbana IL	USA	40 2′ 31″ N, 88 13′ 28″ W	*Miscanthus giganteus*
18	HQ630978	134061	*Chloridium sp1*	MSbale50-42	9/26/08	Urbana IL	USA	40 5′ 38.75″ N, 88 14′ 3.10″ W	*Miscanthus giganteus*
19	HQ630974	134074	*Minimidochium sp1*	MS3p_50-45	9/26/08	Urbana IL	USA	40 2′ 29″ N, 88 13′ 28″ W	*Miscanthus giganteus*
20	HQ630984	135763	*Sporothrix* aff *lignivora*	MSbale50-11	9/26/08	Urbana IL	USA	40 5′ 38.75″ N, 88 14′ 3.10″ W	*Miscanthus giganteus*
21	HQ630968	134075	*Cordyceps* aff *bassiana*	MS3p_50-38	9/26/08	Urbana IL	USA	40 2′ 29″ N, 88 13′ 28″ W	*Miscanthus giganteus*
22	HQ630966	134071	*Gibberella* aff *moniliformis*	MS7p50-29	9/26/08	Urbana IL	USA	40 2′ 34″ N, 88 14′ 17″ W	*Miscanthus giganteus*
23	HQ630977	134060	*Gibberella* aff *avenacea*	MS7p50-6	9/26/08	Urbana IL	USA	40 2′ 34″ N, 88 14′ 17″ W	*Miscanthus giganteus*
24	HQ630964	134070	*Fusarium* aff *aethiopicum*	MS7p50-21	9/26/08	Urbana IL	USA	40 2′ 34″ N, 88 14′ 17″ W	*Miscanthus giganteus*
25	HQ630965	134066	*Fusarium* aff *proliferatum*	MS2-4	9/26/08	Urbana IL	USA	40 2′ 27″ N, 88 13′ 27″ W	*Miscanthus giganteus*
26	HQ630976	135762	*Fusarium* aff *equiseti*	MS6p50-29	9/26/08	Urbana IL	USA	40 2′ 34″ N, 88 13′ 31″ W	*Miscanthus giganteus*

Table 1 Fungi studied with data on source plant, geographic location, GenBank ITS sequence, and CBS accession number *(Continued)*

27	HQ630960	134068	*Hypocrea aff lixii*	MS3p_50-23	9/26/08	Urbana IL	USA	40 2′ 29″ N, 88 13′ 28″ W	*Miscanthus giganteus*
28	HQ630962	134069	*Trichoderma aff spirale*	MSbale50-9	9/26/08	Urbana IL	USA	40 5′ 38.75″ N, 88 14′ 3.10″ W	*Miscanthus giganteus*
29	HQ630959	134067	*Hypocrea aff koningii*	MS5p50-7	9/26/08	Urbana IL	USA	40 2′ 31″ N, 88 13′ 28″ W	*Miscanthus giganteus*
30	HQ630969	134058	*Trichoderma aff atroviride*	MSbale50-8	9/26/08	Urbana IL	USA	40 5′ 38.75″ N, 88 14′ 3.10″ W	*Miscanthus giganteus*

-o-: culture lost. So no CBS number assigned.

We have found (1) that a majority of fungi tested equaled or exceeded the bioconversion abilities of native *T. reesei,* (2) that some of the best of these fungi are in Ascomycota taxa not previously explored for bioconversion, (3) that enzyme activity varies greatly over time and in magnitude among fungi, and (4) that comparing activities of the four enzymes studied here with amount of cell wall removed by fungal solid substrate fermentation suggests that there must be undetected enzyme activities in addition to the four tested here. By focusing on fungi isolated from energy grasses, by documenting total biomass conversion as well as the removal of four key plant cell wall components, and by measuring the activities of four key fungal enzymes, all over an 8-week period, ours is the most thorough examination of the potential contribution of bioprospecting to the biofuel industry. Our positive results provide a strong justification for evaluating the biofuel potential of the fungi that we report here and for further exploration to find fungi with additional, desirable traits.

Results and discussion

Biomass conversion

Over an 8-week period, we found that 25 of the 30 fungi isolated from energy grasses could convert at least 10% of *Miscanthus* biomass. Four of the five species that did not are known for growth on animals (*Exophiala* spp., *Cordyceps* sp., and *Sporothrix* sp.), and likely had been growing on animals collected along with the plant leaves (Figure 1). *Trichoderma reesei,* the industrially most important producer of bioconversion enzymes, converted 12% of the *Miscanthus.* We found 21 other Ascomycota fungi that did as well as *T. reesei,* six of which converted at least 15% *Miscanthus* (Figure 1). The most effective bioconversion fungus isolated by us was a *Chloridium* sp1, which caused 19% Miscanthus weight loss over 8 weeks, followed by *Alternaria* aff. *tenuissima* at 17%. These two fungi were significantly better than *P. placenta* and *T. reesei* and not significantly different than *P. chrysosporium* (20%), the well-studied Basidiomycota wood decay fungus, or *N. crassa* (18.7%), the Ascomycota model for studying fungal bioconversion [23] (Table 2). In total, 6 fungi isolated from decaying energy grasses were not significantly different from *P. chrysosporium,* that is, in addition to *Chloridium* sp1 and *A.* aff. *tenuissima, Bipolaris sp1, Arthrinium* aff. *phaeospermum, Minimidochim sp1,* and *Epicoccum* aff. *nigrum.* Several of the fungi that showed superior bioconversion were members of the Chaetosphaeriales (Sordariomycetes) and Pleosporales (Dothideomycetes), groups of fungi that have not been well-studied for bioconversion of plant biomass and that contain the very effective *Chloridium* sp1 as well as species in the genera, *Alternaria, Bipolaris,* and *Epicoccum.*

Protein titers and activity profiles of cell wall degrading enzymes

Activities of four enzymes, exocellulase, endocellulase, beta-glucosidase, and xylanase, were measured for all 34 fungi after rehydration of lyophilized residue of solid substrate cultures that had been harvested at 0, 1, 2, 4, and 8 weeks after inoculation. Specific enzyme activities are given as μM product/min/mg protein (Figure 2) and colored as a heat map to facilitate comparison among species and time points for a single enzyme, but not among enzymes. Enzyme activity varied over an order of magnitude for exocellulase activities, over two orders of magnitude for endocellulase and β-glucosidase activities and over three orders of magnitude for xylanase activities. For all species, specific enzyme activity was minimal at time 0, and peak enzyme activity could occur at any other time point, depending on the enzyme and fungal species.

Exocellulase activity was highest for *Hypocrea* aff. *koningii,* 23.3 μM glucose/min/mg protein, at week 1, a level that was equaled only by *Dothideomycete* sp. at week 2 and that was twice that of any other fungus at any time. These two species and others that had substantial exocellulase activity (*P. chrysosporium, Penicillium* aff. *minioluteum, A.* aff. *phaeospermum, Cephalosporium* aff. *gramineum,* and *N. crassa*) reached their peak by week 2, in contrast to *E.* aff. *nigrum,* which peaked at week 4 and still showed strong activity at week 8.

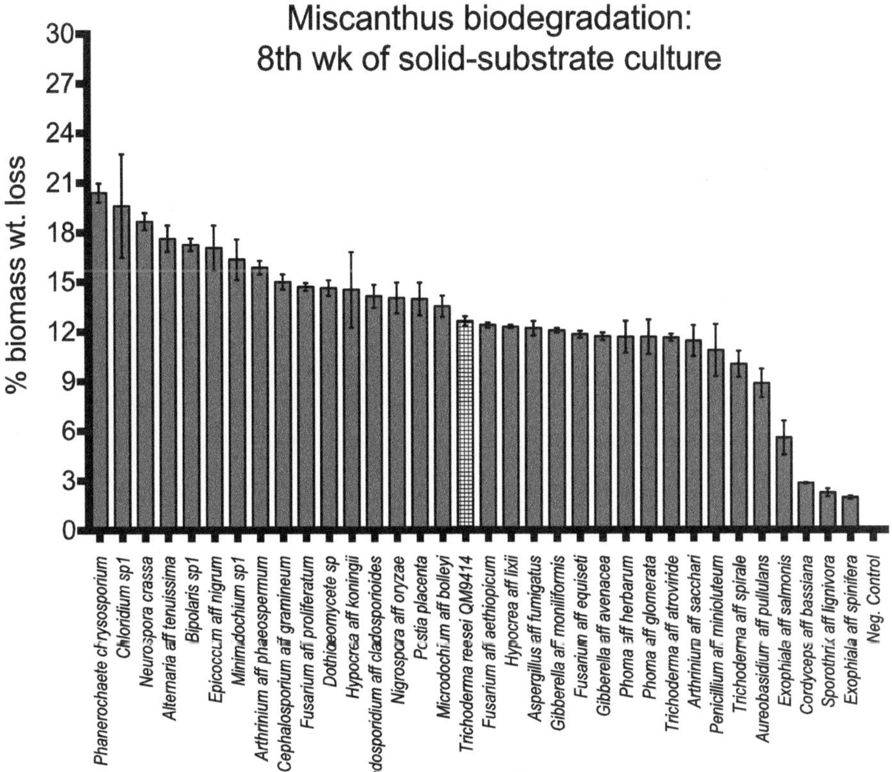

Figure 1 Percent biomass *(Miscanthus)* weight reduction after 8 weeks of solid substrate cultures of fungal species on ground *Miscanthus*. Performance of the industrially important enzyme producer, *Trichoderma reesei*, is shaded differently. Error bars are standard errors ($n = 3$). In addition to 30 species isolated from *Miscanthus* and sugarcane, four well-studied fungi were tested: *Phanerochaetae chrysosporium*, *Neurospora crassa*, *Postia placenta*, and the aforementioned *Trichoderma reesei*.

Endocellulase activity showed a different pattern than exocellulase activity in that it did not peak early and then decline. Instead, most fungi with strong endocellulase activity displayed high activities at weeks 2 through 8 with the highest activities coming at weeks 4 and 8. Fungal species that had better exocellulase activities typically also had higher endocellulase activities, with the notable exception of *Sporothrix* aff. *lignivora*, which showed the highest levels of endocellulase activity seen for any of the fungi, 384 μM glucose/min/mg protein at week 8; interestingly, *S.* aff. *lignivora* lacked exocellulase and xylanase activities. Both of these anomalies are likely a consequence of the low amount of protein secreted by this animal-associated fungus (Figure 3). *E.* aff. *nigrum* again showed consistently strong activity for weeks 2, 4, and 8, and two other Dothideomycetes also achieved high levels of activity, *A.* aff. *tenuissima* and *Dothideomycete* sp. For Sordariomycetes, in addition to the aforementioned *S.* aff. *lignivora*, *A.* aff. *phaeospermum*, *C.* aff. *gramineum*, and *H.* aff. *koningii* all showed high levels of endocellulase activity over extended periods, although *H.* aff. *koningii* was unusual in having the strongest activity at week 1, 234 μM glucose/min/mg protein. Moderate to low levels of endocellulase activity were manifested by cultures of positive control species: *P. chrysosporium*, *P. placenta*, *N. crassa*, and *T. reesei QM9414*, along with *Chloridium sp1* and most species of Hypocreaceae, home to *Trichoderma*, *Gibberella*, and *Fusarium* spp., but not the aforementioned and very active, *H.* aff. *koningii*.

Beta-glucosidase activity showed yet a different pattern of activity, often reaching the highest level in week 1 and maintaining a high level through week 8. In addition, more species achieved the highest levels of enzyme activity for beta-glucosidase than for either exo- or endocellulases. *E.* aff. *nigrum*, *C.* aff. *gramineum*, and *H.* aff. *koningii* showed the highest activities in week 1 (813, 717, and 739 μM pNP/min/mg protein, respectively) and maintained high activities, as did *Dothideomycete* sp., and the Sordariomycete species, *Arthrinium* aff. *sacchari*, *A.* aff. *phaeospermum*, and *Trichoderma* aff. *atroviride*. The four positive control species again had moderate beta-glucosidase activities, except for *P. placenta* at the first week at 427 μM pNP/min/mg protein.

Xylanase activity showed a similar pattern to that of beta-glucosidase. Again, many species reached the highest

Table 2 Matrix of Tukey-Kramer pairwise comparisons of % biomass reduction for all treatments

Fungal isolates	*Trichoderma reesei QM9414*	*Neurospora crassa*	*Phanerochaete chrysosporium*	*Postia placenta*
Chloridium sp1	**6.00**	*0.99*	*0.85*	**7.42**
Alternaria aff *tennuissima*	**3.66**	*4.29*	*2.73*	**5.03**
Bipolaris sp1	*3.52*	*1.49*	*3.34*	*4.94*
Epicoccum aff *nigrum*	*3.30*	*1.71*	*3.55*	*4.72*
Minimidochium sp1	*2.56*	*2.45*	*4.30*	*3.98*
Arthrinium aff *phaeospermum*	*2.02*	*2.98*	*4.83*	*3.45*
Cephalosporium aff *gramineum*	*1.10*	*3.91*	***5.75***	*2.52*
Fusarium aff *proliferatum*	*0.78*	*4.23*	***6.07***	*2.20*
Dothideomycete sp	*0.71*	*4.30*	***6.15***	*2.13*
Hypocrea aff *koningii*	*0.60*	*4.40*	***6.25***	*2.02*
Cladosporidium aff *cladosporioides*	*0.18*	*4.83*	***6.68***	*1.60*
Nigrospora aff *oryzae*	*0.07*	*4.94*	***6.78***	*1.49*
Microdochium aff *bolleyi*	*0.46*	***5.47***	***7.32***	*0.96*
Fusarium aff *aethiopicum*	*1.67*	***6.68***	***8.53***	*0.25*
Hypocrea aff *lixii*	*1.59*	***6.07***	***7.72***	*0.32*
Aspergillus aff *fumigatus*	*1.88*	***6.89***	***8.74***	*0.46*
Gibberella aff *moniliformis*	*2.02*	***7.03***	***8.88***	*0.60*
Fusarium aff *equiseti*	*2.27*	***7.28***	***9.13***	*0.85*
Gibberella aff *avanacea*	*2.42*	***7.42***	***9.27***	*0.99*
Phoma aff *herbarum*	*2.45*	***7.46***	***9.31***	*1.03*
Phoma aff *glomerata*	*2.45*	***7.50***	***9.31***	*1.03*
Trichoderma aff *atroviride*	*2.49*	***7.71***	***9.34***	*1.07*
Arthrinium aff *sacchari*	*2.70*	***8.31***	***9.56***	*1.28*
Penicillium aff *minimoluteum*	*3.30*	***9.20***	***10.16***	*1.88*
Trichoderma aff *spirale*	*4.19*	***10.44***	***11.05***	*2.77*
Aureobasidium aff *pullulans*	***5.43***	***13.96***	***12.29***	*4.01*
Exophiala aff *salmonis*	***8.95***	***16.87***	***15.81***	***7.53***
Cordyceps aff *bassiana*	***11.86***	***17.48***	***18.72***	***10.44***
Sporothrix aff *lignivora*	***12.47***	***17.80***	***19.32***	***11.05***
Exophiala aff *spinifera*	***12.79***		***19.64***	***11.37***
Phanerochaete chrysosporium				**8.28**
Neurospora crassa			*1.85*	**6.43**
Trichoderma reesei QM9414		***5.01***	***6.86***	*1.42*

Pairwise comparisons for all treatments compared to four positive controls for ANOVA with percent weight loss at week 8 as response variable.
Legend: values in italics show no significant pairwise differences; values in bold, column is significantly lower than row; values in bold italics, column is significantly higher than row.

levels of xylanase activity and these high levels (≥2000 μM xylose/min/ mg protein) were reached in week 1 and persisted through week 8. A difference was that the two Ascomycota control species achieved high levels of activity in the first week, *N. crassa* at 2621 μM xylose/min/mg protein and *T. reesei QM9414* at 2873 μM xylose/min/mg protein, and maintained them, but not the Basdiomycota controls, *P. chrysosporium*, and *P. placenta*. Two Dothideomycetes were exceptional, *E.* aff. *nigrum*, with the highest xylanase activity, 4027 μM xylose/min/mg protein, and *Dothideomycete* sp., at 3126 μM xylose/min/mg protein. Other fungi with high and sustained xylanase activity were the Chaetothyriales, *Exophiala* aff. *salmonis*, and the Sordariomycetes, *C.* aff. *gramineum* and *T.* aff. *atroviride*.

Predictors for biomass loss

We analyzed relationships between biomass loss and both enzyme activity and amounts of extracellular protein. As expected, a significant relationship between the four enzyme activities and weight loss of the complex

Species (w/ GenBank Accession #s) / weeks	exocellulase (uM glucose/min.mg protein) 0	1	2	4	8	endocellulase (uM glucose/min.mg protein) 0	1	2	4	8	beta glucosidase (uM pNP/min.mg protein) 0	1	2	4	8	xylanase (uM xylose/min.mg protein) 0	1	2	4	8
Phanerochaete chrysosporium	10.1	5.1	1.4	1.4	1.4	2	32	13	18	21	13	95	45	32	24	39	1232	69	40	360
Postia placenta	0.1	1.7	0.9	2.1	1.2	2	14	0	0	0	36	427	36	50	73	37	660	27	17	1902
Aureobasidium aff pullans (HQ631013)	0.0	1.8	1.1	2.9	0.8	7	74	10	25	125	34	350	359	343	211	0	141	707	1037	1489
Alternaria aff tenuissima(HQ630970)	0.0	1.1	0.7	2.4	4.1	32	35	67	115	186	9	333	342	383	290	14	626	438	2906	1869
Bipolaris sp1(HQ631009)	0.6	1.4	1.9	0.8	1.6	24	11	44	23	22	20	214	334	284	336	43	27	45	73	325
Phoma aff herbarum(HQ630963)	0.0	0.4	0.8	0.5	1.2	9	3	78	48	33	53	23	364	309	250	23	20	796	407	300
Epicoccum aff nigrum(HQ630972)	0.7	3.7	10.9	12.2	7.9	25	79	202	280	269	20	813	642	577	274	61	4027	3270	3236	558
Phoma aff glomerata(HQ630999)	0.0	0.0	0.0	0.0	0.1	16	31	65	46	130	7	160	315	406	454	73	984	844	539	1088
Dothideomycete sp(HQ631008)	0.7	14.6	22.6	5.0	3.4	8	138	229	123	96	81	348	488	387	230	108	2995	3126	2643	1620
Cladosporidium aff cladosporioides(HQ630971)	0.4	0.8	0.6	0.3	1.1	10	0	2	0	15	16	95	94	46	12	41	774	689	769	738
Aspergillus aff fumigatus (HQ631021)	0.2	1.3	1.7	1.7	1.4	1	11	22	19	24	48	118	185	151	122	60	1459	2174	2740	1386
Penicillium aff minioluteum(HQ631007)	1.3	8.4	6.8	3.9	2.9	26	52	17	51	57	75	170	187	226	169	0	2117	1756	1260	1109
Exophiala aff spinifera(HQ631027)	0.2	0.1	0.3	0.2	2.3	6	0	0	0	42	43	34	44	54	79	4	9	52	43	164
Exophiala aff salmonis(HQ630990)	0.0	0.9	0.0	0.0	1.9	0	117	58	159	80	25	26	27	33	29	1	3197	3085	3264	2573
Microdochium aff bolleyi(HQ630981)	0.9	1.2	0.9	0.9	1.8	10	14	24	31	80	9	81	131	189	133	62	23	19	20	140
Nigrospora aff oryzae(HQ630982)	0.1	0.7	1.1	1.8	1.8	0	20	25	39	83	9	38	51	38	39	24	38	0	13	123
Arthrinium aff sacchari(HQ630961)	0.1	2.3	1.9	1.4	3.1	16	58	84	57	77	5	332	393	400	470	18	1011	2361	2051	1785
Arthrinium aff phaeospermum(HQ630967)	0.5	8.4	9.5	5.1	2.6	9	101	162	208	44	4	569	672	500	70	53	1536	1486	3525	2202
Cephalosporium aff gramineum(HQ630973)	0.0	13.2	5.0	4.1	1.4	2	159	134	156	29	5	717	376	335	154	32	2317	2997	2592	2360
Neurospora crassa	0.7	12.4	9.4	4.0	3.2	0	48	39	24	27	7	68	48	17	31	39	2621	2350	1363	1257
Chloridium sp1(HQ630978)	1.0	1.1	0.5	1.2	0.2	0	2	46	13	9	18	18	27	84	122	82	992	723	908	83
Minimidochium sp1(HQ630974)	0.3	0.4	0.4	0.6	0.8	5	0	0	1	18	3	29	24	34	45	21	17	4	24	367
Sporothrix aff lignivora(HQ630984)	0.0	0.0	1.4	0.0	2.4	0	0	176	103	384	34	90	178	229	327	30	19	32	22	152
Cordyceps aff bassiana(HQ630968)	0.0	0.5	3.3	2.2	2.2	0	2	34	65	35	35	59	63	30	75	37	55	16	277	102
Gibberella aff moniliformis(HQ630966)	0.0	0.0	0.1	0.0	2.0	1	0	0	0	4	47	25	20	22	19	23	175	582	973	1014
Gibberella aff avenacea(HQ630977)	0.7	0.3	0.2	0.3	1.0	8	0	0	0	17	109	32	35	32	140	77	280	414	629	1484
Fusarium aff aethiopicum(HQ630964)	0.8	0.9	0.4	0.3	2.2	0	0	0	0	32	9	28	27	14	100	200	576	799	879	667
Fusarium aff proliferatum(HQ630965)	1.1	0.9	0.4	0.9	0.9	17	0	6	5	13	84	54	59	79	37	96	512	441	509	315
Fusarium aff equiseti(HQ630976)	1.2	0.6	0.6	0.3	0.8	4	0	0	0	7	81	81	53	35	52	105	33	0	1	83
Hypocrea aff lixii(HQ630960)	0.2	0.2	0.2	0.1	0.2	1	0	1	1	9	30	14	13	10	11	56	1462	1597	1382	1122
Trichoderma aff spirale(HQ630962)	0.2	0.2	0.2	0.1	0.3	3	2	3	1	8	24	12	12	9	10	21	403	373	264	259
Trichoderma reesei - QM9414	0.3	2.2	1.9	1.3	1.5	0	26	17	0	7	8	32	43	46	50	15	2873	2595	1460	1783
Hypocrea aff koningii(HQ630959)	0.0	23.3	10.7	5.0	2.2	6	234	193	159	47	9	739	379	320	194	45	755	397	2304	1914
Trichoderma aff atroviride(HQ630969)	1.0	1.6	1.4	1.5	1.1	0	21	18	16	22	69	494	493	440	405	131	1411	2665	2974	2895

Figure 2 Heat maps (green = low, yellow = intermediate, red = high) of enzyme activities assessed on crude-cell-free fungal extracts collected periodically from fungal cultures on *Miscanthus*. GenBank accession numbers are given for fungi isolated by us from *Miscanthus* or sugarcane. Names of fungi that showed exceptional enzyme activity are shown in red.

lignocellulosic substrate, *Miscanthus*, was demonstrated by regression analysis ($P < 0.001$, R^2 0.24). However, the 24% correlation between biomass loss and the four enzyme activities combined with the observation that the two best fungi for bioconversion of *Miscanthus*, *P. chrysosporium* and *Chloridium sp1*, never reached the maximum activity for any enzyme, supports the argument that most or all of the fungi studied here must harbor uncharacterized enzymes that are important to bioconversion [4]. This same conclusion has been reached by others who observed an increase in the bioconversion ability of commercial *T. reesei* enzyme preparations upon the addition of secreted proteins from a variety of fungi [11,12,24].

To determine if simple analysis of extracellular protein could predict bioconversion, we examined the correlation between the free protein titers and percent biomass weight loss. The regression model in week 8 was significant ($P < 0.001$, $R^2 = 0.55$). The Pearson correlation is 0.7454 with $P < 0.0001$ and the scatter plot of free protein versus percent biomass weight loss at week 8 reveals a clear association between the variables. Significant correlation between protein concentration profile and percent biomass weight loss was also valid for week 2 ($P < 0.001$, $R^2 = 0.5$) and week 4 ($P < 0.001$, $R^2 = 0.4035$) (Figure 3).

A stepwise regression revealed that the best predictors for percent biomass weight loss at week 2 were free protein concentrations and beta-glucosidase activity. These two factors explained almost 64% of the variance present in percent biomass weight loss at week 2. These two factors were also strong predictors for percent biomass weight loss at weeks 4 and 8.

Removal of plant cell wall components

Removal of four plant cell wall components, glucan, xylan, arabinan, and lignin, were determined from the *Miscanthus* residue that remained following 8 weeks of solid substrate cultures for the 30 fungi showing the most active bioconversion and the four fungi that are well known for bioconversion, *P. chrysosporium*, *N. crassa*, *P. placenta*, and *T. reesei*.

Removal of glucan, a broad category that represents cellulose, was topped by *P. chrysosporium* at 23.3% followed by five other species with more than 18% removal, including *Chloridium* spl at 19.9%, *N. crassa* at 18.1%, and three Dothideomycetes. The two other positive control fungal species, *T. reesei QM9414* and *P. placenta* were able to consume 11.5% and 14.1% glucan by week 8 (Figure 4).

Removal of xylan, one of two polysaccharides representing the hemicelluloses, was the highest for *Bipolaris* spl (30.6%) and at or above 25% for *P. chrysosporium* (25.4%), *Dothideomycete* sp. (25.0%), and *N. crassa* (26.5%). Dropping the amount of removal to 24% included *Chloridium sp1*, *A.* aff. *phaeospermum*, and two more Dothideomycetes spp., *A.* aff. *tenuissima* and *E.* aff. *nigrum*.

Removal of arabinan, the second polymer representing hemicellulose, was almost 50% for *Bipolaris* spl (49.5%) and *Fusarium* aff. *equiseti* (48.8%), followed closely by eight others at more than 45% (Figure 4). The four well-studied fungi removed less than 45% of the arabinan, with *N. crassa* being the highest (43.2%) and *P. chrysosporium* the lowest (36.8%).

Lignin reduction was the highest for the well-studied, wood decay fungus, *P. chrysosporium* (14.7%) followed by *A.* aff. *phaeospermum* at 11.1%. No other fungus removed even 10% of the lignin, including the other Basidiomycota, *P. placenta* (7.6%). The best of the Ascomycota, at between 7% and 8%, were *N. crassa*, *Chloridium sp*l, *Fusarium* aff. *proliferatum*, and *Microdochium* aff. *bolleyi*, all of which were almost twice as good as *T. reesei QM9414* (4.4%). Lignin removal percentages can be difficult to measure for fungi that produce the structurally similar compound, melanin, in their cell walls. Melanin production likely is responsible for the apparent gain in lignin by the Dothideomycetes species, *E.salmonis* aff. *salmonis*, *Dothideomycete sp and Bipolaris sp1*, and the scant reduction by *Epicoccum* aff. *nigrum*.

To make rigorous comparisons of the bioconversion abilities of fungi cultivated from decaying energy grasses to that of four well-studied bioconversion fungi, *T. reesei*, *N. crassa*, *P. chrysosporium* and *P. placenta*, we performed analysis of variance (ANOVA) with percent weight loss as response variable and fungal species as the treatment (Table 2). Tukey-Kramer post hoc tests were used to determine significant difference in mean values of percent biomass weight losses and Dunn-Bonferroni adjustment and Hochberg step-up methods were used to account for Type I errors during multiple pairwise comparisons.

Mean percent weight loss varied significantly across 34 fungal treatments for week 2 ($F = 21.62$, $P < 0.001$), week 4 ($F = 8.62$, $P < 0.001$), and week 8 ($F = 25.55$, $P < 0.001$) weeks. At week 8, none of the fungal species were better at degrading Miscanthus cell walls than *P. chrysosporium* or *N. crassa*, but six were as good as *P. chrysosporium*, 15 were as good as N. crassa, 27 were as good as *T. reesei* QM9414, and 26 were as good as *P. placenta*. One newly isolated species, *Chloridium* spl, bioconverted *Miscanthus* biomass significantly better than *T. reesei* QM9414 and *P. placenta*. The four species of animal associated fungi mentioned above, *Exophiala* aff. *salmonis*, *Cordyceps* aff. *bassiana*, *Sporothrix* aff. *lignivora* and *Exophiala* aff. *spinifera*, showed significantly lower biomass degradation.

Species (w/ GenBank Accession #s) / weeks	% biomass wt loss 0	1	2	4	8	protein (ug/ml) 0	1	2	4	8
Phanerochaete chrysosporium	-0.4	5.4	9.0	13.0	20.4	74	120	170	235	472
Postia placenta	-1.0	2.0	-0.6	0.9	14.0	107	186	111	84	534
Aureobasidium aff pullans (HQ631013)	0.3	3.7	5.6	7.2	8.9	128	179	302	345	378
Alternaria aff tenuissima(HQ630970)	0.1	6.0	8.7	13.4	17.7	179	266	393	413	581
Bipolaris sp1(HQ631009)	-0.3	7.5	9.0	13.6	17.3	100	184	189	295	383
Phoma aff herbarum(HQ630963)	0.5	4.7	9.6	11.5	11.7	88	178	363	468	540
Epicoccum aff nigrum(HQ630972)	-0.2	4.3	7.9	11.5	17.1	48	103	242	279	582
Phoma aff glomerata(HQ630999)	0.4	5.7	7.5	10.5	11.7	101	217	273	359	361
Dothideomycete sp(HQ631008)	-0.2	5.3	7.1	10.0	14.6	114	271	280	315	453
Cladosporidium aff cladosporioides(HQ630971)	-0.8	7.9	10.4	12.7	14.1	149	419	506	519	852
Aspergillus aff fumigatus (HQ631021)	0.7	5.9	7.6	9.8	12.2	142	412	344	221	251
Penicillium aff minioluteum(HQ631007)	-0.6	4.8	6.4	8.3	10.9	167	338	379	351	439
Exophiala aff spinifera(HQ631027)	0.7	1.3	1.7	2.2	2.0	122	150	142	124	118
Exophiala aff salmonis(HQ630990)	1.4	2.0	2.3	3.2	5.6	173	209	208	198	260
Microdochium aff bolleyi(HQ630981)	0.8	5.5	7.5	8.4	13.5	141	371	498	482	656
Nigrospora aff oryzae(HQ630982)	0.3	5.8	7.6	10.6	14.0	151	358	423	567	733
Arthrinium aff sacchari(HQ630961)	0.4	7.4	10.4	10.8	11.4	172	310	358	324	230
Arthrinium aff phaeospermum(HQ630967)	-0.3	3.5	5.9	11.8	15.8	127	165	192	348	624
Cephalosporium aff gramineum(HQ630973)	-0.3	5.2	9.2	11.7	15.0	114	192	417	477	553
Neurospora crassa	-0.1	7.4	10.8	16.2	18.7	112	410	558	671	830
Chloridium sp1(HQ630978)	0.0	6.3	8.6	12.2	19.6	46	248	322	366	556
Minimidochium sp1(HQ630974)	0.0	3.6	6.8	10.6	16.4	150	210	357	449	609
Sporothrix aff lignivora(HQ630984)	0.4	0.6	1.8	4.4	2.3	130	145	132	143	139
Cordyceps aff bassiana(HQ630968)	1.2	2.6	3.9	7.7	2.8	122	145	123	297	121
Gibberella aff moniliformis(HQ630966)	-0.1	5.1	6.9	8.7	12.1	111	349	444	459	536
Gibberella aff avenacea(HQ630977)	0.4	6.7	7.1	9.5	11.7	42	263	423	351	395
Fusarium aff aethiopicum(HQ630964)	0.2	4.2	7.1	9.7	12.4	89	299	433	524	507
Fusarium aff proliferatum(HQ630965)	0.2	6.7	7.9	10.9	14.7	93	457	455	550	626
Fusarium aff equiseti(HQ630976)	0.6	5.8	7.3	11.4	11.9	87	329	437	533	508
Hypocrea aff lixii(HQ630960)	0.4	5.5	7.6	10.4	12.3	130	380	454	527	517
Trichoderma aff spirale(HQ630962)	0.1	4.8	6.4	7.6	10.0	161	327	335	407	493
Trichoderma reesei - QM9414	0.7	6.0	7.5	10.7	12.6	120	258	311	387	441
Hypocrea aff koningii(HQ630959)	-0.4	7.1	11.9	13.1	14.5	56	216	423	510	554
Trichoderma aff atroviride(HQ630969)	0.9	6.0	7.2	9.9	11.7	66	293	324	363	407

Figure 3 Heat map (green = low, yellow = intermediate, red = high) of assessment of extracellular protein secreted by the fungi during 8 weeks of solid substrate cultures on *Miscanthus*. GenBank accession numbers are given for fungi isolated by us from *Miscanthus* or sugarcane. Protein concentrations were determined using the Bradford method with bovine serum albumin as the standard.

Comparison of the amount of bioconversion at 2, 4, and 8 weeks showed that the rate of bioconversion varied by species and that the amount of bioconversion at 8 weeks was better predicted by the amount at 4 weeks than that at 2 weeks. For example, at 2 weeks, the fungi with the most bioconversion were *H.* aff. *koningii*, *Cladosporium* aff. *cladosporioides*, and *Arthrinium* aff. *sacchari*, none of which were among the top five fungal decomposers at 8 weeks. At 4 weeks, the discrepancy was not as great because the top performer, *N. crassa*, and three of the other best performers at 4 weeks, *Chloridium spl*, *P. chrysosporium*, and *Alternaria* aff. *tenuissima*, were among the top five at 8 weeks. In fact, only one of the top five bioconversion fungi at 8 weeks was not among the top five at 4 weeks, *Minimidochium* spl. In addition to *Minimidochium* spl, which increased its bioconversion effort from 11.3% to 17.6% over the last 4 weeks, there were other fungi whose bioconversion increased dramatically over this period, for example, *P. placenta* increased bioconversion of *Miscanthus* from approximately 2% to 14% and *Microdochium* aff. *bolleyi* increased it from 7% to 13%.

With one exception, the fungi that best bioconverted *Miscanthus* were not among the fungi most commonly isolated from *Miscanthus* or sugarcane. The exception was *H.* aff. *koningii*, which was the fungus most commonly isolated (29%) from decaying *Miscanthus* [10]. In contrast, *Chloridium spl* isolates comprised only 1% of fungi isolated from *Miscanthus*. Likewise, *A.* aff. *phaeospermum*, *A.* aff. *tenuissima*, *E.* aff. *nigrum*, and *Minimidochium* spl represented only 2.7%, 2.4%, 1.8%, and 1.2% of fungi isolated from decaying *Miscanthus* samples. With

Species (w/ GenBank Accession #s) / weeks	% reduction (in 8 weeks)			
	glucan	xylan	arabinan	lignin
Phanerochaete chrysosporium	23.3	25.4	36.8	14.7
Postia placenta	14.1	20.5	43.0	7.6
Aureobasidium aff pullans (HQ631013)	1.5	17.8	46.0	-0.6
Alternaria aff tenuissima(HQ630970)	18.8	24.0	42.6	6.5
Bipolaris sp1(HQ631009)	20.6	30.6	49.5	-8.8
Phoma aff herbarum(HQ630963)	12.8	20.1	37.2	2.0
Epicoccum aff nigrum(HQ630972)	16.3	24.0	45.4	1.6
Phoma aff glomerata(HQ630999)	10.0	16.4	35.8	4.6
Dothideomycete sp(HQ631008)	18.0	25.0	45.6	-6.5
Cladosporidium aff cladosporioides(HQ630971)	13.4	22.7	46.1	4.2
Aspergillus aff fumigatus (HQ631021)	11.0	20.4	46.3	-4.9
Penicillium aff minioluteum(HQ631007)	12.3	18.7	35.7	1.5
Exophiala aff spinifera(HQ631027)	-1.2	4.1	13.8	-2.0
Exophiala aff salmonis(HQ630990)	0.6	15.7	27.9	-6.3
Microdochium aff bolleyi(HQ630981)	11.9	14.8	35.0	8.2
Nigrospora aff oryzae(HQ630982)	15.0	18.0	35.3	4.5
Arthrinium aff sacchari(HQ630961)	13.2	16.6	30.0	2.5
Arthrinium aff phaeospermum(HQ630967)	17.7	24.2	46.0	11.1
Cephalosporium aff gramineum(HQ630973)	15.8	18.9	36.2	5.0
Neurospora crassa	18.1	26.5	43.2	7.8
Chloridium sp1(HQ630978)	19.9	24.6	41.6	7.1
Minimidochium sp1(HQ630974)	16.1	20.2	42.0	5.8
Sporothrix aff lignivora(HQ630984)	2.5	2.6	5.8	1.0
Cordyceps aff bassiana(HQ630968)	1.0	0.8	-0.3	-5.7
Gibberella aff moniliformis(HQ630966)	10.6	19.9	43.6	5.1
Gibberella aff avenacea(HQ630977)	11.1	18.5	41.4	0.8
Fusarium aff aethiopicum(HQ630964)	8.5	19.5	45.8	1.6
Fusarium aff proliferatum(HQ630965)	13.6	21.1	39.8	6.9
Fusarium aff equiseti(HQ630976)	12.4	20.9	48.8	2.5
Hypocrea aff lixii(HQ630960)	10.9	20.0	40.9	5.0
Trichoderma aff spirale(HQ630962)	9.0	18.2	45.1	-1.2
Trichoderma reesei - QM9414	11.5	20.5	38.4	4.4
Hypocrea aff koningii(HQ630959)	11.7	20.1	39.4	4.0
Trichoderma aff atroviride(HQ630969)	11.3	18.4	32.3	5.9

Figure 4 Heat map (green = low, yellow = intermediate, red = high) of reduction of *Miscanthus* cell wall components by fungal bioconversion of *Miscanthus* over 8 weeks of solid substrate fermentation. GenBank accession numbers are given for fungi isolated by us from *Miscanthus* or sugarcane.

sugarcane, *Bipolaris sp1* accounted for only 1.5% of isolated strains. Had we had only conducted biomass degradation assays on the top ten species [10] associated with decaying *Miscanthus* and sugarcane, we would not have identified the more efficient biomass degradation activities of these five, moderately represented species.

Five species showed high levels of all four enzyme activities for multiple weeks: two Dothideomycetes, *E.* aff. *nigrum* and *Dothideomycete* sp., and three Sordariomycetes, *A.* aff. *phaeospermum*, *C.* aff. *gramineum*, and *H.* aff. *koningii*. This sustained activity was not seen in the positive control species, where just one activity (exocellulase for *P. chrysosporium*; beta-glucosidase for *P. placenta*, xylanase for *T. reesei QM9414*) or two activities (exocellulase and xylanase for *N. crassa*) were high during the 8 weeks.

Variation in enzyme activity over time has also been reported from other studies where the variation was seen to be as much as twofold after the first 7 days for a selected *Trichoderma* strain by Cianchetta et al. [16], twofold between days 4 and 5 for an *Aspergillus fumigatus* strain obtained from the Amazon forest [25], and twofold between the first and second weeks by strains of *A. fumigatus* and *Myceliophthora* sp. isolated from sugarcane bagasse [14]. In none of these studies was the variation as high as seen here. However, activities were monitored for four times longer in this study than in the previous studies, and, as noted above, the highest levels sometimes were achieved after the longest time interval, that is, 8 weeks.

Conclusion

The most important conclusion from the research presented here is that wild isolates of many fungi recovered from decaying sugarcane or *Miscanthus* were capable of bioconverting ground and alkali-pretreated *Miscanthus* better than *T. reesei*, the fungus that is used to produce most of the enzymes for the commercial deconstruction of plant cell walls. In fact, 21 of the fungi tested did as well as *T. reesei*, 15 did 25% better than *T. reesei*, and one, *Chloridium sp*1 did as well as *P. chrysosporium*, the most active bioconversion fungus of the four well-studied fungi that we included as controls. Another of the four control fungi, *Neurospora crassa*, was among the best fungi at bioconverting *Miscanthus*, validating its use as a model for Ascomycota bioconversion of lignocellulose feed stock for the production of biofuels. Four of the ten best bioconversion fungi isolated from *Miscanthus* or sugarcane are in two taxa of Ascomycota with melanized mycelia, the Chaetospheriales and the Dothideomycetes. Fungi in neither of these taxa have received significant research attention in terms of bioconversion.

Regarding enzyme activity over 8 weeks, the most striking conclusion is that both the level and timing of enzyme activity are quite variable. In terms of activity, for example, xylanase varied over two orders of magnitude in activity among the fungi tested. In terms of timing, the fungi with the most active exocellulases, *Dothideomycete* sp. and *N. crassa*, reached peaks of activity early, after just 1 or 2 weeks, whereas the fungi with the strongest endocellulase activities, *Epicoccum* and *Alternaria*, reached peak activity late, at week 8. With beta-glucosidase, the period of peak activity was reached early and then maintained for a long period, for example, *T.* aff. *atroviride* and *E.* aff. *nigrum* reached peak activity in week 2 and maintained it through week 8; similarly, *H.* aff. *koningii* reached peak activity in week 1 and maintained it through week 4. For xylanase, seven species reached the highest level of activity, six doing so in the eighth week and the seventh, *N. crassa*, achieving the highest level early, at weeks 1 and 2. An important caveat in our measurements of enzyme activity is the contribution of enzyme bound to substrate, which could not contribute to our assays of enzyme activity.

In terms of the potential to discover novel enzymes useful for bioconversion of cellulosic feed stocks, based on statistical analyses, the following species are good candidates for further investigation: *Chloridium sp1*, *Epicoccum* aff. *nigrum*, *Alternaria* aff. *tenuissima*, *Bipolaris sp1*, *Arthrinium* aff. *phaeospermum*, *Minimidochium sp1*, *Cladosporidium* aff. *cladosporioides*, *Microdochium* aff. *bolleyi*, *Nigrospora* aff. *oryzae*, *Dothideomycete* sp., *Fusarium* aff. *proliferatum*, *Aspergillus* aff. *fumigatus*, *H.* aff. *koningii*, and *Cephalosporium* aff. *gramineum*. More importantly, six of these fourteen species: *Epicoccum* aff. *nigrum*, *Dothideomycete* sp., *Alternaria* aff. *tenuissima*, *Arthrinium* aff. *phaeospermum*, *Cephalosproium* aff. *gramineum*, and *H.* aff. *koningii* also showed higher levels of exo- and endocellulase, beta-glucosidase, and xylanase activities across all five time points. Two other fungi are worthy of additional research because they exhibited the highest levels of enzyme activity for at least two enzymes, *N. crassa* and *Trichoderma* aff. *atroviride*.

Another important conclusion regarding unsampled enzyme activity is that the four types of enzymes analyzed here, endocellulase, exocellulase, beta-glucosidase, and xylanase, explained just one quarter of the biomass loss; clearly, other enzymes and processes are playing important roles in biomass conversion. Two measurements explained as much as 64% of the variance in weight loss early in the *Miscanthus* fermentation, that is, at week 2, the amount of secreted protein and the beta-glucosidase activity. In fact, just the concentration of secreted protein correlated more closely with amount of biomass conversion throughout weeks 2 to 8 than summed enzyme activity, again pointing to the action of additional cell wall deconstructing enzymes.

Our final conclusions concern the removal of specific plant cell wall components, that is, glucans, xylans, arabinans, and lignin, by 14 of the best bioconversion fungi. With glucans, *T. reesei* removed less than any of the best 14 fungi. In contrast, four of the wild isolates, plus *N. crassa*, were almost as good as the best fungus, *P. chrysosporium*. With xylans and arabinans, *P. chrysosporium* and *T. reesei* were among the poorest consumers. Instead, *Bipolaris sp1* was best at removing both xylans and arabinans, followed closely by *N. crassa* and additional melanized species. When it comes to lignin, however, *P. crysosporium* is in the lead, having removed 13%, whereas no other species could remove even 10%. A confounding factor when it comes to measuring lignin removal is the production of the structurally similar compound, melanin, by some of the most active bioconversion fungi, including *Bipolaris*, *Epicoccum*, and *Alternaria*, all members of the Dothideomycetes. This production may cause an underestimation of the true amount of lignin removed.

Methods

Fungi

The isolation and identification of fungal isolates used in this study previously were described [10] and cultures have been deposited at Centraalbureau voor Schimmelcultures (CBS) Fungal Biodiversity Center, Utrecht, Netherlands (Table 1). The nomenclatural term, species affinis (abbreviated aff.), is used for taxa with internal transcribed spacer (ITS) sequence identities greater than 97% as compared to named species, and the term species (abbr. sp.) is used for taxa more than 3% distant from any named species.

Substrate and pretreatment

The solid substrate for culturing was ground *Miscanthus* (20 mesh) that had been pretreated with 0.5% *w*/*v* sodium hydroxide (solid to liquid, 1:10) as previously described [10]. Following pretreatment, the *Miscanthus* was recovered via centrifugation, rinsed three times with deionized water, again recovered using centrifugation, and adjusted to pH 4.5 with sulfuric acid in the final rinse. The residue was squeezed to remove excess liquid and then air-dried at room temperature for 48 h before lyophilization and storage at −80°C.

High throughput fungal culture tubes

Miscanthus bioconversion was conducted in round bottom, 15-ml polypropylene tubes [10]. Tubes were weighed, filled with approximately 600 mg pretreated *Miscanthus*, three 5 mm glass beads, and 0.5 ml deionized water, capped and autoclaved at 121°C for 20 min. To determine the initial dry weight of biomass in each tube, the tubes and contents were lyophilized, and this weight was compared to the weight of the empty tube and three 5 mm glass beads.

We chose 30 filamentous fungal isolates for our *Miscanthus* biodegradation study based on their frequency of isolation in decaying *Miscanthus* and sugarcane samples, which included some commonly and rarely isolated species, but no yeasts.

To prepare uniform inocula, fungi were grown in 100 ml of yeast malt (YM) broth as described [10,26]. Fungal colonies were fragmented in a sterile laboratory blender for 1 min and the shredded mycelium was allowed to rejuvenate for 24 h. To minimize nutrient carry over, the fungus was rinsed three times in 100 ml of aqueous NaCl (0.85%) and recovered by centrifugation at each step. Prior to inoculation, the mycelium was resuspended in 50 ml of Vogel's medium [27] with no added sugar.

To start enough solid substrate cultures for three replicates at 0, 1, 2, 4, and 8 weeks (Figure 2) for each fungus, 15 culture tubes were inoculated with 2 ml of suspended mycelium as described [10]. The tubes were plugged with sterile foam and vortexed to mix the biomass and fungal inoculum. Vortexing also spread the mixture along the inner sides of the tube to create a space that provided for air exchange in the central axis of each tube. In addition to testing 30 fungi isolated from *Miscanthus* and sugarcane in the field, we included positive controls with four fungi known to convert biomass, *T. reesei QM9414*, *N. crassa*, *P. chrysosporium*, and *P. placenta*, and a negative control that lacked fungal inoculum. During 8 weeks of solid substrate cultures, we maintained the incubation temperature at 25°C and the relative humidity at 85 ± 5%.

Sampling and analytical assays

We froze and lyophylized three tubes for each fungal species and controls at each sampling time (0, 1, 2, 4, and 8 weeks). Loss of biomass was calculated as the difference between the initial and final dry weights of *Miscanthus* (corrected for the dry weight of added fungal inoculum and assuming that an insignificant amount of fungal biomass was produced during bioconversion) as a percentage of the initial weight and is reported as the mean of the three tubes [10].

Recovery of free sugars and proteins

Following weighing, soluble sugars, organic compounds, and proteins were recovered from the lyophilized *Miscanthus* by adding 10 ml of sterile water to each culture tube, vortexing the tube for 5 min, and centrifuging the tube (2,700 × *g* for 5 min). The supernatant was then filtered (0.22 μm pore size, 25 mm GD/X PES filter membrane, catalog number 6904-2502, Whatman, Piscataway, NJ, USA) into sterile polypropylene tubes and frozen at −80°C. The residues in the culture tubes were also frozen at −80°C.

To analyze total protein (via microwell Bradford Assay) and the activities of four enzymes, xylanase, exocellulase, endocellulase, and b-glucosidase, we used a portion of the filtered, cell-free, supernatant that had been diluted (1:1) in deionized water [23].

Xylanase activity assay

Xylanase activity of the cell-free supernatant (50 μl) was assayed in deep 96 microwell plates with 450 μl of 1% beechwood xylan (Sigma-Aldrich, St. Louis, MO, USA), prepared as 10 g/l in 50 mM sodium acetate buffer at pH 5.0. To aid mixing and reaction, a 3 mm glass bead was added into each of the 96 wells and the sealed plate was shaken at 170 rpm for 20 h in a 37°C incubator. Controls lacked either the substrate or the cell-free supernatant. Specific xylanase activity was determined from the rate of xylose release per unit wt. of protein (μM xylose/min/mg protein) as measured by the dinitrosalicylic acid (DNS) method. The reaction supernatant was recovered by centrifugation (2,500 × *g* for 5 min) and 5 μl were added to 75 μl of DNS reagents for incubation at 99°C for 10 min. The reactions were cooled on ice and diluted with deionized water (1:3) before absorbance was measured at 540 nm. Xylose concentration was determined using a xylose standard curve prepared using xylose standards of 1, 4, 8, 10, 16, and 20 mM.

Exocellulase activity assay

Exocellulase activity of the cell-free supernatant (50 μl) was assayed with 450 μl of 0.5% SigmaCell 20 (Sigma-Aldrich) prepared as 5 g/l in 50 mM sodium acetate buffer at pH 5.0. The reaction conditions were same as described for the xylanase assay. Controls lacked either the substrate or the cell-free supernatant. Specific exocellulase activity was determined from the rate of glucose release per unit wt. of protein (uMglucose/min/mg protein). The reaction supernatant was recovered by centrifugation (2,500 × *g* for 5 min) and 50 μl were added to 150 μl of glucose assay solution (1.5 μl 100 mM o-dianiside, 3 μl 500 U/ml glucose oxidase, 0.3 μl 5,000 U/ml peroxidase and 145.2 μl 50 mM sodium acetate buffer) for incubation at room temperature for 45 min before absorbance was measured at 540 nm. Concentration of glucose was determined by comparison to standard curve prepared from glucose standards of 200, 400, 600, and 1,000 μM.

Endocellulase activity assay

Specific endocellulase activity was measured in the same manner as exocellulase with the exception that the substrate was 0.5% carboxymethyl cellulose (Sigma-Aldrich) prepared as 5 g/l in 50 mM sodium acetate buffer at pH 5.0 and that the enzyme assay plate was incubated at 37°C for 1 h. Released glucose was assayed using glucose oxidase assay as described above.

Beta-glucosidase activity assay

Beta-glucosidase activity of the cell-free supernatant (50 μl) was assayed with 450 μl of 500 μM *p*-nitrophenyl beta D-glucopyranoside (pNPG, Sigma-Aldrich) prepared in 50 mM sodium acetate buffer at pH 5.0. Assays were kept mixed by shaking at 170 rpm for 1 h in a 37°C incubator. Controls lacked either the substrate or the cell-free supernatant. Specific beta-glucosidase activity was determined from the rate of *p*-nitrophenol (pNP) release per unit wt of protein. The reaction supernatant was recovered by centrifugation (2,500 × *g* for 5 min) and 100 μl were mixed with 100 μl of 100 mM sodium bicarbonate before absorbance was measured at 400 nm. Concentration was determined by comparison to *p*-nitrophenol standards of 0, 10, 20, 50, 100, and 200 μM.

Principal biomass component analyses

To prepare biomass for analysis of the glucan, xylan, and lignin fractions remaining after solid substrate cultures, previously frozen residues were thawed and extracted four times at 65°C for 30 min each: twice in 10 ml hot water, once in 10 ml absolute ethanol, and once in 10 ml acetone. The extractive-free residue was air-dried in a chemical hood for 2 days before it was pulverized in a ball mill and dried at 105°C for 16 h. For compositional analysis, the samples were analyzed as outlined in Ibáñez and Bauer [28]. In brief, the pulverized and dried biomass (50 mg) was then incubated at room temperature with 0.5 mL of 72% sulfuric acid in a modified Hungate vial capped with a rubber stopper with vortexing every 15 min. After 1 h, 14 ml of deionized water were added, and the mixture was autoclaved for 60 min (liquid cycle, 121°C) before storage at 4°C overnight to settle the solids. Two milliliters of the clear supernatant was filtered (0.45 μm, PES) and used for high-performance liquid chromatography (HPLC) analysis at 50°C on an HPX-87H (300 × 7.8 mm, Bio-Rad, Hercules, CA, USA) column on an Agilent 1200 series liquid chromatography instrument equipped with a refractive index detector. Elution was performed with 5 mM sulfuric acid at a flow rate of 0.6 ml/min. Glucose, xylose, and arabinose (> = 99%) were obtained from Sigma-Aldrich and linearity of calibration of each standard was tested in the range of 0.01 to 20 mg/ml.

Residues that had not been digested with acid were saved for lignin and ash analyses. The lignin content was determined by the Klason method. Solids were resuspended by vortexing, then filtered through a preweighed glass micro filter after which both the vial, and filter were extensively rinsed with deionized water. The filter and solids were dried at 105°C overnight and weighed after cooling in a desiccator for 30 min. The solids were then ashed by incubation of the filter and content at 575°C (ramp: 105°C for 10 min, 200°C for 10 min, 300°C for 30 min, 575°C for 3 h, cooling to 105° C), cooled in a desiccator for 30 min, and weighed. The percentage of lignin was calculated as the weight of the dry solids minus that of the ash as a percentage of the weight of the initial, dry *Miscanthus* biomass.

Statistical analyses

To compare the biomass degradation ability and extracellular enzyme activity profile of 30 fungal isolates with the four, highly studied species, mean values of the three replicates at each time point were compared. We conducted ANOVA to determine significant differences in data using percent weight loss as the response variable and fungal species as treatments. Tukey-Kramer *post hoc* tests were used to elucidate significant differences in pairwise comparisons. Corrections were made to account for type I errors and *P* values were adjusted using Dunn-Bonferroni and Hochberg step-up methods. Stepwise regressions were used to determine the variables influencing the variation in percent biomass weight loss.

Abbreviations

ANOVA: analysis of variance; CBS: Centraalbureau voor Schimmelcultures; DNS: dinitrosalicylic acid; HPLC: high-performance liquid chromatography; ITS: internal transcribed spacer; pNPG: *p*-nitrophenyl beta D-glucopyranoside; pNP: *p*-nitrophenol; YM: yeast malt.

Competing interests

The authors declare that they have no competing interests.

Authors' contributions

PS designed and performed the study, prepared pure stock cultures of the fungi, analyzed the results, and drafted the manuscript. AI and SB performed biomass component assays and prepared respective results. SG conducted statistical assays of the data. TMS assisted in plant sample collection, isolation, molecular identification, and submission of pure culture of the fungi and yeasts. TDB and JWT coordinated the study and revised the manuscript. All authors approved the final manuscript.

Acknowledgements
The following are acknowledged for helping with this research: The Energy Biosciences Institute for analytical chemistry support; Rafael Silverman y de la Vega, UC Santa Cruz, for help with rDNA sequence corrections, sample preparation for Miscanthus bio-degradation studies, and laboratory assays; Flora Liu for help in enzyme assays; colleagues from the University of Illinois at Urbana-Champaign (Tom Voigt and Rich Pyter) and Louisiana State University at Baton Rouge (Jeff Hoy and M. Catherine Aime) for help in sampling Miscanthus and sugarcane. The Energy Biosciences Institute at the University of California, Berkeley provided funding for this study.

Author details
[1]Department of Plant and Microbial Biology, University of California, Berkeley, CA 94720-3102, USA. [2]Energy Biosciences Institute, University of California, Berkeley, CA 94720, USA. [3]Department of Environmental Science Policy and Management, University of California, Berkeley, CA 94720, USA.

References

1. Farrell AE, Plevin RJ, Turner BT, Jones AD, O'Hare M, Kammen DM. Ethanol can contribute to energy and environmental goals. Science. 2006;311 (5760):506–8.
2. Fargione J, Hill J, Tilman D, Polasky S, Hawthorne P. Land clearing and the biofuel carbon debt. Science. 2008;319(5867):1235–8.
3. Zilberman D, Barrows G, Hochman G, Rajagopal D. On the indirect effect of biofuel. Am J Agric Econ. 2013;95(5):1332–7.
4. Banerjee G, Scott-Craig JS, Walton JD. Improving enzymes for biomass conversion: a basic research perspective. Bioenergy Res. 2010;3(1):82–92.
5. Kazi FK, Fortman J, Anex R, Kothandaraman G, Hsu D, Aden A, et al. Techno-Economic Analysis of Biochemical Scenarios for Production of Cellulosic Ethanol. Techanical Report NREL/TP-6A2-46588 June 2010. Golden, Colorado: National Renewable Energy Laboratory, US DOE; 2010.
6. Klein-Marcuschamer D, Oleskowicz-Popiel P, Simmons BA, Blanch HW. The challenge of enzyme cost in the production of lignocellulosic biofuels. Biotechnol Bioeng. 2012;109(4):1083–7.
7. Lynd LR, Laser MS, Bransby D, Dale BE, Davison B, Hamilton R, et al. How biotech can transform biofuels. Nat Biotech. 2008;26(2):169–72.
8. Pauly M, Keegstra K. Plant cell wall polymers as precursors for biofuels. Curr Opin Plant Biol. 2010;13(3):305–12.
9. Harris PV, Welner D, McFarland KC, Re E, Poulsen JCN, Brown K, et al. Stimulation of lignocellulosic biomass hydrolysis by proteins of glycoside hydrolase family 61: structure and function of a large, enigmatic family. Biochemistry. 2010;49(15):3305–16.
10. Shrestha P, Szaro TM, Bruns TD, Taylor JW. Systematic search for cultivatable fungi that best deconstruct cell walls of *Miscanthus* and sugarcane in the field. Appl Environ Microbiol. 2011;77(15):5490–504.
11. Basso TP, Gallo CR, Basso LC. Cellulolytic activity of isolated fungi from sugarcane bagasse and decayed wood. Pesq Agrop Brasileira. 2010;45(11):1282–9.
12. Berrin JG, Navarro D, Couturier M, Olive C, Grisel S, Haon M, et al. Exploring the natural fungal biodiversity of tropical and temperate forests toward improvement of biomass conversion. Appl Environ Microbiol. 2012;78(18):6483–90.
13. King BC, Waxman KD, Nenni NV, Walker LP, Bergstrom GC, Gibson DM. Arsenal of plant cell wall degrading enzymes reflects host preference among plant pathogenic fungi. Biotechnology for Biofuels. 2011;4:doi:10.1186/754-6834-4-4.
14. Moretti MMS, Bocchini-Martins DA, Da Silva R, Rodrigues A, Sette LD, Gomes E. Selection of thermophilic and thermotolerant fungi for the production of cellulases and xylanases under solid-state fermentation. Braz J Microbiol. 2012;43(3):1062–71.
15. Yang H, Wu H, Wang X, Cui Z, Li Y. Selection and characteristics of a switchgrass-colonizing microbial community to produce extracellular cellulases and xylanases. Bioresour Technol. 2011;102(3):3546–50.
16. Cianchetta S, Galletti S, Burzi PL, Cerato C. Hydrolytic potential of *Trichoderma* sp strains evaluated by microplate-based screening followed by switchgrass saccharification. Enzym Microb Technol. 2012;50(6–7):304–10.
17. Ravalason H, Grisel S, Chevret D, Favel A, Berrin JG, Sigoillot JC, et al. *Fusarium verticillioides* secretome as a source of auxiliary enzymes to enhance saccharification of wheat straw. Bioresour Technol. 2012;114:589–96.
18. Falkoski DL, Guimaraes VM, de Almeida MN, Alfenas AC, Colodette JL, de Rezende ST. *Chrysoporthe cubensis*: a new source of cellulases and hemicellulases to application in biomass saccharification processes. Bioresour Technol. 2013;130:296–305.
19. Ribeiro DA, Cota J, Alvarez TM, Bruchli F, Bragato J, Pereira BMP, et al. The *Penicillium echinulatum* secretome on sugar cane bagasse. Plos One. 2012;7 (12):DOI: 10.1371/journal.pone.0050571.
20. Panagiotou G, Olavarria R, Olsson L. *Penicillium brasilianum* as an enzyme factory; the essential role of feruloyl esterases for the hydrolysis of the plant cell wall. J Biotechnol. 2007;130(3):219–28.
21. McClendon SD, Batth T, Petzold CJ, Adams PD, Simmons BA, Singer SW. *Thermoascus aurantiacus* is a promising source of enzymes for biomass deconstruction under thermophilic conditions. Biotechnology for Biofuels. 2012;5:doi:10.1186/754-6834-5-54.
22. Gao JM, Weng HB, Zhu DH, Yuan MX, Guan FX, Xi Y. Production and characterization of cellulolytic enzymes from the thermoacidophilic fungal *Aspergillus terreus* M11 under solid-state cultivation of corn stover. Bioresour Technol. 2008;99(16):7623–9.
23. Tian CG, Beeson WT, Iavarone AT, Sun JP, Marletta MA, Cate JHD, et al. Systems analysis of plant cell wall degradation by the model filamentous fungus *Neurospora crassa*. Proc Natl Acad Sci U S A. 2009;106(52):22157–62.
24. Couturier M, Navarro D, Olive C, Chevret D, Haon M, Favel A, et al. Post-genomic analyses of fungal lignocellulosic biomass degradation reveal the unexpected potential of the plant pathogen *Ustilago maydis*. BMC Genomics. 2012;13:doi:10.1186/471-2164-13-57.
25. Delabona PD, Pirota RDPB, Codima CA, Tremacoldi CR, Rodrigues A, Farinas CS. Using Amazon forest fungi and agricultural residues as a strategy to produce cellulolytic enzymes. Biomass Bioenergy. 2012;37:243–50.
26. Shrestha P, Khanal SK, Pometto AL, van Leeuwen J. Ethanol production via in situ fungal saccharification and fermentation of mild alkali and steam pretreated corn fiber. Bioresour Technol. 2010;101(22):8698–705.
27. Vogel H. A convenient growth medium for *Neurospora* (medium N). Microb Genet Bull. 1956;13(4):42–3.
28. Ibáñez AB, Bauer S. Downscaled method using glass microfiber filters for the determination of Klason lignin and structural carbohydrates. Biomass Bioenergy. 2014;68:75–81.

11

Elimination of hydrogenase active site assembly blocks H_2 production and increases ethanol yield in *Clostridium thermocellum*

Ranjita Biswas[1,2,4], Tianyong Zheng[2,3], Daniel G Olson[2,3], Lee R Lynd[2,3] and Adam M Guss[1,2,5*]

Abstract

Background: The native ability of *Clostridium thermocellum* to rapidly consume cellulose and produce ethanol makes it a leading candidate for a consolidated bioprocessing (CBP) biofuel production strategy. *C. thermocellum* also synthesizes lactate, formate, acetate, H_2, and amino acids that compete with ethanol production for carbon and electrons. Elimination of H_2 production could redirect carbon flux towards ethanol production by making more electrons available for acetyl coenzyme A reduction to ethanol.

Results: H_2 production in *C. thermocellum* is encoded by four hydrogenases. Rather than delete each individually, we targeted hydrogenase maturase gene *hydG*, involved in converting the three [FeFe] hydrogenase apoenzymes into holoenzymes. Further deletion of the [NiFe] hydrogenase (*ech*) resulted in a mutant that functionally lacks all four hydrogenases. H_2 production in *ΔhydGΔech* was undetectable, and the ethanol yield nearly doubled to 64% of the maximum theoretical yield. Genomic analysis of *ΔhydG* revealed a mutation in *adhE*, resulting in a strain with both NADH- and NADPH-dependent alcohol dehydrogenase activities. While this same *adhE* mutation was found in ethanol-tolerant *C. thermocellum* strain E50C, *ΔhydG* and *ΔhydGΔech* are not more ethanol tolerant than the wild type, illustrating the complicated interactions between redox balancing and ethanol tolerance in *C. thermocellum*.

Conclusions: The dramatic increase in ethanol production suggests that targeting protein post-translational modification is a promising new approach for simultaneous inactivation of multiple enzymes.

Keywords: Cellulosic ethanol, *Clostridium thermocellum*, Hydrogenase maturation, Metabolic engineering

Background

A sustainable future will likely be dependent on large-scale production of fuels, chemicals, and products from renewable resources. One of the most promising approaches to this end is microbially catalyzed conversion of lignocellulosic biomass [1]. Abundant plant biomass resources are available that have the potential to be used as feedstocks [2], but economical bioconversion of plant material into fuel has been elusive [3]. Microbes may ultimately play a central role in the conversion of biomass to fuels and chemicals. While current technologies for biomass fermentation to fuels tend to rely on added cellulolytic enzymes to solubilize hemicellulose and cellulose prior to fermentation, enzyme production represents a significant cost and hinders economic production of biofuels [4]. An alternative approach is to utilize one or more microorganisms to ferment plant cell walls to fuels in one step without added enzymes; this process is called consolidated bioprocessing (CBP). However, no known naturally occurring microbe is capable of robust, high yield, and high titer fuel production from lignocellulose. Thus, genetic modification will be required to create an organism with all of the desired properties for bioconversion of lignocellulosic biomass to fuels.

Clostridium thermocellum, a thermophilic, cellulolytic member of the Firmicutes phylum, is a potential platform to engineer into a CBP organism due to its native ability to efficiently solubilize cellulose and produce ethanol as a fermentation product. However, wild-type

* Correspondence: gussam@ornl.gov
[1]Biosciences Division, Oak Ridge National Laboratory, Oak Ridge, TN 37830, USA
[2]BioEnergy Science Center, Oak Ridge National Laboratory, Oak Ridge, TN 37830, USA
Full list of author information is available at the end of the article

C. thermocellum is limited by its low ethanol yield and titer, producing acetate, lactate, H_2, formate, and free amino acids as additional fermentation products [5], and native strains tolerate low levels of ethanol [6,7]. Recently, methodologies for genetic manipulation have been developed for *C. thermocellum* [8-12], raising the possibility that it can be engineered to economically produce fuels from cellulosic substrates. While these tools are still laborious to use, they have now been successfully used to improve industrially important properties in *C. thermocellum*. For instance, ethanol tolerance has been correlated with mutations in the bifunctional acetaldehyde/alcohol dehydrogenase, *adhE*, in three independent strains of *C. thermocellum* [6,13], and heterologous expression of a mutant *adhE* gene in an otherwise wild-type strain conferred ethanol tolerance [13]. Metabolic flux to major end products has also been blocked via gene deletion, including production of acetic acid [11], lactic acid in an *adhE* mutant [14], and acetic and lactic acid simultaneously [8], increasing ethanol yield. *C. thermocellum* does not encode a pyruvate kinase and is instead thought to divert significant flux from phosphoenolpyruvate → oxaloacetate → malate → pyruvate in what is called a malate shunt, which would generate NADPH while oxidizing NADH to NAD^+, possibly creating a redox imbalance [15-17]. Heterologous expression of pyruvate kinase and disruption of the malate shunt in *C. thermocellum* also substantially improved flux towards ethanol [16], but further increases in ethanol yield are still needed.

While acetic and lactic acids are the primary soluble fermentation products competing with ethanol synthesis, H_2 production is a major sink for electrons that could otherwise be directed toward ethanol production. By using protons as an electron sink rather than glycolytic intermediates, acetyl coenzyme A (acetyl-CoA) becomes available for production of acetate and ATP. While this provides more usable energy for the cell, it decreases the pool of electrons available for reduction of acetyl-CoA.

H_2 production is catalyzed by a class of enzymes called hydrogenases, which broadly fall into three primary categories based on the metals in the active site: [Fe] hydrogenases, which are thus far only found in methanogens, [FeFe] hydrogenases, and [NiFe] hydrogenases. *C. thermocellum* encodes three putative [FeFe] hydrogenases and one ferredoxin-dependent [NiFe] energy-converting hydrogenase (Ech) (Figure 1A, [18,19]). Hydrogenase active sites are complex organometallic catalysts that require a dedicated enzymatic system for post-translational assembly. [FeFe] hydrogenases utilize a single system for active site assembly, consisting of the maturases HydE, HydF, and HydG (Figure 1B, reviewed in [20]). HydF acts as a scaffold upon which the binuclear Fe active site is assembled. HydE likely produces the ligand that bridges the two

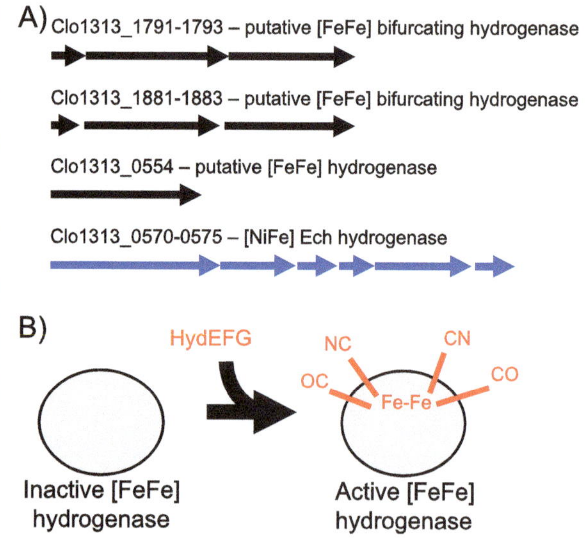

Figure 1 *C. thermocellum* hydrogenases and hydrogenase maturation. A) Chromosomal loci of *C. thermocellum* encoding three [FeFe] hydrogenases (in black) and one [NiFe] hydrogenase (in blue). **B)** Together, HydE, HydF, and HydG assemble the active site of [FeFe] hydrogenases to make active enzymes.

active site Fe molecules, while HydG cleaves tyrosine to generate the -CN and -CO ligands on the active site Fe molecules. Each is required for [FeFe] hydrogenase activity, and thus represents a novel target for simultaneous inactivation of multiple hydrogenases. We hypothesized that targeting electron flux to H_2 can be a fruitful approach to increasing flux to ethanol. Therefore, we targeted inactivation of *hydG* as part of a strategy to eliminate H_2 as a fermentation product and redirect metabolic flux toward ethanol.

Results

Hydrogenase maturase deletion simplifies elimination of H_2 as a fermentation product

To redirect electrons away from H_2 and toward ethanol, we deleted the [FeFe] hydrogenase maturase gene *hydG* to prevent conversion of the hydrogenase apoenzymes into holoenzymes. Because HydEFG is only involved in maturation of the [FeFe] hydrogenases, we further deleted the genes encoding the [NiFe] Ech hydrogenase (Additional file 1). Deletion of *hydG* dramatically decreased H_2 production, with a 15-fold reduction (Figure 2). Further deletion of *ech*, resulting in strain *ΔhydGΔech*, completely eliminated detectable H_2 production and verified that deleting *hydG* simultaneously eliminated all [FeFe] hydrogenase activity.

In anaerobic batch fermentation, the *ΔhydG* mutant, which was later found to also contain a point mutation in the bifunctional acetaldehyde/alcohol dehydrogenase *adhE* (Clo1313_1798) (see below), produced 63% more ethanol than the parent strain (Figure 2). It decreased

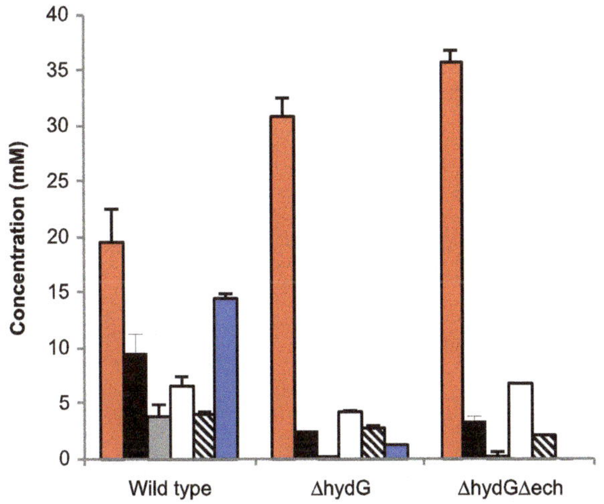

Figure 2 Fermentation products of *C. thermocellum*. The wild-type, *ΔhydG*, and *ΔhydGΔech* strains grown on minimal medium with 5 g/L cellobiose. Red bar, ethanol; black bar, acetate; gray bar, lactate; white bar, formate; diagonal line bar, total amino acids; blue bar, H_2. H_2 concentration is reported as mmol gaseous H_2 per L liquid medium to facilitate comparison to soluble products. Error bars represent one standard deviation.

acetate production by 74%, lactate production was nearly eliminated, formate production decreased by 34% compared to the wild type, and total secreted amino acid levels decreased as well. Further deletion of *ech* in the *ΔhydG* background completely eliminated H_2 production, while ethanol production increased by 90% compared to the wild type and 16% relative to *ΔhydG*. This represents an overall ethanol yield from cellobiose of 64% of the theoretical yield. The maximum optical density was 12% lower in *ΔhydG* (OD600 = 0.70) relative to the wild type (OD600 = 0.80), and the growth rate was also slower (wild type = 0.26 h^{-1}, *ΔhydG* = 0.12 h^{-1}). The growth yield and rate of *ΔhydGΔech* were between these values, with a maximum OD600 = 0.76 and a growth rate of 0.22 h^{-1} (Figure 3).

As *C. thermocellum* is known to divert a significant flux toward production of secreted amino acids, we further examined the abundance of individual amino acids in the supernatant. Among the other secreted amino acids, L-valine was produced in significant amounts in all the strains (Figure 4) as previously observed [21], but was 38% lower in *ΔhydGΔech*. Interestingly, the wild-type strain removed about 1.6 mM of 2.4 mM cysteine in the medium, while *C. thermocellum ΔhydG* and *ΔhydGΔech* did not consume any cysteine from the medium (Figure 4).

Mutation in *adhE* altered cofactor specificity but not ethanol tolerance

During strain validation, a point mutation in the bifunctional acetaldehyde/alcohol dehydrogenase *adhE* (Clo1313_1798) was discovered in *C. thermocellum ΔhydG*, converting aspartate 494 into glycine (D494G). This exact same mutation

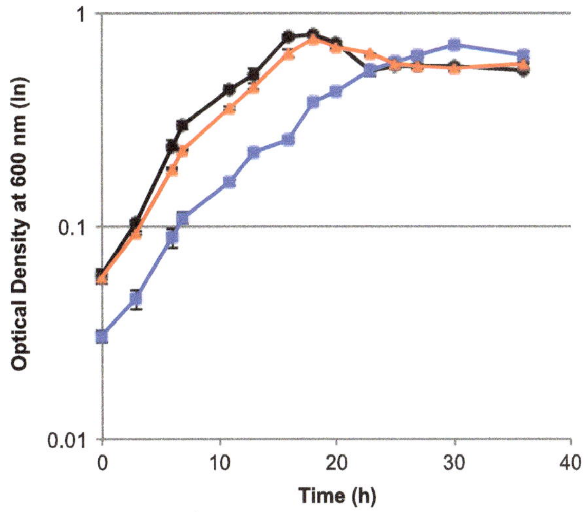

Figure 3 Growth profile of *C. thermocellum* strains on minimal medium. Symbols: Black circles, wild-type *C. thermocellum*; blue squares , *C. thermocellum ΔhydG*; red triangles , *C. thermocellum ΔhydGΔech*. Error bars represent one standard deviation.

was identified in the ethanol-tolerant *C. thermocellum* mutant E50C [6], though the effect of this mutation has not been validated enzymatically or genetically. A different mutant AdhE from the ethanol-tolerant strain *C. thermocellum adhE** (EA) containing two mutations (P704L and H734R) was found to alter coenzyme specificity, eliminating NADH-dependent acetaldehyde reduction but allowing NADPH-dependent acetaldehyde reduction [13]. We therefore tested the effect of the D494G mutation on ADH cofactor specificity in *C. thermocellum ΔhydG*. While the wild-type ADH specific activity was almost exclusively NADH-dependent (7.03 ± 1.06 μmol/min/mg protein with NADH; 0.10 ± 0.08 μmol/min/mg protein with NADPH), ADH specific activity was high with each cofactor in *C. thermocellum*

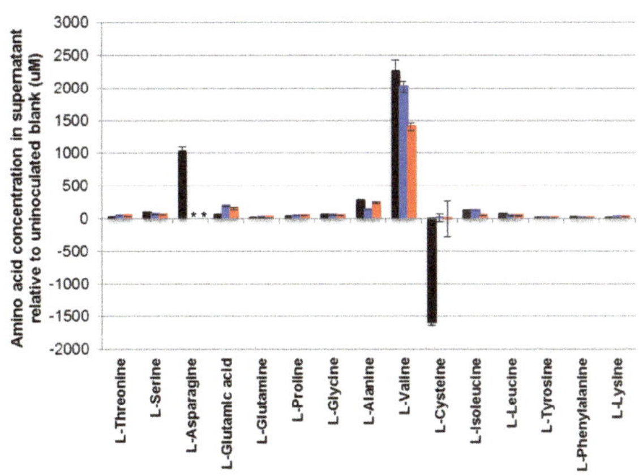

Figure 4 Concentration of amino acids produced by wild-type, *ΔhydG*, and *ΔhydGΔech* strains. *, asparagine quantification for two samples was prevented by interference. Error bars represent one standard deviation. Black bars, wild-type *C. thermocellum*; blue bars, *C. thermocellum ΔhydG*; red bars, *C. thermocellum ΔhydGΔech*.

ΔhydG (6.31 ± 1.27 μmol/min/mg protein with NADH; 5.89 ± 0.89 μmol/min/mg protein with NADPH). As a control, the specific activity of ADH from the *C. thermocellum adhE** (EA) strain containing the P704L and H734R mutations was tested, resulting in 0.006 ± 0.007 μmol/min/mg protein NADH-dependent activity and 0.25 ± 0.005 μmol/min/mg protein NADPH-dependent activity, consistent with previous results [13]. Because previously identified *adhE* mutations in *C. thermocellum*, including D494G, correlated with an ethanol tolerance phenotype [6,13], we examined the effect of added ethanol on the growth of *C. thermocellum ΔhydG*. In rich (Additional file 2, left) and minimal media (Additional file 2, right), *ΔhydG* and *ΔhydGΔech* did not display an ethanol-tolerant phenotype.

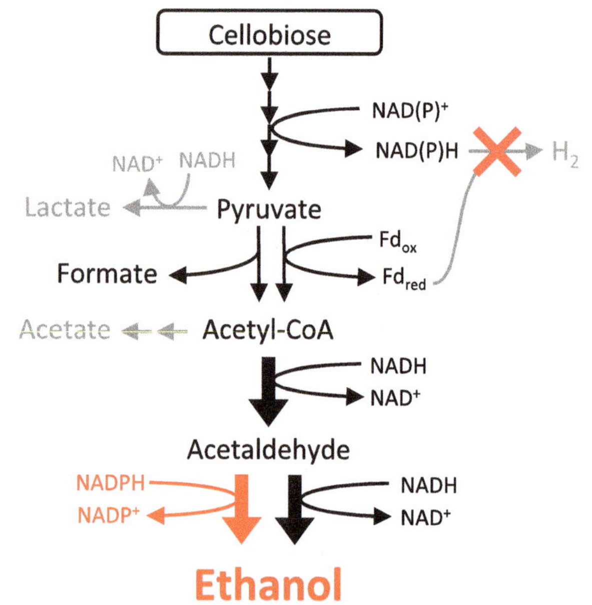

Figure 5 Overview of metabolic changes in *C. thermocellum* ΔhydGΔech. When hydrogenases were inactivated (red X), flux to ethanol increased. Further, the alcohol dehydrogenase mutated, allowing use of NADPH as a cofactor for ethanol production (red pathway). Simultaneously, fluxes to H_2, lactate, and acetate decreased or were eliminated (gray pathways).

Discussion

Genetic manipulation is possible in an increasing number of Firmicutes, and tremendous progress in these systems has been made in recent years. However, depending on the gene and bacterium, the effort to delete a single gene can be time-consuming and a barrier to advanced metabolic engineering approaches. Here, we identified a commonality between multiple desired enzymatic targets and designed a strategy to target them simultaneously. Targeting post-translational modification machinery might be a generally useful approach to streamline metabolic engineering under certain circumstances. Particularly in H_2-producing organisms such as *Clostridium cellulolyticum* (three predicted [NiFe] hydrogenases; four predicted [FeFe] hydrogenases), *Clostridium acetobutylicum* (two predicted [FeFe] hydrogenases), and *Clostridium beijerinckii* (six predicted [FeFe] hydrogenases) [18], this same approach of targeting hydrogenase maturation to divert flux away from H_2 + acetate or butyrate is quite promising. It may also enable fundamental studies of H_2 metabolism in organisms such as sulfate-reducing bacteria [22-24].

Altering electron flux via elimination of H_2 production had a dramatic effect on carbon flux in *C. thermocellum* (Figure 5). Previous *C. thermocellum* metabolic engineering efforts primarily focused on altering carbon flux directly; deletion of phosphotransacetylase (*pta*) nearly eliminated acetate as a fermentation product but had little impact on ethanol yield [11]. Similarly, deletion of *pta* and lactate dehydrogenase together only increased ethanol yield slightly to 27% of the theoretical yield, but strain evolution allowed for increased conversion up to 59% of the theoretical yield [8]. Heterologous expression of a pyruvate kinase and blockage of the malate shunt via deletion of malic enzyme resulted in an ethanol yield of 47% of the theoretical yield [16]. Instead of targeting carbon flux directly, we constrained electron flux by both blocking [FeFe] hydrogenase activity and deleting the Ech hydrogenase. Combined with the spontaneous mutation in *adhE*, this dramatically reduced production of acetate and eliminated H_2, resulting in the highest yet achieved ethanol yield in *C. thermocellum*, 64% of the theoretical maximum without yet evolving the strain for improved performance. The impact of the *adhE* mutation on product yields in these strains is currently unclear, but the evolved strain E50C [6,7] contains the exact same point mutation, and the ethanol yield is comparable to that of the wild type. While *ΔhydG* grew slowly, likely due to difficulty balancing redox reactions, it attained a similar maximum OD to that of the wild type (Figure 3). The metabolic load of producing hydrogenase apoproteins could contribute to the diminished growth rate, especially if they are overproduced in response to the *ΔhydG* mutation, but the hydrogenases were expressed in *ΔhydG* and *ΔhydGΔech* at the same level as in the wild type (CM Wilson, unpublished data), suggesting that metabolic load does not play a significant role in the growth defect. Further deletion of *ech* allowed for a growth rate more similar to that of the wild type while also redirecting more flux to ethanol. This demonstrates that *C. thermocellum* is capable of sufficiently rerouting metabolism to accommodate the lack of H_2 production.

Production of acetic acid plus CO_2 is obligately coupled to production of a more reduced compound, either H_2 or formate in *C. thermocellum*, in order to balance redox reactions. Thus, decreasing or eliminating H_2 production without altering acetate production would result in a redox imbalance, and decreasing flux through this pathway could help alleviate this imbalance. Therefore, the decrease in acetate production seen here is expected upon the elimination of H_2 production. While elimination of the [FeFe] hydrogenases alone increased ethanol yield substantially, the additional removal of the Ech hydrogenase was needed to further increase yield, clearly demonstrating the importance of Ech in this system. This result might be unexpected, because previous proteomics studies showed a lack of Ech protein in the proteome of related strain *C. thermocellum* ATCC27405 [25]. The cause of this discrepancy might well be the difficulty of detecting membrane proteins during proteomics studies, differences between *C. thermocellum* strains ATCC27405 and DSM1313, or differences in physiology due to the deletion of *hydG* in our study.

The hydrogenase mutants also stopped synthesizing lactate via lactate dehydrogenase (*ldh*). In *C. thermocellum* Ldh is known to be allosterically activated by fructose-1,6-bisphosphate (FbP) [26], presenting an alternate mechanism by which lactate production could have been eliminated. Perhaps the hydrogenase deletion strains have low levels of FbP, preventing Ldh from being active. Further, FbP-activated Ldh enzymes from other anaerobic thermophiles are also regulated by other metabolites, such as inhibition by NADPH in *Thermoanaerobacter ethanolicus* [27] and inhibition by pyrophosphate and activation by ATP in *Caldicellulosiruptor saccharolyticus* [28]. While these potential allosteric regulators have not been tested on the *C. thermocellum* Ldh, deletion of the hydrogenases would likely result in an accumulation of reduced ferredoxin via pyruvate: ferredoxin oxidoreductase (PFOR). This reduced ferredoxin plus NADH could then be used as substrates by NfnAB (Clo1313_1848 - 1849) to transfer electrons to NADPH [29], potentially resulting in low levels of NADH and an overabundance of NADPH, leading to inhibition of Ldh. With decreased Ldh activity and an AdhE capable of using both NADH and NADPH, the flux to lactate will decrease and flux to ethanol would then naturally increase to regenerate the cofactors.

Previous characterization of an ethanol-tolerant mutant of *C. thermocellum* revealed that it contained, among many other mutations, two point mutations in *adhE*, resulting in amino acid changes P704L and H734R. This mutation was reconstructed in an otherwise wild-type strain, resulting in altered alcohol dehydrogenase (ADH) enzyme cofactor specificity from NADH-dependent to NADPH-dependent and increased ethanol tolerance [13]. This suggested that an unknown redox imbalance occurs as the ethanol concentration increases. Additional evolved ethanol-tolerant *C. thermocellum* strains E50C and E50A also acquired independent mutations in *adhE*, resulting in amino acid changes to D494G and G553R, respectively [6]. However, the effect of the latter two mutations in an otherwise wild-type strain has not yet been determined. Here, the *hydG* mutant independently acquired the exact same D494G mutation as strain E50C. Unlike the previous P704L-H734R mutant, NADH-dependent activity was not eliminated. Instead, the *hydG* deletion strain contained high levels of both NADH- and NADPH-dependent ADH activities. Interestingly, *ΔhydG* was not ethanol tolerant, illustrating the complexity of the *C. thermocellum* ethanol tolerance phenotype. While it is possible that the D494G mutation does not contribute to the ethanol tolerance phenotype of strain E50C, a more likely explanation is that the additional redox imbalance(s) caused by deletion of *hydG* prevent the D494G mutation from rebalancing metabolism. Regardless, this gain of function mutation that allows use of either NADH or NADPH for reduction of acetaldehyde to ethanol will be useful for further metabolic engineering efforts by minimizing NADH/NADPH redox imbalances. Indeed, *C. thermocellum ΔhydGΔech*, with its high ethanol yield, near wild-type growth rate, and ADH capable of utilizing both NADH and NADPH, is now an ideal platform for further engineering of *C. thermocellum* for production of cellulosic fuels and chemicals.

Conclusions

While *Clostridium thermocellum* degrades cellulose well, efforts to engineer this strain are laborious and time-consuming. In addition to making ethanol, *C. thermocellum* converts cellulosic to lactate, formate, acetate, H_2, ethanol, amino acids, and other products that must be eliminated to achieve high yield. Here, we developed a methodology to simultaneously inactivate multiple enzymes by targeting a common post-translational modification system, simplifying strain construction by minimizing the necessary genetic modifications. We eliminated H_2 production and successfully redirected electron and carbon flux towards desired product. This strain will serve as a platform for further engineering to economically produce fuels and chemicals from lignocellulosic biomass.

Materials and methods

Strains and culture conditions

Standard molecular methods were used for plasmid cloning [30,31]. *Escherichia coli* TOP10 and BL21 were grown in LB medium supplemented with 12 μg ml^{-1} chloramphenicol when appropriate. *Clostridium thermocellum* DSM1313 and mutant strains were grown in a modified DSM122 medium similar to that described by Tripathi *et al.* [11] supplemented with 5 μg ml^{-1} thiamphenicol (TM; Sigma-Aldrich, St. Louis, MO, USA), 50 μg ml^{-1} 5-fluoro-2′-deoxyuridine (FUdR, Sigma-Aldrich, St. Louis, MO, USA), and 500 μg ml^{-1} 8-azahypoxanthine (8AZH; Acros Organics, Pittsburgh, PA, USA) as needed and MTC minimal medium [32] prepared as described in [6]. The modified DSM122 medium composition was (L^{-1}): 3 g sodium citrate tribasic dehydrate, 1.3 g ammonium sulfate, 1.43 g potassium phosphate monobasic, 1.8 g potassium phosphate dibasic trihydrate, 0.5 g cysteine-HCl, 10.5 g 3-morpholino-propane-1-sulfonic acid (MOPS), 6 g glycerol-2-phosphate disodium, 0.41 g sodium acetate, 5 g cellobiose, 4.5 g yeast extract, 0.13 g calcium chloride dehydrate, 2.6 g magnesium chloride hexahydrate, 0.0011 g ferrous sulfate heptahydrate, and 0.0001 g resazurin, adjusted to pH 7.0. The minimal medium consisted of (L^{-1}): 2 g sodium citrate dehydrate, 1.25 g citric acid monohydrate, 1 g sodium sulfate, 1 g potassium phosphate dibasic trihydrate, 2.5 g sodium bicarbonate, 1.5 g ammonium chloride, 2 g urea, 1 g magnesium chloride hexahydrate, 0.2 g calcium chloride dehydrate, 0.1 g ferrous chloride tetrahydrate, 1 g L-cysteine hydrochloride monohydrate, 5 g cellobiose, 0.001 g resazurin, 5 g MOPS, 20 mg pyridoxamine dihydrochloride, 1 mg riboflavin, 1 mg nicotinamide, 0.5 mg DL-thioctic acid, 4 mg 4-amino benzoic acid, 4 mg D-biotin, 0.025 mg folic acid, 2 mg cyanocobalamin, 4 mg thiamine hydrochloride, 0.5 mg $MnCl_2.4H_2O$, 0.5 mg $CoCl_2.6H_2O$, 0.2 mg $ZnSO_4.7H_2O$, 0.05 mg $CuSO_4.5H_2O$, 0.05 mg HBO_3, 0.05 mg $Na_2MoO_4.2H_2O$, and 0.05 mg $NiCl_2.6H_2O$.

Plasmid and strain constructions

Strains were constructed as previously described [8-10] in strain *C. thermocellum Δhpt*, which is a derivative of strain DSM1313 and is herein referred to as wild type. Plasmid pAMG278 (Additional file 3) was used to delete *hydG* (Clo1313_1571), resulting in strain *C. thermocellum Δhpt ΔhydG*, hereafter referred to as *C. thermocellum ΔhydG*. In the *C. thermocellum ΔhydG* background, plasmid pAMG275 (Additional file 3) was used to delete the [NiFe] maturase and hydrogenase enzymes *hypABFCDE* (Clo1313_0564 to 0569) and *echABCDEF* (Clo1313_0570 to 0575), resulting in strain *C. thermocellum Δhpt ΔhydG Δhyp-ech*, hereafter referred to as *C. thermocellum ΔhydGΔech*. All genetic manipulation was carried out using the modified DSM122 medium. Briefly, the plasmids were each isolated from a *dcm*$^-$ strain of *E. coli* (Guss et al., 2012) [9], transformed into *C. thermocellum Δhpt* via electroporation, and plated in medium supplemented with TM. Colonies were picked into liquid medium with TM, grown at 51°C, and plated in medium supplemented with TM and FUdR. Colonies were single colony purified and picked into liquid medium with TM. These cultures were then subcultured without TM and plated with 8AZH to select for the final deletion, followed by single colony purification to obtain pure cultures [8-10]. Deletion strains were confirmed by PCR (Additional file 1; primer sequences in Additional file 4). Strain purity was further confirmed by 16S rRNA gene sequencing.

Fermentation conditions

The inoculum for batch fermentation was prepared by growing the mutants in MTC medium overnight at 55°C in an anaerobic chamber (COY Laboratory Products, Inc., Grass Lake, MI, USA). The fermentation was carbon limited and carried out in 27-ml Balch tubes with 10 ml of MTC media containing 5 g L^{-1} of cellobiose under a N_2 headspace sealed with butyl rubber stoppers. The tubes were inoculated with 0.5% v/v culture and incubated at 55°C. The fermentation products were determined after 53 hours of growth. The final cellobiose concentration was usually less than 0.5 mM, suggesting that fermentation activity was complete. Fermentations were performed at least two times with three independent biological replicates each. Growth rate was calculated using the change in absorbance during mid-log phase, from an OD600 of approximately 0.1 to 0.3.

Ethanol tolerance

Ethanol tolerance was tested in Balch tubes containing minimal medium with 0, 1, 2, 3, 4, and 5% v/v added ethanol, inoculated with 0.5% of overnight grown culture, and incubated at 55°C. Growth was monitored by measuring the optical density at 600 nm on a Unico 1200 spectrophotometer (Unico, Dayton, NJ, USA). Growth experiments were performed at least two times with three independent biological replicates each.

Analytical methods

Fermentation products, including ethanol, acetate, lactate, and formate, were analyzed on a Breeze 2 High Performance Liquid Chromatograph system using an Aminex-HPX-87H column with a 5 mM sulfuric acid mobile phase. H_2 was measured using an Agilent Technologies 6850 Series II Gas Chromatograph (Agilent, Santa Clara, CA, USA) using a thermal conductivity detector at 190°C with a N_2 reference flow and a Carboxen 1010 PLOT (30.0 m × 530 μm I.D.; model Supelco 25467) column.

Enzyme assay of alcohol dehydrogenase (ADH)

50-ml cultures of *C. thermocellum* cells were grown anaerobically to OD_{600} = 0.3 and harvested by centrifugation at 2800 × g for 30 minutes at 4°C and stored at −80°C. Before the assay, the pellet was resuspended anaerobically in 0.5 ml Assay Buffer (0.1 M Tris–HCl, 0.1 mM Fe_2^+, 0.1 mM DTT, pH 7.5). The cells were lysed with 2 μl Ready-Lyse Lysozyme from Epicentre Technologies (Madison, WI, USA), and 1 μl DNase I from New England Biolabs (Ipswich, MA, USA) was added to reduce viscosity. The resulting solution was centrifuged at 17,000 × g for 10 minutes at room temperature, and the supernatant was used as cell-free extracts for enzyme assays. For the ADH (acetaldehyde reduction) reactions, the anaerobic reaction mixture contained 0.25 mM NAD(P)H, 18 mM acetaldehyde, 1 to 10 μl cell-free extract, and 800 μl of Assay Buffer. Decrease in absorbance at 340 nm caused by NAD(P)H oxidation was monitored by an Agilent 8453 UV–vis spectrophotometer with Peltier controlled heating set at 55°C inside an anaerobic chamber. Protein concentration was determined by using the Bradford method [33].

Amino acid analysis

Amino acid analysis was carried out by AminoAcids.com (St. Paul, MN, USA). Samples were prepared for analysis by appropriate dilution with a deproteinizing solution (13.5% w/v 5-sulfosalicylic acid hydrate), L-2-amino-3-guanidinopropionic acid hydrochloride, and eluent lithium buffer. The mixture was vortexed, microcentrifuged, and filtered through a 0.2-μm filter. Amino acid analysis was performed on a Hitachi Model L-8900 HPLC with 10-cm cation exchange columns, four sequential lithium-based eluents (Hitachi and Pickering Laboratories), and lithium hydroxide for column regeneration. Absorbance was measured at 440 and 570 nm following post-column color development by ninhydrin reagent at 135°C. Data are an average of three independent biological replicates. AminoAcids.com reported that an unknown peak overlapped with the asparagine peak, preventing quantification in both the *ΔhydG* and *ΔhydGΔech* samples. As the ninhydrin method detects primary and secondary amines, the interfering peak presumably represents an as yet unidentified amino compound.

Identification of mutation in alcohol dehydrogenase gene in *ΔhydG* strain

The *adhE* region was PCR amplified as 982-bp product using primers XD520 and XD521 (Additional file 3). The primer region was 70 bp away from the D494G mutation [6], which was sequenced and confirmed with primer XD560 (Additional file 4).

Additional files

Additional file 1: Confirmation of *C. thermocellum* deletion mutants. A) Three primer sets were used for confirmation of the *hydG* deletion. Primer set (P1) and (P2) amplified 863-bp and 729-bp fragments of the *hydG*, respectively, which are present in the wild type but absent in *C. thermocellum ΔhydG*. Primer set (P3) amplifies the 3,500-bp region of the wild-type locus, while amplification from *ΔhydG* mutant results in a 2,100-bp fragment. B) Similar primer sets (P4, P5, and P6) are used to confirm deletion of *ech* in the *C. thermocellum ΔhydG* background. P4 and P5 amplify 701-bp and 721-bp regions of *ech*, respectively, which are present in the *ΔhydG* and wild-type strains but absent in the *ΔhydGΔech* strain. Primer set (P6) would amplify a 12,000-bp region of the wild-type locus and a 2,000-bp fragment in the *ΔhydGΔech* strain. C) PCR confirmation of deletion of *hydG* and *ech*. Lane m, DNA ladder with molecular weights noted (in kilobases); lane a, *ΔhydGΔech* template; lane b, *ΔhydG* template; lane c, *C. thermocellum* wild-type template; lane d, no template PCR control.

Additional file 2: Maximum optical density (OD) attained by wild-type and mutant strains of *C. thermocellum*. (Left) rich and (right) minimal medium supplemented with 0 to 5% (v/v) added ethanol. Symbols: Open black squares, *C. thermocellum* wild type; closed gray squares, ethanol-tolerant control *C. thermocellum adhE**(EA); closed blue triangles, *C. thermocellum ΔhydG;* and open red triangles, *C. thermocellum ΔhydGΔech*. Error bars represent one standard deviation.

Additional file 3: Annotated plasmid sequences of pAMG275 and pAMG278.

Additional file 4: List of primers used in this study.

Abbreviations

ADH: alcohol dehydrogenase; 8AZH: 8-azahypoxanthine; CBP: consolidated bioprocessing; FUdR: 5-fluoro-2'-deoxyuridine; MOPS: 3-morpholino-propane-1-sulfonic acid; TM: thiamphenicol.

Competing interests

LRL is a co-founder and employee of Mascoma Corporation, at which DGO was formerly employed. Mascoma has a financial interest in biomass conversion processes.

Authors' contributions

RB participated in the design of the study, participated in strain construction, characterized growth and fermentation, and drafted the manuscript. TZ carried out the enzyme assays. DGO participated in the design of the study. LRL helped conceive of the study and participated in its design. AMG helped conceive of the study, participated in its design and coordination, participated in strain construction, and drafted the manuscript. All authors read and approved the final manuscript.

Acknowledgements

This work was supported by the BioEnergy Science Center, U.S. DOE Bioenergy Research Center supported by the Office of Biological and Environmental Research in the DOE Office of Science. Oak Ridge National Laboratory is managed by UT-Battelle, LLC, for the U.S. DOE under contract DE-AC05-00OR22725. The funders had no role in study design, data collection and analysis, decision to publish, or preparation of the manuscript.

Author details

[1]Biosciences Division, Oak Ridge National Laboratory, Oak Ridge, TN 37830, USA. [2]BioEnergy Science Center, Oak Ridge National Laboratory, Oak Ridge, TN 37830, USA. [3]Thayer School of Engineering at Dartmouth College, Hanover, NH 03755, USA. [4]Current address: Centre for Rural Development and Technology, Indian Institute of Technology Delhi, Hauz Khas, New Delhi 110016, India. [5]One Bethel Valley Road, Oak Ridge, TN 37831-6038, USA.

References

1. Lynd LR, Weimer PJ, van Zyl WH, Pretorius IS. Microbial cellulose utilization: fundamentals and biotechnology. Microbiol Mol Biol Rev. 2002;66:506–77.
2. US Department of Energy. US billion-ton update: biomass supply for a bioenergy and bioproducts industry. Oak Ridge, TN: Oak Ridge National Laboratory; 2011.
3. Blumer-Schuette SE, Brown SD, Sander KB, Bayer EA, Kataeva I, Zurawski JV, et al. Thermophilic lignocellulose deconstruction. FEMS Microbiol Rev. 2014;38:393–448.
4. Lynd LR, Laser MS, Bransby D, Dale BE, Davison B, Hamilton R, et al. How biotech can transform biofuels. Nat Biotechnol. 2008;26:169–72.
5. Ellis LD, Holwerda EK, Hogsett D, Rogers S, Shao X, Tschaplinski T, et al. Closing the carbon balance for fermentation by Clostridium thermocellum (ATCC 27405). Bioresource Technol. 2012;103:293–9.
6. Shao X, Raman B, Zhu M, Mielenz JR, Brown SD, Guss AM, et al. Mutant selection and phenotypic and genetic characterization of ethanol-tolerant strains of Clostridium thermocellum. Appl Microbiol Biotechnol. 2011;92:641–52.
7. Williams TI, Combs JC, Lynn BC, Strobel HJ. Proteomic profile changes in membranes of ethanol-tolerant Clostridium thermocellum. Appl Microbiol Biotechnol. 2007;74:422–32.
8. Argyros DA, Tripathi SA, Barrett TF, Rogers SR, Feinberg LF, Olson DG, et al. High ethanol titers from cellulose using metabolically engineered thermophilic, anaerobic microbes. Appl Environ Microbiol. 2011;77:8288–94.

9. Guss AM, Olson DG, Caiazza NC, Lynd LR. Dcm methylation is detrimental to plasmid transformation in Clostridium thermocellum. Biotechnol Biofuels. 2012;5:30.
10. Olson DG, Lynd LR. Transformation of Clostridium thermocellum by electroporation. Methods Enzymol. 2012;510:317–30.
11. Tripathi SA, Olson DG, Argyros DA, Miller BB, Barrett TF, Murphy DM, et al. Development of pyrF-based genetic system for targeted gene deletion in Clostridium thermocellum and creation of a pta mutant. Appl Environ Microbiol. 2010;76:6591–9.
12. Mohr G, Hong W, Zhang J, Cui GZ, Yang Y, Cui Q, et al. A targetron system for gene targeting in thermophiles and its application in Clostridium thermocellum. PLoS One. 2013;8:e69032.
13. Brown SD, Guss AM, Karpinets TV, Parks JM, Smolin N, Yang S, et al. Mutant alcohol dehydrogenase leads to improved ethanol tolerance in Clostridium thermocellum. Proc Natl Acad Sci U S A. 2011;108:13752–7.
14. Biswas R, Prabhu S, Lynd LR, Guss AM. Increase in ethanol yield via elimination of lactate production in an ethanol-tolerant mutant of Clostridium thermocellum. PLoS One. 2014;9:e86389.
15. Rydzak T, McQueen PD, Krokhin OV, Spicer V, Ezzati P, Dwivedi RC, et al. Proteomic analysis of Clostridium thermocellum core metabolism: relative protein expression profiles and growth phase-dependent changes in protein expression. BMC Microbiol. 2012;12:214.
16. Deng Y, Olson DG, Zhou J, Herring CD, Joe Shaw A, Lynd LR. Redirecting carbon flux through exogenous pyruvate kinase to achieve high ethanol yields in Clostridium thermocellum. Metab Eng. 2013;15:151–8.
17. Raman B, McKeown CK, Rodriguez Jr M, Brown SD, Mielenz JR. Transcriptomic analysis of Clostridium thermocellum ATCC 27405 cellulose fermentation. BMC Microbiol. 2011;11:134.
18. Calusinska M, Happe T, Joris B, Wilmotte A. The surprising diversity of clostridial hydrogenases: a comparative genomic perspective. Microbiol-Sgm. 2010;156:1575–88.
19. Feinberg L, Foden J, Barrett T, Davenport KW, Bruce D, Detter C, et al. Complete genome sequence of the cellulolytic thermophile Clostridium thermocellum DSM1313. J Bacteriol. 2011;193:2906–7.
20. Mulder DW, Shepard EM, Meuser JE, Joshi N, King PW, Posewitz MC, et al. Insights into [FeFe]-hydrogenase structure, mechanism, and maturation. Structure. 2011;19:1038–52.
21. van der Veen D, Lo J, Brown SD, Johnson CM, Tschaplinski TJ, Martin M, et al. Characterization of Clostridium thermocellum strains with disrupted fermentation end-product pathways. J Ind Microbiol Biotechnol. 2013;40:725–34.
22. Caffrey SM, Park HS, Voordouw JK, He Z, Zhou J, Voordouw G. Function of periplasmic hydrogenases in the sulfate-reducing bacterium Desulfovibrio vulgaris Hildenborough. J Bacteriol. 2007;189:6159–67.
23. Casalot L, De Luca G, Dermoun Z, Rousset M, de Philip P. Evidence for a fourth hydrogenase in Desulfovibrio fructosovorans. J Bacteriol. 2002;184:853–6.
24. Casalot L, Valette O, De Luca G, Dermoun Z, Rousset M, de Philip P. Construction and physiological studies of hydrogenase depleted mutants of Desulfovibrio fructosovorans. FEMS Microbiol Lett. 2002;214:107–12.
25. Rydzak T, Grigoryan M, Cunningham ZJ, Krokhin OV, Ezzati P, Cicek N, et al. Insights into electron flux through manipulation of fermentation conditions and assessment of protein expression profiles in Clostridium thermocellum. Appl Microbiol Biotechnol. 2014;98:6497–510.
26. Ozkan M, Yilmaz EI, Lynd LR, Ozcengiz G. Cloning and expression of the Clostridium thermocellum L-lactate dehydrogenase gene in Escherichia coli and enzyme characterization. Can J Microbiol. 2004;50:845–51.
27. Bryant FO. Characterization of the fructose 1,6-bisphosphate-activated, L (+)-lactate dehydrogenase from Thermoanaerobacter ethanolicus. J Enzyme Inhib. 1991;5:235–48.
28. Willquist K, van Niel EW. Lactate formation in Caldicellulosiruptor saccharolyticus is regulated by the energy carriers pyrophosphate and ATP. Metab Eng. 2010;12:282–90.
29. Wang S, Huang H, Moll J, Thauer RK. NADP+ reduction with reduced ferredoxin and NADP+ reduction with NADH are coupled via an electron-bifurcating enzyme complex in Clostridium kluyveri. J Bacteriol. 2010;192:5115–23.
30. Maniatis T, Fritsch EF, Sambrook J. Molecular cloning: a laboratory manual. Cold Spring Harbor, NY: Cold Spring Harbor Laboratory; 1982.
31. Shanks RMQ, Kadouri DE, MacEachran DP, O'Toole GA. New yeast recombineering tools for bacteria. Plasmid. 2009;62:88–97.
32. Hogsett D. Cellulose hydrolysis and fermentation by Clostridium thermocellum for the production of ethanol [PhD thesis]. Hanover, NH: Dartmouth College, Thayer School of Engineering; 1995.
33. Bradford MM. A rapid and sensitive method for the quantitation of microgram quantities of protein utilizing the principle of protein-dye binding. Anal Biochem. 1976;72:248–54.

12

Highly efficient enzymatic biodiesel production promoted by particle-induced emulsification

Juan Mangas-Sánchez* and Patrick Adlercreutz

Abstract

Background: At present, the conversion of oils to biodiesel is predominantly carried out using chemical catalysts. However, the corresponding lipase-catalysed process has important advantages, which include mild reaction conditions and the possibility of using cheap, low quality feedstocks with a high free fatty acid content. Further increases in the efficiency of the enzymatic process are desired to make it even more attractive and suitable for large-scale applications.

Results: Herein, we present a simple and efficient two-phase lipase-catalysed system for the preparation of biodiesel in which different parameters (biocatalyst composition, ethanol concentration and the presence of additives) were optimised in order to obtain the maximum productivity starting from triolein with a high free oleic acid content. In the two-phase system, the enzyme tolerated high-ethanol concentrations, which made it possible to reach high conversions. The addition of silica particles increased the reaction rate substantially. It was suggested that such particles can catalyse acyl migration as a step to the full conversion to glycerol and biodiesel. However, in the system studied here, the effect of the particles was shown to be due to the formation of smaller and more uniform emulsion droplets leading to better mass transfer between the two phases. Particles of widely different size had positive effects, and the highest rate was obtained with silica particles derivatised with phenyl groups. The optimal conditions were applied to the solvent-free ethanolysis of rapeseed oil, and a yield of 96% was reached in 5 h. Under the mild conditions used, chemical catalysts were inefficient.

Conclusions: Triacylglycerol oils with a high free fatty acid content can be efficiently converted to ethyl esters using *Thermomyces lanuginosus* lipase as the catalyst in an aqueous/organic two-phase system. Fast mass transfer can be achieved using silica particles, which helped to decrease the size of the emulsion droplets and thus led to a more efficient process. The high-ethanol concentration tolerated by the lipase in this system made it possible to reach almost quantitative yields.

Keywords: Biodiesel, Lipase, Transesterification, *Thermomyces lanuginosus*, High fatty acid content, Ethanolysis, Pickering emulsion and two-phase system

Background

Biodiesel is a mixture of esters that can be produced by the transesterification of triglycerides (oils) with short-chain alcohols, obtaining glycerol as the major by-product [1-3]. Biodiesel production has gained importance in recent years for its ability to replace fossil fuels since it can be blended with conventional diesel fuel. Additionally, its use does not require any important technical modifications on the engine, and it provides environmental benefits since it is a renewable source of energy [4]. Global biodiesel production was less than 1 billion litres in 2000, which has increased to almost 15 billion litres in 2009 [5]. The EU is the world's main producer, with a market share of around 55%. This production is highly influenced by domestic fuel policy [6], and in this context, several targets have been established for the use of biodiesel in fuels [7].

Industrial biodiesel production is still mainly carried out via base-catalysed transesterification using sodium or potassium hydroxides with an excess of methanol [8-11]. Alkoxides are also often used since less water is released in the reaction. These processes present several drawbacks such as the need to remove inorganic salt in the downstream process, the high temperature required and undesirable side reactions. Also, these systems have been found to be inefficient when a high free fatty acid (FFA) content is present in the starting material [12],

* Correspondence: juan.mangas@biotek.lu.se
Department of Biotechnology, Lund University, P.O. Box 124, Lund SE-221 00, Sweden

which restricts the use of conventional chemical pathways to a highly pure feedstock. For ecological and economical reasons, the development of new methodologies has attracted attention in recent years. In this context, the use of lipases may become an alternative solution [13-16]. Immobilised lipase-catalysed transesterification in the presence of an organic solvent has been studied in detail. The possibility of reusing the enzyme and the simplicity of the downstream process makes this approach attractive [17]. While the use of biocatalytic methods is well-established for the preparation of pharmaceutical products, much higher productivities are required for commodity chemicals due to the price of biocatalysts. Concerning biodiesel production, lipase immobilisation adds an extra cost to the process and partial inactivation may also occur. Therefore, the catalyst has to be reused many times in order to be economically feasible [18]. For these reasons, the use of liquid lipase preparations has attracted interest, but the optimisation of these processes is needed in order to establish them as a realistic industrial alternative.

Two-phase systems are of particular interest for reactions with both non-polar and polar substrates and products. Lipases specifically evolved for the hydrolysis of fats and oils; they are activated in biphasic mixtures in a process called interfacial activation [19]. Most lipases have a lid covering the lipophilic active site. Once the lipase comes into contact with a non-polar surface, the lid opens, exposing the lipophilic area where the active site is situated. This part is oriented towards the organic layer, such that the lipid can easily enter the active site and the hydrophilic products can diffuse into the aqueous layer. For that reason, two-phase reaction media seem to be the most suitable environment for these particular enzymes [20]. In this sense, we have recently developed an efficient methodology for the chemoenzymatic preparation of 2-monoolein and 1,2-diolein in a two-phase system [21]. Nevertheless, these systems present mass transfer limitation issues, especially when working with surfactant-like species such as diacyl and monoacylglycerides (DAGs and MAGs). To overcome this limitation, the use of strategies which can favour mass transfer in the complex system is important.

In the current study, aqueous-organic two-phase systems for biodiesel production were developed. In the development work, a substrate containing triolein rich in free oleic acid was used to simulate a low-quality oil. Since the substrate contained no fatty acids other than oleic acid, it was possible to quantify all intermediate acylglycerol products accurately, thus facilitating the optimization of the system in a rational way. Finally, the best reaction conditions were also applied to the solvent-free conversion of rapeseed oil to biodiesel.

Results and discussion

Enzyme screening

Enzymatic ethanolysis of triglycerides (TAGs) for the preparation of biodiesel constitutes a complex system because it includes two liquid phases and several enzyme-catalysed reactions. Lipases express different specificities for each reaction involved [16], and therefore, the screening of lipases in the reaction system is important. Initially, both the organic and aqueous phase were analysed, but it was found that the amounts of acylglycerols, biodiesel and FFA in the aqueous phase were negligible. Lipases from *Rhizomucor miehei* and *Pseudomonas fluorescens* converted only small amounts of TAG under the conditions used (Figure 1). The lipase from *Pseudomonas* sp. converted most of the TAG, but only minor amounts of biodiesel were formed; instead, DAG,

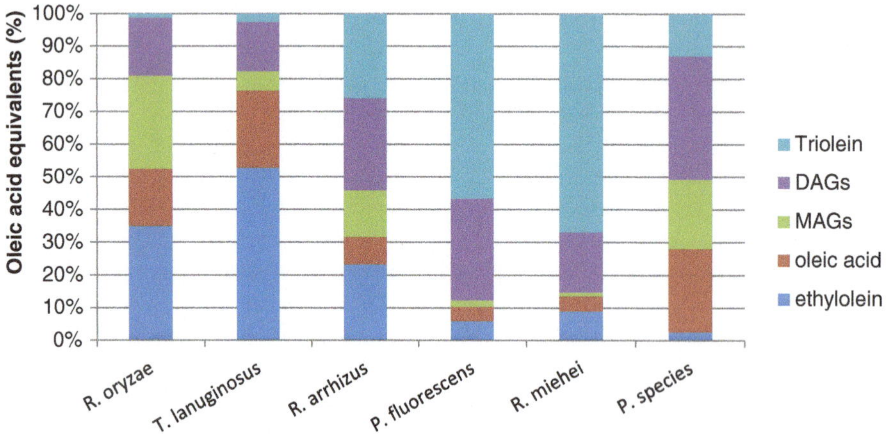

Figure 1 Product composition after 6 h using different lipases. Ethanolysis of triolein catalysed by different lipases at 37°C and 1,000 r.p.m using 400 mM ethanol and 75 mg enzyme/g substrate for solid preparations and 100 μL enzyme/g substrate for liquid formulations. DAGs, diacylglycerols; MAGs, monoacylglycerols.

MAG and FFA were the main products. *Rhizopus oryzae*, *Rhizopus arrhizus* and *Thermomyces lanuginosus* lipases produced biodiesel yields >23%. *R. oryzae* and *R. arrhizus* lipases express strong 1,3-specificity and promoted the accumulation of high 2-MAG concentrations. *T. lanuginosus* lipase was efficient in converting all the acylglycerols and was found to be the most suitable biocatalyst for biodiesel preparation. Using this lipase, 97.5% conversion of TAG was reached after 6 h, providing a 53% yield of biodiesel.

Biocatalyst composition

The biodiesel yield improved with an increasing amount of TLL up to 100 μL of enzyme per gram of substrate (Figure 2). It has been shown that combinations of lipases can favour the process since different lipases have different substrate specificity [16,22]. As a complement to TLL, which is efficient in converting TAG, *Candida antarctica* lipase B (CAL-B) was chosen since this enzyme can efficiently convert DAG and MAG [16]. However, under the conditions used, the addition of CAL-B did not improve the biodiesel yield. It should be pointed out that the observed yields are higher than those obtained using immobilised TLL in a low-water system with a similar substrate (Figure 2, [16]), which shows that TLL is more efficient in biodiesel production in a two-phase system than in a pure organic medium. This agrees with previous observations that TLL is very efficient in aqueous/organic two-phase systems rather than in pure organic media compared to CAL-B [23]. However, a drawback of the two-phase system is that the high-water content causes a significant hydrolysis reaction, which competes with biodiesel formation.

Effect of ethanol concentration

In order to suppress the hydrolysis reaction and favour the alcoholysis reaction, the alcohol concentration in the reaction mixture can be increased, although it is known that alcohols can also inhibit and inactivate lipases [24,25]. A range of ethanol concentrations between 400 mM and 3 M was tested. The initial rate was highest at ethanol concentrations of 1.6 and 2.0 M (0.17 μmol min^{-1} $μL^{-1}$ enzyme), while after 6 h, the highest biodiesel yield was observed with 2.5 M ethanol (Figure 3). After 24 h, equilibrium was reached in most cases, and the yield increased with an increasing ethanol concentration, with the exception of the highest one (Table 1). It should be noted that the FFA concentrations decreased towards the end of the reaction, which indicates that the FFA formed under kinetic control was esterified in the later part of the reaction, when equilibrium was approached. An additional experiment was conducted using the same experimental set-up with 100 mM oleic acid and 2 M ethanol as the substrates. After just 1 h, 95% conversion was observed, showing that the enzyme was able to convert FFA into biodiesel efficiently. It seemed that the enzyme tolerated high-ethanol concentrations in the two-phase system quite well and a concentration of 2 M was chosen for the rest of the experiments.

Effects of added particles

Due to the high 1,3-specificity expressed by many lipases, including TLL, the accumulation of 2-MAG is expected to constitute a bottleneck in the process. To promote the overall conversion to glycerol and biodiesel, the addition of substances able to catalyse the isomerisation of 2-MAG to the thermodynamically more stable 1 (3)-MAG, which is a better substrate for lipases, has been applied successfully. In a typical example, soybean oil is converted to biodiesel using a mixture of immobilised

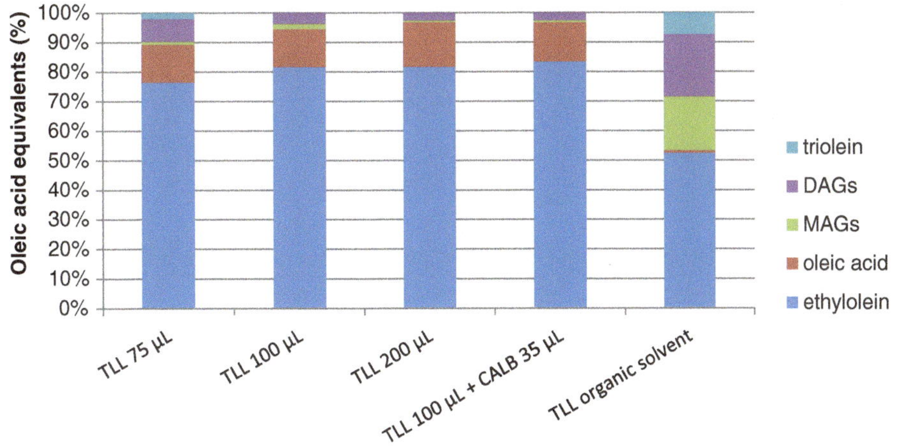

Figure 2 Product composition after 24 h using different enzyme amounts and combinations. Ethanolysis of triolein catalysed by different lipase types and amounts at 37°C, 1,000 r.p.m. and 400 mM ethanol concentration. DAGs, diacylglycerols; MAGs, monoacylglycerols; TLL, *T. lanuginosus* lipase.

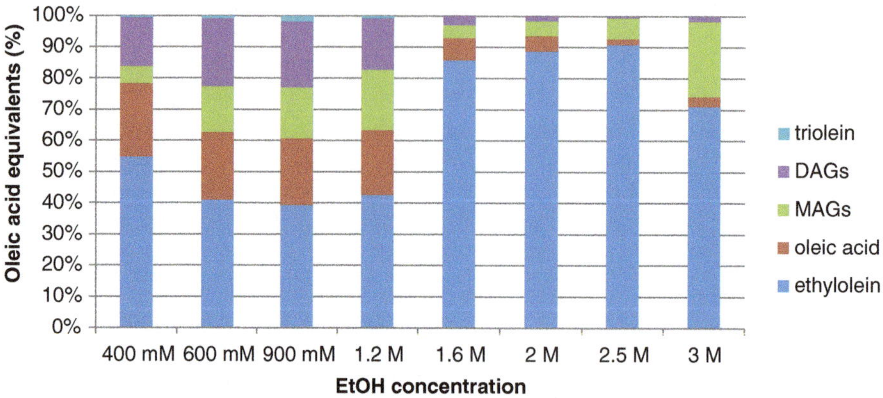

Figure 3 Product composition after 6 h using different ethanol concentrations. Ethanolysis of triolein catalysed by TLL lipase using different ethanol concentrations and 100 μL of TLL/g substrate at 37°C and 1,000 r.p.m. DAGs, diacylglycerols; MAGs, monoacylglycerols.

lipase and silica gel as the catalyst [26]. Another approach is to immobilise the lipase on an anion exchange resin, which works both as a support for the lipase and as an acyl migration catalyst [27]. Since silica is the most commonly used acyl migration catalyst, the addition of 4% *w*/*v* silica (35- to 70-μm particle size) was evaluated in the reaction system used in the present study; it was found to increase the rate of biodiesel formation significantly. In order to confirm that it was acyl migration that caused the increased conversion rate, 2-monoolein was synthesised separately and its isomerisation to 1(3)-monoolein was studied under different conditions. Acyl migration was promoted by silica in pure heptane, in agreement with previous observations (Figure 4). However, surprisingly enough, no acyl migration was observed in the water/heptane two-phase system in the presence or absence of silica (Figure 4). In order to determine the mechanism of the silica-induced increase in the biodiesel production rate, the reaction mixtures were subjected to microscopic investigation. It was found that the presence of silica particles caused the formation of an emulsion with smaller and more uniform droplets than in the absence of silica particles (Figure 5). It is likely that the increased interfacial area caused the observed increase in the reaction rate, by speeding up the mass transfer of substrates and products between the phases. Alternatively, the increased rate could have been due to increased interfacial activation of TLL due to the increased interfacial area.

It is known that small solid particles can act as amphiphilic molecules, adsorbing on droplets and preventing flocculation and coalescence and thus creating so-called Pickering emulsions [28-32]. However, this requires much smaller particles and more power input than the shaking that was used in the present study. To further study the effects of particles on the studied biodiesel process, the addition of smaller silica particles was evaluated. Furthermore, since it has been shown that not only the size but also the surface character of the particles can affect the stability of the emulsion [33], all sizes of silica particles were evaluated in a phenyl-derivatised form as well. All types of particles increased the biodiesel formation rate and the phenyl-modified particles were somewhat more efficient (Figure 6). After a 3-h reaction time, all acylglycerols had been converted in the reactions involving Phenyl 35–70 and Phenyl 0.007 (Figure 7), and in the continued reaction, a further increase in yield was observed due to esterification of the originally formed FFA. After 6 h, yields over 95% were obtained regardless of the particle size and hydrophobic character.

All particles caused the formation of emulsions with a similar droplet size (Figure 5). Since the effect was observed not only with nanoparticles, but also with

Table 1 Product composition using different amounts of ethanol

	400 mM	600 mM	900 mM	1.2 M	1.6 M	2 M	2.5 M	3 M
Ethylolein	72.1 ± 2.4	73.3 ± 3.1	78.5 ± 2.2	91.0 ± 2.7	98.3 ± 1.4	98.0 ± 1.6	99.7 ± 0.2	99.4 ± 0.3
Oleic acid	24.6 ± 2.2	15.5 ± 0.7	11.7 ± 2.1	5.9 ± 3.6	1.7 ± 1.4	2.0 ± 1.6	0.3 ± 0.2	0.6 ± 0.3
MAGs	0.8 ± 0.2	3.5 ± 0.7	4.7 ± 0.7	3.1 ± 0.6	0	0	0	0
DAGs	2.5 ± 0.8	7.7 ± 0.6	5.1 ± 0.1	0	0	0	0	0
Triolein	0	0	0	0	0	0	0	0

Product composition (oleic acid equivalents, %) after 24 h at 37°C and 1,000 r.p.m. using different ethanol concentrations and 100 μL of TLL/g substrate. DAGs, diacylglycerols; MAGs, monoacylglycerols.

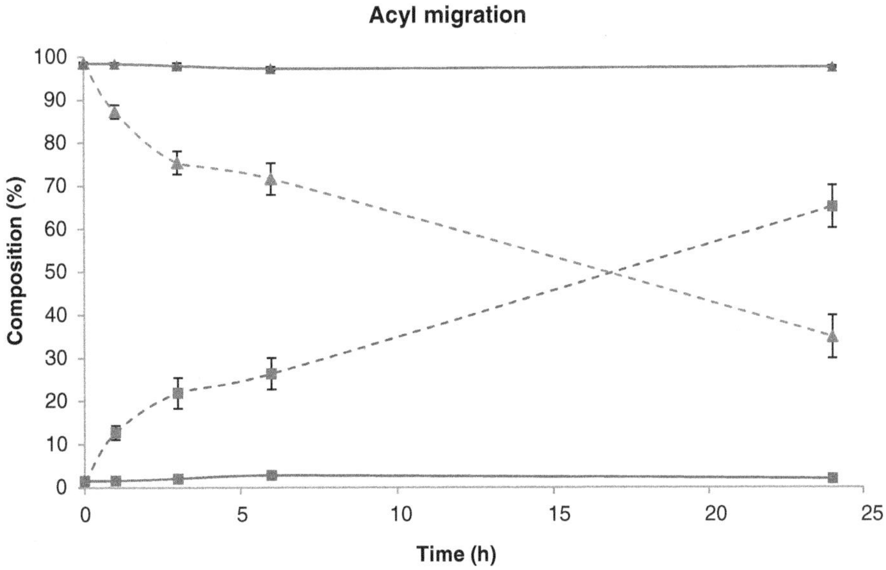

Figure 4 Acyl migration study. The 2-monoolein (triangles) and 1-monoolein (squares) as a function of time using 4% *w/v* silica gel (35- to 70-μm particle size) as acyl migration promoter in the two-phase system (*n*-heptane/water 20%, solid lines) and in pure *n*-heptane (dotted lines).

particles much larger than the liquid droplets, it is clear that this is not a typical example of a Pickering emulsion. Furthermore, the power input was most probably not high enough to create a Pickering emulsion, which normally requires ultrasound, high-pressure homogenizers or other devices producing high-shear stress [31]. However, our emulsions induced by particles without the use of high-energy emulsification equipment may be quite useful for biodiesel production because of the considerably increased reaction rate. The fact that slightly higher rate enhancement was observed with phenyl-derivatised silica particles indicates that an additional activation

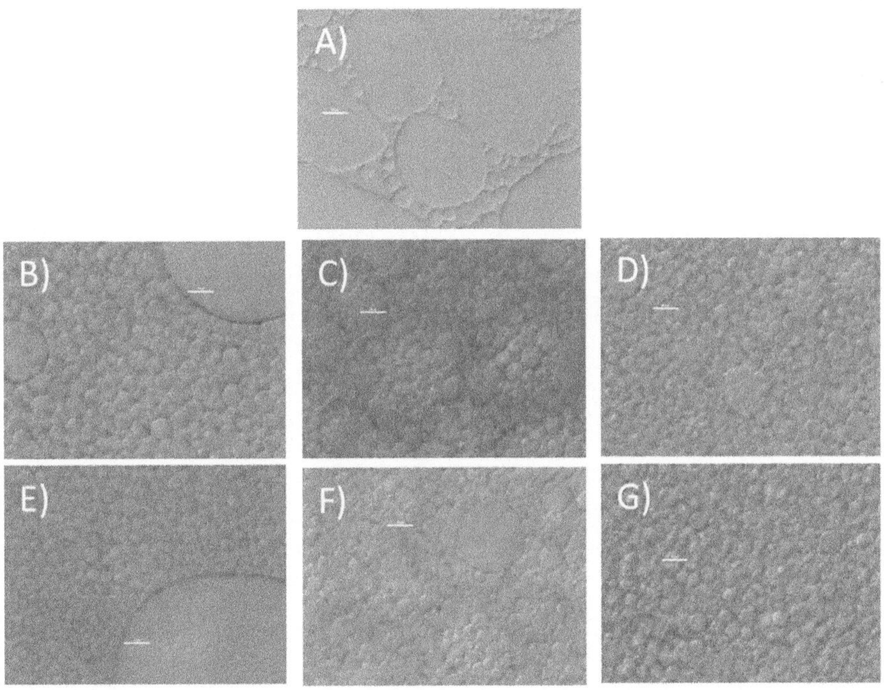

Figure 5 Emulsion pictures using different conditions. (A) No additive, **(B)** non-modified silica gel 35 to 70 μm, **(C)** non-modified silica gel 0.2 to 0.3 μm, **(D)** non-modified silica gel 0.007 μm, **(E)** phenyl silica gel 35 to 70 μm, **(F)** phenyl silica gel 0.2 to 0.3 μm, **(G)** phenyl silica gel 0.007 μm. Used was 4% *w/v* of the different additives and 100 μL of TLL/g substrate. Scale bars: 10 μm. The large structures in (B) and (E) are silica particles.

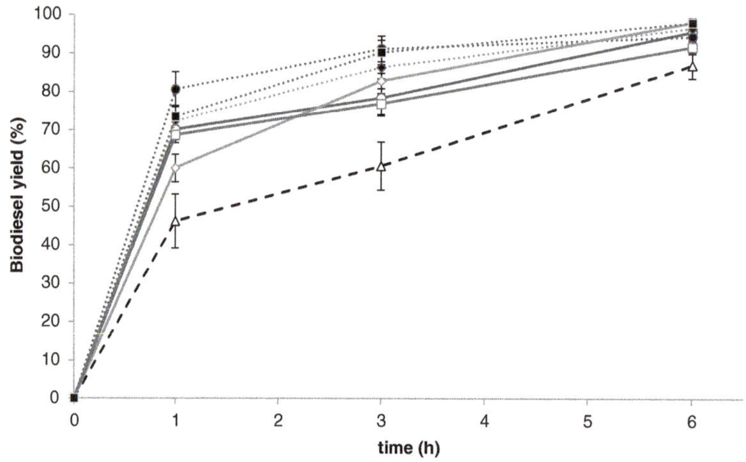

Figure 6 Biodiesel batch production (yield) using different solid particles. No additive (dashed line), phenyl-modified silicas (dotted lines, closed symbols) and non-modified silicas (solid lines, open symbols). Particle size: 35 to 70 μm (circles), 0.2 to 0.3 μm (diamonds) and 0.007 μm (squares). Used was 4% *w/v* of the different additives and 100 μL of TLL/g substrate.

mechanism might have been in operation. It is possible that the lipase to some extent was adsorbed on the hydrophobic surface of the derivatised particles, thereby causing opening of the lid and interfacial activation, as previously observed after adsorption of lipases on hydrophobic materials. However, since almost equal activation was observed with underivatised silica particles, it appears that the main rate-enhancing effect of the particles was due to the reduction of droplet size.

Comparison with chemical catalysis

Methanolysis of vegetable oils in the presence of a homogeneous base catalyst is nowadays the dominant process in industrial biodiesel production. A comparison between the biocatalytic method developed in this work and the homogeneous catalyst under the same conditions (pressure, temperature and time) was carried out. The most common acid/base homogeneous catalysts used in industry, for example, NaOH, NaOMe, KOH and H_2SO_4 were included in the comparison at a concentration of 2% (*w*/*v* or *v*/*v*). No reaction was detected in any case after 24 h. These results can be explained by the high free fatty acid content of the starting material and also by the mild reaction conditions used.

Biodiesel preparation from rapeseed oil

Once the optimised conditions were set, these were applied to a real feedstock material. Rapeseed oil was selected as the substrate, and the reaction was carried out under solvent-free conditions, starting from 300 mg of oil, 0.4 mL of water, 8 mg of phenyl silica (35- to 70-μm particle size) and 100 μL of enzyme/g substrate at

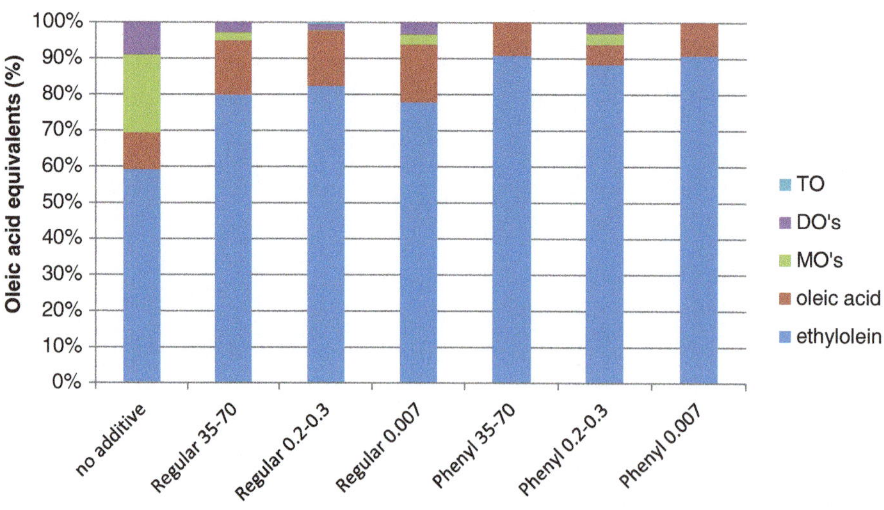

Figure 7 Product composition after 3 h using different silica gels. Ethanolysis of triolein catalysed by TLL using regular silica or phenyl silica with different particle size (given in microns) as additives and 100 μL of TLL/g substrate at 37°C and 1,000 r.p.m. DO, dioleins; MO, monoolein; TO, triolein.

37°C and 1,000 r.p.m. The process was monitored by GC, and after 5 h, 96% biodiesel yield was observed, with 2% MAGs and 2% DAGs remaining.

Conclusions

We have developed a very effective method for the enzymatic preparation of biodiesel, which is also useful for oils with a high free fatty acid content. The method uses a liquid formulation of *T. lanuginosus* lipase in an aqueous/organic two-phase system. The enzyme has shown an excellent tolerance to ethanol, which not only made the process faster but also allowed for the total conversion of the free fatty acid into biodiesel with a short-reaction time. Triacylglycerol conversion was very fast, while the main bottleneck in the process was the conversion of DAG and MAG. The use of solid silica particles led to changes in the emulsion structure which permitted a larger surface area in the interphase, overcoming mass transfer limitations and thus increasing the rate of the process. Surprisingly, the silica particles did not cause acyl migration as observed in other biodiesel production systems. The emulsion-enhancing effects of particles of widely varying size has not been described before, at least not in the biodiesel production field, and is clearly different from the creation of Pickering emulsions, which requires small particles and high-energy input. This work demonstrates the potential of liquid lipase preparations for biodiesel production starting from high free fatty acid feedstocks such as waste oils. Under the best conditions, almost quantitative yields were achieved after only 6 h of reaction. These results constitute a good starting point for efficient and cheap biodiesel production, and the next step would be to adapt the methodology to readily available cheap waste oils.

Materials and methods

Chemicals

Lipases from *T. lanuginosus* (2,200 U/mL, Sigma), *R. oryzae* (Aldrich, product number 80612), *R. arrhizus* (Sigma, product number 62305), *P. fluorescens* (Aldrich), *R. miehei* (Sigma), *Pseudomonas species* (Aldrich) and *Candida antarctica* lipase B (Sigma), *p*-nitrophenyl butyrate and silica gels of different particle sizes (silica gel 35 to 79 μm, silica fumed 0.2 to 0.3 μm and silica fumed 0.007 μm) were obtained from Sigma-Aldrich Co. (Steinheim, Germany); cyclohexane and tetradecane were purchased from Merck (Darmstadt, Germany); *n*-heptane, ethanol (99.5%) from VWR (Stockholm, Sweden); and *N*-methyl-*N*-trimethylsilylheptafluorobutyramide (MSHFBA, Machery-Nagel) from ScantecLab (Partille, Sweden). Triolein was obtained from Larodan Fine Chemicals (Malmö, Sweden), perfluorodecyltriethoxysilane from Alfa Aesar and phenyltrimethoxysilane and *n*-octyltriethoxysilane were acquired from Acros Organics. Rapeseed oil was purchased in a local supermarket. Other chemicals were of analytical grade. A high oleic acid content starting material was used in order to simulate a low-quality oil. The composition (molar ratio) of the starting material was 54% triolein, 24% DAGs and 22% oleic acid.

GC analysis

The samples were analysed by GC using a Varian gas chromatograph (430-GC-FID, Agilent Technologies Inc., Santa Clara, CA) equipped with a flame-ionisation detector and a FactorFour™ capillary column (VF = 1 ms, length 15 m and ID 250 μm, Varian, Agilent Technologies Inc., Santa Clara, CA). Helium was used as the carrier gas, and the temperature of both the detector and injector was 350°C. The starting temperature of the column was 180°C. This was maintained for 2.5 min, then increased by 10°C/min to a temperature of 340°C and was then maintained for 26 min.

Samples were withdrawn from the organic phase and derivatised by silylation by adding an equal volume of MSHFBA and incubating the mixture at room temperature for 30 min. An equal volume of dry ethanol was then added to stop the reaction. The samples were diluted 30 times with an internal standard solution (tetradecane) in cyclohexane to obtain a sample in which the final solution of tetradecane was 9 mM, suitable for GC. The concentrations of the compounds were calculated using response factors obtained from a standard curve.

Alcoholysis reactions

The standard reaction conditions were 265 mg of the starting material (mainly triolein, see the section on 'Chemicals') in 1.6 mL of *n*-heptane with different ethanol concentrations in a 4.5-mL vial with septum. The reaction was started by adding the biocatalyst (different amounts) in 0.4 mL of H_2O. The reactions were carried out in a thermomixer at 1,000 r.p.m. Samples from the organic phase were regularly withdrawn with a Hamilton syringe and analysed by GC. The concentration of each component was calculated as a percentage of the initially available oleic acid equivalents, bearing in mind that diacylglycerols and triacylglycerols contain two and three fatty acid moieties, respectively. All experiments were carried out in duplicate. Error bars are omitted in some charts in order to have a better understanding of the results since standard deviations are in all cases below 4%.

Silica modification

Silica gels of different particle sizes were modified with phenyl moieties in order to increase their hydrophobicity. The most often used method for preparing hydrophobic inorganic support materials is to silanise the support. To that purpose, 200 mg of silica (30- to 75-,

0.2- to 0.3- and 0.007-μm particle sizes) were suspended in 1.5 mL of dry toluene, and the mixture was kept at 130°C for 6 h. Then, 0.5 mL of phenylmethoxysilane was added, and the mixture was stirred overnight at 70°C. The suspension was cooled and the solvent was removed by decantation. The preparation was washed with fresh toluene, and MTBE and was finally air-dried at room temperature.

Microscopy

The structure of the emulsions was studied by interference microscopy. Images were taken using a Nikon Optishot-2 microscope at × 400 magnification with Digital Sight DS 2Mv camera and processed using NIS-elements 3.1D.

Abbreviations

CAL-B: *Candida antarctica* lipase B; DAGs: diacylglycerols; GC: gas chromatography; MAGs: monoacylglycerols; MSHFBA: *N*-methyl-*N*-trimethylsilylheptafluorobutyramide; MTBE: methyl *tert*-butyl methyl ether; TLL: *Thermomyces lanuginosus* lipase.

Competing interests

The authors declare that they have no competing interests.

Authors' contributions

JMS designed the study, performed the experiments, analysed the results and drafted the manuscript. PA participated in the design of the study and commented and revised the manuscript. Both authors approved the final manuscript.

Acknowledgements

This work was financially supported by the Swedish Research Council Formas (Grant no. 213-2012-820).

References

1. Atadashi IM, Aroua MK, Aziz AA. Biodiesel separation and purification: a review. Renew Energy. 2011;36:437–43.
2. Fjerbaek L, Christensen KV, Norddahl B. A review of the current state of biodiesel production using enzymatic transesterification. Biotech Bioeng. 2009;102:1298–312.
3. Saleh J, Tremblay AY, Dubé MA. Glycerol removal from biodiesel using membrane separation technology. Fuel. 2010;89:2260–6.
4. Yusuf NNAN, Kamarudin SK, Yaakub Z. Overview on the current trends in biodiesel production. Energy Convers Manag. 2011;52:2741–51.
5. Sorda G, Banse M, Kemfert C. An overview of biofuel policies across the world. Energy Policy. 2010;38:6977–88.
6. Hervé G, Agneta F, Yves. D. Biofuels and world agricultural markets: outlook for 2020 and 2050, Economic effects of biofuel production. Dr. Marco Aurelio Dos Santos Bernardes (Ed.); 2011. ISBN: 978-953-307-178-7, InTech, doi: 10.5772/20581. Available from: http://www.intechopen.com/books/economic-effects-of-biofuel-production/biofuels-and-world-agricultural-markets-outlook-for-2020-and-2050.
7. Bozbas K. Biodiesel as an alternative motor fuel: production and policies in the European Union. Renew Sustan Energ Rev. 2008;12:542–52.
8. Srivastava A, Prasad R. Triglycerides-based diesel fuels. Renew Sustan Energ Rev. 2000;4:111–33.
9. Zhang Y, Dube MA, Mclean DD, Kates M. Biodiesel production from waste cooking oil: 1. Process design and technological assessment. Bioresour Technol. 2003;89:1–16.
10. Meher LC, Sagar DV, Naik SN. Technical aspects of biodiesel production by transesterification-a review. Renew Sustan Energ Rev. 2006;10:248–68.
11. Stamenkovic OS, Velickovic AV, Veljkovic VB. The production of biodiesel form vegetable oils by ethanolysis: current state and perspectives. Fuel. 2011;90:3141–55.
12. Atadashi IM, Aroua MK, Abdul Aziz AR, Sulaiman NMN. Production of biodiesel using high free fatty acid feedstocks. Renew Sustan Energ Rev. 2012;16:3275–85.
13. Narwal SK, Gupta R. Biodiesel production by transesterification using immobilized lipase. Biotechnol Lett. 2013;35:479–90.
14. Ranganathan SV, Narasimham SL, Muthukumar K. An overview of enzymatic production of biodiesel. Biores Technol. 2008;99:3975–81.
15. Bajaj A, Lohan P, Jha PN, Mehrotra R. Biodiesel production through lipase catalyzed transesterification: an overview. J Mol Catal B Enz. 2010;62:9–14.
16. Šinkūnienė D, Adlercreutz P. Effects of regioselectivity and lipid class specificity of lipases on transesterification, exemplified by biodiesel production. J Am Oil Chem Soc. 2014;91:1283–90.
17. Nordblad MVT, Silva L, Nielsem PM, Woodley JM. Identification of critical parameters in liquid enzyme-catalyzed biodiesel production. Biotech Bioeng. 2014;111:2446–53.
18. Tufvesson P, Lima-Ramos J, Nordblad M, Woodley JM. Guidelines and cost analysis for catalyst production in biocatalytic processes. Org Proc Res Dev. 2011;15:266–74.
19. Drauz K, Gröger H, May O. Enzyme catalysis in organic synthesis. 1st ed. Weinheim: Wiley-VCH; 2012.
20. Faber K. Biotransformations in organic chemistry. 6th ed. Berlin: Springer; 2011.
21. Mangas-Sánchez J, Serrano-Arnaldos M, Adlercreutz P. Effective and highly selective lipase-mediated synthesis of 2-monoolein and 1,2-diolein in a two-phase system. J Mol Catal B Enz. 2015;112:9–14.
22. Huang Y, Zheng H, Yan YJ. Optimization of lipase-catalyzed transesterification of lard for biodiesel production using response surface methodology. Appl Biochem Biotech. 2010;160:504–15.
23. Salis A, Svensson I, Monduzzi M, Solinas V, Adlercreutz P. The atypical lipase B from *Candida antarctica* is better adapted for organic media than the typical lipase from *Thermomyces lanuginosa*. Biochim Biophys Acta. 2003;1646:145–51.
24. Shimada Y, Watanabe Y, Sugihara A, Tominaga Y. Enzymatic alcoholysis for biodiesel fuel production and application of the reaction to oil processing. J Mol Catal B: Enzym. 2002;17:133–42.
25. Adlercreutz P. Immobilisation and application of lipases in organic media. Chem Soc Rev. 2013;42:6406–36.
26. Du W, Xu YY, Liu DH, Li DH. Study on acyl migration in immobilized lipozyme TL-catalyzed transesterification of soybean oil for biodiesel production. J Mol Catal B: Enzym. 2005;37:68–71.
27. Wang YD, Shen XY, Li ZL, Li X, Wang F, Nie XA, et al. Immobilized recombinant Rhizopus oryzae lipase for the production of biodiesel in solvent free system. Journal of Molecular Catalysis B-Enzymatic. 2010;67:45–51.
28. Binks BP, Lumsdon SO. Influence of particle wettability on the type and stability of surfactant-free emulsions. Langmuir. 2000;16:8622–31.
29. Dickinson E. Food emulsions and foams: stabilization by particles. Current Opinion in Colloid and Interface Science. 2010;15:40–9.
30. Binks BP. Particles as surfactants-similarities and differences. Current Opinion in Colloid and Interface Science. 2002;7:21–41.
31. Chevalier Y, Bolzinger M-A. Emulsions stabilized with solid nanoparticles: Pickering emulsions. Colloids and Surfaces A: Physicochem Eng Aspects. 2013;439:23–34.
32. Wu C, Bai S, Ansorge-Schumacher MB, Wang D. Nanoparticle cages for enzyme catalysis in organic media. Adv Mater. 2011;23:5694–9.
33. Aveyard R, Binks BP, Clint JH. Emulsions stabilised solely by colloidal particles. Adv Colloid Interface Sci. 2003;100–102:503–46.

13

Evolved strains of *Scheffersomyces stipitis* achieving high ethanol productivity on acid- and base-pretreated biomass hydrolyzate at high solids loading

Patricia J Slininger[1*], Maureen A Shea-Andersh[1], Stephanie R Thompson[1], Bruce S Dien[1], Cletus P Kurtzman[2], Venkatesh Balan[3], Leonardo da Costa Sousa[3], Nirmal Uppugundla[3], Bruce E Dale[3] and Michael A Cotta[1]

Abstract

Background: Lignocellulosic biomass is an abundant, renewable feedstock useful for the production of fuel-grade ethanol via the processing steps of pretreatment, enzyme hydrolysis, and microbial fermentation. Traditional industrial yeasts do not ferment xylose and are not able to grow, survive, or ferment in concentrated hydrolyzates that contain enough sugar to support economical ethanol recovery since they are laden with toxic byproducts generated during pretreatment.

Results: Repetitive culturing in two types of concentrated hydrolyzates was applied along with ethanol-challenged xylose-fed continuous culture to force targeted evolution of the native pentose fermenting yeast *Scheffersomyces (Pichia) stipitis* strain NRRL Y-7124 maintained in the ARS Culture Collection, Peoria, IL. Isolates collected from various enriched populations were screened and ranked based on relative xylose uptake rate and ethanol yield. Ranking on hydrolyzates with and without nutritional supplementation was used to identify those isolates with best performance across diverse conditions.

Conclusions: Robust *S. stipitis* strains adapted to perform very well in enzyme hydrolyzates of high solids loading ammonia fiber expansion-pretreated corn stover (18% weight per volume solids) and dilute sulfuric acid-pretreated switchgrass (20% w/v solids) were obtained. Improved features include reduced initial lag phase preceding growth, significantly enhanced fermentation rates, improved ethanol tolerance and yield, reduced diauxic lag during glucose-xylose transition, and ability to accumulate >40 g/L ethanol in <167 h when fermenting hydrolyzate at low initial cell density of 0.5 absorbance units and pH 5 to 6.

Keywords: Lignocellulose, Biofuel, Adaptation, Yeast, *Pichia stipitis*, Fermentation

Background

An estimated 1.3 billion dry tons of lignocellulosic biomass could be available annually to support ethanol production at a level that would allow the U.S. to reduce its petroleum consumption by 30% [1]. The fibrous, plant cell-wall material that is characteristic of lignocellulosic feedstocks is highly recalcitrant and difficult to deconstruct into fermentable sugars. The chemical pretreatment required to open the structure of plant biomass to enzymatic hydrolysis results in solutions rich in glucose and xylose. However, pretreatment generates fermentation-inhibiting byproducts, including acetic acid, furfural, hydroxymethyl furfural (HMF), and others. Traditional industrial yeasts do not ferment xylose and are not able to survive, grow, or ferment in toxic concentrated hydrolyzates which contain high sugar concentrations (>100 g/L) to support ethanol accumulation (>40 g/L) high enough for economical recovery [2,3].

Pichia stipitis is known to ferment D-xylose to ethanol more efficiently than other native yeasts previously

* Correspondence: Pat.Slininger@ars.usda.gov
[1]Bioenergy Research Unit, National Center for Agricultural Utilization Research, USDA-ARS, 1815 N. University, Peoria, IL 61604, USA
Full list of author information is available at the end of the article

described [4]. The species, represented by type strain NRRL Y-7124 (CBS 5773), was recently renamed *Scheffersomyces stipitis* [5] and is particularly useful because it has strong NADH-linked as opposed to NADPH-linked aldose reductase activity providing for a more favorable cofactor balance in the conversion of xylose to xylulose [6] and a high ethanol yield. Strain NRRL Y-7124 was selected for our adaptation study because it ferments hexoses and xylose to economically recoverable concentrations of ethanol exceeding 40 g/L with almost no accumulation of xylitol byproduct [7-9]. In nutritionally optimized media, *S. stipitis* strain NRRL Y-7124 is able to produce over 70 g/L ethanol in 40 h (1.75 g/L/h) from 150 g/L sugars at a yield of 0.41 ± 0.06 g/g in high density fermentations (6 g/L cells) [7,10,11]. Given appropriate nitrogen levels and sources, it is also relatively resistant to fermentation inhibitors ethanol, furfural, and HMF [12]. *S. stipitis* is one of the most viable native pentose-fermenting yeasts available for commercial scale-up [13]. For industrial application, sugar uptake rate in biomass hydrolyzates needs improvement, including reducing the effects of diauxy and improving ethanol and inhibitor tolerance. To advance the science and application of *S. stipitis,* our objective was to apply appropriate selective pressure to guide its evolution toward an industrially robust derivative that is tolerant of diverse lignocellulosic hydrolyzates.

Two different types of industrially promising hydrolyzate were selected for application in our adaptation process, namely ammonia fiber expansion-pretreated corn stover hydrolyzate (AFEX CSH) [14,15] and enzymatically saccharified dilute acid-pretreated post-frost switchgrass (SGH). In contrast to AFEX CSH, SGH is characterized by high contents of furan aldehydes and acetic acid and very low levels of available nitrogen needed to support yeast growth and fermentation. Dilute acid-pretreated switchgrass hydrolyzate liquor (PSGHL) is the liquor in association with the hydrolyzed solids. It is separable from the solids by filtration or centrifugation and is characteristically rich in xylose but low in glucose, a feature which may make PSGHL a useful enrichment medium to force selection for improved xylose utilization in hydrolyzates, a failing point for many types of native and engineered yeast tried in the past. The low available N content of the switchgrass hydrolyzates provided opportunity to explore the utility of nitrogen supplementation in the process of screening and ranking improved strains. Both AFEX CSH and PSGHL were used as challenging selective media applied in sequence and in parallel to force the evolution of *S. stipitis* toward derivatives with enhanced ability to grow and ferment in diverse hydrolyzates. The repetitive culturing and retrieval of functional populations from increasingly concentrated hydrolyzate environments was the general strategy to be accomplished in microplates employing a dilution series of 12% glucan AFEX CSH or PGSHL prepared at 20% solids loading. This strategy utilized natural selection and enrichment to recover spontaneous hydrolyzate inhibitor-tolerant derivatives of strain NRRL Y-7124.

An additional feature of this research was the application of ethanol-challenged continuous culture to further enhance and stabilize AFEX CSH-adapted populations. Ethanol-challenged xylose growth and fermentation was targeted to enrich for populations able to resist ethanol toxicity, grow, and survive on xylose as a sole C source and able to induce xylose-specific enzymes allowing fermentation of xylose to ethanol, even in the presence of high concentrations of ethanol. Recent investigations [11] have shown that ethanol concentrations around 15 to 50 g/L progressively repressed enzyme inductions specific to xylose utilization. The resulting populations enriched in ethanol-tolerant derivatives of the AFEX CSH tolerant population were next subjected to further enrichment on PSGHL to explore this approach to broadening the functionality of strains in various types of hydrolyzates.

Once adaptations were completed, isolates were evaluated. Optimally adapted individuals from populations occurring at various phases of adaptation were obtained by enrichment under target stress conditions followed by dilution plating to skim the most prevalent populations from which to pick colonists. Selected colonists were then ranked using dimensionless relative performance indices to determine best overall performance considering xylose uptake rate and ethanol yield on various hydrolyzates with or without nutrient supplementations. To improve the performance of *Scheffersomyces* strains, several researchers have applied various adaptation procedures [16-20]. Although the potential utility of adaptation to improving the functionality of *Scheffersomyces* strains in lignocellulosic hydrolyzates is suggested by the literature, application of adapted strains of *S. stipitis* to hydrolyzates with economical ethanol production has not been reported. The evolved strains reported in this paper are significantly improved over the parent NRRL Y-7124 in accordance with high solids loading hydrolyzate screening targets and are able to produce >40 g/L ethanol in AFEX CSH and in appropriately soy nitrogen-supplemented SGH. Given the sequenced genome of a second *S. stipitis* strain, NRRL Y-11545 [21], these novel strains are candidates for future studies to uncover the genetic changes that have occurred and the mechanisms underlying hydrolyzate tolerance and improved fermentation.

Results and discussion

Figure 1 provides a flow diagram of the adaptation processes used to obtain the tolerant strains reported and indicates the screening numbers and points of isolation for superior ranking strains characterized either by consistently high xylose consumption rates and high ethanol

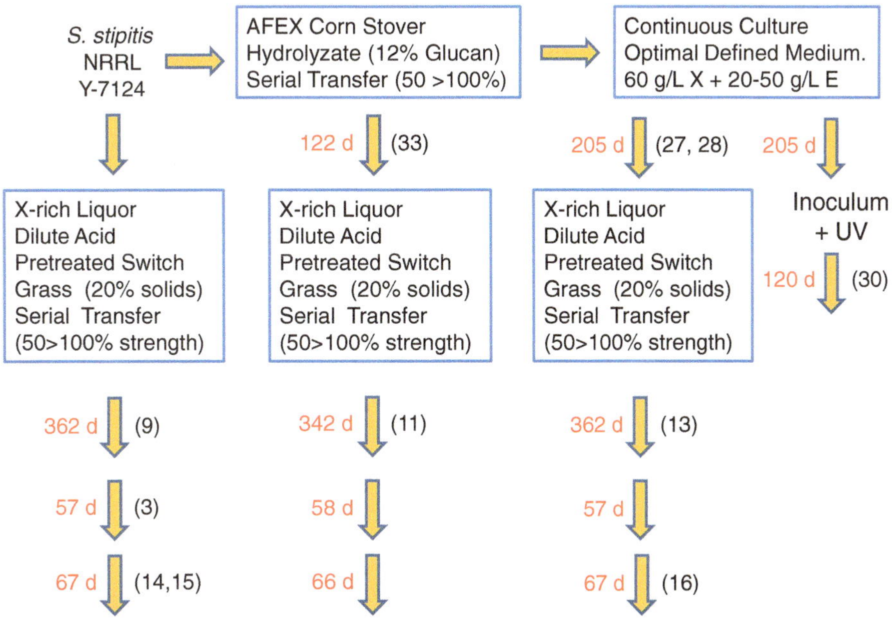

Figure 1 *Scheffersomyces stipitis* adaptation flow chart. The diagram shown indicates the order of the stresses applied during the adaptation process and the points of recovery of superior isolates (numbers in parenthesis). See also Table 2 isolate key as reference for strain identities. To provide time orientation, the numbers in red indicate the number of days in each phase of adaptation. For the serial transfer phases in AFEX CSH and xylose-rich PSGHL, each day of adaptation represents approximately two to four generations. For the continuous culture phase (205 days total), the dilution rate D was variable at approximately 0 to 0.1 h^{-1} during 125 days of operation with pH-actuated feeding. In the next 80 days, operation was at a continuous flow with D at 0.012 h^{-1}, providing a generation time (ln 2/D) of 58 h or one generation per 2.4 days at steady state. Next, a sample of the adapted population from the 205-day continuous culture was mutagenized with UV light and inoculated to a continuous culture operated with D at 0.012 h^{-1}.

yields across all hydrolyzate formulations (Table 1) or by extremely high, consistent performance on at least one type of hydrolyzate. On the basis of identical nucleotide sequences for domains D1/D2 of the nuclear large subunit rRNA gene, all superior tolerant isolates were identified as *S. stipitis* [22] and deposited in the ARS Patent Culture Collection with the accession numbers listed in Table 2 along with adaptation stresses applied during strain evolution and screening numbers used in graphs. Results will be noted and discussed in order of the adaptation sequence.

Adaptation to enzyme saccharified AFEX CSH in microplate batch cultures

After the parent strain NRRL Y-7124 was exposed to increasing concentrations of 12% glucan AFEX CSH for several weeks, an adapted population was stored as a glycerol stock, and when a single colony isolate (colony 5) was cultivated on 6% glucan AFEX CSH in comparison with the parent strain, a significantly enhanced performance was observed as shown in Figure 2. Superior performance features of the adapted population isolate, colony 5, included faster glucose and xylose volumetric uptake rates (that is, 27% and 67% increases, respectively), more complete xylose uptake (2 g/L *versus* 7 g/L), and higher ethanol production rate (16% increase) and accumulation (7.7% increase).

When single colony isolates of the adapted population were obtained and compared on optimal defined medium (ODM) with 66 g/L glucose and 87 g/L xylose against the parent NRRL Y-7124, all isolates showed similar performance to the adapted population and significantly less diauxic lag compared to the parent. Figure 3 shows relative performances and indicates that the adapted population and superior isolated single-cell clones (colonies 1, 5, and 7 (not shown)) were significantly improved in their ability to rapidly switch to xylose fermentation once glucose had been depleted. Both clones 1 and 5 shown had consumed all xylose by 200 h and made approximately 57 g/L ethanol, while the parent had 7 g/L xylose remaining even at 300 h and had accumulated only about 44 g/L ethanol. Prior research [11] has shown that the induction of enzymes specific for xylose metabolism is repressed in the parent NRRL Y-7124 when ethanol concentration exceeds 15 g/L. In the fermentations of Figure 3, ethanol reached nearly 30 g/L by the time glucose was depleted, and the adapted population and clones were not repressed in xylose utilization in contrast to the parent strain, which was severely crippled in its ability to use xylose after the glucose had been consumed. Exposure to increasing

Table 1 Compositions of hydrolyzates used in cultivations[a]

Component[b]	AFEX-pretreated corn stover hydrolyzate				Dilute acid-pretreated switch grass hydrolyzates (20% solids)							
	6% Glucan		12% Glucan		PSGHL		SGH		SGH-N1		SGH-N2	
	Mean	s	Mean	s	Mean	s	Mean	s	Mean	s	Mean	s
Glucose (g/L)	58.9	9.0	107.4	16.7	7.9	2.5	69.2	3.2	67.4	6.5	64.2	1.3
Xylose (g/L)	34.2	7.3	48.7	15.8	52.1	5.1	48.6	2.7	45.3	3.5	47.4	0.8
Arabinose (g/L)	4.3	0.6	9.5	0.4	7.6	0.9	6.1	0.3	4.4	0.7	7.6	0.2
Galactose (g/L)	3.1	0.3	5.7	0.6	3.0	0.7	5.2	0.3	5.2	0.3	5.2	0.3
Fructose (g/L)	4.0	1.7	8.2	3.0	0.9	0.4	0.7	0	0.7	0	0.7	0
Mannose (g/L)	1.0	0.1	2.0	0.2	8.9	5.1	0	0	0	0	0	0
Acetic acid (g/L)	1.8	0.2	4.7	1.0	6.1	2.8	5.4	0.4	4.3	0.6	5.8	0.3
HMF (mM)	0.3	0.5	1.1	1.5	2.8	3.1	1.8	0.2	1.1	1.3	3.6	0.5
Furfural (mM)	0.2	0.1	0.4	0.0	24.4	7.6	18.3	4.5	24.5	4.5	19.1	0.8
PAN (mg N/L)	318.7	49.9	493.2	69.1	33.7	15.6	69.9	14.7	188.0	4.0	173.0	22.0
Urea (mg N/L)	83.1	29.3	105.5	9.1	0.7	0.7	7.0	2.7	1101.0	–[c]	962.0	105.0
Ammonia (mg N/L)	1,193.4	289.7	2,707.6	385.6	25.0	21.0	23.0	24.6			369.0	63.0

[a]Values are reported in terms of mean and standard deviation(s) across hydrolyzates used in experiments reported.
[b]Abbreviations: HMF = hydroxymethylfurfural; PAN = primary amino nitrogen; PSGHL = dilute acid pretreated switchgrass hydrolyzate liquor; SGH = switchgrass hydrolyzate; or with nutrient supplements -N1 or -N2, as described in 'Methods' section.
[c]No values of *s* since N content was calculated based on urea addition.

concentrations of AFEX-pretreated corn stover hydrolyzate, that is, increasing concentrations of sugars and inhibitors, led to an adapted population better able to ferment this hydrolyzate not only because of enrichment of the population in members more resistant to the inhibitory environment of the hydrolyzate but also because of enrichment of the population in members that were less susceptible to ethanol-associated repression of enzymes specific to xylose metabolism, thus avoiding extended diauxic lag.

Acetic acid arises as a product of hemicellulose hydrolysis during pretreatment, and impaired abilities of engineered *S. cerevisiae* strains to shift from glucose to xylose utilization in the presence of 7.5 to 15 g/L acetic acid have been documented [23]. Consequently, the sensitivity of strain Y-7124 and its adapted derivatives to acetic acid was studied in ODM. After exposure to relatively low levels of acetic acid (2 to 4.7 g/L in dilutions of 12% glucan AFEX CSH) during the adaptation process, colony 5 was able to ferment both glucose and xylose in ODM containing 2 to 10 g/L acetic acid more efficiently than the parent strain NRRL Y-7124 (Figure 4), particularly due to reduced diauxic lag and more rapid xylose fermentation. Even at 10 g/L acetic acid, colony 5 continued

Table 2 Summary of superior tolerant *Scheffersomyces stipitis* strains for fermentation of hydrolyzates of plant biomass

Accession number (NRRL)	Screening number	Isolate designation	Adaptation stress[a]	Preferred hydrolyzate[a]
Y-50871	33	Colony 5	AFEX CSH	All; AFEX CSH
Y-50872		Colony 1	AFEX CSH	NR
Y-50873		Colony 7	AFEX CSH	NR
Y-50861	27	2A.1.53R-E20-C1	AFEX CSH > E	All; AFEX CSH
Y-50862	28	2A.1.53R-E30-C3	AFEX CSH > E	All; SGH-N2
Y-50864	30	2A.30R2-E40-C5	AFEX CSH > E (UV)	SGH-N2; AFEX CSH
Y-50857	13	2A.1.53R S100E40-1	AFEX CSH > E > PSGHL	All
Y-50860	16	2A.1.53R-1	AFEX CSH > E > PSGHL	SGH-N1 or -N2
Y-50865	11	Colony 5 GP-6	AFEX CSH > PSGHL	SGH-N2 or -N1
Y-50874	3	Y-7124 S90E40-1	PSGHL	All
Y-50863	9	Y-7124 GP-5	PSGHL	SGH-N1 or -N2
Y-50859	14	Y-7124-6	PSGHL	All
Y-50858	15	Y-7124-10	PSGHL	All

[a]Abbreviations are as follows: AFEX CSH = ammonia fiber explosion-pretreated corn stover enzyme hydrolyzate; E = ethanol-fed continuous culture; UV = ultraviolet light-treated inocula for E; PSGHL = dilute acid-pretreated switchgrass hydrolyzate liquor; NR = not ranked.

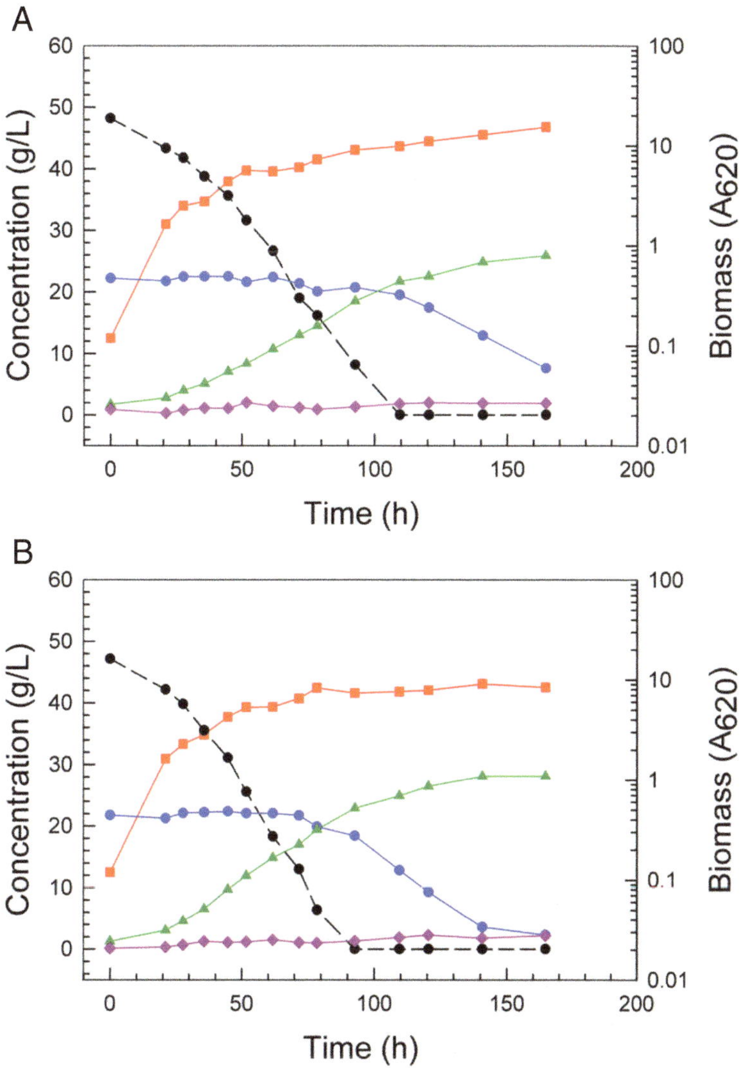

Figure 2 Improved batch fermentation of 6% glucan AFEX CSH. *Scheffersomyces stipitis* NRRL Y-7124 parent strain fermentation of 6% glucan AFEX CSH (**A**) is compared with adapted colony 5 fermentation of 6% glucan AFEX-pretreated corn stover hydrolyzate (**B**). Symbols designate biomass (red square), glucose (black circle with dashed line), xylose (blue circle with solid line), ethanol (green triangle), and xylitol (purple diamond).

to ferment xylose to ethanol, while the parent strain was not able to do so. Despite the strong impact of acetic acid on fermentation, the growth of both parent and colony 5 was relatively unimpaired across the concentrations used, although the numbers of viable colony forming units fell more in colony 5 than parent cultures, either due to greater cell flocculation or death or both.

To validate the application of colony 5, we applied high cell densities for rapid ethanol production from AFEX CSH. Replicate colony 5 populations grown on 6% glucan AFEX CSH and repitched to A_{620} of 40 in higher glucan hydrolyzates were tested and found to allow the accumulation of over 40 g/L ethanol on circa 8% glucan hydrolyzates (Figure 5). Although full accumulation of the ethanol required over 188 h, around 85% of it was accumulated in the first 48 h. Further testing of the adapted strain in comparison to its parent in populations repitched to A_{620} of 40 in 6% glucan hydrolyzate and later fed with an equal volume of 12% glucan hydrolyzate indicated a significant improvement in the adapted strain over the parent with respect to xylose uptake rate, ethanol productivity, and maximum ethanol accumulation (Figure 6) during both the initial batch and fed-batch phases of the cultivation. The adapted isolate was able to consume both glucose and xylose more quickly than the parent strain, reducing the sugar consumption time by 25% from 200 h to just under 150 h. In addition, ethanol accumulation of the adapted strain was improved by about 30% over that of the parent. However, xylose utilization was not complete and the adapted strain cell viability fell near the end of glucose consumption/beginning of xylose uptake, suggesting a

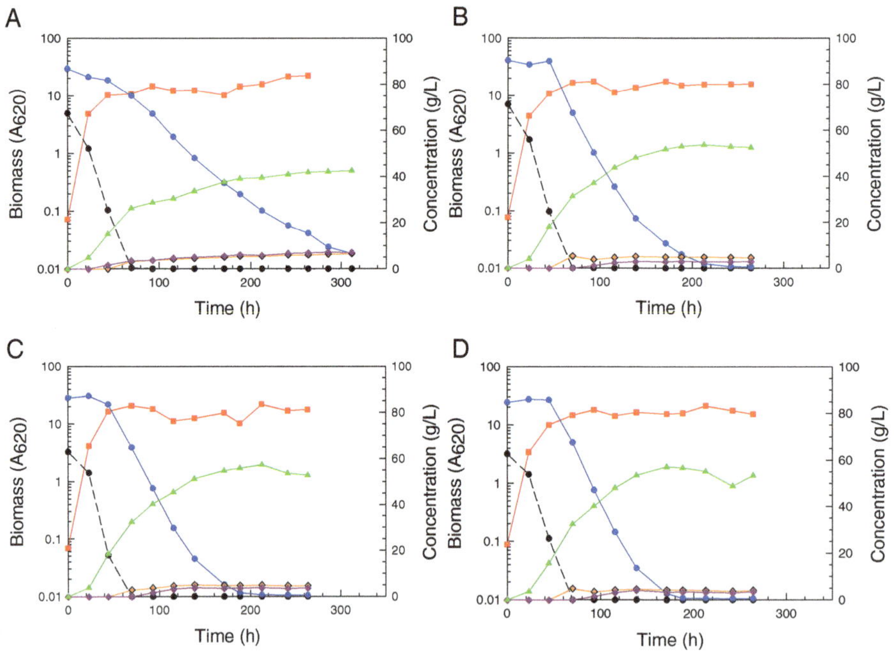

Figure 3 Reduced diauxic lag in defined medium with mixed sugars. Fermentation performances are compared in ODM with 66 g/L glucose and 87 g/L xylose for parent strain *S. stipitis* NRRL Y-7124 (**A**), the AFEX CSH adapted population derived from Y-7124 (**B**), single-cell colony 1 isolated from the adapted *S. stipitis* population (**C**), single-cell colony 5 isolated from the adapted population (**D**). Symbols designate biomass (red square), glucose (black circle with dashed line), xylose (blue circle with solid line), ethanol (green triangle), xylitol (purple diamond), and adonitol (gold diamond with black edge).

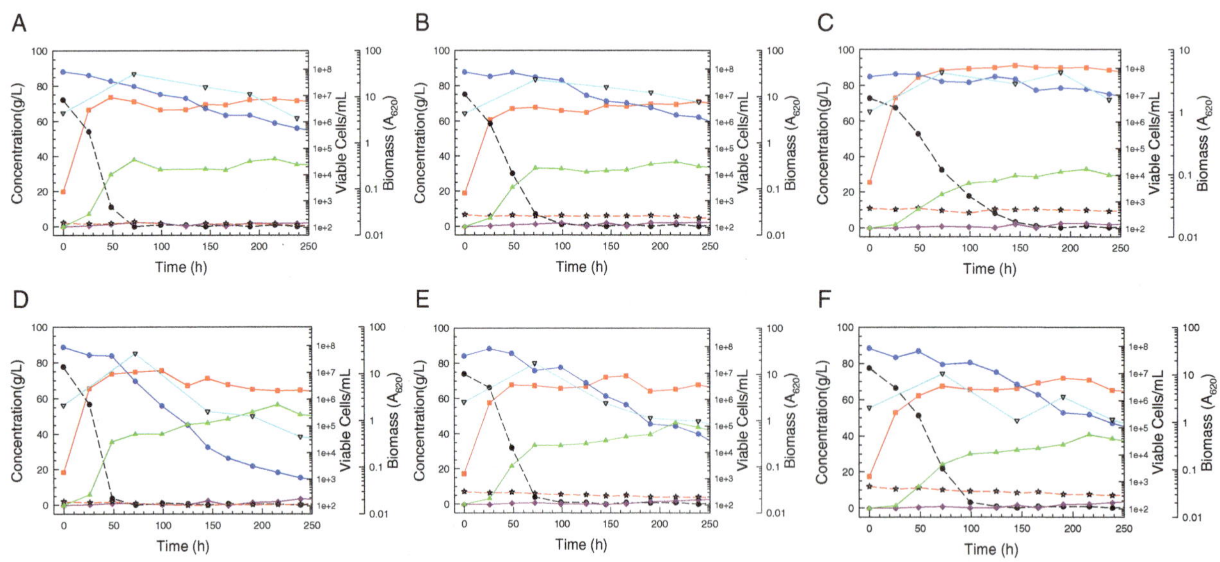

Figure 4 Reduced sensitivity to acetic acid in optimal defined medium. Sensitivity of *S. stipitis* NRRL Y-7124 in ODM to increasing acetic acid concentrations at 2 (**A**), 6 (**B**, and 10 g/L (**C**) is compared to relative tolerance of adapted colony 5 at the same acetic concentrations ((**D**, (**E**) and (**F**), respectively). The following time course data are shown: biomass (red squares), glucose (black circles and dashed line), xylose (blue circles and solid line), xylitol (purple diamonds), acetic acid (red stars), ethanol (green triangles), viable cells (inverted aqua triangles).

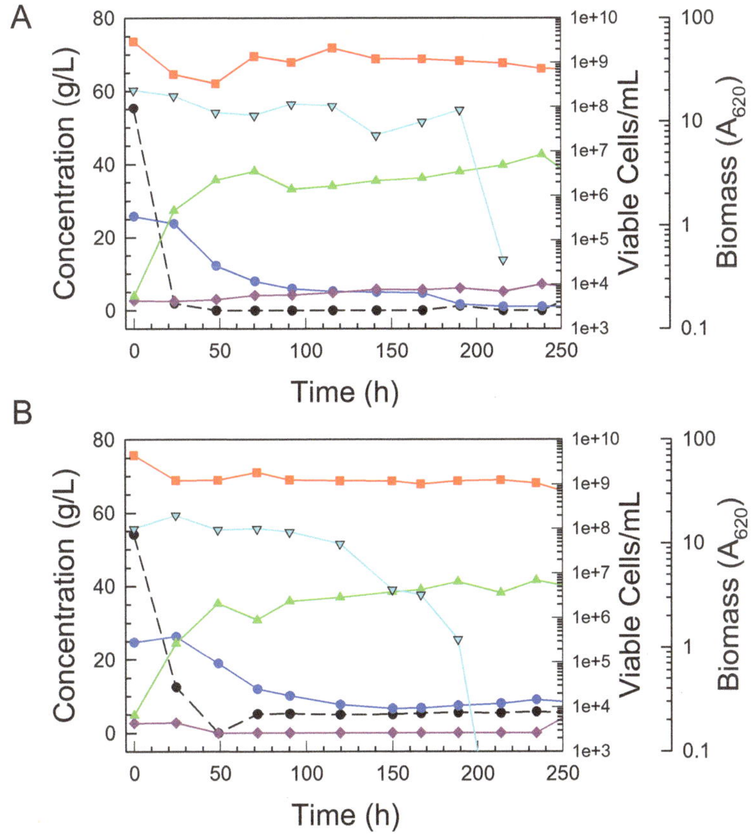

Figure 5 Fermentation of 8% glucan AFEX CSH to > 40 g/L ethanol with a high cell density of adapted strain. Fermentation batches (**A**) and (**B**) are shown for 8% glucan AFEX-pretreated corn stover hydrolyzate that was inoculated with a large population of *S. stipitis* AFEX CSH-tolerant colony 5 repitched from a 6% glucan batch growth during xylose utilization. Time courses of biomass (red squares), glucose (black circles and dashed line), xylose (blue circles and solid line), xylitol (purple diamonds), ethanol (green triangles), and viable cells (aqua inverted triangles) are shown.

potential problem combating deleterious impacts of ethanol and inhibitors from the 12% glucan AFEX CSH feed while trying to metabolize the xylose. It is notable that when compared to simple batch operation, the fed-batch process ethanol productivity and ethanol accumulation of colony 5 were lower by 30% and 22%, respectively, based on final volume, even though the fed-batch process was fed more sugar overall, 105 g/L compared to 80 g/L. Thus, the fed-batch mode of operation did not overcome the inhibitor challenges accompanying the abrupt addition of an equal volume of concentrated 12% glucose AFEX CSH hydrolyzate to the first stage 6% glucan AFEX CSH.

Application of continuous culture to improve xylose utilization with increasing ethanol

To improve ethanol tolerance and resistance to repression of enzymes specific to xylose utilization, colony 5 was inoculated to a continuous culture operated at a low dilution rate with high ethanol concentrations and xylose in the feed. The goal was to select and enrich for yeast cells with improved growth and fermentation on xylose as sole carbon source in the presence of high ethanol concentrations. Cells were exposed to >15 g/L ethanol, which is associated with repression of xylose-specific enzymes in the parent strain NRRL Y-7124 [11]. In Figure 7a,b, the progressive improvement in initial growth rate on xylose in the presence of 40 g/L ethanol is seen for two enriched populations, 2A.1.53R and 2A.1.30R2, which had been collected at earlier and later times, respectively, during operation and then frozen in glycerol. Population 2A.1.30R2 was obtained from the continuous culture following reinoculation with UV-mutagenized populations and further enrichment under 40 to 50 g/L ethanol challenge. These data suggest that the ethanol resistance loss shown by colony 5 was alleviated by the continuous culture process and UV exposure. In addition, the ability of glucose-induced cells to switch to xylose metabolism in the presence of 40 and 45 g/L ethanol was shown to be further improved over that of the colony 5 inoculum and far better than that of the parent control (Figure 7a,b). The enriched populations

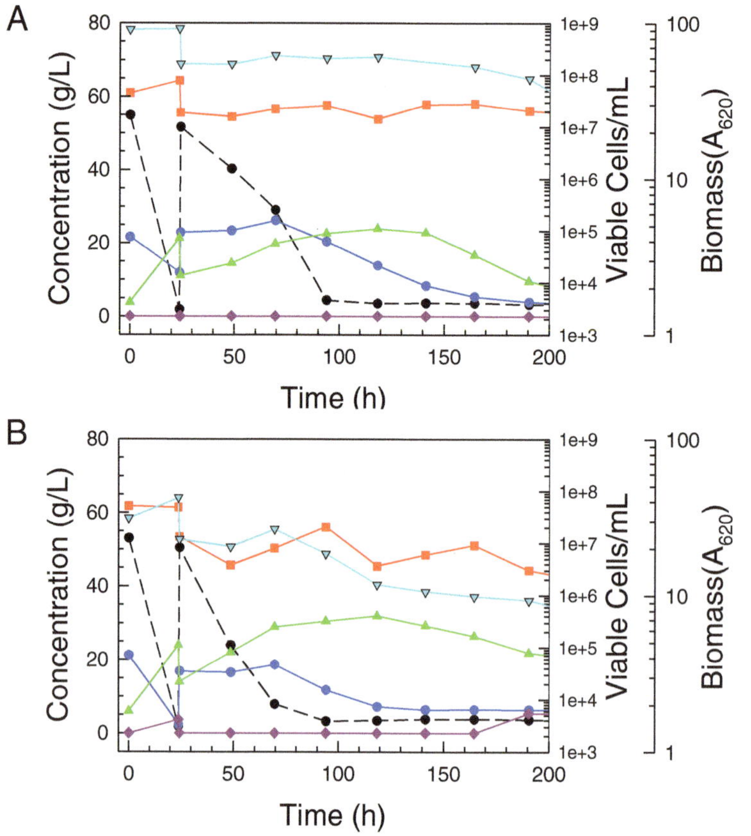

Figure 6 Comparative fed-batch fermentations. High density cultures of *S. stipitis* NRRL Y-7124 parent (**A**) and hydrolyzate-tolerant Colony 5 (**B**) repitched from 6% glucan during xylose uptake into fresh 6% glucan hydrolyzate were then fed at 24 h with an equal volume of 12% glucan hydrolyzate. Time courses of biomass (red squares), glucose (black circles and dashed line), xylose (blue circles and solid line), xylitol (purple diamonds), ethanol (green triangles), and viable cells (aqua inverted triangles) are shown.

2A.1.30R2 and 2A.1.53 R were successful at using xylose even in the presence of >40 g/L ethanol when yeast had been cultivated on glucose and transferred in high density to ODM with xylose and ethanol (Figure 7b). Three single-cell isolates (2A.1.53R-E20-C1, 2A.1.53R-E30-C3, and 2A.30R2.E40-C5) were recovered from these populations, which showed superior ability to ferment hydrolyzates in subsequent screens.

Adaptation to PSGHL in microplate batch cultures

To further broaden hydrolyzate tolerance, colony 5 and the ethanol-tolerant derivative population 2A.1.53R were further challenged by repetitive serial transfer to increasing concentrations of xylose-rich PSGHL in microplates. Additionally, strain NRRL Y-7124 was subjected to direct adaptation in PSGHL as a control relative to sequential application of different adaptation challenges. Using dilution plating, over 150 isolates were recovered either directly from adaptation microplates or from frozen glycerol stock populations after streaking to YM agar and selective enrichment with hydrolyzate and/or ethanol challenges. Four superior hydrolyzate-fermenting isolates were identified from NRRL Y-7124 adaptation to PSGHL (Y-7124 GP-5, Y-7124 S90E40-1, Y-7124-6, and Y-7124-10) and one each was derived from colony 5 and population 2A.1.53R (colony 5 GP-6 and 2A.1.53R-1, respectively) when the following screening and ranking processes were applied (Table 2).

Primary screen of isolates on xylose-rich PSGHL

Isolates from all phases of the adaptation process were screened in PSGHL as the primary elimination point for those cultures not able to ferment xylose well. The isolates chosen for representation in Figure 8 are among the best performing on PSGHL out of the approximately 150 ranked in this primary screen and in all secondary phases of performance ranking described below. In order to indicate improvement relative to the parent strain, the performance of each isolate was expressed as the ratio of isolate kinetic parameter value to parent strain NRRL Y-7124 kinetic parameter value. Ratio values of 'one' occurred if the isolate performance, based on either yield or xylose uptake rate, was equivalent to the parent. Ethanol yield per initial sugar available and xylose

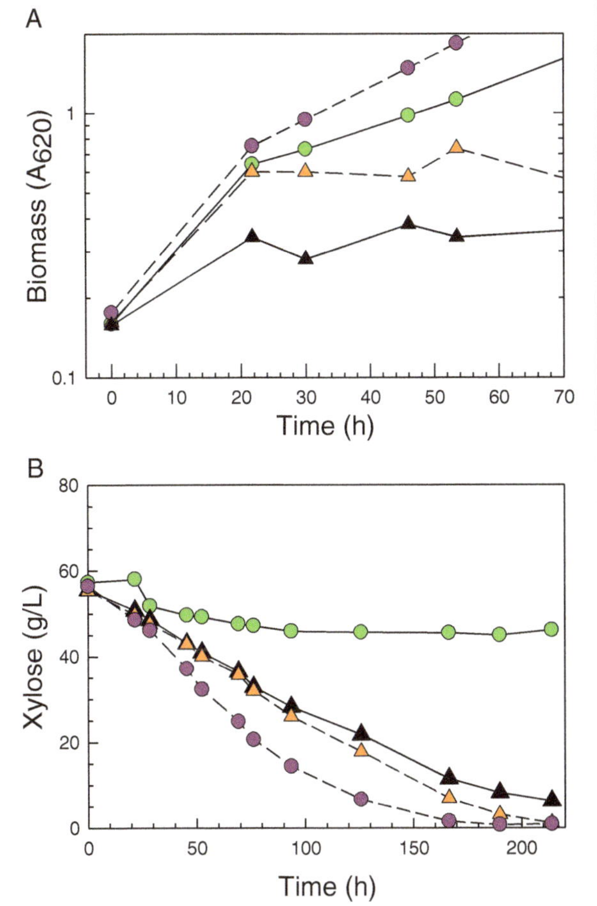

Figure 7 Ethanol-resistant derivatives of colony 5. Hydrolyzate-tolerant colony 5 was further developed by continuous culture selection on ODM containing xylose as sole carbon source and high levels of ethanol. (**A**) The growth rate on ODM + 60 g/L xylose +40 g/L ethanol of two derivative glycerol stock populations obtained early in the selection process (2A.1.53R, orange triangle and dashed line) and after UV irradiation of continuous culture inocula (2A.1.30R.2, purple circle and dashed line) are shown in comparison with the NRRL Y-7124 parent strain (green circle with solid line) and AFEX CSH-tolerant colony 5 (black triangle with solid line). (**B**) Xylose uptake by dense populations of glucose-grown yeast ($A_{620} = 50$) in ODM with 40 g/L ethanol indicated that all adapted strains surpassed the unadapted parent in the ability to induce xylose metabolism.

uptake rate ratios were used to rank relative performances of isolates.

Some general trends in isolate responses to harshness of the primary hydrolyzate screen were notable. As the harshness of the hydrolyzate environment was increased with inhibitor and glucose concentrations, the performance ratios became progressively smaller as shown in Figure 8a-d, respectively. Figure 8a,b provides a summary of top isolate performances on 60% and 75% strengths of PSGHL with ODM and ODM + YM nutrient supplements, respectively. In the 60% strength PSGHL, five of seven top isolates exposed to PSGHL selection pressure performed many times better than NRRL Y-7124 (isolate 1), and four isolates performed slightly better than colony 5 (isolate 33), which had evolved during exposure to AFEX CSH but had no previous selective exposure to PSGHL. However, in the harsher 75% strength of PSGHL, only three isolates significantly surpassed both the parent and colony 5 (isolate 33) despite improved nutrient richness.

In a related experiment, isolates obtained from the continuous culture challenged with xylose growth in the presence of ethanol (isolates 27, 28, and 30) were screened on similar PSGHL media that were also supplemented with 75 g/L glucose (Figure 8c,d). The glucose addition was chosen to heighten the ethanol repression challenge to xylose utilization in order to detect isolates with the conserved reduced diauxy feature characteristic of the parent colony 5 (isolate 33) from which they arose, as well as general ethanol resistance to cell damage. However, these isolates had not previously been exposed to PSGHL, and, as shown in Figure 8c,d, they struggled to surpass par with the NRRL Y-7124 parent even when the nutrient environment was enriched with YM components. Thus, Figure 8 indicates that the relative ranking of isolates on a particular culture medium, such as PSGHL, could vary due to the prior adaptation medium exposure, the strength of the hydrolyzate, and the nutrients applied.

Impact of nutrient environment on differentiation of isolates in a secondary screen on enzyme-saccharified hydrolyzates

When isolates passing the primary screen were submitted to a secondary screen on unamended SGH without any nitrogen supplementation, all isolates performed poorly, and there was no significant variation ($P > 0.05$). Across the 33 isolates tested, including the parent NRRL Y-7124, the kinetic parameter means and standard deviations were 0.19 ± 0.1 g/g for ethanol yield per initial sugar and 0.10 ± 0.03 g/L/h, for xylose uptake rate.

Past studies on defined media have shown that optimal xylose fermentation by *S. stipitis* strain NRRL Y-7124 requires C:N 57 to 126:1, and 20% to 60% of available nitrogen must be in the form of amino acids while 80% to 40% must be urea [10]. Higher amino acid contents may be required in the presence of hydrolyzate inhibitors, especially during xylose fermentation [10,24]. The C:N ratio of the typical SGH used in this study was approximately 600:1, and nitrogen starvation would result if not supplemented. Since nitrogen availability and quality have been shown to impact the growth and fermentation capacity of the parent NRRL Y-7124, SGH was supplemented with both urea and amino acids. SGH-N1 was fortified to 42:1 C:N (approximately 15% of N from PAN and approximately 85% as urea) with complex nitrogen sources and growth factors including casamino acids, tryptophan,

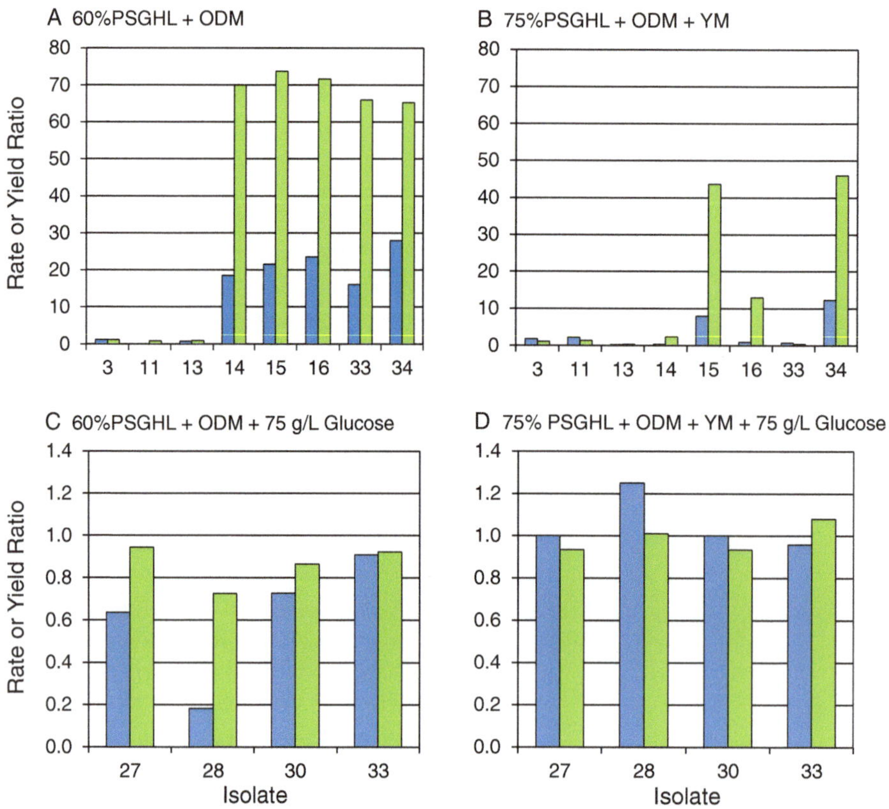

Figure 8 Ratio of performance improvement of tolerant isolate compared to parent. The performances of superior tolerant isolates are summarized relative to the control parent strain NRRL Y-7124 for each formulation of PSGHL (**A-D**). Performances were assessed in terms of xylose uptake rate (blue bars representing ratios of isolate to parent) and ethanol yield per sugar supplied (green bars representing ratios of isolate to parent). Isolates from ethanol-challenged continuous cultures on xylose were tested in this screen using PSGHL supplemented with glucose (C and D) in order to allow formation of significant ethanol to more strongly challenge induction of xylose utilization enzymes and metabolism of xylose.

cysteine, and vitamins. The SGH-N2 was fortified to 37:1 C:N with urea and soy flour as the lowest cost commercial source of PAN and vitamins, providing approximately 12% of N from PAN and 88% as urea. Thus, the two SGH nitrogen environments -N1 and -N2 were close, but below the ideal. The 6% glucan AFEX CSH had the optimal mix of ammonia and amino acids but a C:N at approximately half of the optimal 60:1 feed in defined media.

While overall C:N and PAN:ammonia content are easily measurable, the hydrolyzate environments are complex and likely to vary in amino acid, growth factor, and mineral profiles, which may potentially impact fermentations. The strategy for the secondary isolate screen then was to apply all three nutritional environments in search of isolates with the nutritional flexibility to perform well and consistently despite the potential for nutritional variability to occur under commercial circumstances. A two-way ANOVA testing the impact of isolates × hydrolyzate types on yield (data not shown) and xylose uptake rate (Figure 9) was carried out on data collected from hydrolyzate screen conducted in duplicate. When superior isolates passing the primary screen on PSGHL were evaluated on the three hydrolyzate types (SGH-N1, SGH-N2, and AFEX CSH), the overall mean xylose uptake rate across isolates varied significantly between hydrolyzate types ($P < 0.001$): 0.24 g/L/h for SGH-N1, 0.19 g/L/h for SGH-N2, and 0.12 g/L/h for AFEX CSH. This interaction is evident in the data presented in Figure 9. The relative xylose consumption rates among isolates were significantly dependent on the hydrolyzate type used in the screen, such that there was a nearly significant dependence on isolate ($P = 0.058$) and a strongly significant interaction of isolate × hydrolyzate type impacting xylose uptake rate ($P < 0.001$). In contrast, the ethanol yield (data not shown) was not significantly impacted by hydrolyzate type ($P = 0.967$) or isolate ($P = 1$). The overall yield mean and standard deviation were 0.30 ± 0.03 g ethanol/g initial ethanol per g initial sugar supplied.

Ranking of single-cell isolates from AFEX CSH, ethanol, and PSGHL adaptation phases

Relative performance indices (RPIs) were calculated in order to rank isolates in sets within a series of different experiments based on their relative performance in a

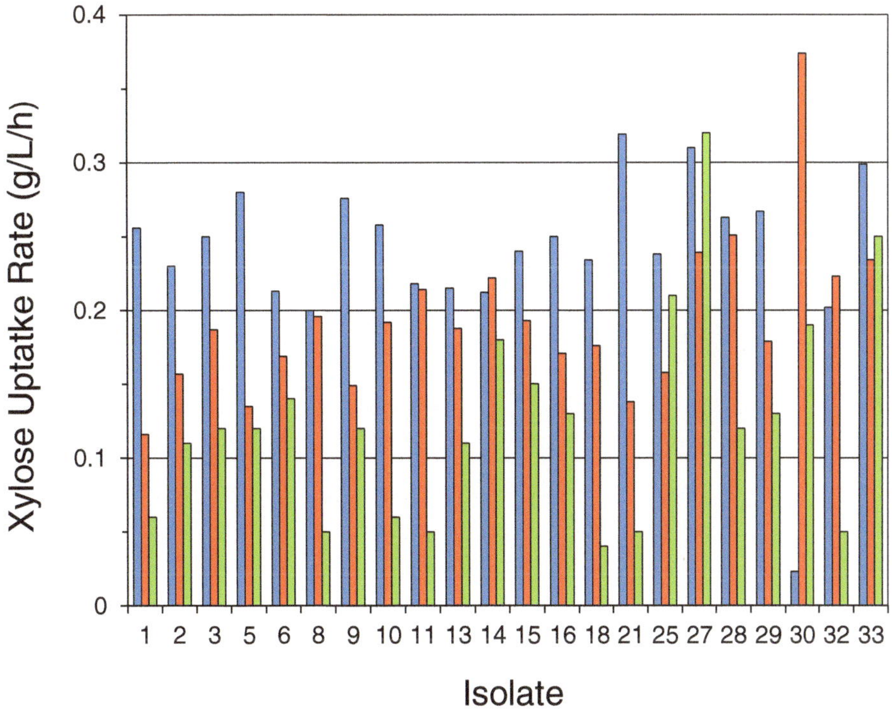

Figure 9 Significant dependence of xylose uptake rate on the interaction of isolate with hydrolyzate type ($P < 0.001$). Isolates (identified along x-axis) were tested in a secondary screen on two nutrient formulations of switchgrass hydrolyzate, SGH-N1 (blue bars) and SGH-N2 (red bars), and additionally AFEX-pretreated corn stover hydrolyzate AFEX CSH (green bars) without nutrient supplementation.

variety of hydrolyzates and nutritional environments tested. RPI is a dimensionless value that is useful in combining data sets to use in overall ranking and/or statistical analysis of subjects submitted to various testing procedures. In this case, yeast isolates were ranked based on ethanol yield per initial sugar available and xylose uptake rate within various experiment sets. Given that the parameters calculated for the performance of each isolate on each hydrolyzate were normally distributed across the group of isolates tested, the value of $F = (X - X_{avg})/s$ ranges from −2 to +2. Here, X can designate yield (Y) or rate (R) observed per isolate, and X_{avg} and s are the average and standard deviation, respectively, of all values observed for the group of isolate treatments within a given experiment, such as the testing of isolates on AFEX CSH. $RPI = (2 + F) \times 100/4$, such that the value of RPI ranges from approximately 0 to 100 percentile from lowest to highest value, respectively, of rate or yield. RPI averages for each isolate within a given hydrolyzate type trial were calculated as $RPI = (RPI_Y + RPI_R)/2$, where yield and rate contributions (subscripted Y and R, respectively) are given equal weighting in this application. (In general, pending the application and relative importance, different weights could be rationally assigned to the yield and rate contributions to the overall RPI average.) Additionally in this study, $RPI_{overall}$ was computed across the n types of hydrolyzates tested: $RPI_{overall} = [(RPI_Y + RPI_R)_1 + (RPI_Y + RPI_R)_2 + \ldots (RPI_Y + RPI_R)_n]/2n$. During primary ranking of the approximately 150 single-colony isolates adapted to xylose-rich PSGHL, the $RPI_{overall}$ was first calculated for rates and yields across the two PSGHL formulations applied in the screen, that is, 60% PSGHL and 75% PSGHL. During secondary screening of the top 20% of isolates from the primary screen, the $RPI_{overall}$ was next similarly calculated for each of the isolates performing on three enzyme-saccharified pretreated hydrolyzate formulations including AFEX CSH, SGH-N1, and SGH-N2, and this ranking parameter was applied to further winnow the list of superior isolates.

Figure 10a shows RPI results within the three different hydrolyzate types for a selection of the better isolates among the 33 tested in the secondary screen. A two-way ANOVA indicated significant variation in RPI due to isolate ($P = 0.003$); but since RPI values were scaled within each hydrolyzate type relative to the data set mean and standard deviation, the mean RPI did not significantly vary among the three hydrolyzates tested ($P = 0.27$), which had mean RPIs ranging from 50 to 55. However, the interaction of isolate × hydrolyzate type was strongly significant ($P < 0.001$), and the relative ranking of isolates depended upon the type of hydrolyzate. This variation in ranking may have come about because of variations in isolate nutritional requirements or inhibitor sensitivities, but it also may have arisen due to instability of certain isolates and inconsistencies even within the same type of hydrolyzate.

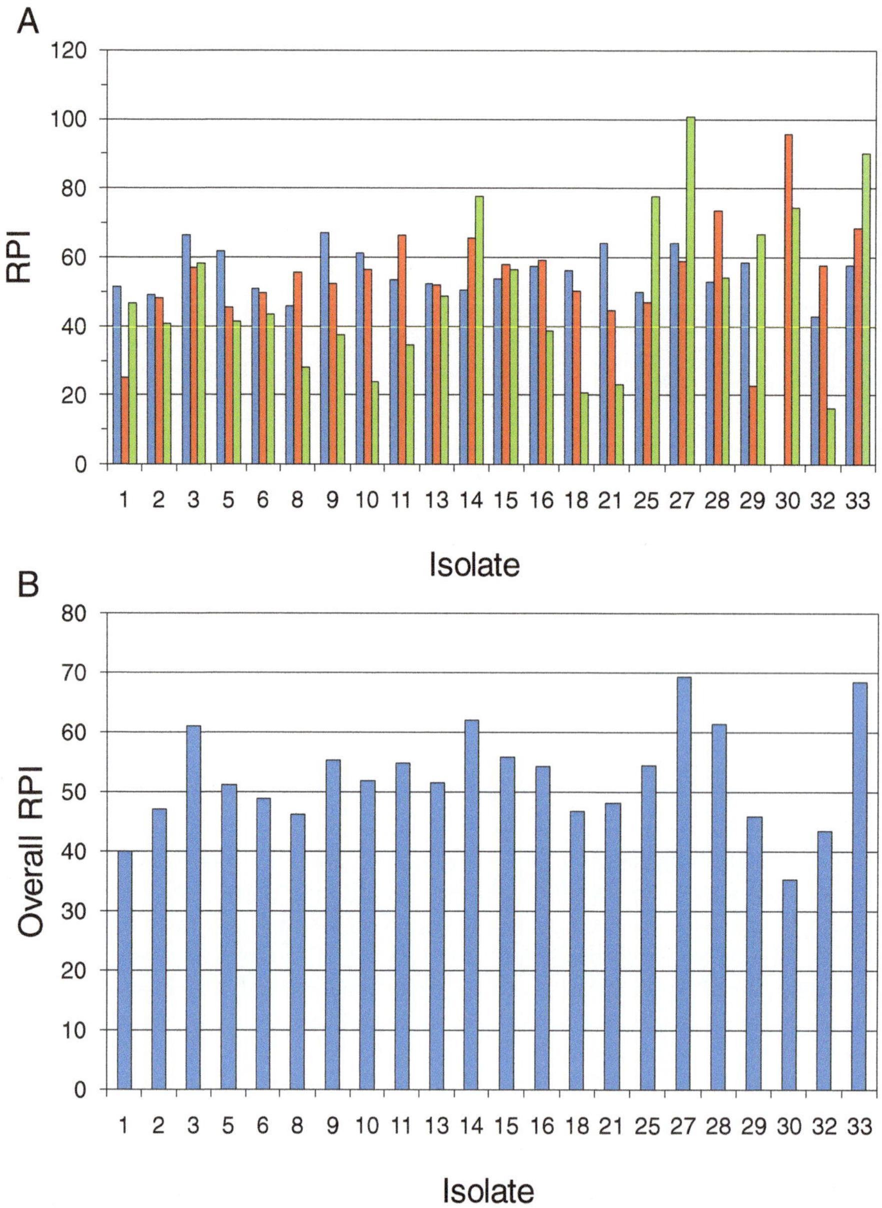

Figure 10 Isolate ranking based on RPI. The relative performance index (RPI) concept was applied to the performance results of the secondary screen in order to rank 33 isolates within each hydrolyzate type based on xylose uptake rate and ethanol yield per sugar supplied. (**A**) The relative ranking of any given isolate depended on the hydrolyzate type ($P < 0.001$): SGH-N1 (blue bars), SGH-N2 (red bars), and AFEX CSH (green bars). (**B**) The overall RPI calculated across all hydrolyzate types (light blue bars) indicated superior strains with most robust performance across different hydrolyzate conditions.

The identification of isolates with highest overall RPI and lowest relative standard deviation among rankings on different hydrolyzates and replicates was a goal of our analysis. Such isolates possess broad inhibitor tolerance, nutritional diversity, and genetic stability - all characteristics useful toward commercial robustness. Five such superior isolates with RPI >60 are indicated in Figure 10b: 3, 14, 27, 28, and 33. Another strategy was to identify any isolates that were specialists or that consistently ranked highly RPI >55 within a hydrolyzate type or overall with a low relative standard deviation: 11 (SGH), 13, 15, 16 (SGH), and 30 (SGH-N2) as indicated in Table 3 and Figure 10. Most superior isolates fell within statistics group A or B (as shown in Table 3), but isolate 30 was in group D since it was very good in SGH-N2 and AFEX CSH but very poor in SGH-N1. Figure 11 indicates the improvement of each of the superior isolates over the unadapted parent NRRL Y-7124 and shows that isolate

Table 3 Summary of relative performance indexes ($RPI_{overall}$) of isolates

		SGH-N1 and SGH-N2				SGH-N1, SGH-N2, and AFEX CSH			
Screen number	Isolate designation	$RPI_{overall}$	s	Rel. s (%)	Statistic group	$RPI_{overall}$	s	Rel. s (%)	Statistic group
1	Y-7124	38.3	23.6	61.5	D	40.0	23.0	57.3	C
2	Y-7124 S80E40-2	48.6	24.7	50.8	C	47.0	22.2	47.1	C
3	*Y-7124 S90E40-1*	*61.8*	*15.6*	*25.2*	*A*	*61.1*	*14.7*	*24.0*	*A*
4	Colony 5 S90E40-5	40.1	12.6	31.5	C				
5	Colony 5 S100E40-5	53.6	11.8	22.0	B	51.1	12.3	24.0	B
6	2A.1.53R S90E40-4	54.0	19.0	35.2	B	48.8	12.4	25.5	C
7	2A.1.53R S100E40-5	52.3	16.6	31.7	B				
8	Y-7124 GP-1	50.7	23.9	47.1	B	46.2	23.1	50.1	C
9	*Y-7124 GP-5*	*59.8*	*15.7*	*26.3*	*B*	*55.3*	*17.6*	*31.9*	*B*
10	Colony 5 GP-2	58.9	10.1	17.2	B	51.9	17.4	33.5	B
11	*Colony 5 GP-6*	*59.9*	*15.4*	*25.8*	*B*	*54.8*	*17.7*	*32.3*	*B*
12	2A.1.53R S90E40-2	34.2	17.1	50.1	D				
13	*2A.1.53R S100E40-1*	*52.2*	*10.5*	*20.1*	*B*	*51.5*	*9.5*	*18.4*	*B*
14	*Y-7124-6*	*58.2*	*16.2*	*27.9*	*B*	*62.1*	*17.0*	*27.3*	*A*
15	*Y-7124-10*	*57.6*	*6.6*	*11.4*	*B*	*55.9*	*6.7*	*11.9*	*B*
16	*2A.1.53R-1*	*58.3*	*10.2*	*17.4*	*B*	*54.4*	*14.0*	*25.8*	*B*
17	2A.1.53R-6	49.4	13.1	26.5	C				
18	Colony 5-3	53.3	11.7	22.0	B	46.7	17.2	36.8	C
19	Colony 5-4	41.6	16.7	40.1	C				
20	Y-7124 GP3-1	32.0	18.2	56.8	D				
21	Y-7124 GP3-5	54.4	18.9	34.7	B	48.1	21.2	44.1	C
22	Colony 5 25%-2 N	46.8	29.3	62.6	C				
23	2A.1.53R 25%-1 N	47.8	13.8	28.9	C				
24	2A.1.53R 25%-2	51.5	16.7	32.4	B				
25	2A.44R-E20-C1	48.6	15.4	31.7	C	54.4	18.4	33.8	B
26	2A.44R-E40-C2	55.0	29.7	54.0	B				
27	*2A.1.53R-E20-C1*	*61.5*	*23.7*	*38.6*	*A*	*69.3*	*27.7*	*39.9*	*A*
28	*2A.1.53R-E30-C3*	*63.3*	*17.6*	*27.9*	*A*	*61.4*	*16.2*	*26.4*	*A*
29	2A.30R2-E30-C5	40.5	52.9	130.6	C	45.8	48.5	105.9	C
30	*2A.30R2-E40-C5*	*25.5*	*81.4*	*319.3*	*D*	*32.3*	*74.7*	*231.4*	*D*
31	3A.1.57-E20-C1	32.2	13.7	42.5	D				
32	3A.1.57-E30-C1	50.2	11.2	22.3	B	43.4	17.7	40.8	C
33	*Colony 5*	*63.0*	*14.8*	*23.5*	*A*	*68.4*	*17.4*	*25.4*	*A*

Isolates in italics are considered superior based on high overall RPI across hydrolyzate types, low relative standard deviation (Rel. s), and/or exceedingly high RPI on at least one hydrolyzate type as seen in Figure 10A.
Strains are ranked based on ethanol yield and xylose uptake rate in screens on dilute acid-pretreated switchgrass enzyme hydrolyzates with two different nutrient supplements (SGH-N1 and SGH-N2) or over all three hydrolyzates including AFEX-pretreated corn stover enzyme hydrolyzates.

abilities were best separated by fermentation on AFEX CSH and SGH-N2. AFEX CSH supported the highest dynamic range of improvement in xylose uptake rate among isolates tested, but SGH-N2 supported the highest range of improvement in the ethanol yield per initial sugar supplied. Isolate abilities were not very distinguishable from one another on SGH-N1 (the least cost effective) perhaps because it was the least challenging since it was the most nutritionally fortified with added yeast extract, malt extract, amino acids, vitamins, and minerals.

Comparative kinetic performances of isolates recovered after varied adaptation stress

Reduction of diauxic lag

Significant ethanol production can occur on AFEX CSH prior to xylose uptake due to the presence of approximately

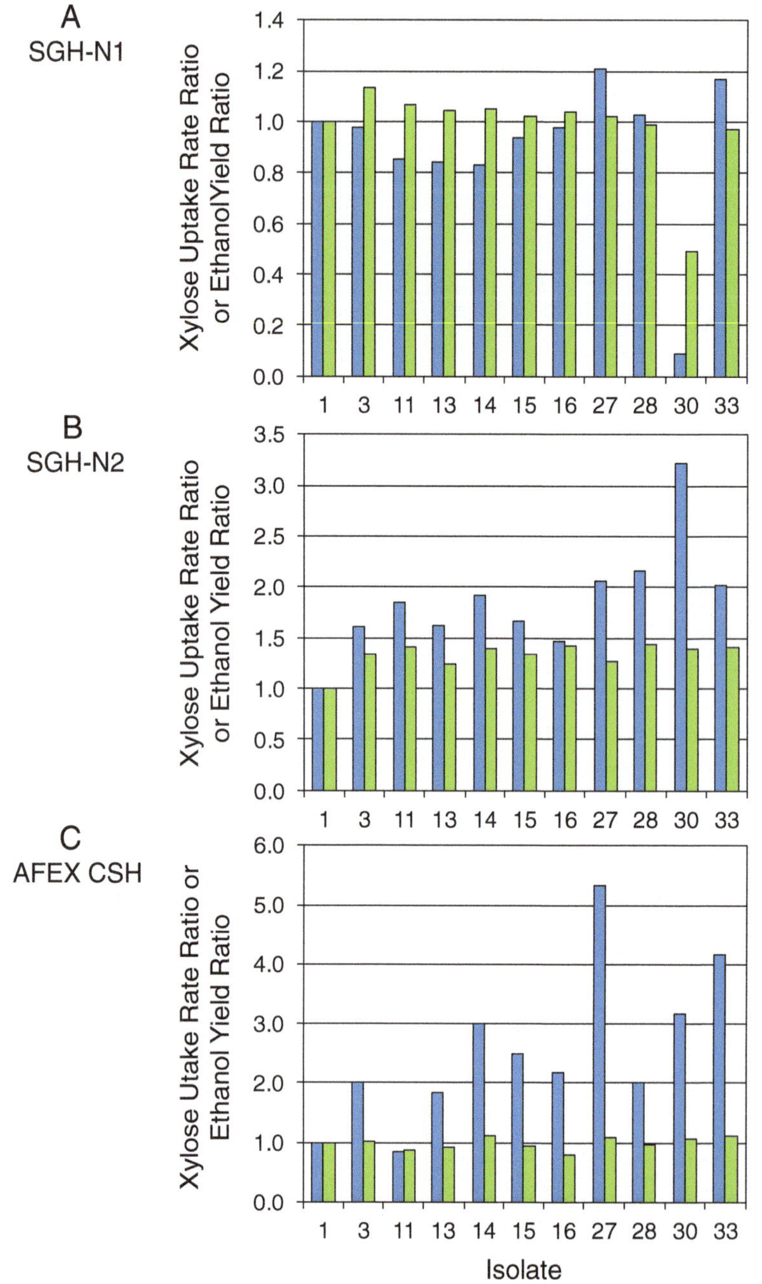

Figure 11 Performance summary of superior adapted isolates of *S. stipitis*. Isolates were screened on three hydrolyzate types, including switchgrass hydrolyzate with two nutrient formulations and unsupplemented AFEX-pretreated corn stover hydrolyzate: (**A**) SGH-N1, (**B**) SGH-N2, and (**C**) AFEX CSH. The improvement in superior adapted isolates over the parent strain was hydrolyzate dependent and is indicated in terms of the xylose uptake rate ratio (dark blue bars) or the ethanol yield ratio (light green bars), which are ratios of adapted strain to parent strain NRRL Y-7124 kinetic parameter values.

60 g/L glucose in this hydrolyzate, and this presents a challenge to induction of enzymes for xylose utilization. After exposure to AFEX CSH selection pressure, colony 5 displayed significantly reduced diauxic lag (Figure 3). Consequently, the induction of xylose utilization in cultures inoculated to low initial cell densities ($A_{620.0}$ of 0.1) on ODM with 75 g/L each of xylose and glucose were evaluated on a selection of the superior strains in order to check the occurrence of the reduced diauxy phenotype in PSGHL-adapted isolates and its retention in derivatives of colony 5 (AFEX CSH) obtained after exposure to ethanol-challenged continuous cultures and/or PSGHL. The data summarized in Table 4 indicated that while glucose uptake rate remained the same among all isolates, the reduced

Table 4 Comparative performance of superior isolates on ODM + 74.5 g/L xylose + 77.6 g/L glucose inoculated to low cell density (A_{620} = 0.1)[a]

			Sugar uptake rate (g/L/h)				Ethanol productivity (g/L/h)				Yield per initial sugar (g/g)					
Adaptation stress	Isolate		Glucose		Xylose		Glucose		Xylose		Ethanol		Biomass		Xylitol	
AFEX CSH	33	Colony 5	1.08	A	0.38	A	0.48	A	0.14	AB	0.37	A	0.016	BC	0.018	B
AFEX CSH > E	28	2A.1.53R-E30-C3	1.09	A	0.38	A	0.48	A	0.15	A	0.37	A	0.020	AB	0.018	B
AFEX CSH > E (UV)	30	2A.1.30R2-E40-C5	1.10	A	0.38	A	0.48	A	0.13	AB	0.35	AB	0.022	A	0.029	A
AFEX CSH > E > PSGHL	13	2A.1.53R-S100E40-1	0.96	A	0.30	C	0.41	AB	0.12	ABC	0.32	BC	0.014	BC	0.011	C
AFEX CSH > E > PSGHL	16	2A.1.53R-1	1.05	A	0.27	C	0.41	AB	0.09	C	0.29	C	0.012	CD	0.015	BC
AFEX CSH > PSGHL	11	Colony 5-GP6	0.91	A	0.27	C	0.37	B	0.11	BC	0.30	C	0.010	D	0.010	C
PSGHL	3	Y-7124-S90E40-1	1.00	A	0.38	A	0.41	AB	0.15	A	0.35	AB	0.013	CD	0.026	A
PSGHL	14	Y-7124-6	0.97	A	0.34	B	0.38	B	0.13	AB	0.32	BC	0.017	BC	0.018	B
PSGHL	15	Y-7124-10	1.08	A	0.29	C	0.44	AB	0.11	BC	0.30	C	0.009	D	0.012	C
Wild	1	Y-7124	1.05	A	0.29	C	0.48	A	0.11	BC	0.34	ABC	0.019	AB	0.010	C

[a]With the exception of glucose uptake rate, all parameters varied significantly among isolates based on one-way ANOVA ($P < 0.01$). Within columns, values with no letters in common are significantly different at $P < 0.05$ (Student-Newman-Keuls pairwise comparison method).

diauxy trait was evidenced by faster xylose uptake rates (and higher ethanol productivities on xylose) in colony 5 recovered from AFEX CSH adaptation, both AFEX CSH > E isolates with or without UV exposure and two of three PSGHL-evolved strains. However, the trait was notably lost from all AFEX CSH treatments later exposed to repetitive culturing in PSGHL. Unlike AFEX CSH, PSGHL is rich in xylose and high in furfural and acetate but poor in glucose, and so during exposure, there would be little selective pressure favoring reduced diauxy due to <5 g/L ethanol formed prior to xylose uptake. This situation may have led to loss of the trait after extended repetitive culturing of AFEX CSH +/− E derivatives in PSGHL. However, two of three PSGHL stressed isolates gained a similar reduced diauxy phenotype. Repeated exposure to xylose among the other inhibitors present, such as acetic acid and furan aldehydes, may have benefited the evolution process toward reduced diauxy but likely by different mechanisms than during exposure to AFEX CSH. For example, during PSGHL exposure, furfural and HMF could potentially compete with xylose for reducing equivalents [25-27] while acetate would cause cell damage, forcing the need for cells to metabolize xylose to accommodate repair of cell damage.

Acetic acid tolerance

Parent strain NRRL Y-7124 and hydrolyzate-tolerant derivatives showed little difference in their abilities to grow on xylose or glucose in the presence of acetic acid at 6 to 15 g/L (Table 5). Specific growth rates in the absence of acetic acid were similar, ranging from 0.23 to .28 h^{-1} on glucose and from 0.22 to .26 h^{-1} on xylose. However, the results of a three-way ANOVA conducted on the ratio of specific growth rate in the presence of inhibitory acetic acid to that in the absence of acetic acid (μ_i/μ_o) showed that cells growing on xylose were significantly more inhibited by acetic acid (growth rates reduced to 68% by 15 g/L acetic acid) than cells growing on glucose (growth rates reduced to just 79%). Isolate 30 (2A.1.30R2-E40-C5) was significantly more resistant to acetic acid on xylose (average $\mu_i/\mu_o = 1.1$) than all other strains and among the most resistant to acetic acid on glucose (average $\mu_i/\mu_o = 0.93$).

Fermentation of 75 g/L xylose in ODM by large populations was significantly more inhibited as acetic acid was increased from 5 to 15 g/L. The level of inhibition was significantly higher when the cell populations were glucose grown rather than xylose grown, suggesting the increased difficulty of glucose-using cells to switch to xylose utilization when under stress by acetic acid (Table 6). The overall impact of acetic acid across isolates was reflected by significant differences in xylose uptake rate, ethanol productivity, and ethanol yield. When strain populations were pre-grown on xylose, the xylose uptake and ethanol production rates and yields were higher than the parent NRRL Y-7124 for all strains that had been developed with AFEX CSH stress. The three strains that had received only PSGHL adaptation stress fermented xylose more slowly than the parent strain NRRL Y-7124 (Table 6A). On the contrary, when the cell populations were grown on glucose, all strains that had been exposed to only PSGHL adaptation stress fermented xylose significantly faster and at significantly higher ethanol yield than when they were grown on xylose (Table 6B).

Furfural tolerance

The furfural content of the PSGHL used in adaptation cultures was around 24 mM with little accompanying HMF (Table 1). Furfural is typically reduced enzymatically with NADH cofactors by yeasts to less toxic furan

Table 5 Comparative average resistances of superior isolates to 6, 10, and 15 g/L acetic acid in ODM with 50 g/L sugar[a,b]

Adaptation stress	Isolate		Average μ_o (h^{-1})[c]				Average ratio μ_i/μ_o^c			
			Glucose		Xylose		Glucose		Xylose	
AFEX CSH	33	Colony 5	0.23	AB	0.23	A	0.86	AB	0.90	B
AFEX CSH > E	27	2A.1.53R-E20-C1	0.28	A	0.26	A	0.72	AB	0.80	B
AFEX CSH > E	28	2A.1.53R-E30-C3	0.26	AB	0.24	A	0.75	AB	0.68	B
AFEX CSH > E (UV)	30	2A.1.30R2-E40-C5	0.24	AB	0.21	A	0.93	AB	1.10	A
AFEX CSH > E > PSGHL	13	2A.1.53R S100E40-1	0.26	AB	0.25	A	0.84	AB	0.79	B
AFEX CSH > E > PSGHL	16	2A.1.53R-1	0.25	AB	0.24	A	0.88	AB	0.75	B
AFEX CSH > PSGHL	11	Colony 5-GP6	0.22	B	0.25	A	1.02	A	0.84	B
PSGHL	3	Y-7124 S90E40-1	0.25	AB	0.22	A	0.66	B	0.83	B
PSGHL	15	Y-7124-10	0.25	AB	0.24	A	0.97	A	0.85	B
PSGHL	14	Y-7124-6	0.23	AB	0.23	A	0.84	AB	0.81	B
Wild	1	Y-7124	0.24	AB	0.25	A	0.96	AB	0.87	B
Significance			$P = 0.039$		$P = 0.223$		$P < 0.01$		$P < 0.001$	

Acetic acid (g/L)	Average ratio μ_i/μ_o[d]			
	Glucose		Xylose	
6	0.98	A	0.98	A
10	0.80	B	0.89	B
15	0.79	B	0.64	C
Significance	$P < 0.001$		$P < 0.001$	

[a]Abbreviations are the following: μ_o = initial specific growth rate in the absence of acetic acid; μ_i = the specific growth rate in the presence of the inhibitory acetic acid.
[b]Within columns, values with no letters in common are significantly different at $P < 0.05$ (Student-Newman-Keuls pairwise comparison method).
[c]Average across acetic acid concentrations for a particular isolate.
[d]Average across isolates for a particular acetic acid concentration.

methanol [25-27]. When the parent NRRL Y-7124 and hydrolyzate-tolerant derivatives were challenged to grow in ODM + 50 g/L glucose (or xylose) amended with 25 mM distilled furfural, all were able to survive and begin growing within 32.8 to 42.1 h (data not shown). Past reports have suggested that xylose and low level furan utilization may be compatible due to the competition for NADH limiting xylitol accumulation [28]. Statistical analysis of our results indicated some benefit of xylose shortening lag phase as a general trend across all isolates, but only by approximately 2.5 h. Considering glucose as the growth substrate most likely to be encountered during detoxification lag, one exceptional isolate, 2A.1.30R2-E40-C5 (AFEX CSH > E (UV)) had a significantly shorter lag phase of 32.7 h compared with that of the parent NRRL Y-7124 at 37.6 h. It is also notable that the growth of isolate 2A.1.30R2-E40-C5 was not reduced in the presence of up to 15 g/L acetic acid (Table 5). These attributes may give this strain a competitive advantage in hydrolyzate fermentations as seen later in Figure 12B. Interestingly, strain 2A.1.30R2-E40-C5 was exposed to AFEX CSH, high levels of ethanol with xylose in extended continuous cultures, and UV irradiation during its development, but its exposures to furfural and acetate were only low to moderate during the hydrolyzate phase, suggesting that adaptive changes were generally useful to coping with stress from inhibitors. Additionally, the finding that most adapted isolates were in general not especially faster at detoxifying furfural in defined medium compared with parent NRRL Y-7124 suggests that other attributes and mechanisms were involved in aiding their ability to cope in hydrolyzates.

Hydrolyzate utilization

When moderately high cell densities were applied initially to inoculate larger flask cultures with SGH-N2, almost all of the adapted strains ranked as superior in our screen were consuming sugars more quickly and producing ethanol more quickly on a volumetric rate basis (Table 7A) compared to the control, wild type strain Y-7124. To reduce the influence of cell biomass variations due to growth advantages of some strains, especially during glucose consumption, rates were normalized based on the average absorbance reading during glucose or xylose uptake (Table 7B). This procedure further enhanced statistical separation among strains, especially with respect to xylose uptake and ethanol productivity on xylose. This procedure indicated NRRL Y-7124 S90E40-1 as having

Table 6 Impact of acetic acid on xylose fermentation by large isolate populations ($A_{620} = 50$)

A. Comparative fermentation of 75 g/L xylose in ODM with 5 to 15 g/L acetic acid by isolates precultured on xylose

Adaptation stress	Isolate		Xylose uptake rate (g/L/hA)[a,b]		Ethanol productivity (g/L/hA)[a,b]		Ethanol yield per initial sugar (g/g)[b]	
AFEX CSH	33	Colony 5	0.090	A	0.0231	AB	0.26	B
AFEX CSH > E	28	2A.1.53R-E30-C3	0.078	AB	0.0197	ABC	0.23	BCD
AFEX CSH > E (UV)	30	2A.1.30R2-E40-C5	0.071	BC	0.0155	ABCDE	0.20	DE
AFEX CSH > E > PSGHL	13	2A.1.53R S100E40-1	0.069	BC	0.0249	A	0.23	BCD
AFEX CSH > E > PSGHL	16	2A.1.53R-1	0.078	AB	0.0175	ABCD	0.25	BC
AFEX CSH > PSGHL	11	Colony 5-GP6	0.066	BC	0.0141	BCDE	0.21	CD
PSGHL	3	Y-7124 S90E40-1	0.043	D	0.0080	DE	0.14	F
PSGHL	14	Y-7124-6	0.053	CD	0.0105	CDE	0.16	EF
PSGHL	15	Y-7124-10	0.045	D	0.0061	E	0.30	A
Wild	1	Y-7124	0.060	C	0.0148	BCDE	0.19	DE
	Acetic acid (g/L)							
	5		0.083	A	0.0223	A	0.29	A
	10		0.065	B	0.0152	B	0.22	B
	15		0.048	C	0.0088	C	0.14	C

B. Comparative fermentation of 75 g/L xylose in ODM with 5 to 15 g/L acetic acid by isolates precultured on glucose

Adaptation stress	Isolate		Xylose uptake rate (g/L/hA)[a,b]		Ethanol productivity (g/L/hA)[a,b]		Ethanol yield per initial sugar (g/g)[b]	
AFEX CSH	33	Colony 5	0.014	C	0.0027	B	0.09	C
AFEX CSH > E	28	2A.1.53R-E30-C3	0.018	C	0.0039	B	0.09	C
AFEX CSH > E (UV)	30	2A.1.30R2-E40-C5	0.029	BC	0.0094	AB	0.16	BC
AFEX CSH > E > PSGHL	13	2A.1.53R S100E40-1	0.021	C	0.0047	B	0.18	B
AFEX CSH > E > PSGHL	16	2A.1.53R-1	0.019	C	0.0044	B	0.16	BC
AFEX CSH > PSGHL	11	Colony 5-GP6	0.014	C	0.0049	B	0.15	BC
PSGHL	3	Y-7124 S90E40-1	0.058	A	0.0165	A	0.26	A
PSGHL	14	Y-7124-6	0.044	AB	0.0108	AB	0.27	A
PSGHL	15	Y-7124-10	0.050	A	0.0141	A	0.23	A
Wild	1	Y-7124	0.018	C	0.0050	B	0.13	BC
	Acetic acid (g/L)							
	5		0.043	A	0.0126	A	0.24	A
	10		0.028	B	0.0071	B	0.17	B
	15		0.014	C	0.0032	C	0.10	C

[a]Rates are normalized relative to population density in absorbance units (A) at 620 nm.
[b]Parameter variations based on two-way ANOVA (isolate × acetic) were significant ($P < 0.001$). Within columns, values with no letters in common are significantly different at $P < 0.05$ (Student-Newman-Keuls pairwise method).

particularly high fermentation capacity on a specific rate basis although its ability to grow and accumulate biomass in the full strength hydrolyzate was weaker than for other strains. The ethanol yields of 0.31 to 0.34 g/g initial sugar did not vary significantly among strains (Table 7C). In Figure 12A, corresponding to the kinetic data of Table 7, time courses are shown for the parent NRRL Y-7124 and selected adapted strains representing the most successful under these conditions: colony 5 (AFEX CSH), Y-7124-6 (PSGHL), and 2A.53R-E30-C3 (AFEX CSH > E). Highest ethanol accumulations reached 39 g/L for adapted strains compared to 36 g/L for the parent strain.

A similar flask experiment was run to compare kinetics in cultures inoculated to a low initial cell density (Figure 12B). In this situation, the ability of strains to grow in the hydrolyzate and transition to fermentation was tested. Time courses of the control and selected adapted strains are shown in Figure 12B, and here the parent strain performed particularly poorly relative to all of the adapted strains because it suffered a 48-h

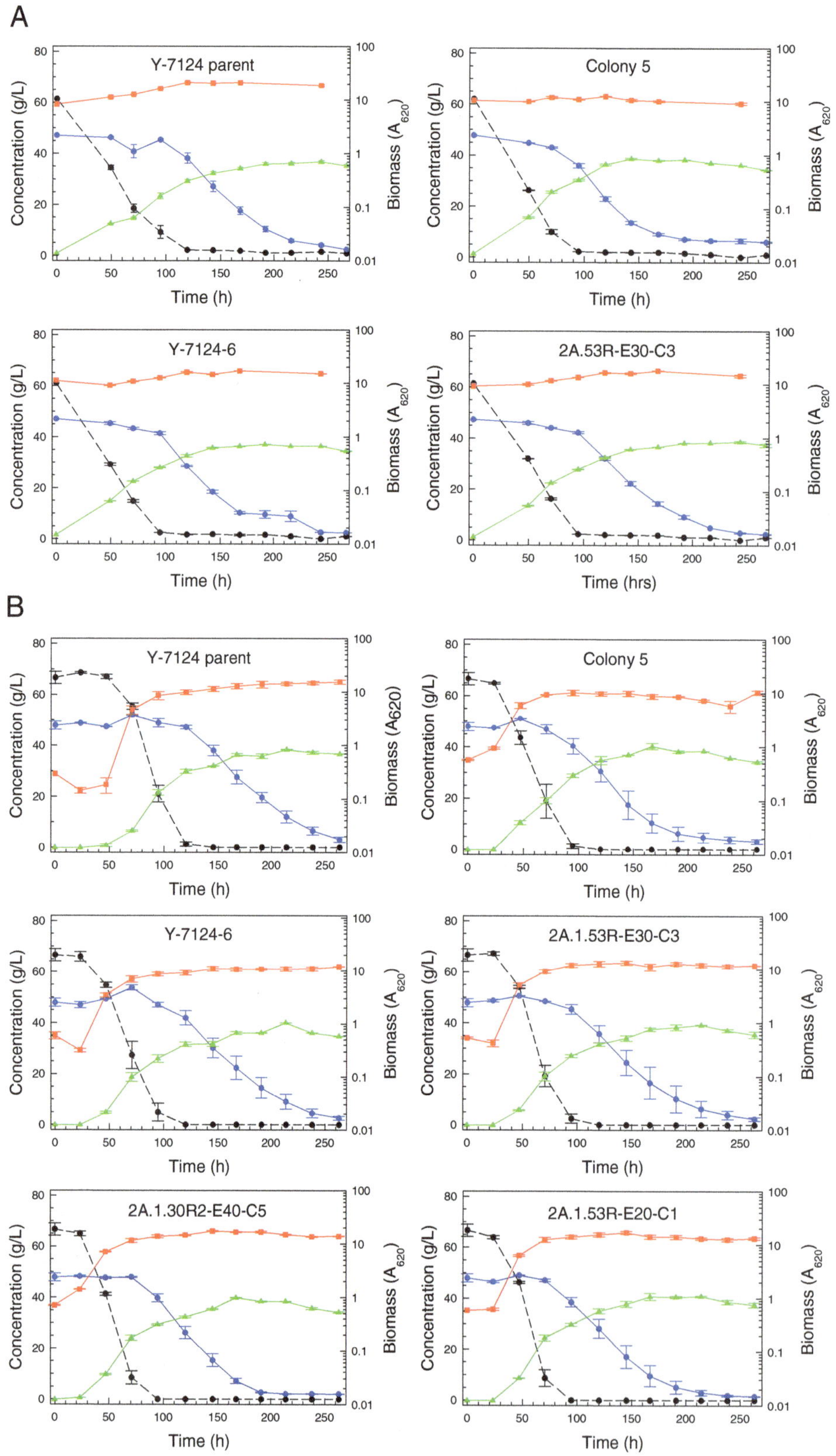

Figure 12 (See legend on next page.)

(See figure on previous page.)
Figure 12 Comparative SGH fermentations of superior adapted isolates of *S. stipitis*. Superior adapted isolates and their parent strain NRRL Y-7124 are compared fermenting enzymatic hydrolyzates of dilute acid-pretreated switchgrass (20% solids loading) at 25°C and initial pH 6.2 at high initial cell density (**A**) or low initial cell density (**B**). Time courses of biomass (red squares), glucose (black circles and dashed line), xylose (blue circles and solid line), and ethanol (green triangles) are shown. Error bars represent the range about the mean value marked by symbols.

lag period before growth began whereas the lag for adapted strains was much shorter at 24 h or less. Highest ethanol accumulations reached were 42 g/L at 167 h for adapted strains compared to 38 g/L at 213 h for the parent NRRL Y-7124. For the adapted strains, 85% of maximum ethanol accumulation was reached by 120 h. These kinetics are the most favorable ever reported for high solids loading hydrolyzates inoculated with non-engineered yeast strains at pH 5 to 6 without prior detoxification.

Table 7 Comparative kinetics of isolates on switchgrass hydrolyzate SGH-N2 inoculated to $A_{620} = 8.4 +/- 2.5$[a]

A. Volumetric rates during glucose or xylose consumption

Adaptation stress	Isolate		Sugar uptake rate (g/L/h)				Ethanol productivity (g/L/h)					
			Glucose		Xylose		Glucose		Xylose		Overall	
AFEX CSH	33	Colony 5	0.65	A	0.19	D	0.32	A	0.174	A	0.26	A
AFEX CSH > E	28	2A.1.53R-E30-C3	0.63	A	0.27	B	0.29	A	0.076	C	0.16	D
AFEX CSH > E > PSGHL	13	2A.1.53R S100E40-1	0.43	C	0.27	B	0.20	C	0.089	C	0.15	E
AFEX CSH > E > PSGHL	16	2A.1.53R-1	0.57	AB	0.23	C	0.27	AB	0.083	C	0.17	D
PSGHL	3	Y-7124 S90E40-1	0.65	A	0.16	E	0.30	A	0.140	B	0.24	B
PSGHL	14	Y-7124-6	0.62	A	0.24	C	0.28	AB	0.090	C	0.18	C
PSGHL	15	Y-7124-10	0.63	A	0.20	D	0.30	A	0.066	C	0.16	D
Wild	1	Y-7124	0.51	B	0.30	A	0.24	B	0.059	C	0.15	E

B. Normalized rates per unit absorbance (620 nm) during glucose or xylose consumption

Adaptation stress	Isolate		Sugar uptake rate (g/L/hA)				Ethanol productivity (g/L/hA)					
			Glucose		Xylose		Glucose		Xylose		Overall	
AFEX CSH	33	Colony 5	0.060	D	0.017	B	0.029	C	0.0161	B	0.024	B
AFEX CSH > E	28	2A.1.53R-E30-C3	0.056	D	0.017	B	0.025	CD	0.0048	DF	0.012	C
AFEX CSH > E > PSGHL	13	2A.1.53R S100E40-1	0.040	E	0.019	B	0.018	D	0.0063	CDF	0.012	C
AFEX CSH > E > PSGHL	16	2A.1.53R-1	0.082	C	0.026	A	0.039	B	0.0094	C	0.022	B
PSGHL	3	Y-7124 S90E40-1	0.152	A	0.025	A	0.072	A	0.0216	A	0.043	A
PSGHL	14	Y-7124-6	0.057	D	0.016	B	0.026	C	0.0059	CDF	0.014	C
PSGHL	15	Y-7124-10	0.089	B	0.023	A	0.042	B	0.0076	CD	0.020	B
Wild	1	Y-7124	0.037	E	0.014	B	0.017	D	0.0028	F	0.009	D

C. Yields and residuals per total initial sugar

Adaptation stress	Isolate		Yields or residuals per total initial sugar (g/g)							
			Ethanol		Biomass		Xylitol		Residual xylose	
AFEX CSH	33	Colony 5	0.34	A	0.0030	D	0.025	A	0.055	B
AFEX CSH > E	28	2A.1.53R-E30-C3	0.34	A	0.0134	B	0.020	A	0.023	C
AFEX CSH > E > PSGHL	13	2A.1.53R S100E40-1	0.31	B	0.0067	D	0.024	A	0.060	B
AFEX CSH > E > PSGHL	16	2A.1.53R-1	0.34	A	0.0045	D	0.019	A	0.096	A
PSGHL	3	Y-7124 S90E40-1	0.32	B	0.0060	D	0.023	A	0.113	A
PSGHL	14	Y-7124-6	0.33	AB	0.0091	C	0.021	A	0.024	C
PSGHL	15	Y-7124-10	0.33	AB	0.0050	D	0.029	A	0.073	B
Wild	1	Y-7124	0.33	AB	0.0165	A	0.026	A	0.023	C

[a]With the exception of xylitol yield, all parameters varied significantly among isolates based on one-way ANOVA ($P < 0.001$). Within columns, values with no letters in common are significantly different at $P < 0.05$ (Student-Newman-Keuls pairwise comparison method).

Conclusions

The application of single-phase and multi-phase adaptation schemes led to the recovery of significantly improved strains of *S. stipitis* with genetic stability indicated by performance consistency in defined and hydrolyzate-challenged culture media. The AFEX CSH adaptation was sufficient to cause significant improvement of the performance of *S. stipitis* in enzyme hydrolyzates of both AFEX-pretreated corn stover and dilute acid-pretreated switchgrass. This improvement, especially with respect to reduced diauxic lag during sugar transition, was further stabilized, and ethanol tolerance was enhanced and broadened by a second phase of adaptation in ethanol-challenged continuous culture. While exposure to only PSGHL was also sufficient to cause significant improvement of the performance of *S. stipitis* in enzyme hydrolyzates of both AFEX-pretreated corn stover and dilute acid-pretreated switchgrass, the exposure to AFEX CSH and/or ethanol-challenged continuous culture followed by PSGHL did not appear to generate additional performance advantages over the simple exposure to one hydrolyzate or the other. The reverse order of PSGHL exposure followed by AFEX CSH was not investigated, however. Preferred hydrolyzate applications for the top isolates are summarized in Table 2.

The competitive advantages of adapted strains were strongly related to significant improvements in sugar uptake rates and especially xylose uptake rate. Improvements in sugar uptake rate translated to increased ethanol productivity, particularly on xylose. However, adapted strains often demonstrated only modest improvements over the parent strain in maximal ethanol accumulation and yield per initial sugar. Based on the results of studies in a defined medium optimized for *S. stipitis* (ODM), the mechanisms of improved hydrolyzate performance varied among strains. Most strains, except those exposed to both AFEX CSH and PSGHL types of hydrolyzate, demonstrated reduced diauxic lag during transition from use of glucose to xylose. All strains showed better xylose fermentation performance in the presence of acetic acid, but relative performance depended on the level of prior induction on xylose. One strain (2A.1.30R2-E40-C5) showed significantly better growth resilience to acetic acid in the 6 to 15 g/L range on xylose, and the same strain showed shorter lag time in the presence of 25 mM furfural. However, in general, low cell densities of adapted strains were as susceptible to furfural-induced growth lag as the parent strain.

The incorporation of a variety of nutrient supplements into high solids loading hydrolyzates, including low cost commercial sources, was key to controlling the dynamic range of isolate performances for improved statistical separation during screening. The importance of nutrients to successful fermentation of concentrated hydrolyzates is perhaps due to an elevated need for amino acids and perhaps other nutrients required for cell maintenance, redox balancing, and repair. Additionally, the variation of supplements and hydrolyzate types during the screening process allowed identification of the most robust isolates with ability to function consistently in a variety of culture environments, as might be encountered in an industrial setting. The resulting new strains of *S. stipitis* NRRL Y-7124 were characterized by a greatly reduced growth lag and economically harvestable accumulations of ethanol at faster rates than previously reported for non-genetically engineered yeast fermenting hydrolyzates at pH 5 to 6 without prior overliming or other detoxification measures. The evolved yeast strains will support lower cost production of renewable ethanol from agricultural biomass, reducing dependence on fossil fuels.

Our studies so far provide a valuable set of *S. stipitis* isolates related by evolution under different selection pressures and an associated catalog of their kinetic attributes pertinent to hydrolyzate fermentation to ethanol biofuel. Future studies utilizing genome sequencing tools are expected to reveal how the genetic structure of the various superior isolates (Table 2) has changed from one phase of evolution to the next, following the flow diagram of selection pressure phases traversed by the evolving *S. stipitis* population in Figure 1. This figure essentially provides a useful road map notating which of the top isolates (Table 2) were obtained from each phase, such that it may be possible to associate adaptation method with specific genetic changes obtained. Fermentation performance characteristics indicated in the figures and tables may also be logically associated with the genetic changes that have occurred in each isolate. Such future work promises valuable new knowledge to help with the continued research and engineering of next-generation yeast biocatalysts and processes for converting lignocellulosic biomass to biofuels.

Methods

Stock cultures

A lyophilized culture of the parent strain *Scheffersomyces stipitis* NRRL Y-7124 (CBS 5773) was acquired from the ARS Culture Collection (National Center for Agricultural Utilization Research, Peoria, IL). Stock cultures of the strain NRRL Y-7124 and its derivatives were maintained in 10% glycerol at −80°C. Glycerol stocks were used to inoculate yeast malt peptone dextrose (YM) agar plates which were incubated 48 to 72 h at 25°C [11]. Developed plates were stored up to a week at 4°C prior to use as liquid preculture inocula.

Optimal defined medium (ODM)

The following defined medium composition optimized for ethanol production from high xylose concentration feeds [10] was used in all precultures and growth cultures for inhibitor tolerance bioassays: *purines/pyrimidines* - 10 mg/L adenine, cytosine, guanine, uracil, and thymine; *macro-minerals* - 1 g/L K_2HPO_4, 1 g/L KH_2PO_4, and 0.75 g/L $MgSO_4 \cdot 7H_2O$; *trace minerals* - 10 mg/L NaCl, 50 mg/L $FeSO_4 \cdot 7H_2O$, 5.5 mg/L $ZnSO_4 \cdot 7H_2O$, 1.6 mg/L $CoCl_2 \cdot 6H_2O$, 12.5 mg/L $MnCl_2$, 5 mg/L $(NH_4)_6 (Mo_7O_{24}) \cdot 4H_2O$, 8 mg/L $CuSO_4 \cdot 5H_20$, 27.5 mg/L $CaCl_2$ H_2O, 250 mg/L EDTA; *vitamins* - 0.5 mg/L biotin, 0.5 mg/L thiamin, riboflavin, calcium pantothenate, niacin, pyridoxamine, thioctic acid, 0.05 mg/L folic acid, and B_{12}; *carbon and nitrogen sources* - the medium was originally optimized to accommodate efficient conversion of 150 g/L xylose with 0.15 *M* N supplied 80% by 3.56 g/L urea and 20% by amino acids (10 g/L Difco Vitamin-Assay Casamino acids (Difco, Corpus Christi, TX, USA) (product 228830) + 0.1 g/L D,L-tryptophan (Sigma-Aldrich T3300, Sigma-Aldrich, St. Louis, MO, USA) + 0.4 g/L L cysteine (Sigma-Aldrich C7352)) [10]. In certain instances as designated, it was applied at one-third of the sugar and nitrogen source loadings - that is, with 50 g/L xylose or glucose and 0.05 *M* N as amino acids and/or urea to maintain C:N at 33:1, which is near the optimal 37:1 ratio.

Hydrolyzates

The comparative compositions of AFEX CSH at 6% and 12% glucan levels, PSGHL, and the enzyme saccharified switchgrass hydrolyzate (SGH) are given in Table 1. Hydrolyzate preparation details are given below. To study the impact of nutrient supplementation, SGH was amended with the following nutrients, then filer sterilized for use in isolate performance screening: SGH-N1(nutrient level 1) = SGH + 6.66 g/L casamino acids, 0.066 g/L tryptophan, 0.266 g/L cysteine + 2.36 g/L urea + ½ of liquid vitamin stock for ODM + ½ of dry $MgSO_4$ for ODM added prior to pH adjustment to 5.6 +/− 0.1 followed by filter sterilization of the finished hydrolyzate. SGH-N2 (nutrient level 2) = SGH supplemented with soy flour (ADM Toasted Nutrisoy Flour, Product Code 063160) and then 2.36 g/L urea to yield the nutrient levels set forth in Table 2, and pH was adjusted to 5.75. This mixture of soy and urea nitrogen sources provided the optimal approximately 20% primary amino nitrogen and approximately 80% nitrogen as urea noted previously for *S. stipitis* NRRL Y-7124 when the ODM sugar loading was approximately 100 g/L sugars [10].

Preparation of AFEX CSH at 6% and 12% glucan

Corn stover harvested in September 2008 was obtained from Arlington Research Station located in WI, USA. The corn hybrid used in this study was Pioneer 36H56 (triple stack - corn borer/rootworm/Roundup Ready (Monsanto, Creve Coeur, MO, USA)) variety. The biomass size reduction was done first by using hammer mills and dried at room temperature until the moisture content of the biomass was <10% (dry weight basis). Then, further fine milling was done using a Thomas Model 4 Wiley® Mill (Swedesboro, NJ, USA) to 4 mesh size (0.5 cm) and stored at 4°C in zip-lock bags until further use. AFEX pretreatment was carried out using a 5-gallon high pressure stainless steel batch reactor purchased from Parr Company (Moline, IL, USA) at Michigan Biotechnology Institute (MBI) pretreatment facility (Lansing, MI, USA) [14]. About 750 g of biomass was pretreated in a batch process. Biomass moisture content was raised to 60% by spraying de-ionized water and loaded into the reactor. The reactor was then charged with nitrogen followed by pumping of anhydrous liquid ammonia using an ammonia delivery system (comprised of an ammonia pump and a flow meter) into the reactor at 1:1 ammonia to biomass ratio. The reactor was then heated up using a heating mantle until the temperature of the biomass reached 100°C (approximately 300 psi). This condition was maintained for a period of 30 min, and the ammonia was released by venting. The pretreated biomass was then transferred to a plastic tray and dried in the hood overnight to remove residual ammonia present in the biomass. The AFEX-treated biomass was then packed in a plastic zip-lock bag and stored at 4°C until further use.

AFEX-pretreated biomass was hydrolyzed at high solids loading (6% and 12% glucan loading) using commercial enzymes supplied by Novozymes (Franklinton, NC, USA) and Genencor (Palo Alto, CA, USA) at 30 mg/g of glucan enzyme loading (70% Ctec2, 15% Htec2, and 15% Multifect Pectinase). Enzyme hydrolysis was done under sterile conditions using 2-L baffled shake flasks at 50°C, 250 rpm for 168 h. The pH was maintained at 4.8 using 3 M HCL. Biomass was loaded in two (6%) to three (12%) batches during hydrolysis to overcome mixing problems caused due to high viscosity during the initial stages of hydrolysis [15]. After the completion of hydrolysis, the hydrolyzate slurry was transferred to 1-L centrifuge tubes and spun at 6,000 rpm for 30 min to remove the solids from the liquid using Beckman Avanti centrifuge system (Brea, CA, USA). The hydrolyzed sugar stream (supernatant liquid) was sterile filtered using a 0.2 μm Millipore (Billerica, MA, USA) sterile-cup membrane filtration system and stored at 4°C until further use.

Preparation of PSGHL

Switchgrass hydrolyzates were prepared from Kanlow N1 baled post-frost from Mead, NE, USA, that was milled to pass through a 2-mm screen. Switchgrass was pretreated at the 20% solids level by mixing 20 g dry weight of biomass with 80 mL of 0.936% (v/v) sulfuric acid solution and 0.3 g Pluronic F-268. Each of 12 closed stainless steel vessels was loaded with reactants and mounted in a

Mathis Labomat IR Dyer Oven where they were rotated at 50 rpm (1 min right then 1 min left) and heated to 160°C, held for 15 min, and then cooled at 40°C. To prepare PSGHL, the pretreatment reaction products were centrifuged for 45 min at 7,000 rpm and sterile filtered through 0.2 μm Nalgene filter units. Supernates were combined and adjusted with $Ca(OH)_2$ to pH 6.0 to 6.5, and the resulting PSGHL was filter sterilized and refrigerated at approximately 4°C until use.

Preparation of SGH

To prepare SGH, switchgrass was pretreated in the Labomat oven at the 20% solids level as described above. After pretreatment, the product was adjusted to pH 4.5 by adding 7.14 mL of 15% $Ca(OH)_2$ solution and 4.5 mL of 1 M citric acid buffer directly into each vessel and then tumbling 15 min in the Labomat. Pretreatment hydrolyzates were transferred to 250 mL pyrex bottles for saccharification. To each bottle, 2.7 mL of CTec and 0.5 mL of HTec enzymes (Novozyme) were added. Tightly capped bottles were incubated approximately 72 h at 50°C and 175 rpm. Resulting hydrolyzates were sterile filtered through 0.2 μm Nalgene filter units and refrigerated at 4°C until use.

AFEX CSH serial transfer culture adaptation process

A preculture of strain NRRL Y-7124 was inoculated by loop transfer of cells from YM agar to 75 mL ODM + 150 g/L xylose to challenge growth under osmotic stress. Precultures in 125-mL flasks with Bellco silicon sponge closures (Bellco, Vineland, NJ, USA) were incubated 24 h at 25°C with shaking (150 rpm, 1″ orbit). Frozen aliquots of 6% to 12% glucan AFEX-pretreated corn stover hydrolyzate were thawed in cold water and used at pH 5 to prepare a dilution series in 96-well microplates. Plates were filled with 50 μL per well and eight wells per dilution, then inoculated with a few microliters of preculture per well to allow for an initial absorbance (620 nm) $A_{620.0} \geq 0.1$. Plates were statically incubated in a plastic box with a wet Wypall for humidity at 25°C for 24 to 48 h. Using the most concentrated hydrolyzate dilution that visibly grew (that is, $A_{620} > 1$), 1 to 5 μL was transferred to each well of a new hydrolyzate dilution series ($A_{620.0} \geq 0.1$). Cell growth was monitored by culture absorbance (620 nm) using a plate-reading spectrophotometer (Biotek PowerWave XS; Biotek, Winooski, VT, USA). An uninoculated dilution series served as a control and blank. Glycerol stocks of adaptation cultures were prepared regularly for subsequent isolation of improved strains or use in reinoculating continuing hydrolyzate dilution series by mixing 200 μL of the greatest hydrolyzate concentration colonized with an appropriate glycerol solution to suspend cells in 10% glycerol in cryovials for freezing at −80°C.

For isolation of single tolerant colonists, selected glycerol stocks of adaptation cultures were streaked to YM agar and used to inoculate 50 μL of 3% glucan hydrolyzate (pH 5) in each of three microplate wells to $A_{620,0} = 0.1$. The 96-well microplates were incubated as before at 24 h and 25°C. Colonized culture wells were pooled and dilution plated to YM agar or 6% glucan AFEX CSH agar. Selected single colonies were picked from the highest dilution plated after 24 to 48 h incubation at 25°C and then restreaked to YM plates for incubation 24 h prior to glycerol stock preparation.

Batch and fed-batch culture evaluations of AFEX CSH utilization

For evaluation in 6% glucan AFEX CSH batch cultures, cells from 48-h plates streaked from glycerol stocks were suspended in buffer to $A_{620} = 10$ and 1 μL used to inoculate each of four wells of 50 μL 3% glucan hydrolyzate (12% glucan AFEX CSH at pH 5 diluted 1:3 with sterile water) to $A_{620} = 0.2$. Microplates were developed 24 h and then two wells were transferred to inoculate precultures of 25 mL of pH 5 6% glucan AFEX CSH/50-mL flasks with Bellco silicon sponge closures. The precultures were incubated for 24 h at 25°C, approximately 150 rpm (1″ orbit) and then used to inoculate similar 25-mL growth cultures to $A_{620} = 0.1$. The cultures were incubated as for precultures and sampled daily (0.2 mL) for monitoring biomass accumulation (A_{620}) and concentrations of sugars and fermentation products (HPLC).

The performance of AFEX 6% glucan CSH-grown populations repitched to 8% glucan hydrolyzates was also studied. Inocula for 30-mL growth cultures were prepared as described above for AFEX 6% glucan CSH batch cultures. Growth cultures were inoculated to an absorbance of 0.1 and incubated in 50-mL flasks with silicon sponge closures (Bellco). Cultures were sampled daily for A_{620} and HPLC analyses. When the majority of the xylose had been consumed, the cells from growth cultures were harvested by centrifugation and repitched to A_{620} of 40 in 4.2 mL of 8% glucan AFEX CSH in 50-mL flasks with septum caps, vented with 3/8″ 26 G needles. All cultures were incubated at initial pH 5, 25°C, 150 rpm, 1″ orbit. The daily samples were plated for viable cells, A_{620} was measured, and the remaining sample was centrifuged to collect supernate for HPLC analyses.

In later studies, repitched AFEX 6% glucan CSH cultures were fed with 12% glucan hydrolyzate. Following the above procedure, cells were harvested by centrifugation and resuspended to an A_{620} of 50 in 4.2 mL of 6% glucan AFEX CSH (pH 5) in a 50-mL flask with vented septum cap. All flasks were incubated as before. After approximately 24 h and significant sugar consumption in the 6% glucan hydrolyzate, the cultures were then fed with 4 mL of 12% glucan hydrolyzate (pH 5). The cultures were sampled prior to the feed and thereafter.

Evaluation of diauxic lag on ODM with mixed sugars and tolerance of acetic acid

Precultures of 75 mL ODM with 150 g/L xylose in 125-mL flasks with silicone sponge closures (Bellco) were inoculated by loop transfer from YM glycerol streaks. The 24-h precultures were used to inoculate similar 75-mL test cultures but with ODM containing 75 g/L of glucose and 75 g/L of xylose. All flask cultures had initial pH 6.5 and were shaken at 25°C, 150 rpm (1″ orbit). The impact of 0 to 15 g/L acetic acid on strain fermentation of glucose and xylose was also tested using a similar protocol except that the initial pH was set at 6.0 ± 0.2 in test cultures to buffer the impact of acetic acid consumption.

Continuous culture selection for ethanol-challenged xylose utilization

The continuous culture feed medium was ODM with 60 to 100 g/L xylose, 20 to 50 g/L ethanol at pH 6.3 ± 0.2. AFEX-tolerant colony 5 was precultured in 75 mL of the ethanol-free feed medium in 125-mL flasks at 25°C, 150 rpm (1″ orbit) for 24 h. The continuous culture was initiated with colony 5 preculture by inoculating 100 mL of ODM + 100 g/L xylose + 20 g/L ethanol to $A_{620.0}$ approximately 0.5. The 100-mL culture holding volume was maintained at 25°C in a jacketed 100-mL spinner flask (Bellco) stirred at 200 rpm and outfitted with a sterilizable pH electrode. Temperature was controlled with a refrigerated circulating water bath. For the first 125 days of cultivation, the feed medium was dosed using a pH actuated pump such that when the culture fermentation was sufficient to drop the pH to 5.4, the feed medium at 100 g/L xylose and 50 g/L ethanol would dose to prevent the pH from dropping lower. A continuously pumping effluent pump drawing from the culture surface maintained the fermentor volume. Thus, the ethanol concentration of the culture rose at an artificially high rate in response to fermentation progress. Samples (1 to 2 mL) were taken from continuous cultures every 48 to 72 h and analyzed for A_{620}, cell viability, sugars, and ethanol. Effluent was collected and measured at sample times. Based on effluent collection rates, the dilution rate varied from 0 to 0.01 h^{-1}. Glycerol stocks were saved on a regular basis by isolating from viability spread plates allowing formation of 30 to 100 colonies, which would be a sampling of the most prevalent, robust colonists at that point in the enrichment process. This plate was flooded with approximately 5 mL of 10% glycerol to prepare duplicate cryovials. On occasion, it was necessary to restart the continuous culture using the most recent glycerol stock.

Once pH-actuated continuous fed cultures were able to grow solely on xylose in the presence of up to 28 g/L ethanol, the remainder of the continuous culture selection process (next 80 days) was carried out at a dilution rate of approximately 0.012 h^{-1} using ODM + 60 g/L xylose + 30 to 50 g/L ethanol. The selection culture was restarted from the current, most resistant glycerol stock population streaked to YM and transferred to a preculture of ethanol-free ODM with 60 g/L xylose for incubation as above. The 100-mL holding volume of ODM with 60 g/L xylose and 20 g/L ethanol was inoculated to A_{620} 0.5, and the population was allowed to grow batch-wise to stationary phase; then the feed medium flow was started. Over time, ethanol concentration in the feed was raised as yeast tolerance improved and advancing populations were captured in glycerol stocks for use as described above.

During the last 6 months of operation, ultraviolet (UV) irradiation was used approximately monthly to induce further mutations in the glycerol stock populations used to restart continuous cultures which were fed as before. Colonies from plates streaked with glycerol stocks were resuspended in 10 mL of ODM with 60 g/L xylose (as used for precultures) and transferred to a common sterile flask. The combined cell suspension at approximately 5×10^8 viable cells/mL was used to cover the bottom of four to five petri plates. Each open plate was placed below the UV light source in a biological safety cabinet and exposed for 45 min. The excess cell suspension remaining after filling plates and a post-irradiation sample were dilution plated to allow estimating the kill rate at approximately 97%. The UV-exposed cultures (approximately 30 mL) were transferred to a foil-covered 50-mL flask to prevent photo reactivations and to preserve mutations as cultures were incubated at 25°C and 150 rpm for 24 h while viable cell counts returned to approximately 1×10^8 viable cells/mL for continuous culture inoculation to approximately 1×10^7 viable cells/mL.

Performance of adapted population growth and fermentation of xylose with ethanol present

In order to focus isolation efforts, selected glycerol stock cultures were screened to identify those with best growth and fermentation of xylose in the presence of ethanol. First, xylose uptake by glucose-grown high cell densities was evaluated in the presence of ethanol. Precultures in ODM with 150 g/L glucose were incubated as described above for 96 h prior to use as inocula for the test flask cultures in order to produce large populations requiring enzyme induction for xylose utilization. Test cultures in ODM +60 g/L xylose + 30 to 45 g/L ethanol were inoculated to A_{620} of 40 and incubated at 25°C, 150 rpm (1″ orbit) in 50-mL flasks with silicon sponge closures.

Growth on xylose in the presence of ethanol was also evaluated. Precultures were inoculated by loop transfer to 75 mL of ODM with 150 g/L xylose in 125 mL flasks and incubated as previously described. Test cultures were inoculated to an A_{620} of 0.1 in 25 mL of ODM + 60 g/L xylose + 30 to 45 g/L ethanol in 125-mL flasks. Flasks were incubated at 25°C, 300 rpm, 1″ orbit, and sampled.

Isolation of single-cell colonies utilizing xylose in the presence of ethanol

For each glycerol stock showing superior ability to grow on and ferment xylose in the presence of ethanol, 1-mL precultures on PSGHL mixed 1:1 with ODM + 50 g/L xylose (no ethanol) were inoculated by picking from glycerol stock streaks. Precultures were contained in 96-well, deep-well plates with low evaporation covers (Duetz clamping system, Applikon Biotechnology, Delft, The Netherlands) and incubated 48 h at 25°C, 400 rpm, 1″ orbit. Precultures were used to inoculate 16 × 1 mL replicate cultures to A_{620} = 0.5 in 1:1 PSGHL:ODM + 50 g/L xylose with 20, 30, or 40 g/L ethanol for enrichment of tolerant colonists. Enrichment cultures were incubated similarly to precultures. Harvesting from highest ethanol concentration allowing growth and xylose use, each cell line was plated to YM agar to obtain single colonies. Ten colonies per cell line were picked and streaked to new YM agar plates for glycerol stock preparation.

PSGHL serial transfer culture adaptation process

Precultures of NRRL Y-7124 (parent), 5MSU colony 5, and 2A.1.53R in 75 mL of ODM + 150 g/L xylose were prepared and incubated as described above for the AFEX CSH hydrolyzate adaptation process. PSGHL was diluted with water to provide a series of increasing concentrations in 96-well microplates for each of the three cell lines. Each 50 μL micro-culture of the dilution series was initiated to A_{620} ~ 0.1 to 0.5 with precultures. The PSGHL adaptation was carried out using the same procedure as the AFEX CSH hydrolyzate adaptation detailed above, and glycerol stocks of progressive populations were prepared. Single-cell isolates were also obtained directly from final (480 day) adaptation plates by dilution plating each of the three cell lines to YM agar since all were growing in the full strength PSGHL. Ten large colonies were picked for each of the three cell lines and then bar streaked to YM plates for glycerol stocks.

Glycerol stocks of the three cell lines at two earlier points of adaptation (360 and 420 days) were streaked to YM agar, precultured 24 h on ODM + 50 g/L xylose, then challenged in 50 μL of 50% PSGHL per microplate well incubated statically 48 h from initial A_{620} ~ 0.2, then spread onto PSGHL gradient plates ranging from 0% to 50% strength hydrolyzate (delivering approximately 300 to 400 viable cells per plate), and ten single colonies picked per cell line from the highest hydrolyzate concentration area of the gradient [29]. Picked colonies were streaked to YM for glycerol stock preparation. Alternatively, cell lines were propagated from glycerol stocks as for selective plating on gradient agar, but instead of the gradient agar plates, 96-well microplates with a range of hydrolyzate concentrations from 50% to 100% strength and ethanol from 10 to 40 g/L were inoculated to A_{620} ~ 0.2. Microplates were developed 72 to 96 h with the following conditions: 50 μL/well, static incubation, 25°C, humidified box. Via dilution plating on YM agar, ten single colonies were isolated from wells of the harshest hydrolyzate-ethanol combinations showing growth, and glycerol stocks were prepared as above.

Performance ranking screens to select best isolates from AFEX CSH, ethanol, and dilute acid SGH adaptation phases

Deep-well plate screen of PSGHL performance as primary elimination of inferior isolates

Five sets of thirty isolates were screened along with NRRL Y-7124 parent (control) to choose six top strains from each set of 30. For higher throughput, the screen was carried out in Applikon Duetz System deep-well plates filled with 1 mL per well and covered with stainless steel lids with black silicone low evaporative seals. Plates were incubated in a New Brunswick Innova 42R shaker (Eppendorf, Hamburg, Germany) at 25°C and 400 rpm (1″ shaker orbit). Isolates were picked from glycerol streaks to duplicate wells of ODM + 50 g/L xylose and incubated 48 h. All deep-well plate filling patterns were designed to allow separation of different isolates by open wells. The 50% PSGHL was prepared by mixing PSGHL 1:1 with ODM + 10 g/L glucose + 50 g/L xylose. A 50-μL volume of ODM precultures (A_{620}@ ~ 10) was transferred to each of the two 50% PSGHL wells for each of the isolates and controls to obtain an A_{620} of approximately 0.5. Cells of each isolate were harvested at 72 h from the 50% PSGHL challenge precultures and used to inoculate 50 μL per 1,000 μL to A_{620} ~ 0.5 in five deep wells for each of two test media: 60% PSGHL + ODM nutrients and 75% PSGHL + ODM + YM nutrients. In the two test media using the indicated partial strength of PSGHL, the nutrients (excluding sugars) were at half of the strength as standard for ODM (when designed for use with 50 g/L sugar). When present, YM nutrients were also used at half of the standard strength of 3 g/L yeast extract, 3 g/L malt extract, and 5 g/L peptone. For each sampling, a well was transferred to centrifuge tubes to obtain supernate (7,000 rpm, 15 min) for ethanol, glucose, and xylose high throughput analyses. Biomass was measured in 96-well plates (200 μL/well) with a Biotek Powerwave spectrophotometer. Within each of the five sets of isolates tested, relative performance indexes were calculated and used to rank each strain based on ethanol yield and xylose uptake rate on both test media.

Comparison of top PSGHL performers on SGH at different nutrient levels

The top 32 isolates performing in the deep-well plate screen of PSGHL and the parent strain NRRL Y-7124 control were next screened twice in 16-mL flask cultures on SGH and SGH amended with two levels of nitrogen, SGH-N1, and SGH-N2. Isolates were picked from glycerol streaks to duplicate deep wells of 1 mL ODM + 50 g/L xylose as before and incubated 48 h in the Duetz System. Then, 50 μL of ODM precultures was transferred to 50% SGH challenge cultures which were incubated in the Duetz System. The 50% SGH was prepared by mixing SGH 1:1 with sugarless ODM + 50 g/L xylose (pH 5.6). Isolates were harvested after 72 h from the 50% SGH challenge precultures (A_{620} ~ 10). For each of the isolates, a 16-mL aliquot of SGH, SGH-N1, or SGH-N2 was inoculated with the cell pellet (15 min, 4,900 rpm) from three wells of challenge culture to yield initial test culture A_{620} ~ 2.0. Test cultures were incubated at 25°C, 180 rpm (1″ orbit) in 25-mL flasks with Bellco silicone sponge closures. Flasks were sampled, and samples were analyzed per the PSGHL screen.

Comparison of top SGH performers on AFEX CSH

The top 21 isolates performing in the deep-well plate screen of SGH and the parent strain NRRL Y-7124 control were next similarly screened twice in 16 mL flask cultures on 6% glucan AFEX CSH at pH 5.2.

Comparative kinetics of superior isolates

Diauxy during glucose and xylose fermentation at low cell density ($A_{620.0}$ 0.1) on ODM was evaluated as previously described for AFEX CSH isolates. Additionally, the impact of acetic acid on fermentation of ODM with mixed sugars and diauxy at high cell density was also evaluated. Precultures were inoculated by loop transfer from YM glycerol streaks to 200 mL ODM with 150 g/L xylose or glucose at pH 6 in 300-mL flasks with silicone sponge caps (Bellco). Flasks were incubated at 150 rpm (1″ orbit) for 96 h at 25°C. Test cultures were inoculated to an A_{620} of 50 using cell pellets from precultures in 12 mL ODM + 75 g/L xylose with 0 to 15 g/L acetic acid at pH 6.0. The 12-mL treatments were distributed 1 mL per well to 12 wells of a deep-well plate and incubated in the Duetz System.

To study impact of inhibitors on growth, liquid precultures were inoculated by sterile loop in 20-mL cultures of ODM with 50 g/L xylose (or glucose) at pH 6.0 in 50-mL flasks with silicone sponge caps. Flasks were incubated at 150 rpm (1″ orbit) for 24 h at 25°C. Test cultures were inoculated to an A_{620} of 0.1 in duplicate wells containing 0.8-mL volumes of ODM + 50 g/L xylose (or glucose) +/− acetic acid or furfural inhibitors at pH 6.0 in 48-well MTP flower plates (P/N MTP-48-BOHS, M2P Labs, Baesweiler, Germany). Test cultures were incubated at 25°C, 1,100 rpm in a Biolector Instrument where light scattering was monitored.

For larger-scale hydrolyzate fermentations in flasks for comparative kinetics, liquid precultures were inoculated by sterile loop to 75-mL cultures of ODM + 50 g/L xylose in 125-mL flasks (silicone sponge caps) and incubated 48 h, 150 rpm (1″ orbit), 25°C. The 48-h precultures were used to inoculate 75-mL challenge cultures of 1:1 SGH-N2:water (pH 6.2) to A_{620} ~ 0.5, and they were incubated similarly in 125-mL flasks. Challenge cultures were harvested at approximately 72 to 96 h as required to obtain populations in the midst of xylose consumption. Test cultures were inoculated at A_{620} 8.4 ± 2.5 to 75 mL SGH-N2 (pH 6.2) incubated in 125-mL flasks with silicone sponge caps at 25°C, 150 rpm (1″ orbit). For low initial cell density experiments, test cultures were inoculated to A_{620} 0.5 in 23 mL SGH-N2 per 50-mL flask.

Analyses

Cell biomass was measured by culture absorbance at 620 nm in 1-cm cuvettes using a Genesys 2 spectrophotometer (0.167 g/L biomass per unit absorbance) or in microplates using a BioTek Powerwave XS plate reader (200 μL/well). All absorbance data were reported in terms of Genesys 2 absorbance units (equivalent to microplate absorbances/0.438). Quantitations of sugars, ethanol, furfural, HMF, and acetic acid in culture samples were by Waters HPLC system equipped with a Biorad HPX-87H Aminex ion exclusion column (Bio-Rad, Hercules, CA, USA) fitted with a Micro-guard Cation H Micro-Guard Cartridge (125–0129) precolumn. Samples (10 μL) injected to the precolumn were isocratically eluted at 60°C with acidified water (15 mM HNO_3) at 0.6 mL/min to achieve component separations which were monitored by both refractive index and ultraviolet absorbance detectors (215 nm). For hydrolyzate compositional analysis, Biorad Aminex HPX-87P carbohydrate analysis column (125–0098) (with Deashing cartridge (125–0118) and Carbo-P Micro-Guard Cartridge (125–0119)) was used at 80°C with water mobile phase. For higher throughput analyses of isolate screenings in deep-well micro-plates, a Biorad Fast Acid Analysis column (125–0100) at 0.6 mL/min acidified water mobile phase was used to assay ethanol, while glucose and xylose were analyzed in microplates using a YSI 2900 Biochemistry Analyzer.

For available nitrogen, enzyme-based test kits were used to assay primary amino nitrogen, ammonia, and urea (Megazyme International Ireland Ltd., Wicklow, Ireland).

Analysis of variance (ANOVA) and Student-Newman-Keuls (SNK) pairwise comparison analyses were performed using Sigmastat 3.5 (Systat Software, Inc., San Jose, CA, USA) at significance criterion $P \leq 0.05$.

Abbreviations

A: spectrophotometric absorbance; AFEX: ammonia fiber expansion; AFEX CSH: AFEX-pretreated corn stover hydrolyzate; AFEX CSH > E: adaptation stress was AFEX CSH followed by ethanol; C:N: carbon to nitrogen ratio; E: ethanol; F: statistical parameter dependent on observed data, data average, and standard deviation that is used in the calculation of RPI; HMF: hydroxymethyl furfural; NRRL Y-7124: the accession number of the parent strain of *Scheffersomyces stipitis* used in evolution studies, which is deposited in the Agricultural Research Service Culture Collection at the National Center for Agricultural Utilization Research at Peoria, IL; ODM: optimal defined medium for *S. stipitis* NRRL Y-7124 ethanol production; PAN: primary amino nitrogen; PSGHL: dilute acid-pretreated switchgrass hydrolyzate liquor; RPI: relative performance index based on fermentation rate (RPI_R) or yield (RPI_Y); $RPI_{overall}$: average relative performance based on both rate and yield parameter data; SGH: enzymatically saccharified dilute acid-pretreated post-frost switchgrass; SGH-N1: SGH fortified to 42:1 C:N with urea, casamino acids, tryptophan, cysteine, and vitamins; SGH-N2: SGH fortified to 37:1 C:N with soy flour and urea; *S. stipitis*: *Scheffersomyces stipitis*; UV: ultraviolet irradiation, used where specified to help induce mutations; YM: yeast malt peptone dextrose agar; μ: specific growth rate.

Competing interests

The authors declare that they have no competing interests.

Authors' contributions

PJS, BSD, VB, and LdCS together conceptualized developing novel pentose-fermenting yeast strains tolerant of hydrolyzate through the repetitive application of multiple hydrolyzates as growth and fermentation substrates to challenge the parent strain *Scheffersomyces stipitis* NRRL Y-7124, induce mutations, and naturally enrich for tolerant derivatives. BED, VB, LdCS, and NU prepared the needed AFEX-pretreated corn stover hydrolyzates using their novel technologies for pretreatment and enzyme hydrolysis. BSD optimized pretreatment and saccharification technologies to prepare dilute acid-pretreated switchgrass hydrolyzate as a second inhibitory medium for adaptation. PJS, SRT, and MAS devised the culture adaptation strategies, nutrient supplementation, isolation strategies, and performance evaluation methodologies required to develop and identify improved evolved strains. MAC participated in discussions about the research and contributed to setting up the high throughput analytical equipment required to rapidly screen the performances of a large number of isolates. CPK provided expertise on the life cycle of *S. stipitis* which influenced our strategy of cultivation and adaptation, and he also contributed taxonomic evaluations throughout the adaptation process. PJS deposited strains in the ARS Culture Collection, Peoria IL, and drafted the manuscript. Others edited the draft. All authors read and approved the final manuscript.

Acknowledgements

We would like to express our sincere appreciation to Drs. Kenneth Vogel, Robert Mitchell, and Gautam Sarath, Grain, Forage, and Bioenergy Research Unit, Agricultural Research Service, Lincoln, NE, for their kind supply of switchgrass for this project. We also thank the U.S. Department of Energy for funding to VB and NU through the DOE Great Lakes Bioenergy Research Center (GLBRC) Grant DE-FC02-07ER64494.

Disclaimer

The mention of trade names or commercial products in this article is solely for the purpose of providing specific information and does not imply recommendation or endorsement by the U.S. Department of Agriculture.

Author details

[1]Bioenergy Research Unit, National Center for Agricultural Utilization Research, USDA-ARS, 1815 N. University, Peoria, IL 61604, USA. [2]Bacterial Foodborne Pathogens and Mycology Research, National Center for Agricultural Utilization Research, USDA-ARS, 1815 N. University, Peoria, IL 61604, USA. [3]DOE Great Lakes Bioenergy Research Center, Michigan State University, East Lansing, MI 48824, USA.

References

1. U.S. Department of Energy. U.S. billion-ton update: biomass supply for a bioenergy and bioproducts industry (R.D. Perlack and B.J. Stokes (Leads)). In: Book U.S. billion-ton update: biomass supply for a bioenergy and bioproducts industry (R.D. Perlack and B.J. Stokes (Leads)). Oak Ridge, TN: U.S. Department of Energy, Oak Ridge National Laboratory; 2011. p. 227.
2. Ladisch MR, Dyck K. Dehydration of ethanol: new approach gives positive energy balance. Science. 1979;205:898–900.
3. Cardona Alzate CA, Sanchez Toroa OJ. Energy consumption analysis of integrated flowsheets for production of fuel ethanol from lignocellulosic biomass. Energy. 2006;31:2447–59.
4. Prior BA, Kilian SG, DuPreez JC. Fermentation of D-xylose by the yeasts *Candida shehatae* and *Pichia stipitis*. Process Biochem. 1989;24:21–32.
5. Kurtzman CP, Suzuki M. Phylogenetic analysis of ascomycete yeasts that form coenzyme Q-9 and the proposal of the new genera *Babjeviella, Meyerozyma, Millerozyma, Priceomyces and Scheffersomyces*. Mycoscience. 2010;5:2–14.
6. Bruinenberg PM, De Bot PHM, VanDijken JP, Scheffers WA. NADH-linked aldose reductase: the key to anaerobic alcoholic fermentation of xylose by yeasts. Appl Microbiol Biotechnol. 1984;19:256–60.
7. Slininger PJ, Bothast RJ, Okos MR, Ladisch MR. Comparative evaluation of ethanol production by xylose-fermenting yeasts presented high xylose concentrations. Biotechnol Lett. 1985;7:431–6.
8. Slininger PJ, Bothast RJ, Ladisch MR, Okos MR. Optimum pH and temperature conditions for xylose fermentation by *Pichia stipitis*. Biotechnol Bioeng. 1990;35:727–31.
9. Slininger PJ, Branstrator LE, Lomont JM, Dien BS, Okos MR, Ladisch MR, et al. Stoichiometry and kinetics of xylose fermentation by *Pichia stipitis*. Annals New York Acad Sci. 1990;589:25–40.
10. Slininger PJ, Dien BS, Gorsich SW, Liu ZL. Nitrogen source and mineral optimization enhance D-xylose conversion to ethanol by the yeast *Pichia stipitis* NRRL Y-7124. Appl Microbiol Biotechnol. 2006;72:1285–96.
11. Slininger PJ, Thompson SR, Weber S, Liu ZL. Repression of xylose-specific enzymes by ethanol in *Scheffersomyces (Pichia) stipitis* and utility of repitching xylose-grown populations to eliminate diauxic lag. Biotechnol Bioeng. 2011;108:1801–15.
12. Slininger PJ, Gorsich SW, Liu ZL. Culture nutrition and physiology impact inhibitor tolerance of the yeast *Pichia stipitis* NRRL Y-7124. Biotechnol Bioeng. 2009;102:778–90.
13. Agbogbo FK, Coward-Kelly G. Cellulosic ethanol production using the naturally occurring xylose-fermenting yeast, Pichia stipitis. Biotechnol Lett. 2008;30:1515–24.
14. Balan V, Bals B, Chundawat S, Marshall D, Dale BE. Lignocellulosic biomass pretreatment using AFEX. In: Mielenz JR, editor. Biofuels: Methods and protocols, Methods in Molecular Biology. Volume 581. New York, NY: Humana Press, a part of Springer Science+Business Media, LLC, Springer; 2009. p. 61–77.
15. Jin M, Gunawan C, Uppugundla N, Balan V, Dale BE. A novel integrated biological process for cellulosic ethanol production featuring high ethanol productivity, enzyme recycling, and yeast cells reuse. Energy Environ Sci. 2012;5:7168–75.
16. Nigam JN. Development of xylose-fermenting yeast *Pichia stipitis* for ethanol production through adaptation on hardwood hemicellulose acid prehydrolysate. J Appl Microbiol. 2001;90:208–15.
17. Nigam JN. Ethanol production from wheat straw hemicellulose hydrolysate by *Pichia stipitis*. In: *Book Ethanol production from wheat straw hemicellulose hydrolysate by Pichia stipitis* (Editor ed.^eds.), vol. 87. 2001. p. 17–27.
18. Hughes SR, Gibbons WR, Bang SS, Pinkelman R, Bischoff KM, Slininger PJ, et al. Random UV-C mutagenesis of *Scheffersomyces* (formerly *Pichia*) *stipitis* NRRL Y-7124 to improve anaerobic growth on lignocellulosic sugars. J Indus Microbiol Biotechnol. 2012;39:163–73.
19. Bajwa PK, Shireen T, D'Aoust F, Pinel D, Martin VJJ, Trevors JT, et al. Mutants of the pentose-fermenting yeast *Pichia stipitis* with improved tolerance to inhibitors in hardwood spent sulfite liquor. Biotechnol Bioeng. 2009;104:892–900.
20. Bajwa PK, Pinel D, Martin VJJ, Trevors JT, Lee H. Strain improvement of the pentose-fermenting yeast *Pichia stipitis* by genome shuffling. J Microbiol Methods. 2010;81:179–86.
21. Jeffries TW, Grigoriev I, Grimwood J, Laplaza JM, Aerts A, Salamov A, et al. Genome sequence of the lignocellulosic-bioconverting and xylose-fermenting yeast *Pichia stipitis*. Nat Biotechnol. 2007;25:319–26.

22. Kurtzman CP, Robnett CJ. Identification and phylogeny of ascomycetous yeasts from analysis of nuclear large subunit (26S) ribosomal DNA partial sequences. Antonie Van Leeuwenhoek. 1998;73:331–71.
23. Casey E, Sedlak M, Ho NWY, Mosier NS. Effect of acetic acid and pH on the cofermentation of glucose and xylose to ethanol by a genetically engineered strain of *Saccharomyces cerevisiae*. FEMS Yeast Res. 2010;10:385–93.
24. Wang X, Jin M, Balan V, Jones AD, Li X, Li B-Z, Dale BE, Yuan Y-J. Comparative metabolic profiling revealed limitations in xylose-fermenting yeast during co-fermentation of glucose and xylose in the presence of inhibitors. Biotechnology and Bioengineering 2013:n/a-n/a.
25. Liu ZL, Slininger PJ, Dien BS, Berhow MA, Kurtzman CP, Gorsich SW. Adaptive response of yeasts to furfural and 5-hydroxymethylfurfural and new chemical evidence for HMF conversion to 2,5-bis-hydroxymethylfuran. J Indus Microbiol Biotechnol. 2004;31:345–52.
26. Liu ZL, Slininger PJ, Gorsich SW. Enhanced biotransformation of furfural and hydroxymethylfurfural by newly developed ethanologenic yeast strains. Appl Biochem Biotechnol. 2005;121–124:451–60.
27. Wiegert B, Klein C, Rizzi M, Lauterbach C, Dellweg H. Xylose fermentation by yeasts. 8. Influence of furfural on the aerobic growth of the yeast *Pichia stipitis*. Biotechnolgy Letters. 1988;10:895–900.
28. Hahn-Hagerdal B, Jeppsson H, Skoog K, Prior BA. Biochemistry and physiology of xylose fermentation by yeasts. Enzyme Microb Technol. 1994;16:933–43.
29. Syzbalski W, Bryson Y. Genetic studies on microbial cross resistance to toxic agents, I. Cross resistance of Escherichia coli to fifteen antibiotics. J Biotechnol. 1952;64:489–99.

14

An integrated approach for efficient biomethane production from solid bio-wastes in a compact system

Haoyu Wang[1,2,3*], Yu Tao[1,2], Margarida Temudo[4], Margot Schooneveld[4], Henk Bijl[4], Nanqi Ren[1], Monika Wolf[5], Cornelia Heine[5], Anne Foerster[5], Vincent Pelenc[5], Joris Kloek[4], Jules B van Lier[2,3] and Merle de Kreuk[2*]

Abstract

Background: Solid bio-wastes (or organic residues) are worldwide produced in high amount and increasingly considered bioenergy containers rather than waste products. A complete bioprocess from recalcitrant solid wastes to methane (SW2M) via anaerobic digestion (AD) is believed to be a sustainable way to utilize solid bio-wastes. However, the complex and recalcitrance of these organic solids make the hydrolysis process inefficient and thus a rate-limiting step to many AD technologies. Effort has been made to enhance the hydrolysis efficiency, but a comprehensive assessment over a complete flow scheme of SW2M is rare.

Results: In this study, it comes to reality of a complete scheme for SW2M. A novel process to efficiently convert organic residues into methane is proposed, which proved to be more favorable compared to conventional methods. Brewers' spent grain (BSG) and pig manure (PM) were used to test the feasibility and efficiency. BSG and PM were enzymatically pre-hydrolyzed and solubilized, after which the hydrolysates were anaerobically digested using different bioreactor designs, including expanded granular sludge bed (EGSB), continuously stirred tank reactor (CSTR), and sequencing batch reactor (SBR). High organic loading rates (OLRs), reaching 19 and 21 $kgCOD \cdot m^{-3} \cdot day^{-1}$ were achieved for the EGSBs, fed with BSG and PM, respectively, which were five to seven times higher than those obtained with direct digestion of the raw materials via CSTR or SBR. About 56% and 45% organic proportion of the BSG and PM can be eventually converted to methane.

Conclusions: This study proves that complex organic solids, such as cellulose, hemicellulose, proteins, and lipids can be efficiently hydrolyzed, yielding easy biodegradable/bio-convertible influents for the subsequent anaerobic digestion step. Although the economical advantage might not be clear, the current approach represents an efficient way for industrial-scale treatment of organic residues with a small footprint and fast conversion of AD.

Background

The growing worldwide energy demands and concomitant fossil fuels constraints have led to the decades' pursuit of alternative energy from renewable sources. Methane is an energy-rich component that is formed as the end product during the anaerobic decomposition of organic matter, such as domestic slurries and residues coming from food-processing manufactories. Among many different materials that can be used for biogas production, lignocellulose-rich materials, such as plant wastes, and protein-rich materials, such as animal manure, are highly promising due to their high methane potential [1-3].

It is estimated that the world lignocellulosic biomass fixes tenfold the solar energy amount per year compared to the total yearly energy demand of all humans [4]. Therefore, in principle, lignocellulosic biomass could play an increasingly important role in the world future energy production. Brewers' spent grain (BSG) is largely produced along with the increasing production of beer in recent years. About 15- to 20-kg BSG waste are generated from 1 hL of produced beer, and worldwide, about 1.85 billion hL of beer is produced annually [5]. Animal manure can be considered

* Correspondence: wanghaoyuwh@163.com; M.K.deKreuk@tudelft.nl
[1]State Key Laboratory of Urban Water Resource and Environment, Harbin Institute of Technology, 150090 Harbin, China
[2]Section of Sanitary Engineering, Department of Water Management, Delft University of Technology, 2628 CN, Delft, The Netherlands
Full list of author information is available at the end of the article

an even more important energy-containing organic waste, which is also largely produced worldwide. For example, China produces about two billion tons of livestock and poultry manure annually [6]. The number for USA has exceeded one billion tons since 2005 [7]. Anaerobic digestion (AD) of manure can improve the fertilizer value due to enhanced nutrient availability and reduction of the number of pathogens [8]. Moreover, application of manure digestion leads to the recovery of methane as energy source and, when properly applied, reduction of untended emissions. In Europe, AD is regarded the most favorable way for bioenergy production in terms of CO_2 emission reduction among the various biofuel production possibilities [9]. In addition, considerable amounts of nutrients such as nitrogen, phosphorous, and potassium are mineralized during AD, which can be subsequently reused for agriculture purposes [10].

A bottleneck of applying AD on plant and livestock wastes is the slow rate of hydrolysis because of the complex and recalcitrance of certain components in these materials. Both macroscopic-scale factors, such as tissue compositional heterogeneity and mass transfer limitations, and microscopic-scale factors, such as lignin-carbohydrate cross-linking and cellulose crystallinity, contribute synergistically to the recalcitrance [11]. Direct hydrolysis of this biomass by anaerobic hydrolyzing bacteria is inefficient and regarded as the rate-limiting step in the traditional AD processes [12]. Traditional methanization approaches, such as the use of continuous stirred tank reactor (CSTR), require long residence times in order to meet the slow (rate limiting) hydrolysis step and to prevent the loss of slowly growing microorganisms. A solution is to enhance the hydrolysis step by physicochemical pretreatment, breaking the crystalline structures and promoting access to enzymes for hydrolysis [13-15]. In addition, the solid residues left after pretreatment can be exposed to selective hydrolytic enzyme(s), yielding considerable amounts of protein, glucose, xylose, arabinose, and other compounds from cellulose and hemicellulose, as well as hydrolyzed proteins, into the liquid stream. An integrated approach combining such enzymatic hydrolysis and AD is a promising way for energy recovery from solid substrates. For example, Nkemka et al. performed the digestion of the hydrolysate from pretreated and pre-hydrolyzed wheat straw in an upflow anaerobic sludge blanket (UASB) reactors [16]. Some other previous studies focused more on applying enzymatic hydrolysis to commercial crops for a biorefinery purpose, such as switch grass and corn stover, for bioethanol production [13,17-19]. AD of hydrolysates, for instance, agricultural lignocellulosic wastes, and animal manure is often not considered in these studies.

In this study, two types of biomass, BSG and pig manure (PM), are used to produce methane through a novel approach. In this two-step approach, raw materials are hydrolyzed via a multienzyme pretreatment to convert the chemical oxygen demand (COD)-containing components into a soluble form. After this, the soluble COD is anaerobically converted in high-efficient expanded granular sludge bed (EGSB) reactors to harvest methane. A comprehensive assessment was performed to describe the feasibility, productivity, stability, and energy yield of the mentioned approach by comparing it to CSTRs and sequencing batch reactors (SBRs), which are two traditional anaerobic reactor designs. Our current study includes the research on COD yield after pre-hydrolysis, methane production, volatile fatty acid (VFA) accumulation and utilization, and the impact of salinity and pH.

Results

Enzymatic hydrolysis

Multistep enzymatic hydrolysis was applied to hydrolyze raw BSG and PM at a pilot scale (Figure 1). As the raw BSG and PM contain many proteins and lignocellulose, protease, cellulase, and hemicellulase were applied as specific enzymes for hydrolysis. The lysing enzyme was specially used in the PM hydrolysis process to dissolve the cell walls of microorganisms. There were three steps processing the raw BSG and PM: (1) thermochemical pretreatment, which broke down the structure of raw materials to increase the solubilization yield; (2) enzymatic hydrolysis, which was performed by various steps of enzymes under different pHs and temperatures; (3) filtration, which separated the liquid and solid fraction in the last step. The filtered liquid was used as influent (after dilution) for the anaerobic digestion setups and is addressed as hydrolysates. The elemental characterization of raw materials and BSG/PM hydrolysates are shown in Table 1.

Thermochemical pretreatment

The lignocellulosic materials were constructed by the lignin-carbohydrate complexes, in which the biodegradable cellulose and hemicellulose were partially blocked by lignin [20]. Thereof, an appropriate pretreatment is very important to enhance the conversion efficiency from lignocelluloses to saccharides. Temperature and pH are both critical parameters to a successful pretreatment before enzymatic hydrolysis [21,22]. In this study, a series of batch tests was designed (Figure 2a) to compare the effect of different times under different temperatures (4 h 70°C, 1 h 90°C, 4 h 90°C, and 20 min 120°C) and pH conditions (pH 1.5, 4.0, 6.6, and 11.5) on the hydrolysis efficiency of BSG, that is, solubilization yields. Results of solubilization yield (Figure 2b) clearly show that 40% to 50% of the organic dry matter can be solubilized at mild pH conditions (pH 4 and 6.6), whereas 60% to 70% can be solubilized at more extreme pH conditions (pH 1.5 and 11.5). The enzymatic hydrolysis contributes 10% to 20% to the 4-h 70°C pretreatment, while there was 15% to 40% more

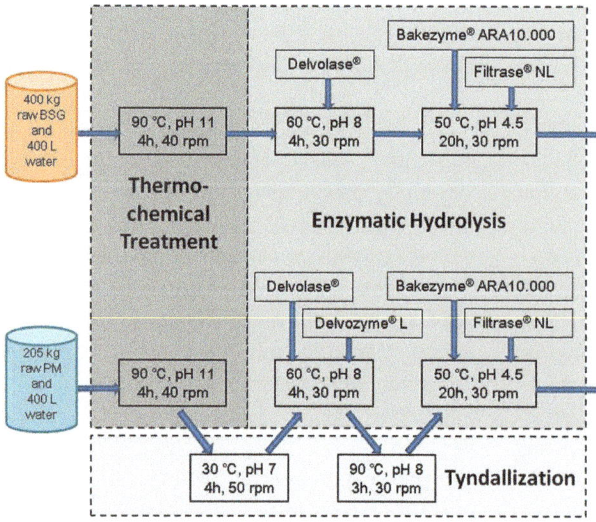

Figure 1 Integrated enzymatic hydrolysis process scheme of raw BSG and PM. Each of Bakezyme®, ARA10.000, and Filtrase® NL was mixed in a total volume of 8 kg solution. BSG, brewers' spent grain; PM, pig manure.

Table 1 Characterization of raw BSG, raw PM, and BSG/PM hydrolysates

Content (g · kg^{-1})	Raw BSG	Raw PM	BSG hydrolysates	PM hydrolysates
Dry matter	199	301	74	58
Organic dry matter	191	233	63	40
Ash	9	78	12	19
Protein	57	48	N.A.	7
Lipids	15	8	N.D.	N.D.
Lignin	29	88	N.D.	N.D.
Total carbohydrates[a]	84	67	15	2
Monomeric glucose	N.D.[b]	N.D.	4.59	0.87
Monomeric xylose	N.D.	N.D.	6.95	0.85
Monomeric arabinose	N.D.	N.D.	3.30	0.56
COD	108	N.A.[c]	100	41
TN	9.15	N.A.	3.54	2.77
Organic nitrogen	N.A.	N.A.	N.A.	1.17
Ammonia-N	0.05	5.00	0.24	1.60
Phosphorus	1.18	N.A.	0.48	0.82
Sulfur	0.59	N.A.	0.27	0.71
Sodium	0.02	N.A.	4.93	3.70
Chloride	0.01	N.A.	2.70	N.A.
Calcium	0.66	N.A.	0.18	2.23
Magnesium	0.36	N.A.	0.12	1.18
Potassium	0.06	N.A.	0.04	1.09

[a]Carbohydrates in raw BSG and raw PM are mainly polysaccharides and so no glucose, xylose, and arabinose monomers were detectable in this study despite of the fact that these sugars were present as building blocks in the polysaccharides. It is clear that all carbohydrates in the hydrolysates of BSG and PM were monosaccharides after degradation. [b]Not detectable. [c]Data not available. BSG, brewers' spent grain; PM, pig manure.

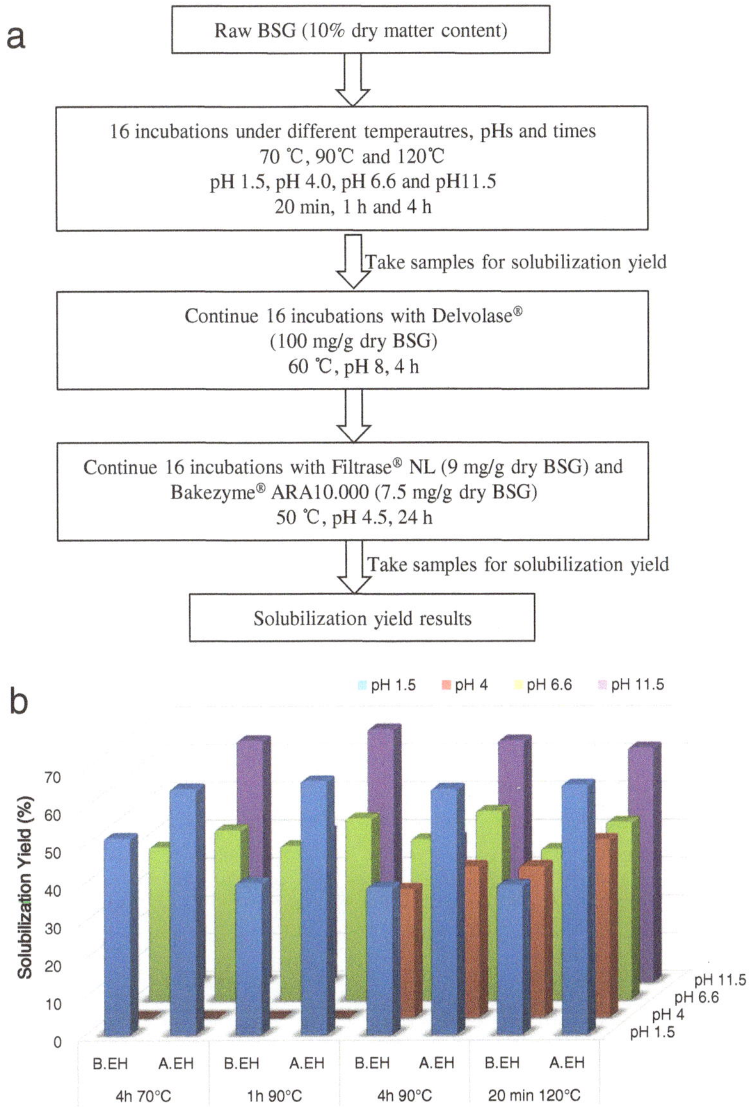

Figure 2 Comparison of solubilization yields under different combinations of temperatures, pHs, and times. (**a**) Scheme of solubilization yields of raw BSG enzymatic hydrolysis process. (**b**) Solubilization yields responding to different conditions. The results of test 4 h 70°C at pH 4 are not available. The solubilization yield was determined using the organic dry matter content of the supernatant and the total slurry after pretreatment (see 'Calculation of solubilization yield'). *N.A., data not available; A.EH, after enzymatic hydrolysis; B.EH, before enzymatic hydrolysis; BSG, brewers' spent grain.

solubilization derived from enzymatic hydrolysis at 4-h 90°C and 20-min 120°C pretreatments. The extreme pH conditions further benefit enzymatic hydrolysis with a higher improvement on solubilization yields compared to neutral pH conditions. The batch tests proved that the sole thermochemical method (for example, 70°C to 120°C, either pH <2 or pH >11) could contribute to a hydrolysis efficiency in a range of 34% to 52%, while a further enzymatic process can enhance this efficiency to 67%. Considering that an alkaline condition is optimal to the subsequent use of protease, the conditions of pH 11, 4 h, and 90°C were selected for the pilot-scale thermochemical pretreatment (Figure 1).

Enzymatic hydrolysis

Enzymatic hydrolysis of BSG Six hundred ten kilograms of liquid BSG hydrolysates was obtained out of a total weight of 800 kg of raw BSG after filtration (Figure 1). The obtained hydrolysates were used to feed the different anaerobic reactor systems, that is, the EGSB, CSTR, and SBR. After the enzymatic hydrolysis, about 70% of the organic compounds from the raw BSG, calculated based on COD, were present in the liquid hydrolysates and 30% was lost in the residual solid. The concentrations of the main components from the raw and liquid hydrolyzed BSG are listed in Table 1. Notably, both lipids and lignin were not detected in the BSG hydrolysates. Results indicate that more recalcitrant matter can be well separated from the hydrolysates, which can greatly benefit the downstream anaerobic digestion process.

Enzymatic hydrolysis of PM In total, 496 kg of liquid PM hydrolysates were obtained from raw PM (Figure 1). The dry matter and organic dry matter concentrations of liquid hydrolysates were only about 47% and 41% of the raw PM (Table 1). More than 91% of carbohydrates and 63% of proteins were converted during the hydrolysis process. The protein concentration decreased by 85%, and lipids and lignin were not detected anymore after the enzymatic hydrolysis. Similar to BSG, the coupling of enzymatic hydrolysis and filtration resulted in liquid section that contains less recalcitrance, likely enhancing the efficiency of anaerobic digestion process.

Methane yield

BMP test

The biological methane potential (BMP) test is used to show the potential methane yield of organic matter, following standardized protocols [23,24]. In our study, the BSG hydrolysates had the highest biogas-production potential value, reaching 810 $NmL \cdot gODM^{-1}$, followed by suspended BSG, which has a biogas-production potential value 680 $NmL \cdot gODM^{-1}$. The raw BSG has the lowest production potential value of 450 $NmL \cdot gODM^{-1}$, which is only 55% of the hydrolysate value. PM hydrolysates also had the highest biogas-production potential value, reaching 485 ± 24 $NmL \cdot gODM^{-1}$ compared to suspended PM and raw PM, which had a biogas-production potential value of 308 ± 10 $NmL \cdot gODM^{-1}$ and 153 ± 8 $NmL \cdot gODM^{-1}$, respectively.

Comparison of digestion performance between different reactor configurations

The performance of CSTR, SBR, and EGSB bioreactors was compared with the purpose to determine the most optimal reactor configuration for digesting hydrolysates.

The maximum organic loading rates (OLRs), that is, the ones that could be reached before reactor perturbation, are shown in Table 2. The CSTR results clearly show that pre-hydrolyzed BSG was methanized at two-fold higher OLRs compared to the raw BSG. Moreover, results also show that the EGSB reactors fed by hydrolysates were able to run in stable at OLRs as high as 11 $kgCOD \cdot m^{-3} \cdot day^{-1}$ within 3 months after startup and reach to 21 $kgCOD \cdot m^{-3} \cdot day^{-1}$ after 9-month acclimation. The EGSB reactor was characterized by the highest methane production rate, as well as the highest methane yield and the shortest applied hydraulic retention time (HRT). Apparently, the EGSB was the most efficient reactor for digesting BSG hydrolysates compared to CSTR and SBR. In addition to the BSG hydrolysates treatment, the EGSB was also most efficient when treating PM hydrolysates (Table 2). The maximum OLR for the EGSB

Table 2 Comparison of methane yield in different reactors treating BSG and PM

Substrate	Reactor type	Operational time (days)	OLR[a] ($kgCOD \cdot m^{-3} \cdot day^{-1}$)	OLR[a] ($kgODM \cdot m^{-3} \cdot day^{-1}$)	Methane yield ($L \cdot kgCOD^{-1}$)	Methane yield ($L \cdot kgODM^{-1}$)	Methane production rate ($mL \cdot L^{-1} \cdot day^{-1}$)	HRT (days)	Soluble COD removal (%)
BSG									
Raw BSG	CSTR (5.0 L)	120	4.0	3.0	153	205	617	15.0	N.A.[b]
Suspended BSG[c]	CSTR (5.0 L)	120	5.3	4.0	150	258	802	15.0	N.A.
BSG hydrolysates	CSTR (5.0 L)	120	6.3	4.5	196	295	1,224	10.0	82
BSG hydrolysates	SBR (5.0 L)	80	7.9	5.6	193	292	1,521	12.1	89
BSG hydrolysates	One-stage EGSB (3.8 L)	80	11.5	N.D.[d]	260	N.D.	3,079	2.5	88
BSG hydrolysates	Two-stage EGSB (3.8 L)	120	19.0	N.D.	253	N.D.	4,864	1.5	87
PM									
PM hydrolysates	SBR (5.0 L)	80	3.0	3.3	83	140	249	12.0	91
PM hydrolysates	EGSB (3.8 L)	280	21.0	N.D.	275	N.D.	5,456	1.5	93

[a]The OLR values were the maximum values that were achievable by each reactor, meanwhile the reactors were under stable operational under such OLR conditions. [b]Data not available. [c]The mixture of the solid and liquid fraction of hydrolyzed BSG. [d]Not detectable. BSG, brewers' spent grain; COD, chemical oxygen demand; EGSB, expanded granular sludge bed; OLR, organic loading rate; PM, pig manure.

was in this case seven times higher than the maximum OLR that could be applied to the SBR.

Methane yield from hydrolysates in EGSBs

Accumulating VFAs in reactor effluents is generally associated with instability of the AD process [25,26]. A rapid VFA increase might be followed by a subsequent period with low methane production rates [27]. Long-term VFA accumulation and concomitant lack of methane production can even cause a serious drop in pH and may lead to biomass washout and deterioration of the AD process. In our present study, the maximum applicable OLR was searched for, imposing rapid OLR increases to the system. As a consequence, VFA accumulation appeared in all EGSB reactors (Figure 3). The EGSB process stability could be easily recovered in all reactors after temporary

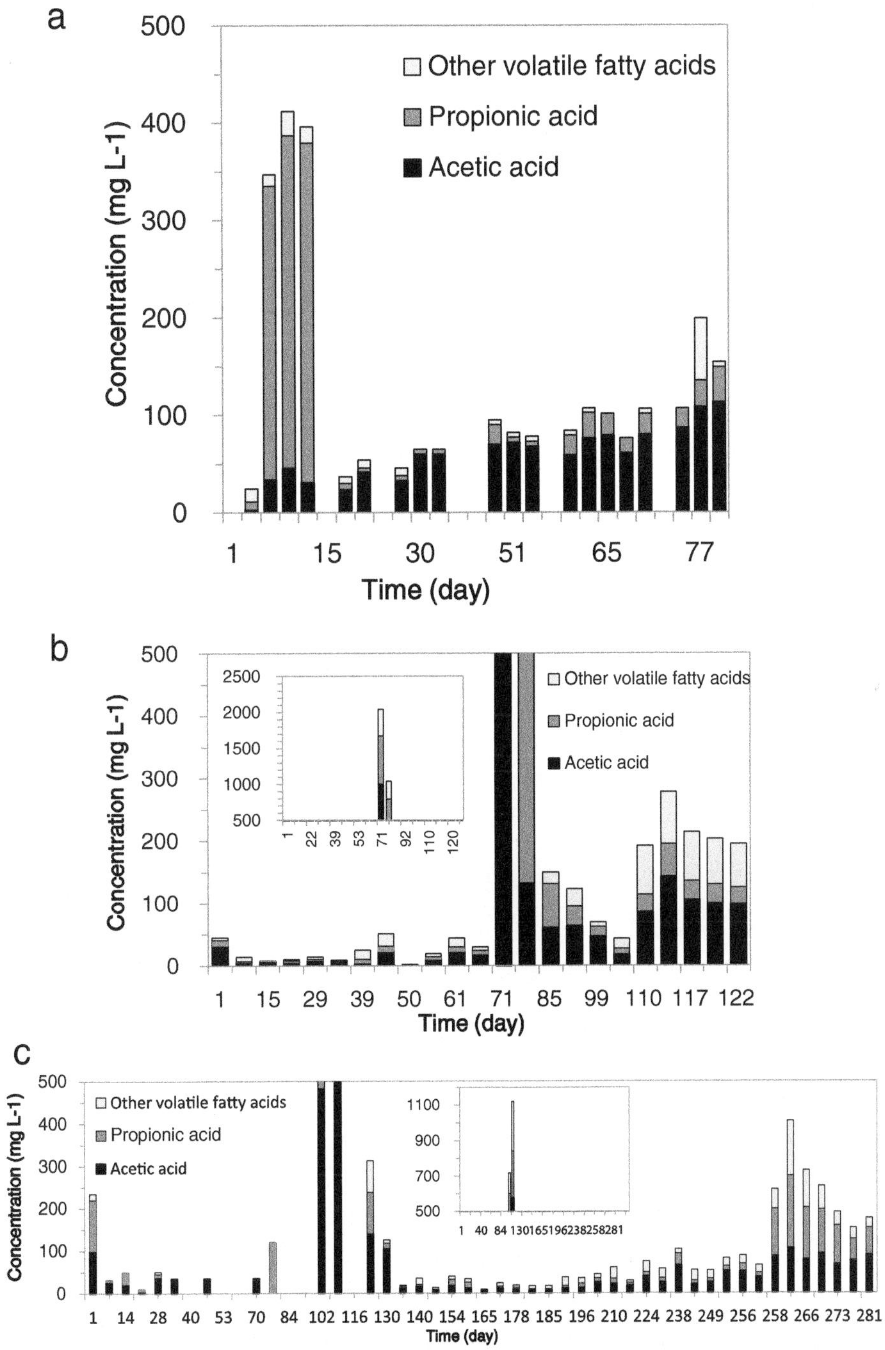

Figure 3 VFA concentrations in the one-stage BSG-EGSB (**a**), two-stage BSG-ESGB (**b**), and PM-EGSB (**c**).

OLR decrease, when VFAs was indeed observed in the effluent.

The average methane yield, that is, gCH_4-COD/ $gCOD_{depleted}$, for the one-stage BSG-EGSB, two-stage BSG-ESGB, and PM-EGSB was about 79%, 78%, and 81%, respectively. The methane yield of the one-stage BSG-EGSB fluctuated between 60% and 95% in the first month of operation (Figure 4b). A relatively low methane yield, that is, less than 60% was observed in the two-stage BSG-ESGB 1 week after the start-up, which also appeared after the high and sudden OLR increase at days 64 to 71 (Figure 4b). The contribution of the added enzymes to the overall COD of hydrolysates was approximately less than 3.5% and 4% for BSG and PM, respectively, while the released nitrogen only accounted for about 1.2% and less than 1% of the total nitrogen of the BSG and PM hydrolysates, respectively.

The seeding sludge for the three EGSBs was characterized by a high specific methanogenic activity (SMA) of 0.73 to 1.2 $gCOD_{CH4} \cdot gVSS^{-1} \cdot day^{-1}$. Nonetheless, reactors were started at low OLRs of 0.1 to 0.2 $gCOD \cdot gVSS^{-1} \cdot day^{-1}$, in order to facilitate the adaptation of microbial communities. The applied OLRs at the start of the continuous flow experiment were less than a quarter of the biogas producing capacity of the inoculum. Subsequently, OLRs were increased stepwise, according to the performance of each EGSB. The monitored indicators that were used for assessing the reactor stability were methane COD conversion efficiency and VFA accumulation. The inoculums could easily adapt to BSG hydrolysates, resulting in a more rapid increase in OLR applied to the BSG-EGSBs compared to the PM-EGSB (Figure 4a).

Large amount of aceticlastic methanogens were expected in the seeding sludge as it showed very high values in the SMA tests, in which acetate was used as the sole carbon source. In order to understand if the applied sludge loading rates (expressed as $gCOD \cdot gVSS^{-1} \cdot$

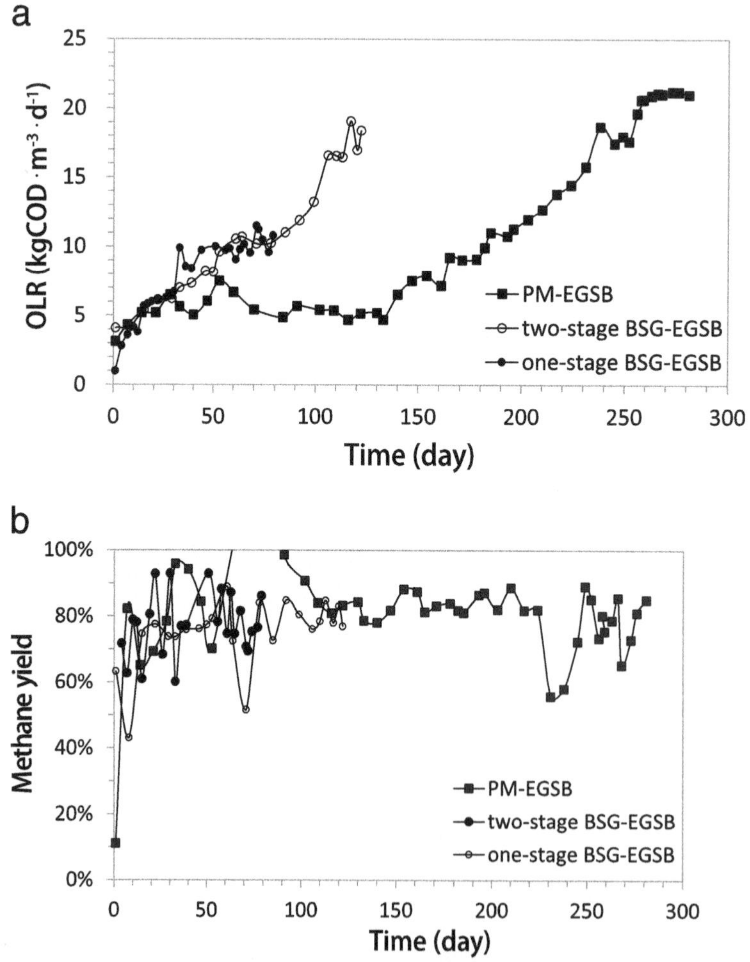

Figure 4 Organic loading rates (OLRs) and methane yields of the one-stage BSG-EGSB (white circle), two-stage BSG-ESGB (black circle), and PM-EGSB (black square). BSG, brewers' spent grain; EGSB, expanded granular sludge bed; PM, pig manure.

day^{-1} of biodegradable hydrolysates) were appropriate, SMA tests were applied to both BSG-EGSB and PM-EGSB sludge at different periods, more (Table 3). The results showed that the applied sludge loading rates were higher than the corresponding SMAs of the PM reactor sludge, but were close to the corresponding SMAs of the BSG reactor sludge, except for the starting period of the two-stage BSG-EGSB (Table 3). Interestingly, the observed seemingly *overloading* of PM biodegradable hydrolysates did not lead to unstable methane production or reactor perturbation. It is very likely because the sludge SMA values were assessed with acetate as the sole substrate, which missed the potential contribution of hydrogenotrophic methanogens. As a matter of fact, by combining the results of 454-pyrosequencing and real-time quantitative polymerase chain reaction (see Additional file 1: Supplementary Material), we found in the PM-EGSB that the quantity of hydrogenotrophic methanogens was sometimes two orders of magnitude higher than the amount of aceticlastic methanogens; accordingly, the relative abundance of hydrogenotrophic methanogens was also much higher than that of aceticlastic methanogens (Additional file 1: Table S1), while such differences were small or reversed in both one-stage and two-stage BSG-EGSBs. These results indicate that hydrogenotrophic methanogens very likely played a more important role in the PM reactor compared to the BSG ones. However, such metabolic route cannot be revealed by a standard SMA test because syntrophic associations between acetogens and hydrogenotrophic methanogens are not maximized when acetate is used as the sole carbon source in such tests. These results reminded us of the possible population shift after a period feeding of substrates such as PM hydrolysates, and such change in microbial community may lead to biased (or confusing) results from an unchanged analytical method.

VFA accumulation was observed in the one-stage BSG-EGSB immediately following the start-up (Figure 3a). However, there was no such VFA accumulation after the start-up of the two-stage BSG-ESGB. The more stable process performance might be attributed to a higher Archaea/bacteria ratio in the second (EGSB) stage of the two-stage process compared to a single-stage process. However, severe VFA accumulation was observed in the two-stage BSG-ESGB reactor with maximum total VFA concentrations reaching about 2.0 g · L^{-1} during the week of day 70 (Figure 3b), corresponding to a simultaneous drop in methane yield (Figure 4b). The PM-EGSB also experienced severe VFA accumulation with total VFA concentrations exceeding 1.0 g · L^{-1} between days 100 to 120 (Figure 3c). Slightly accumulating VFAs were observed in the PM-EGSB during days 218 to 237 (Figure 3c) followed by a VFA accumulation up to 0.4 g · L^{-1} as total VFA, from day 258 onwards.

Discussion

High-rate biomethanation of organic residues can be achieved by coupling a separate, enzymatic pre-hydrolysis step to a high-rate anaerobic reactor system. BSG and PM were selected to test the feasibility, productivity, stability, and energy yield of the current approach. Firstly, BSG and PM were treated by enzymatic hydrolysis, whereafter the

Table 3 Specific methane activity (SMA) of the EGSBs fed with BSG hydrolysates or PM hydrolysates during different periods

Date	SMA[a] ($gCOD_{CH4}$ · $gVSS^{-1}$ · day^{-1})	Biomass-based OLR ($gCOD_{hydrolysates}$ · $gVSS^{-1}$ · day^{-1})	Biomass-based OLR (g biodegradable $COD_{hydrolysates}$ · $gVSS^{-1}$ · day^{-1})	Overloading rate
One-stage BSG-EGSB				
Inoculum	0.73 ± 0.08	N.A.[b]	N.A.	N.A.
Day 79	0.49 ± 0.01	0.67	0.48	Not overloading
Two-stage BSG-EGSB				
Inoculum	0.73 ± 0.08	N.A.	N.A.	N.A.
Day 8	0.44 ± 0.01	0.20	0.14	Not overloading
Day 122	0.39 ± 0.02	0.63	0.45	15%
PM-EGSB				
Inoculum	0.73 ± 0.08	N.A.	N.A.	N.A.
Day 8	0.20 ± 0.06	0.47	0.43	115%
Day 105	0.22 ± 0.01	0.33	0.30	36%
Day 140	0.27 ± 0.03	0.42	0.39	44%
Day 263	0.32 ± 0.02	0.76	0.70	119%

[a]Measured with acetate as the substrate. [b]Data not available. BSG, brewers' spent grain; COD, chemical oxygen demand; EGSB, expanded granular sludge bed; OLR, organic loading rate; PM, pig manure.

solubilized hydrolysates was processed in anaerobic CSTRs, SBRs, and EGSBs for biogas production.

One of the critical parameters before/during enzymatic pretreatment is pH. Our results show that the extreme pH conditions of pH <2 or pH >11 could further enhance the enzymatic pre-hydrolysis, measured as solubilization yield. Results also showed that the solubilization yields at neutral pH conditions are quite limited. Our observations are in line with other studies reporting that extreme pH conditions contribute to yield considerable percentages of sugars from hemicelluloses and cellulose and favored the subsequent enzymatic hydrolysis [13,28,29]. Compared to acid pretreatments, a high pH condition could be more preferable because it requires lower temperatures [29] and is more efficient in removing lignin-like materials [30].

Solid waste streams are generally digested using large CSTR type of digester systems that are usually operated with hydrolysis as the rate-limiting step. Solubilizing these solid waste streams creates the possibility to use high-rate wastewater treatment reactors to digest the hydrolysates. In fact, all types of bioreactors can be used. Of the various high-rate systems, EGSB reactors are characterized by a very compact configuration, a small footprint, and an advanced gas/liquid/solid separation device. Moreover, an EGSB reactor is well accepted for industrial applications and, therefore, can be easily scaled up for methane recovery from BSG and PM hydrolysates. In our present research, the solubilized BSG hydrolysates showed a four- to fivefold higher biogas production efficiency in a two-stage EGSB system, compared to the direct treatment of suspended BSG hydrolysates and raw BSG solid waste.

A high treatment efficiency and a high methane yield on hydrolyzed BSG/PM are the two main observations of our current study. Firstly, the EGSBs were characterized by a stable treatment performance applying extreme OLRs, as high as 19 $kgCOD \cdot m^{-3} \cdot day^{-1}$ for pre-acidified BSG hydrolysates and 21 $kgCOD \cdot m^{-3} \cdot day^{-1}$ for PM hydrolysates. The applied OLR could be rapidly increased to 11.5 $kgCOD \cdot m^{-3} \cdot day^{-1}$ within 3 months for the one-stage BSG-EGSB, with a high and stable methane yield. Secondly, the methane yield of the EGSBs is about 80%, meaning that 80% of organic matters in BSG and PM hydrolysates are eventually converted to methane. Considering the high organic yield of the pre-hydrolysis process, that is, about 70% for BSG and 45% for PM, the overall methane yield from raw BSG and PM, including losses during pretreatment, were 56% and 36%, respectively. The enzymatic pretreatment saves considerable time for hydrolysis compared to direct biological hydrolysis by anaerobic bacteria; the solubilized hydrolysates are easier for acidogens to utilize, which will result in more compact anaerobic reactor systems with higher loading potentials than when applying direct digestion. Meanwhile, the current method can maximize the organic conversion efficiency, especially when pretreatment is further improved, minimizing COD losses.

In this study, ordinary anaerobic granules from a food-processing factory were used to inoculate all reactors. It is notable that the CSTR might perform better on digesting the raw BSG or PM if it was inoculated with some types of *more appropriate* seeding sludge. However, the success of our inoculation was a more representative confirmation to the proposed method than inoculating specialized sludge.

A potential downside of the used pH control conditions during pre-hydrolysis is the introduction of considerable amounts of Cl^- and Na^+ into the liquid hydrolysates. The hydrolysate characterization (Table 1) shows that the contents of sodium and chloride in the BSG hydrolysates liquor are 4.93 and 2.70 $g \cdot kg^{-1}$, respectively, which are much higher than their concentrations in raw BSG solids (0.02 and 0.01 $g \cdot kg^{-1}$, respectively). Also in PM hydrolysates, the measured salinity was very high, that is, 13 $g \cdot L^{-1}$, which is distinctly higher than generally found in anaerobic digestion reactors. High salinity, for example, exceeding 10 $g \cdot L^{-1}$, may negatively impact the anaerobic digestion process [31,32] and may lead to weak sludge granules [33,34]. The strength and size of these granules is essential for operating a high-efficiency anaerobic reactor system such as UASB and EGSB [35]. Considering these negative potentials, alternative acids and bases should be tested to overcome salinity-derived problems and even favor the subsequent anaerobic digestion process.

Another negative issue is the high mineral content in pig manure. It is notable that the calcium and magnesium concentrations in PM hydrolysates are 2.23 and 1.18 $g \cdot kg^{-1}$, which are 12.5 and 9.5 times higher than that in BSG hydrolysates. Bivalent cations can potentially precipitate inside reactors during anaerobic digestion, depending on operational temperatures and pH conditions. In our study, we observed such precipitates on the surface of PM-EGSB granules (Figure 5a) and the energy-dispersive X-ray (EDX) analysis proved that the major mineral elements of these precipitates were sodium and calcium (Figure 5b). The accumulation of precipitates inside an EGSB reactor may cause high total suspended solid (TSS) concentrations and low ratios of volatile suspended solids (VSS) to TSS.

Although this study proves that the concept of high-efficiency methanization of BSG and PM is technically feasible using this integrated approach, the enzymatic hydrolysis processes might not have apparent advantage from the economical perspective. An in-depth economical analysis on the selection of enzymes and operational optimization is necessary, and some alternatives in making this process more economically attractive are needed.

Conclusions

This paper presents a novel way to methanize two common organic residues, namely BSG and PM, at high

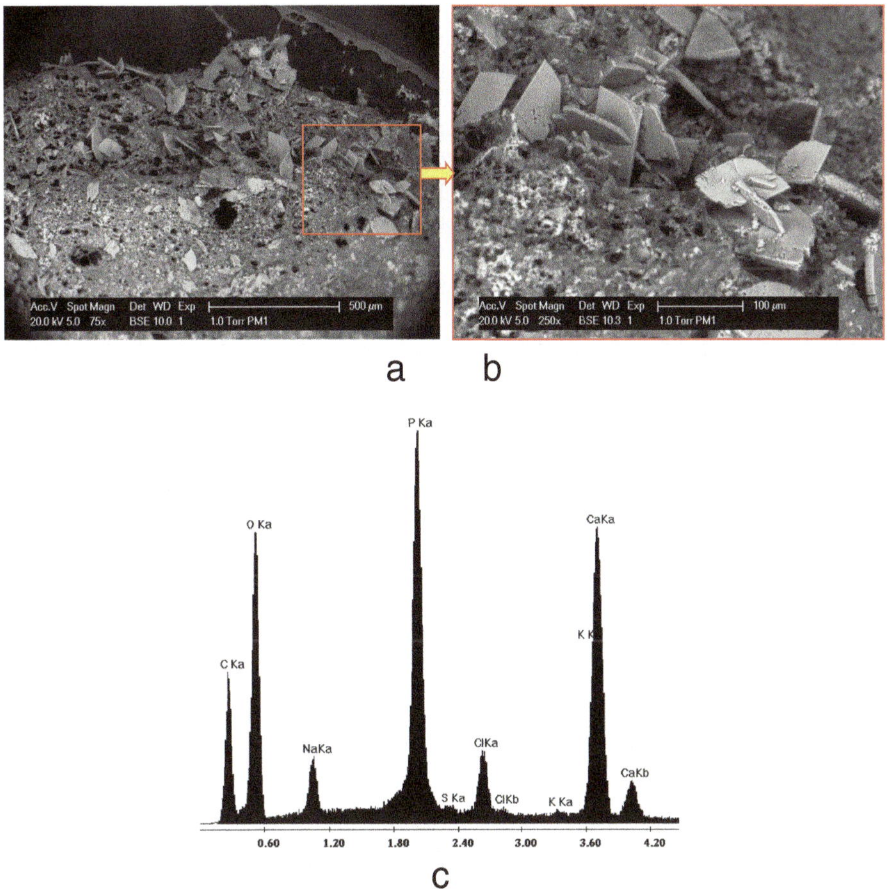

Figure 5 Scanning electron microscopic photo of a PM-EGSB granule (**a**, **b**) and energy-dispersive X-ray analysis (**c**) on the precipitates on the granule.

efficiency. Firstly, the enzymatic pretreatment helps to break down the rigid solid matrix and to convert the large-molecule organic matter, such as cellulose, hemicellulose, and protein, into small monomers, and the non-hydrolyzed residue, such as lignin, is separated from the hydrolysates by solid–liquid separation. The organic-rich hydrolysate liquor could be transformed into methane via anaerobic digestion in CSTR, SBR, or EGSB reactors. The whole process that is catalyzed by enzymes is shown to have a high yield and a high efficiency. In this proof of concept study, about 56% and 45% of the total organic matter from BSG and PM were eventually converted to methane. Solubilized hydrolysate-fed EGSBs performed stable at OLRs of 19 $kgCOD \cdot m^{-3} \cdot day^{-1}$ (BSG) and 21 $kgCOD \cdot m^{-3} \cdot day^{-1}$ (PM), which is five to seven times higher than conventional reactor systems and methods, applying CSTRs or SBRs directly fed with raw organic solids. Our study demonstrates that the separation of the enzymatic hydrolysis step and the methanation step provides an optimal control and selection of conditions to efficiently treat the organic solids. The proposed technology represents a promising technique for the industrial-scale treatment of organic solids with a high energy yield and a high efficiency but using only a small footprint of AD.

Materials and methods

Raw BSG and PM

Raw wet BSG was obtained from a brewery plant in The Netherlands. Raw BSG is a wet slurry, mainly consisting of organic dry matter, consisting of protein, lipids, lignin, and carbohydrates, and an inert ash fraction (Table 1). PM was obtained from a manure trader in The Netherlands. The compositions of raw PM and BSG are listed in Table 1. The main differences between the hydrolysates are the elevated ammonia nitrogen concentration in raw PM (5 $gNH_4^+\text{-}N \cdot kg^{-1}$), as well as the elevated salinity in PM.

Enzymatic hydrolysis

Enzymatic hydrolysis setup

The enzymatic hydrolysis of BSG and PM was carried out in a stainless steel tank with a working volume of 1,500 L. The reactor was equipped with a cooling/heating jacket, to control the temperature of the slurry (30°C to 95°C). The slurry was stirred with a variable speed anchor type mixer.

Enzymes

The enzymes that were used for the hydrolysis of raw PM and BSG were provided by DSM (Delft, The Netherlands). Delvolase® (DSM, Delft, The Netherlands) is an alkaline protease derived from bacteria *Bacillus licheniformis*. The optimum conditions of Delvolase® are pH 8 to 10 at 55°C to 60°C. Delvozyme® L is a purified enzyme extracted from egg white and is able to lyse cell walls of vegetated bacteria, which shares the same optimum conditions with Delvolase®. Filtrase® NL (DSM, Delft, The Netherlands) is a liquid fungal beta-glucanase, cellulase, and xylanase from *Talaromyces emersonii*. Bakezyme® ARA10.000 (DSM, Delft, The Netherlands) is a type of hemicellulase commonly used in bakery process. The optimum conditions for Filtrase® NL and Bakezyme® ARA10.000 match well at pH 4.5 around 50°C.

Enzymatic hydrolysis of BSG

The hydrolysis process of BSG is shown in Figure 1a. Firstly, a thermochemical pretreatment was applied to the raw BSG slurry to improve the solubilization yield of organic matters under the conditions of pH 10.7, 90°C for 4 h. And then several steps of enzymatic hydrolysis were applied under the optimum conditions of each enzyme. The pH was adjusted by adding NaOH/HCl. Delvolase® was applied in the first place, which can hydrolyze both native and denatured proteins after the thermochemical pretreatment. After 4 h, Filtrase® NL and Bakezyme® ARA10.000 were added together to break intricate 3D networks of cellulose and hemicellulose. Since the optimum pH of protease (Delvolase®) is higher, it was added prior to cellulase/hemicellulase to avoid too many pH adjustments. Twenty hours later, the pH of slurry was neutralized and then the slurry was filtrated by a multifilament cloth (Sefar Tetex, Bern, Switzerland) to get the liquid fraction for the subsequent AD reactors.

Enzymatic hydrolysis of PM

The raw PM contained many microorganisms (such as pathogens and spores), which could overgrow, utilizing the hydrolyzed carbohydrates and decreasing the biomethane potential of the hydrolysates. Hence, a tyndallization was applied to inactivate the microorganisms in PM (Figure 1b). Most cells were inactivated after the thermochemical treatment but spores can still survive. An incubation at pH 7 and 30°C for 4 h was applied to germinate the live spores, and another heat treatment was applied at 90°C for 3 h. Additionally, the pH of PM was adjusted to 3.5 at the end to prevent the growth of microorganisms. The total amount of active cells decreased from more than 1×10^8 CFU/mL in the raw slurry to less than 120 CFU/mL in the PM hydrolysates. The left hydrolysis process of PM was similar to BSG's process.

Anaerobic digestion setups

Continuously stirred tank reactors

Three identical CSTR reactors (CSTR1, CSTR2, CSTR3) had a working volume of 5 L. They were fed with raw BSG (CSTR1), suspended BSG (CSTR2), and solubilized BSG (CSTR3) hydrolysates, respectively. The operation temperature was 35 ± 1°C. The pH of all CSTRs was controlled in a range of 7.1 to 7.5. The HRT of CSTR1 and CSTR2 was set to 15 days and CSTR3 had an HRT of 10 days.

Sequencing batch reactor

Two identical SBRs had a working volume of 5 L. Solubilized BSG and PM hydrolysates were fed to each SBR. The operation temperature was 35 ± 1°C and the HRT was set to 12 days. The pH of the two SBRs was controlled between 7.2 and 7.7.

One-stage EGSB

Three identical glass-made EGSB reactors (manufactured by Louwers, Enschede, The Netherlands) with a working volume of 3.8 L were used in this study. Two EGSBs were fed with BSG hydrolysates (BSG-EGSB) and another one was fed with PM hydrolysates (PM-EGSB). The operational temperature was maintained at 35 ± 1°C by a water bath (Tamson Instruments, Bleiswijk, The Netherlands). Peristaltic pumps (Watson Marlow, Cornwall, United Kingdom) were used to feed the reactor with a constant influent flow rate to maintain the HRT to about 1.5 days. An effluent recirculation was used to supply a constant upflow velocity of $8\ m \cdot h^{-1}$. The biogas was collected via a three-phase separator from top of the EGSB, and the biogas flow was constantly measured by a milligas counter (type MGC-1 PMMA, Ritter, Schwabmünchen, Germany). Temperature, oxidation-reduction potential (ORP), and pH were online monitored by sensors (Mettler Toledo, Greifensee, Switzerland), and the pH was controlled to be always above 6.9 by adding 0.1 mM NaOH. LabView software was used to control the pumps and data collection.

Two-stage EGSB

A 1.8-L (working volume) pre-acidification bottle was set prior to one BSG-EGSB and the combined system was named two-stage BSG-EGSB. The sludge that was washed out from the one-stage BSG-EGSB was inoculated to the pre-acidification bottle. The HRT was controlled as 8 h by a peristaltic pump and the bottle was operated under ambient temperature (18°C to 24°C). A magnetic stirrer was used to mix the bulk liquid, and some biomass was inevitably pumped into EGSB along with the pre-acidified hydrolysates. The online monitor system for temperature, pH, and ORP was the same as the ones equipped for EGSBs.

Inoculum

Anaerobic granules from a full-scale upflow anaerobic sludge bed that treated potato processing wastewater (Germany) were inoculated to all the lab-scale anaerobic digestion reactors (CSTRs, SBRs, and EGSBs) and were used for the BMP tests.

Analytical methods

Basic analysis

TSS, total solids (TS), VSS, and volatile solids (VS) were measured twice per week according to standard methods [36]. COD, ammonia nitrogen and total nitrogen (TN) were measured by the corresponding testing kits (product numbers 1145410001, 114559, and 114763, Merck, Darmstadt, Germany) and spectrophotometer (Spectroquat TR420/NOVA60, MERCK, Merck, Darmstadt, Germany). Each analysis was performed in duplicate. The samples for soluble COD measurement were filtered through a 0.45-μm fiberglass filter (Spartan 30, Whatman, GE Healthcare, Buckinghamshire, United Kingdom) before analysis. The salinity was tested using a conductivity meter. VFAs (C2-acetic acid, C3-propionic acid, C4-butyric acid, iC4-isobutyric acid, C5-valeric acid, and iC5-isovaleric acid) were quantified by gas chromatography (GC, HP7890 Agilent Technologies, Palo Alto, CA, USA) equipped with a flame ionization detector. The GC was fitted with a capillary column (19095 N-123 HP INNOWX). The temperature of column, the injector port, and the detector was 70°C, 250°C, and 300°C, respectively. The carrier gas was nitrogen at a flow rate of 10 $mL \cdot min^{-1}$ and a split flow of 40 $mL \cdot min^{-1}$. The methane content in biogas was measured using a 7890A gas chromatograph (Agilent Technology, USA) with a thermal conductivity detector and a 45- to 60-mesh matrix molecular sieve 5A column (Sigma-Aldrich, St. Louis, MO, USA). Helium gas was the carrier gas at a flow rate of 30 $mL \cdot min^{-1}$. The temperature of the injection inlet, oven, and detector was 100°C, 60°C, and 105°C, respectively.

Calculation of solubilization yield

The solubilization yield was determined using the organic dry matter content of the supernatant and the total slurry after pretreatment, using the following equation:

$$\text{Solubilization Yield } \% = (ODM_S/ODM_T) \times F \times 100\%$$

in which ODM_S is the organic dry matter content of the supernatant (%), ODM_T is the organic dry matter content of the total slurry (%), *F* is the correction factor for the pellet volume, which is calculated as:

$$F = (1\text{–}ODM_T)/(1\text{–}ODM_S)$$

Characterization of liquid hydrolysates solution and EGSB effluent

Protein characterization was a combination of precipitation of proteins using trichloroacetic acid (TCA) to remove disturbing substances and allow determination of the protein concentration with the colorimetric Biuret reaction. The standardization was performed using bovine serum albumin (BSA). Carbohydrates and lignin content were determined as described in Sluiter et al. [37].

Sludge characterization

Specific methanogenic activity SMA was used to estimate the capacity of methanogenic microorganisms to convert acetate into CH_4 in the anaerobic system. In this study, the SMA of the EGSB sludge was determined using an Automated Methane Potential Test System (AMPTS, Bioprocess Control, Lund, Sweden). The synthetic media for the control group consisted of a mixture of macronutrients, trace elements, and phosphate buffer solution. This medium was also used for the SMA test itself, supplemented with sodium acetate ($2\ g \cdot L^{-1}$) as substrate. The inoculum amount was determined by setting an inoculum VSS to substrate COD ratio (I/S) of 2:1. Both the reactor sludge and control group were analyzed in triplicate. The interpretation of SMA was expressed as $kgCOD_{CH4}$ per $kgVSS_{sludge}$ per day, in order to compare with other studies.

BMP test The BMP was determined using 15-mL hydrolysates added to a 50-mL batch digestion vial. In two independent experiments, 4 replicates of 15-mL hydrolysates were incubated during 21 days at 39°C. Increasing gas pressure in the head space during digestion was measured once per day in the first week, each second day in the second week, and each third day in the last week. Normalized gas production in consideration of methane content that was measured by GC was calculated and compared to corresponding controls, which were the incubations without hydrolysates.

ESEM and EDX element analyzing system ESEM and EDX analysis were applied to observe the surface of PM-EGSB granules and analyze the elemental constitution of precipitates on such granules. The granule samples were freshly sampled and fixed by 2% (*v*/*v*) glutaraldehyde for 2 h before ESEM observation. For ESEM observation, the samples were mounted on a 1-cm^2 metal support and kept in place with adhesive tape and observed with a Philips XL30 Series ESEM (Philips, Amsterdam, The Netherlands). There was no need to apply a gold sputter coating on the sample, since by using the ESEM, there is a reduced build-up of static electricity. The EDMA-3 system (SUTW 3.3 EDX window and 128.0-eV EDX resolution) was applied to analyze the key elements of precipitates.

Additional file

Additional file 1: Supplementary materials.

Abbreviations
AD: anaerobic digestion; BMP: biological methane potential; BSG: brewers' spent grain; CFU: colony forming units; COD: chemical oxygen demand; CSTR: continuously stirred tank reactor; EDX: energy-dispersive X-ray; EGSB: expanded granular sludge bed; ESEM: environmental scan electron microscopy; GC: gas chromatograph; HRT: hydraulic retention time; ODM: organic dry matter; OLR: organic loading rate; ORP: oxidation-reduction potential; PM: pig manure; SBR: sequencing batch reactor; SMA: specific methanogenic activity; TN: total nitrogen; TS: total solids; TSS: total suspended solids; VFA: volatile fatty acid; VS: volatile solids; VSS: volatile suspended solids.

Competing interests
The authors declare that they have no competing interests.

Authors' contributions
HYW, YT, MT, JK, JBvL, and MdK designed the experiment. MS and HB developed the method for pre-hydrolyses, prepared the hydrolysates, and measured the composition of raw materials and hydrolysates. MdK designed the EGSB reactors. HYW and YT operated the EGSBs and did the measurements and analysis of this part. MW, CH, AF, and VP designed the experiments and operated the CSTRs and SBRs. YT applied microbial ecology experiments. HYW, YT, MT, MS, MW, VP, JK, JBvL, and MdK contributed to discuss and analyze the data. HYW, YT, and MdK wrote the paper and MT, MS, NQR, JK, and JBvL critically revised the paper. JK, JBvL, and MdK managed the project. All authors have read and approved the final manuscript.

Acknowledgements
The authors would like to express their gratitude for the PhD Fellowship awards provided by the China Scholarship Council to Haoyu Wang (No. 2011612233) and Yu Tao (No. 2011612235).

Author details
[1]State Key Laboratory of Urban Water Resource and Environment, Harbin Institute of Technology, 150090 Harbin, China. [2]Section of Sanitary Engineering, Department of Water Management, Delft University of Technology, 2628 CN, Delft, The Netherlands. [3]UNESCO-IHE Institute for Water Education, 2601 DA, Delft, The Netherlands. [4]DSM Biotechnology Center, 2600 MA, Delft, The Netherlands. [5]Biopract GmbH, 12489 Berlin, Germany.

References
1. Ding SY, Liu YS, Zeng Y, Himmel ME, Baker JO, Bayer EA. How does plant cell wall nanoscale architecture correlate with enzymatic digestibility? Science. 2012;338:1055–60.
2. DeMartini JD, Pattathil S, Miller JS, Li HJ, Hahn MG, Wyman CE. Investigating plant cell wall components that affect biomass recalcitrance in poplar and switchgrass. Energy Environ Sci. 2013;6:898–909.
3. Xie S, Wu G, Lawlor PG, Frost JP, Zhan X. Methane production from anaerobic co-digestion of the separated solid fraction of pig manure with dried grass silage. Bioresour Technol. 2012;104:289–97.
4. Perlack RD, Wright LL, Turhollow A, Graham RL, Stokes B, Erbach DC. Biomass as feedstock for a bioenergy and bioproducts industry: the technical feasibility of a billion-ton annual supply. US Dept Energy, US Depart Agr. 2005. 1–78. DOE Accession #ADA436753.
5. FAO: http://faostat.fao.org/site/636/DesktopDefault.aspx?PageID=636#ancor. 2011.
6. Liu ZG, Zhou XF, Zhang YL, Zhu HG. Enhanced anaerobic treatment of CSTR-digested effluent from chicken manure: the effect of ammonia inhibition. Waste Manag. 2012;32:137–43.
7. USDA. National Program 206: Manure and Byproduct Utilization. http://www.ars.usda.gov/research/programs/programs.htm?np_code=206&docid=13337 2008.
8. Xie S, Lawlor PG, Frost JP, Hu Z, Zhan X. Effect of pig manure to grass silage ratio on methane production in batch anaerobic co-digestion of concentrated pig manure and grass silage. Bioresour Technol. 2011;102:5728–33.
9. Van Foreest F. Perspectives for biogas in Europe. Report NG 70. Oxford: The Oxford Institute for Energy Studies; 2012. 54 p.
10. Pabon CP, Slingerland MA, vanLier JB, Rabbinge R. Anaerobic digestion as a key technology for biomass valorization: contribution to the energy balance of biofuel chains. In: Wellinger A, Murphy J, Baxter D, editors. The biogas handbook: science, production and applications. Cambridge, UK: IEA Bioenergy, Woodhead Publishing; 2013. Chapter 7: 166–188. ISBN: 978-0-85709-498-8.
11. Chundawat SPS, Donohoe BS, Sousa LD, Elder T, Agarwal UP, Lu FC, et al. Multi-scale visualization and characterization of lignocellulosic plant cell wall deconstruction during thermochemical pretreatment. Energy Environ Sci. 2011;4:973–84.
12. Nizami AS, Korres NE, Murphy JD. Review of the integrated process for the production of grass biomethane. Environ Sci Technol. 2009;43:8496–508.
13. Jae JH, Tompsett GA, Lin YC, Carlson TR, Shen JC, Zhang TY, et al. Depolymerization of lignocellulosic biomass to fuel precursors: maximizing carbon efficiency by combining hydrolysis with pyrolysis. Energy Environ Sci. 2010;3:358–65.
14. Gao X, Kumar R, Singh S, Simmons BA, Balan V, Dale BE, et al. Comparison of enzymatic reactivity of corn stover solids prepared by dilute acid, AFEX™, and ionic liquid pretreatments. Biotechnol Biofuel. 2014;7:71.
15. Yan L, Zhang L, Yang B. Enhancement of total sugar and lignin yields through dissolution of poplar wood by hot water and dilute acid flowthrough pretreatment. Biotechnol Biofuel. 2014;7:76.
16. Nkemka VN, Murto M. Biogas production from wheat straw in batch and UASB reactors: the roles of pretreatment and seaweed hydrolysate as a co-substrate. Bioresour Technol. 2013;128:164–72.
17. Jin MJ, Gunawan C, Uppugundla N, Balan V, Dale BE. A novel integrated biological process for cellulosic ethanol production featuring high ethanol productivity, enzyme recycling and yeast cells reuse. Energy Environ Sci. 2012;5:7168–75.
18. Cateto C, Hu G, Ragauskas A. Enzymatic hydrolysis of organosolv Kanlow switchgrass and its impact on cellulose crystallinity and degree of polymerization. Energy Environ Sci. 2011;4:1516–21.
19. Olcay H, Subrahmanyam AV, Xing R, Lajoie J, Dumesic JA, Huber GW. Production of renewable petroleum refinery diesel and jet fuel feedstocks from hemicellulose sugar streams. Energy Environ Sci. 2013;6:205–16.
20. Mussatto SI, Fernandes M, Milagres AMF, Roberto IC. Effect of hemicellulose and lignin on enzymatic hydrolysis of cellulose from brewer's spent grain. Enzyme Microb Technol. 2008;43:124–9.
21. Faulds CB, Robertson JA, Waldron KW. Effect of pH on the solubilization of brewers' spent grain by microbial carbohydrases and proteases. J Agric Food Chem. 2008;56:7038–43.
22. Pierre G, Sannier F, Goude R, Nouviaire A, Maache-Rezzoug Z, Rezzoug SA, et al. Evaluation of thermomechanical pretreatment for enzymatic hydrolysis of pure microcrystalline cellulose and cellulose from Brewers' spent grain. J Cereal Sci. 2011;54:305–10.
23. Angelidaki I, Alves M, Bolzonella D, Borzacconi L, Campos JL, Guwy AJ, et al. Defining the biomethane potential (BMP) of solid organic wastes and energy crops: a proposed protocol for batch assays. Water Sci Technol. 2009;59:927–34.

24. Pabón-Pereira CP, Castanares G. vanLier JB. An Oxitop® protocol for screening plant material for its biochemical methane potential (BMP). Water Sci Technol. 2012;66:1416–23.
25. Ho DP, Jensen PD, Batstone DJ. Methanosarcinaceae and acetate oxidising pathways dominate in high-rate thermophilic anaerobic digestion of waste activated sludge. Appl Environ Microbiol. 2013;79(20):6491–500.
26. Lee YH, Pavlostathis SG. Decolorization and toxicity of reactive anthraquinone textile dyes under methanogenic conditions. Water Res. 2004;38:1838–52.
27. Gao DW, Tao Y, An R, Fu Y, Ren NQ. Fate of organic carbon in UAFB treating raw sewage: impact of moderate to low temperature. Bioresour Technol. 2011;102:2248–54.
28. Gao XD, Kumar R, DeMartini JD, Li HJ, Wyman CE. Application of high throughput pretreatment and co-hydrolysis system to thermochemical pretreatment. Part 1: dilute acid. Biotechnol Bioeng. 2013;110:754–62.
29. Li H, Gao X, Demartini JD, Kumar R, Wyman CE. Application of high throughput pretreatment and co-hydrolysis system to thermochemical pretreatment. Part 2: dilute alkali. Biotechnol Bioeng. 2013;110(11):2894–901.
30. Gupta R, Lee YY. Investigation of biomass degradation mechanism in pretreatment of switchgrass by aqueous ammonia and sodium hydroxide. Bioresour Technol. 2010;101:8185–91.
31. Yang J, Spanjers H, Jeison D, van Lier JB. Impact of Na+ on biological wastewater treatment and the potential of anaerobic membrane bioreactors: a review. Cri Rev Environ Sci Technol. 2013;130301143417004. doi:10.1080/10643389.2012.694335
32. Mccarty PL, Mckinney RE. Salt toxicity in anaerobic digestion. J Water Pollut Con F. 1961;33:399–415.
33. Ismail SB, de La Parra CJ, Temmink H, van Lier JB. Extracellular polymeric substances (EPS) in upflow anaerobic sludge blanket (UASB) reactors operated under high salinity conditions. Water Res. 2010;44:1909–17.
34. Rinzema A, Vanlier J, Lettinga G. Sodium inhibition of acetoclastic methanogens in granular sludge from a UASB reactor. Enzyme Microb Technol. 1988;10:24–32.
35. Hulshoff Pol LW, de Castro Lopes SI, Lettinga G, Lens PN. Anaerobic sludge granulation. Water Res. 2004;38:1376–89.
36. APHA. Standard methods for the examination of water and wastewater. 18th ed. Washington D.C.,USA: American Public Health Association; 2005.
37. Sluiter A, Hames B, Ruiz R, Scarlata C, Sluiter J, Templeton D, et al. Determination of structural carbohydrates and lignin in biomass. Technical report NREL/TP-510-42618. 2011.

Exploitation of algal-bacterial associations in a two-stage biohydrogen and biogas generation process

Roland Wirth[1†], Gergely Lakatos[2†], Gergely Maróti[2], Zoltán Bagi[1], János Minárovics[4], Katalin Nagy[4], Éva Kondorosi[2], Gábor Rákhely[1,3] and Kornél L Kovács[1,3,4*]

Abstract

Background: The growing concern regarding the use of agricultural land for the production of biomass for food/feed or energy is dictating the search for alternative biomass sources. Photosynthetic microorganisms grown on marginal or deserted land present a promising alternative to the cultivation of energy plants and thereby may dampen the 'food or fuel' dispute. Microalgae offer diverse utilization routes.

Results: A two-stage energetic utilization, using a natural mixed population of algae (*Chlamydomonas* sp. and *Scenedesmus* sp.) and mutualistic bacteria (primarily *Rhizobium* sp.), was tested for coupled biohydrogen and biogas production. The microalgal-bacterial biomass generated hydrogen without sulfur deprivation. Algal hydrogen production in the mixed population started earlier but lasted for a shorter period relative to the benchmark approach. The residual biomass after hydrogen production was used for biogas generation and was compared with the biogas production from maize silage. The gas evolved from the microbial biomass was enriched in methane, but the specific gas production was lower than that of maize silage. Sustainable biogas production from the microbial biomass proceeded without noticeable difficulties in continuously stirred fed-batch laboratory-size reactors for an extended period of time. Co-fermentation of the microbial biomass and maize silage improved the biogas production: The metagenomic results indicated that pronounced changes took place in the domain Bacteria, primarily due to the introduction of a considerable bacterial biomass into the system with the substrate; this effect was partially compensated in the case of co-fermentation. The bacteria living in syntrophy with the algae apparently persisted in the anaerobic reactor and predominated in the bacterial population. The Archaea community remained virtually unaffected by the changes in the substrate biomass composition.

Conclusion: Through elimination of cost- and labor-demanding sulfur deprivation, sustainable biohydrogen production can be carried out by using microalgae and their mutualistic bacterial partners. The beneficial effect of the mutualistic mixed bacteria in O_2 quenching is that the spent algal-bacterial biomass can be further exploited for biogas production. Anaerobic fermentation of the microbial biomass depends on the composition of the biogas-producing microbial community. Co-fermentation of the mixed microbial biomass with maize silage improved the biogas productivity.

Keywords: Microalgae, Biogas, Biohydrogen, Algal bacterial co-culture, Metagenomics

* Correspondence: kovacs.kornel@brc.mta.hu
†Equal contributors
[1]Department of Biotechnology, University of Szeged, Közép fasor 52, H-6726 Szeged, Hungary
[3]Institute of Biophysics, Biological Research Center, Hungarian Academy of Sciences, Temesvári krt. 62, H-6726 Szeged, Hungary
Full list of author information is available at the end of the article

Introduction

Biomass utilization for energy generation is commonly regarded as a major contributor to the achievement of renewable energy production targets [1-4]. Energy carriers from biomass are currently predominantly produced through the use of terrestrial plants [5]. The intensive exploitation of land for the cultivation of crops destined for biofuel production, however, may exert a negative impact on the global supply and the price of food and feed [6].

The search for alternative biomass sources still continues. Economically and environmentally friendly solutions should be found. Huge energetic and biorefinery opportunities are offered by the conversion of solar energy via the use of photosynthetic microorganisms. Hence, the interest in photosynthetic microorganisms (and especially microalgae) is growing worldwide. The microalgae are a large and diverse group of microscopic, photoautotrophic, or photoheterotrophic organisms, which grow profusely in both salt and fresh natural waters [7]. Microalgae are able to double their biomass much faster than terrestrial plants, and they therefore produce more biomass per hectare than higher plants do [8]. The relatively small land area needed to cultivate microalgae may be arable or marginal land, which further decreases the competition for agricultural land and smothers the 'food or fuel' dispute [7]. Microalgae can be harvested practically all year round, hence improving the biomass production efficacy and eliminating numerous storage problems. Cultivation is possible in closed photobioreactors or in open ponds. Open systems are usually considered to be economical, while closed systems are more efficient from the aspect of biomass production and control of the cultivation parameters [9,10]; either concept may therefore be competitive in diverse applications [11]. Additional beneficial features of a microalgal biomass include versatility and the variety of utilization for energetic purposes such as biohydrogen (bioH_2), bioethanol, biodiesel, and biogas production [12-14], besides biorefinery applications [14-16].

The important properties of a microalgal biomass to be used in anaerobic digestion (AD) include high contents of lipids and/or carbohydrates and a lack of recalcitrant lignin [12]. The lipid and carbohydrate content amounts up to 50% of the biomass dry weight in some strains [10,17]. Research on the AD of algal biomass started more than 50 years ago [18]. Until recently, only a few studies followed up this line of research [19-24]. Levels of biogas productivity from various fresh and salt water algal strains have been compared under mesophilic conditions [25]. The biogas potential was found to depend strongly on the species and on the cell disruption method applied. The CH_4 content of the gas evolved from the microalgae was 7% to 13% higher than that from maize silage [25]. A closed-loop system to convert the algal biomass to biogas and electricity has been tested [26]. The microbial communities thriving in anaerobic digesters fed with algal biomass have not been investigated extensively. The archaeal community formed during microalgal fermentation was recently analyzed by next-generation sequencing [27].

Some microalgae, such as the most extensively studied green microalga *Chlamydomonas reinhardtii*, have the noteworthy ability to produce H_2 via a photosynthetic water-splitting reaction coupled with the dark hydrolysis of storage materials [28-30]. Sulfur deprivation becomes a standard method through which to switch the algal metabolism from photoautotrophy to dark heterotrophic H_2 generation. The two-step process during which the cells undergo major metabolic and biochemical changes demands considerable energy input both by the process operators and by the algae.

Naturally formed, mixed algal-bacterial microbial communities have been observed to have beneficial effects on algal growth [31-34]. The mutualistic relationship involves supplying the algae with important growth factors, notably vitamin B12, by the bacterial partner in exchange for organic nutrients [35-39]. Little is known about H_2 production by algal-bacterial systems [40]. A recent study proposed that by consuming the O_2 generated photosynthetically by the algae, the bacteria maintain an anaerobic environment suitable for algal bioH_2 production [41]. This may eliminate the need for the sulfur-deprivation step [28-30].

In this study, we modeled a two-stage biorefinery process, that is, H_2 production in the first stage by an algal-bacterial mixed biomass grown under nonsterile photoheterotrophic conditions, with biogas generation from the residual biomass in the second stage. The composition of the microalgal-bacterial mixture was monitored during the process by using next-generation DNA sequencing technology.

Results and discussion

H_2 production by the mixed algal-bacterial system

H_2 accumulated in the reactor headspace and concomitantly O_2 disappeared in time when a mixture of *Scenedesmus* sp. and *Chlamydomonas* sp. was cultivated under nonsterile conditions together with their natural mutualistic bacterial partners (AB + S culture), which consumed the O_2 produced by the algae. The results were compared with the H_2 evolution by a mixture of the pure cultures of the two microalgae supplemented with hydrogenase-deficient *Escherichia coli* cells (AE + S culture) and by sulfur-deprived, bacterium-free algal cultures (A-S culture) (Figure 1). Striking differences were observed in terms of accumulated H_2 yields and the commencement and duration of H_2 evolution.

In the headspace of the growing algal-bacterial culture, the O_2 level decreased from 21% to 4.5% in 12 h (Figure 1B). The low O_2 level allowed H_2 evolution by the algal biomass

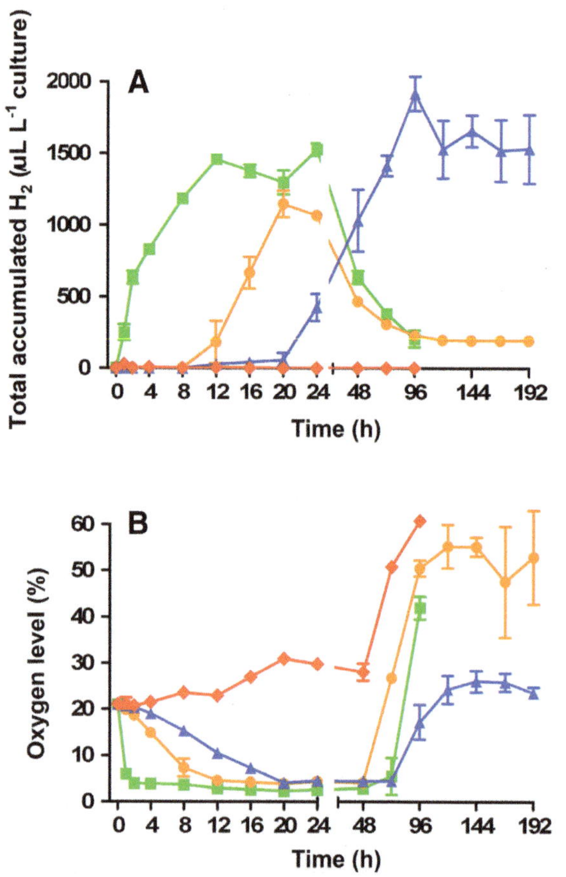

Figure 1 H_2 accumulation (A) and O_2 content (B) in the headspaces of the various cultures in time. Orange circles: mixed algal-bacterial co-culture (AB + S); green squares: algal-bacterial mixture with added *E. coli* ΔhypF (AE + S); blue triangles: sulfur-deprived bacterium-free co-culture of *Chlamydomonas* sp. and *Scenedesmus* sp. (A-S); red diamonds: bacterium-free co-culture of *Chlamydomonas* sp. and *Scenedesmus* sp. without sulfur deprivation (A + S).

after 8 h and 1.15 ± 0.09 mL H_2 L^{-1} was produced during the next 16 h, confirming earlier observations in similar systems (Figure 1A) [41].

The mutualistic bacteria were eliminated from the algal culture by photoautotrophic cultivation on minimal medium supplemented with rifampicin. H_2 production was not observed of the bacterium-free algal culture (A + S), because O_2 was not consumed by the mutualistic bacteria and the biosynthesis of the O_2 sensitive hydrogenases was repressed (Figure 1A,B). The facultative anaerobic wild-type *E. coli* tends to consume O_2 when it is available. Under anaerobic conditions, *E. coli* evolves H_2 by using its own hydrogenases [42]. In order to eliminate the contribution of H_2 production by *E. coli*, a pleiotropic hydrogenase mutant (Δ*hypF*) strain was used in these experiments so that only the facultative anaerobic property of this bacterium is functioning. Addition of *E. coli* Δ*hypF* cells and acetate to the pure algal culture (AE + S) efficiently reduced the level of O_2 from 21% to 4% in 2 h. Pronounced H_2 production accompanied this condition (1.52 ± 0.04 mL H_2 L^{-1}) (Figure 1A). The bacterial cell number in the spontaneously formed algal-bacterial culture (AB + S) was markedly lower than in the algal-*E. coli* Δ*hypF* co-culture (AE + S), which may explain why H_2 generation by the AE + S started earlier than without the O_2 scavenger *E. coli* strain (Figure 1). These data were compared with the H_2 production by the mixture of the pure algal strains using the photoheterotrophic TRIS-acetate-phosphate medium (TAP) and employing the sulfur-deprivation method [43,44]. The sulfur-deprived pure *Scenedesmus* sp. and *Chlamydomonas* sp. mixture (A-S culture) became anaerobic after 20 h as opposed to the 2 to 8 h in the case of AB + S and AE + S. H_2 evolution starts when anaerobic conditions are established; therefore, the difference in time required to reach anaerobicity is critical for the efficacy of the process. Additional benefits from practical aspect are the lower cost of alga production under nonsterile conditions and the elimination of labor- and cost-intensive transfer of algae into the sulfur-deficient medium.

The highest level of H_2 generation by the A-S (1.91 ± 0.12 mL H_2 L^{-1}) was reached after 4 days (Figure 1A), which exceeded the H_2 production of the AE + S culture only by about 20%. In view of the exceptionally thick cell walls of the *Scenedesmus* strains, the H_2 productivity may have been partly diffusion-limited in the mixed algal culture, which may explain the lower H_2 yield of A-S relative to the pure culture of sulfur-deprived *Chlamydomonas* sp. 549 strain (2.63 ± 0.04 mL H_2 L^{-1}) reported earlier [41]. Taken together, these experiments demonstrated that algal-bacterial natural mutualistic consortia are superior to the bacterium-free sulfur-deprived algal cultures from the aspect of H_2 evolution.

There are two possible reasons why the H_2 production ceased after about 24 h in the algal-bacterial co-cultures cultivated in TAP medium (see the 'Materials and methods' section). First, the H_2 yield depends on the H_2 partial pressure in a closed system [45]. Removal of the product H_2 from the headspace allows the extension of the production time, leading to sustainable H_2 evolution (data not shown). Secondly, in separate experiments, we have demonstrated that the depletion of acetate also results in a rapid loss of the mutualistic bacteria [41]. This can be remedied by the systematic addition of acetic acid to the system. Acetate is a low-value commodity produced in a number of anaerobic fermentative processes. The limiting factors of this bioH_2 production methodology appear to be relatively easy to overcome. H_2 production by algae under nonsterile conditions may make this approach economically viable on a large scale.

Biogas production from algal-bacterial mixed biomass

The levels of biogas production from the various biomass substrates were determined after a 1-month of start-up and stabilization phase, that is, in weeks 1 to 4 of the experiment. During this time, the reactors were fed with the AB + S substrate to ensure that all the remaining and digestible biomass from the inoculum (containing pig slurry and maize silage) had been degraded and did not contribute to the biogas formation. Gas production data were collected during weeks 5 to 9, when the evolved gas was produced from the AB + S biomass. Biogas generation from the algal-bacterial mixture was compared with co-fermentaion of the alga-rich biomass and maize silage, and reactors fed with maize silage were used as controls. The CH_4 concentration in the gas made from the AB + S biomass substrate was 58% to 61%, which is comparable to previous findings [25,26,46]. The biogas CH_4 content from maize silage alone was 50% to 52%, as found previously [47]. The co-fermentation of algal-bacterial biomass with maize silage, in a ratio of 1:1, on the basis of organic dry matter (oDM), yielded a CH_4 content of 54% to 57%, an intermediate value between those for maize silage and the algal-bacterial biomass. The daily average generated biogas volumes were as follows: from maize silage 3.20 L day^{-1}, from co-fermentation 3.15 L day^{-1}, and from algal-bacterial mixture 2.20 L day^{-1}. Figure 2 shows the specific average CH_4 production values (mL) calculated for g oDM^{-1}.

For the appreciation of the potential value of the AB + S biomass as biogas substrate, its advantages relative to the widely used maize silage have to be taken into account. Most importantly, the AB + S biomass can be cultivated under nonsterile conditions on lands not useful for agricultural production and can be continuously harvested during extended period of the year. Although several technical issues related to the large-scale production of AB + S biomass for energetic purpose remain to be elaborated, this material may effectively replace a large portion of maize silage in the biogas reactors.

VOAs/TAC ratio indicated stable operation

The ratio of the volatile organic acids (VOAs) and the total alkaline capacity (TAC) is an appropriate measure of the functional stability of the anaerobic digestion process [48,49]. A VOAs/TAC ratio below 0.1 means that the reactor needs feeding, whereas at a ratio ≥0.5 the biomass input is excessive and the process is out of balance. During the experiments, the average content of VOAs was 1.5 g L^{-1} and the average TAC was between 9 and 10 g $CaCO_3$ L^{-1} in all cases. Figure 3 shows the weekly measured VOAs/TAC ratios.

A constant value of VOAs/TAC is a reliable indicator of a stable fermentation process. The organic loading rate was on the low side and allowed stable and balanced operation.

NH_4^+ accumulation

From the decomposition of nitrogen-containing compounds, ammonia (NH_3) is formed, which is present in the aqueous medium in the form of ammonium ion (NH_4^+) [50]. Values above 3,000 mg NH_4^+ L^{-1} may have a negative effect on the methanogenic community [51,52]. During the anaerobic fermentation, slight fluctuations in the weekly NH_4^+ concentrations were observed. In the case of using the algal-bacterial mixture, the NH_4^+ content

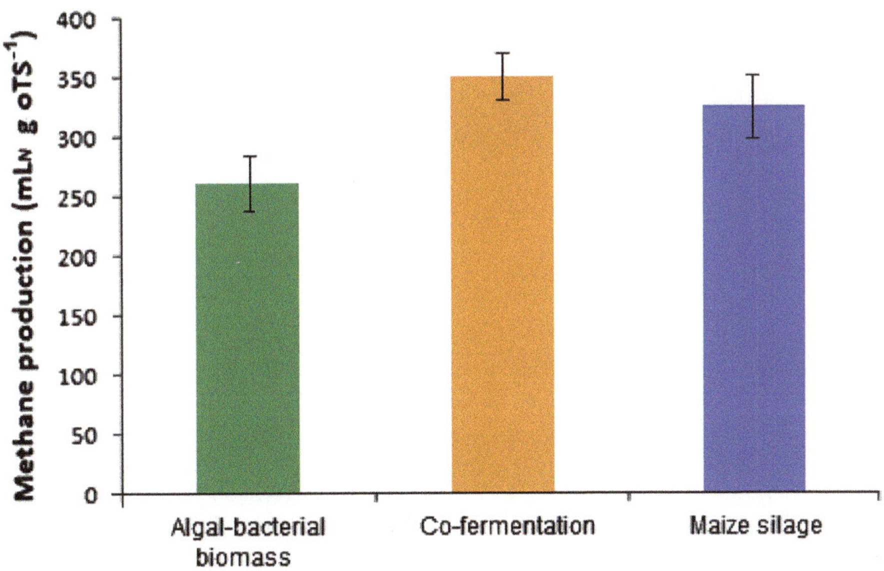

Figure 2 Specific CH_4 production from the various biomasses.

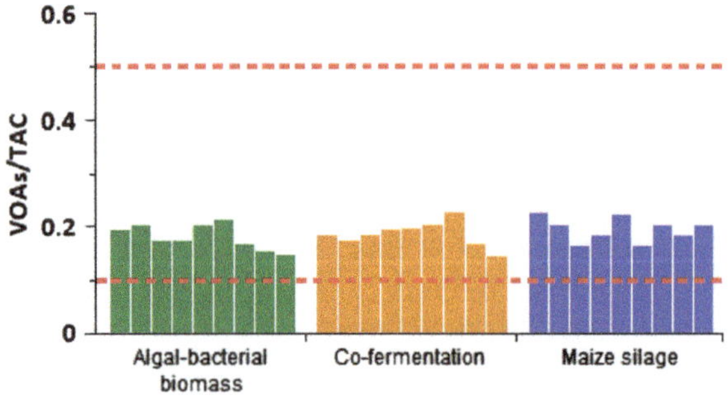

Figure 3 Weekly measured VOAs/TAC ratios. The area between the dashed red lines indicates the optimum range.

tended to increase but remained under the critical 3,000 mg NH_4^+ L^{-1} level (Figure 4). Co-fermentation efficiently balanced this elevated NH_4^+ level.

The effect of the C/N ratio

The ideal C/N ratio for AD is 20 to 30 [53,54], because the microbes in the anaerobic reactor can utilize carbon (C) 20 to 30 times faster than nitrogen (N) [54]. The risk of C starvation increases if the C/N ratio is lower than 20; the methanogens are inhibited by the high NH_3 accumulation, making the AD process vulnerable. At the other end of the spectrum, if the C/N ratio exceeds 30, the concentration of volatile fatty acids escalates, leading to process inhibition. The C/N ratios of the substrates used in this work are presented in Table 1. During the fed-batch continuous AD of microalgae and their mutualistic bacterial flora (AB + S), the nitrogen content increased. The initial C/N ratio of the AB + S biomass was low, 5.3. The nitrogen content increased as the fermentation progressed (Figure 5), accompanied by a slight but persistent free N concentration increase. Co-fermentation of the algal-bacterial biomass with maize silage, which had a C/N ratio of 45.3, led to a less pronounced N accumulation, indicating a buffering effect of the maize silage. In the reactors fed with maize silage alone, the N level remained nearly constant (Figure 5).

Olsson *et al.* reported that feeding AD reactors with a high proportion of microalgal biomass in co-fermentation with waste water sludge had a negative effect under both thermophilic (55°C) and mesophilic (37°C) conditions, possibly because of the high N content of the microalgal biomass [55]. Co-fermentation of a microalgal biomass with waste paper improved the AD performance [56], presumably in consequence of the higher C/N ratio of the mixed substrate and the induction of cellulase biosynthesis by the paper sludge. In our case, co-fermentation of the algal-bacterial biomass with the cellulose-rich maize silage likewise enhanced the biogas productivity.

Microbial community

The composition of the microbial community was established at four time points: at the start of feeding with the selected substrate (start), 1 week later (week 1), when the system was working at full capacity (week 5), and at the end of the process (week 9). The microbial community compositions of the substrates were determined separately.

Microbiological compositions of the substrates

The microbial flora of the maize silage included representatives of the genera *Lactobacillus* and *Acetobacter*, as expected (Figure 6A). Lactobacilli produce lactate from mono- and disaccharides [57]. Upon ensilaging, the accumulating acid decreases the pH and preserves the green plant material. Members of the genus *Acetobacter* primarily contribute to acetate production [58].

The mixture of *Chlamydomonas sp.* and *Scenedesmus sp.* microalgae was cultivated under nonsterile conditions and contained copious amounts of the mutualistic bacteria (Figure 6B). *Rhizobium* species predominated in the bacterial population. *Rhizobium* is well known for its syntrophic interaction with plants and mutualism has also been observed in the cases of several microalgal

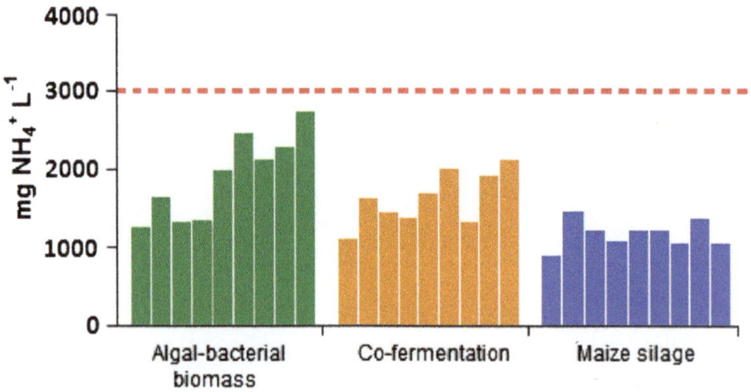

Figure 4 Weekly measured NH_4^+ concentrations. The dashed red line indicates the highest value recommended by the various studies.

Table 1 The initial substrate compositions

Substrate	Wet mass N (mg g^{-1})	Wet mass C (mg g^{-1})	C/N ratio	TS (%)	oDM (%)
Maize silage	4.35	196.86	45.3:1	41.19	94.59
Algal-bacterial mix	18.65	98.33	5.3:1	30.30	97.71

TS = total solids, oDM = organic dry material.

species [36,39]. The major probable driving force behind this association is vitamin B_{12}, which the algae needs for growth but cannot synthesize. *Rhizobium* is there to supply the algae with vitamin B_{12} in exchange for fixed carbon. The growth rate and the resistance to environmental stresses improve as a result of the algal-bacterial interactions [36,39]. Other forms of mutualism between microalgae and bacteria have also been recognized [31-34].

The biogas-producing microbial community

The distribution of the microbial taxa in the biogas-producing microbial community at the beginning of the experiments was very similar to that found in earlier studies on reactors fed with pig manure and maize silage [59], in good agreement with starting the reactors with inocula from a mesophilic industrial biogas facility digesting such substrates. These results may therefore be regarded as an internal control validating the metagenome sequencing method. In the following detailed analysis of the metagenomic results, the unidentified sequences are disregarded.

Microbial community of maize silage AD (domain Bacteria)

Only relatively minor and trivial rearrangements occurred in the relative distribution of the bacterial taxa during the experimental period (Figure 7). This is not surprising in view of the fact that the reactors were sustained on pig manure and maize silage prior to the start of the experiment. In the domain *Bacteria*, the most abundant strains belong in the phylum *Firmicutes*. Pronounced changes were seen in the phylum *Proteobacteria*. Some of the *Proteobacteria* were apparently displaced by *Firmicutes* and *Bacteroidetes*. In the phylum *Firmicutes*, the orders *Clostridiales* and *Bacteroidales* predominated (Figure 8). Among the *Clostridiales*, the genus *Clostridium* increased in abundance, followed by the genus *Bacillus*. In the order *Bacteroidales*, the genus *Bacteroides* predominated (data not shown).

Microbial community of co-fermentation (domain Bacteria)

Co-fermentation of the algal-bacterial mixture with maize silage provoked major changes in the composition of the bacterial community within a week as compared with the AD of maize silage (Figures 7 and 8). At the starting time point, there was no difference between the reactors fed with the various substrate compositions, indicating that the same microbial community was established during the start-up phase and the initial conditions were therefore identical. Supplying the reactors with a 1:1 mixture of microbial biomass and maize silage instigated a rearrangement within the biogas-producing microbial community. Representatives of the phylum *Proteobacteria* gradually predominated in the community, and within the taxon, the orders *Rhizobiales* and *Burkholderiales* prevailed (Figure 8). At higher resolution, a marked accumulation of the genera *Rhizobium* and *Burkholderia* was evident as the experiment progressed, although the phylum *Proteobacteria* displayed a diverse representation at the start. At the same time, members of the phylum *Firmicutes* and to a lesser degree those belonging to the phylum *Bacteroidetes* lost their significance within the AD community. The majority of bacteria belonging in these taxa have gained a reputation as outstanding cellulose degraders and H_2 producers, both of these metabolic activities being crucial for efficient biogas production from plant biomass.

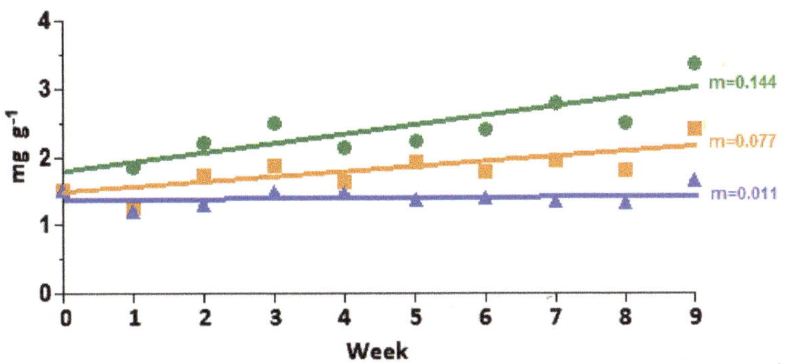

Figure 5 Changes in N content during the AD of various substrates. Green: AB + S, orange: co-fermentation, blue: maize silage.

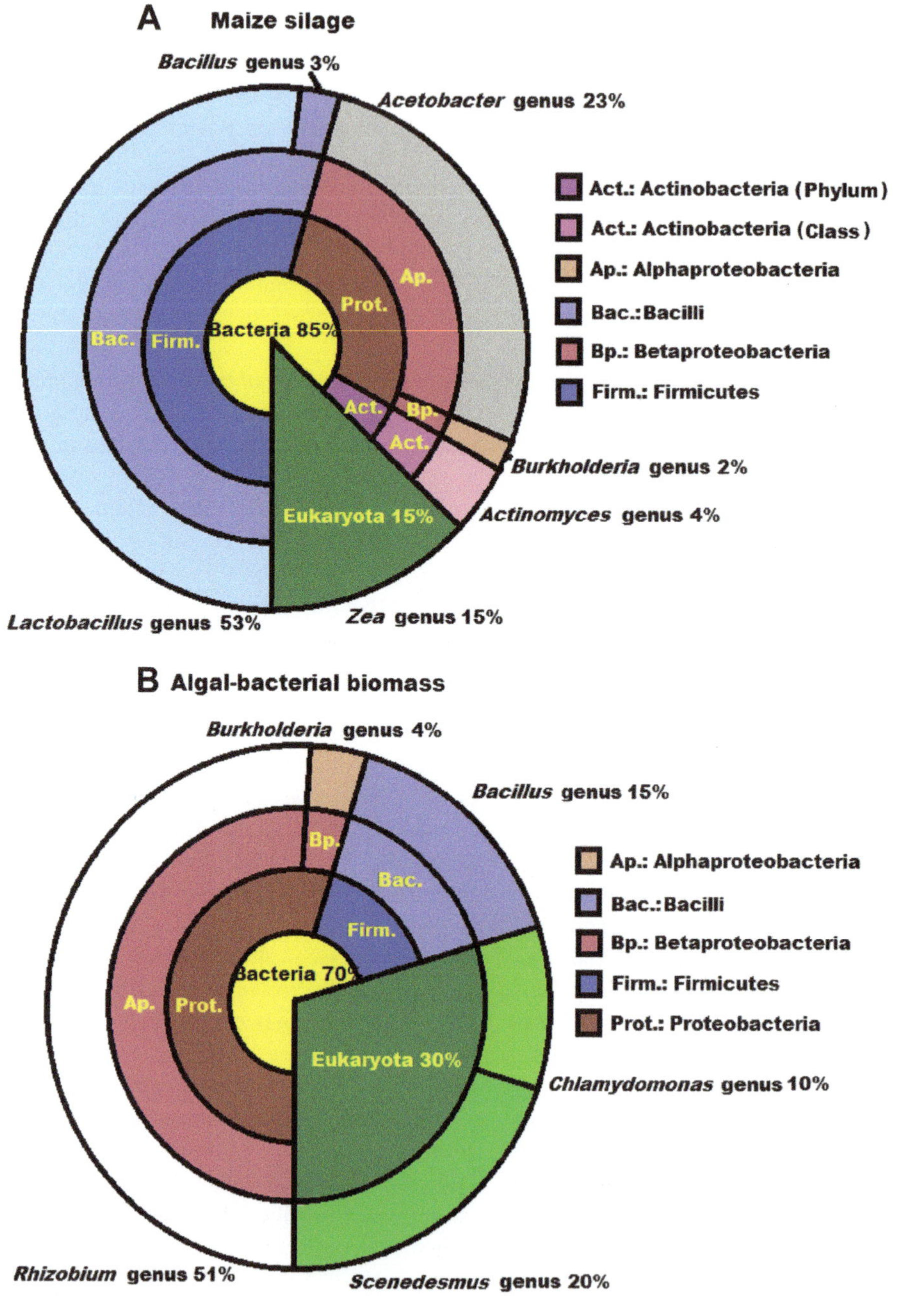

Figure 6 Microbial compositions of the substrates: (A) Maize silage, (B) AB + S. The communities at domain, phylum, class, and genus levels are indicated.

Microbial community of microalgal-bacterial fermentation (domain Bacteria)

A noteworthy fast response by the biogas-producing microbial community was observed when the substrate added to the reactors was changed from the mixture of pig slurry and maize silage to the algal-bacterial biomass. The reaction was less pronounced, but similar when the reactors were fed with a 1:1 mix of plant and microbial biomasses, as discussed above. The main outcome of this reorganization was the predominance of the phylum *Proteobacteria*, which surpassed the phyla *Firmicutes* and *Bacteroidetes*. The genera *Rhizobium* and *Burkholderia* were introduced into the reactor with the substrate (Figure 6) and accumulated in time (Figure 8), in spite of the relatively low daily organic loading rate. Either the decomposition was too slow to convert the total administered bacterial biomass to biogas, or the *Rhizobia* multiplied faster than their anaerobic degradation. *Rhizobium* species survive in free living form under anaerobic

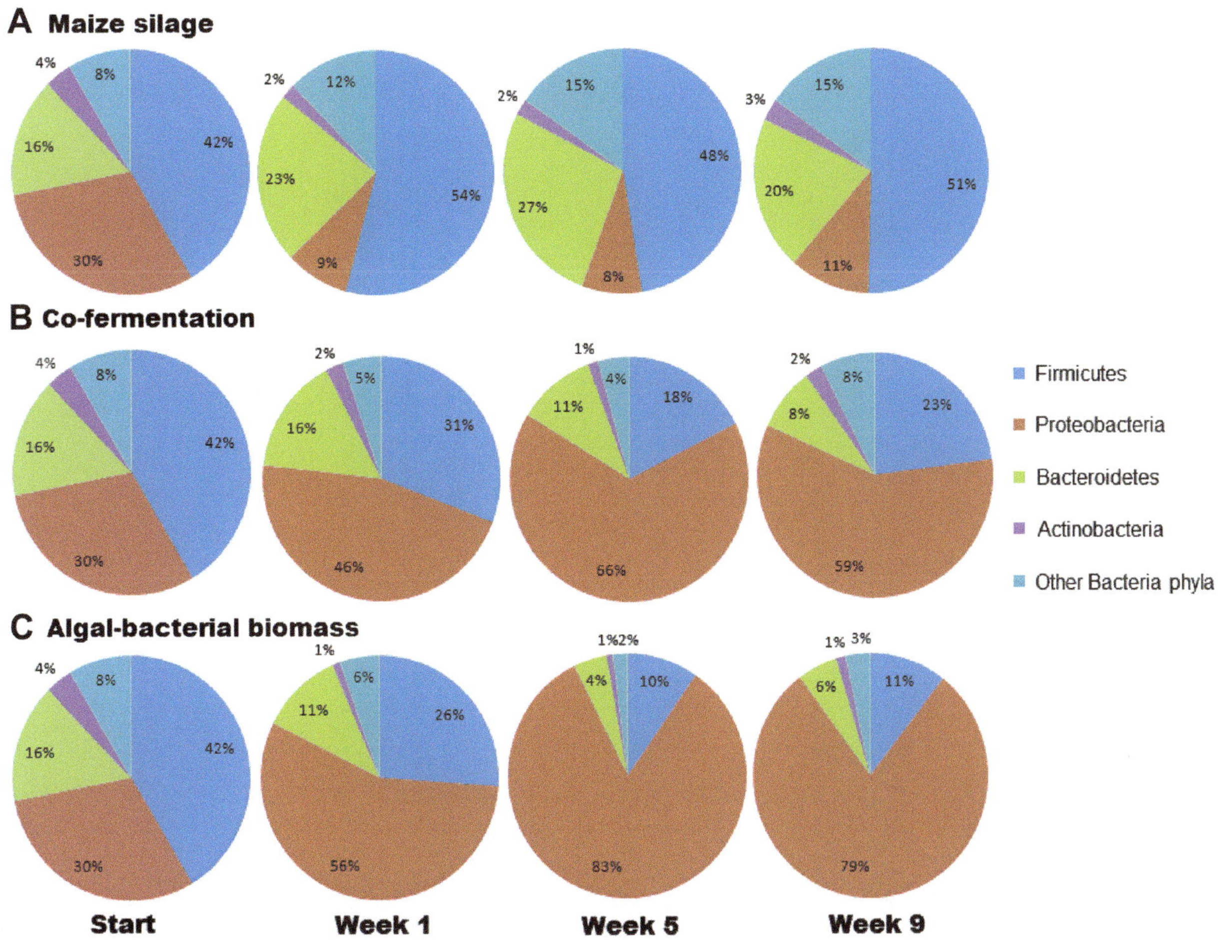

Figure 7 Changes in the domain Bacteria of the microbial community at phylum level. (A) Maize silage, **(B)** co-fermentation, and **(C)** algal-bacterial biomass.

conditions, taking advantage of their nitrate respiration capability [60,61], but it is unlikely that their growth rate exceeds that of the anaerobic degradation by the biogas microbial community. The substrate was stored at −20°C for about 3 months before being fed into the reactors. It seems likely that the build-up of *Proteobacteria* in time is due to their relatively slow decomposition under the AD conditions. In this respect, it is noteworthy that the relative abundance of eukaryotic sequences in the reactors also increased in time (Figure 9). The eukaryotic DNA accumulation from the algal biomass was twice that from the maize, suggesting that the algal cell wall may be more resistant than that of the maize silage to microbial degradation.

This implies that the biogas potential of the algal biomass is higher than that of the bacterial biomass, although a correct mass balance is difficult to achieve because of the complexity of the organic materials in the reactor.

The domain Archaea

In the domain *Archaea*, a microbial composition was found that was distinct from those observed in previous studies in reactors fed with 'conventional' substrates [59,62-66]. The class *Methanomicrobia* represented the domain *Archaea* in great abundance. The *Methanomicrobia* are able to operate all three routes of methanogenesis [67]. The order *Methanomicrobiales* was the most prevalent from the start, and at higher resolution, the members of the genus *Methanosarcina* predominated. Seasonal changes or other uncontrolled factors may also be responsible for these alterations in the AD communities [68-70]. At any rate, the genus *Methanosarcina* remained predominant in all fermentations tested in this study (Figure 10). Interestingly, in a previous study, involving the use of next-generation sequencing *mcrA* genes, the order *Methanosarcinales* was also found to be predominant in the AD of a mixture of waste water sludge and a nonsterile, unidentified algal biomass [27]. In the Archaeal community converting that substrate to biogas, the acetotrophic genus *Methanosaeta* (order *Methanosarcinales*) was identified as the prevailing taxonomic unit. Members of the genus *Methanosaeta* were present in our study too, although in less abundance.

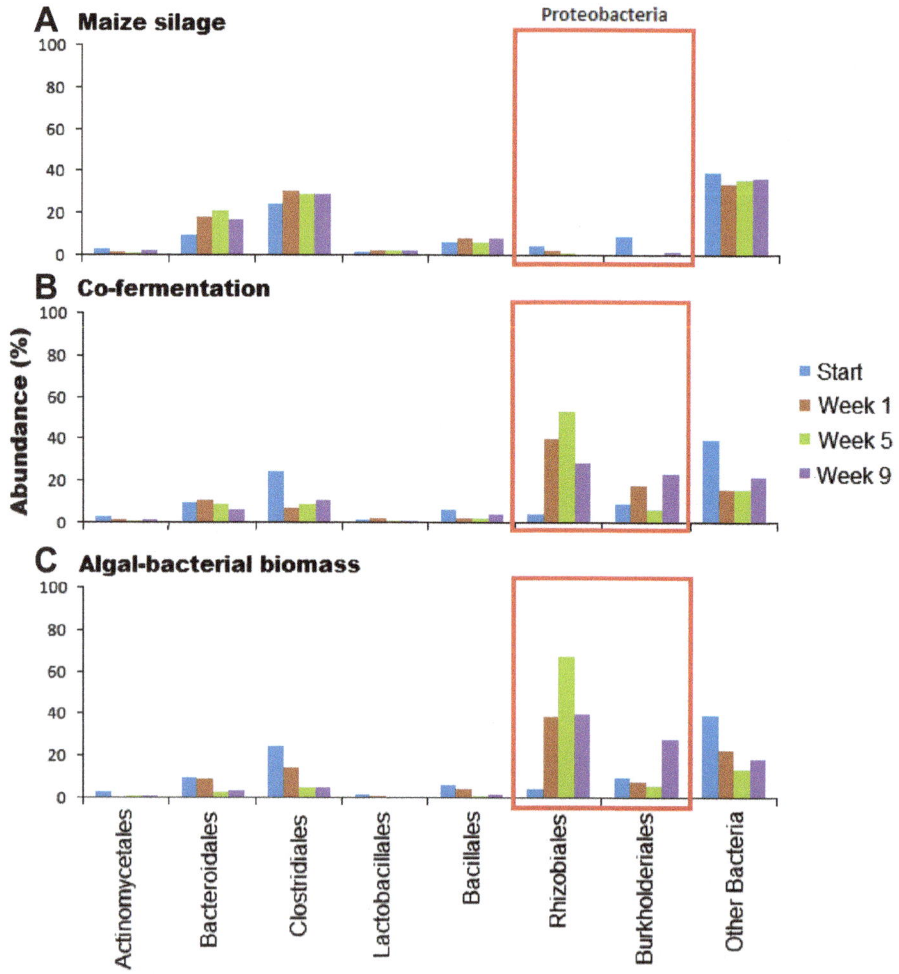

Figure 8 Changes in the domain Bacteria of the microbial community at the order level. (A) Maize silage, **(B)** co-fermentation, and **(C)** algal-bacterial biomass.

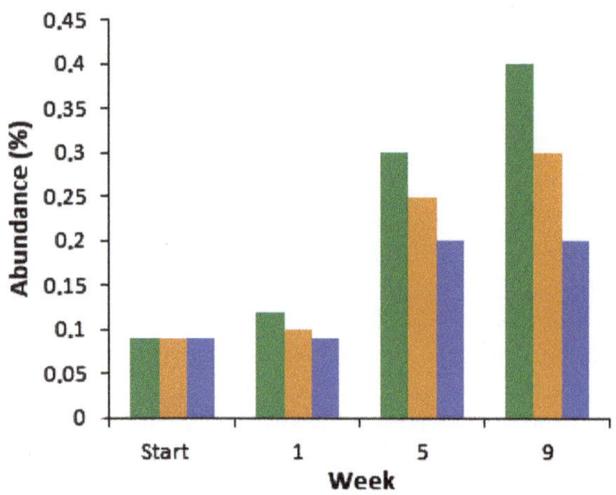

Figure 9 Eukaryotic sequences in the reactors. Green: AB + S, orange: co-fermentation, blue: maize silage.

Conclusions

A combination of bioH_2 and biogas production by a mixture of nonsterile microalgae and natural bacterial flora was demonstrated. In a closed system, the mutualistic bacteria consumed the O_2 evolved by the algae and created a sufficiently anaerobic environment for algal H_2 evolution without damaging the photosynthetic apparatus of the algae. With the help of the bacterial partners, the algae succeeded in capturing light energy by photosynthetic water splitting and evolved H_2 at the same time without the need for further manipulation of the system, such as sulfur deprivation.

H_2 production through the use of a mixture of microalgae and syntrophic bacteria started earlier than the H_2 evolution following sulfur deprivation, although sulfur-deprived *C. reinhardtii* produced bioH_2 for a longer period of time.

AD and biogas evolution from the nonsterile microalgal-bacterial biomass yielded a gas enriched in CH_4 relative to

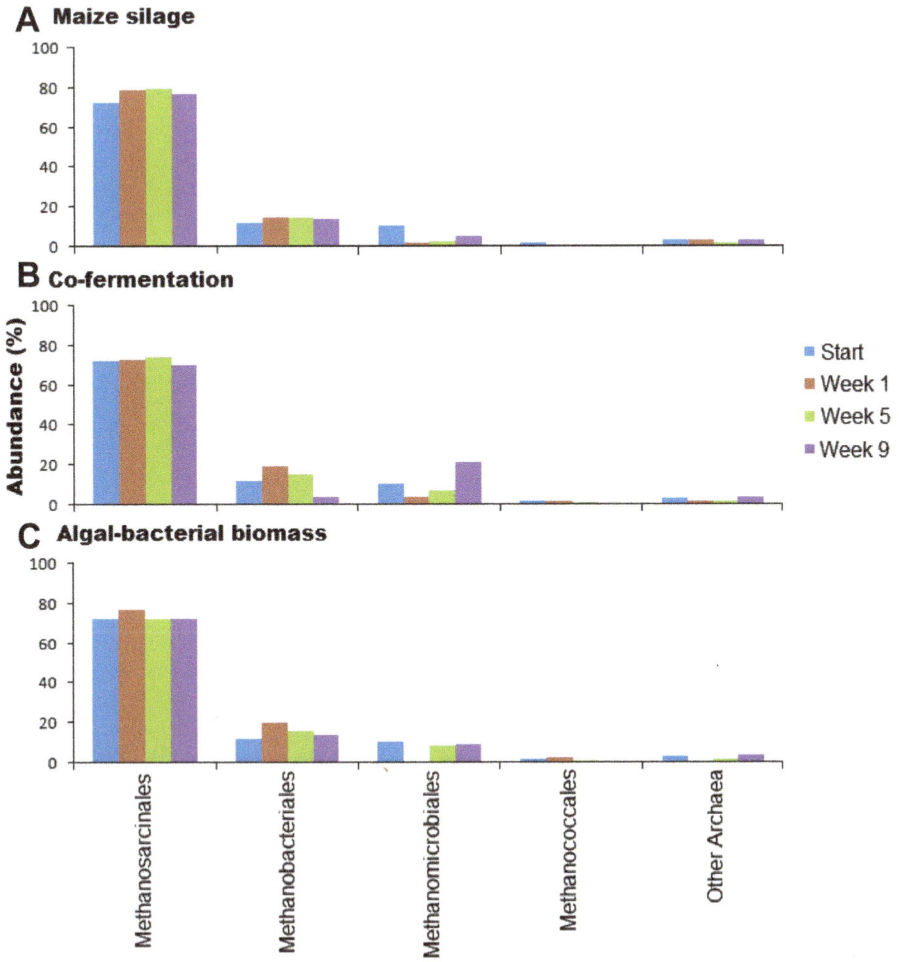

Figure 10 Distribution of the domain Archaea in the microbial community at the order level. (A) Maize silage, **(B)** co-fermentation, and **(C)** algal-bacterial biomass.

the commonly used maize silage. The specific biogas production estimated on the basis of the organic material input, however, was smaller than that from maize silage. The addition of maize silage to the algal-bacterial mixed biomass increased the C/N ratio considerably and improved the balanced digestibility of the microbial biomass.

The metagenome analysis of the microbial communities present in the AD reactors revealed the persisting impact of the microalgae and their bacterial companions on the composition of the AD microbial community within a few days. The large amount of bacteria belonging in the genera *Rhizobium* and *Burkholderia*, dosed together with the microalgal biomass, significantly changed the bacterial community composition. Co-fermentation of the algal-bacterial biomass with maize silage compensated somewhat for the *Rhizobium* and *Burkholderia* predominance due to the 50% lower loading of the microbial biomass on an organic dry matter basis. In the control reactors fed with maize silage, the microbial taxa belonging in the phyla *Firmicutes* and *Bacteroidetes* persisted.

Interestingly, the pronounced alterations observed in the domain *Bacteria* did not affect the composition of the domain *Archaea*. The order *Methanosarcinales* predominated in the Archaeal community regardless of the substrate composition.

Materials and methods

Cultivation of pure and mixed cultures

The *Chlamydomonas* sp. and *Scenedesmus* sp. algae and their mutualistic bacteria (AB + S culture) were obtained as algal strain 810 from the Mosonmagyaróvár Algal Culture Collection (MACC) of Hungary. The purified algal mixture was maintained and cultivated on TP (TRIS-phosphate) medium supplemented with rifampicin. The TP medium is a modified TAP (TRIS-Acetate-Phosphate) medium where acetate is replaced with HCl. The TAP and TP plates were incubated under 50 μmol m^{-2} s^{-1} light intensity at 25°C. Algae used for H_2-evolution experiments were harvested as fresh cultures grown on TP-agar plates supplemented with rifampicin and transferred into liquid TAP medium

[41,71,72]. The algal stock solutions were equally distributed into 40-mL Hypo-Vial bottles, resulting in a final volume of 35 mL and a final optical density (OD_{750}) of 0.7. *E. coli ΔhypF,* a hydrogenase-deficient strain, was grown on LB (Luria-Bertani medium) plates at 30°C in the dark.

The original algal-bacterial culture, used for H_2 production experiments, was pre-grown in TAP medium. The medium of the pre-grown stock culture was changed to fresh TAP medium by centrifugation. It was diluted to OD_{750} of 1.2. Bottles were sealed with butyl rubber stoppers and aluminum caps. All experiments were performed in at least three parallel repetitions.

Algal-bacterial culture scaling up for biomass production

For biomass production, an unpurified culture of a *Scenedesmus* sp. and *Chlamydomonas* sp. mixture, as obtained from the MACC collection, was cultivated in 13-L polycarbonate vessels under 50 μmol m^{-2} s^{-1} light intensity at 25°C for 5 days before harvesting. The biomass yield was approximately 2 g L^{-1}. The microbial biomass was harvested by using a cross-flow centrifuge and the harvested biomass was stored at −20°C until utilization.

Gas chromatographic analyses

The H_2 and O_2 levels in the headspace of the Hypo-Vial bottles were measured by gas chromatography. An Agilent 7890A gas chromatograph (Agilent Technologies, Santa Clara, USA), equipped with a thermal conductivity detector and an Agilent HP-Molsieve column (length 30 m, diameter 0.320 mm, film 12.0 μm; Agilent Technologies, Santa Clara, USA) was used in splitless mode. Linde HQ argon 5.0 (Linde Group, Munich, Germany) was used as carrier and reference gas. The temperatures of the injector, TCD detector, and column were kept at 150°C, 160°C, and 60°C, respectively. The column pressure was 47.618 psi. The flow rate of the column was 12 mL min^{-1}. Samples (50 μL) were analyzed. Three biological replicates were used for the measurements. A H_2 calibration curve was used to determine accurate gas volumes. Serial dilutions of pure H_2 gas were prepared in 25-mL gas-tight vials, and identical volumes were injected into the gas chromatograph: data from three replicates were used to draw the H_2 calibration curve.

Anaerobic fermentation and biogas analysis

Anaerobic fermentations were carried out in 5-L continuously stirred tank reactors [73], and the fed-batch operational mode was used. The reactors were operated by using a pig manure and maize silage mixture [59] until the biogas production stabilized. This start-up period lasted for 4 to 5 weeks. Feeding with the various substrate compositions was started thereafter. One of the reactors was fed with a AB + S loading of 1 g oDM L^{-1} day^{-1}, an identical reactor was supplied with AB + S and corn silage (0.5 + 0.5 g oDM L^{-1} day^{-1}), and the control received only corn silage (1 g oDM L^{-1} day^{-1}). The initial parameters of the substrates are summarized in Table 1. Heating was maintained by means of an electronically heated jacket which surrounded the cylindrical apparatus. Temperature was measured with a bimetallic sensor and was maintained constant at 37°C ± 1.0°C. The pH was between 7 and 8, and the redox potential was < −500 mV. The generated gas and its quality were measured daily after the 1-month start-up (weeks 1 to 4) and stabilization phase on the designated substrate. Gas volumes were measured with thermal mass flow devices (DMFC; Brooks Instrument, Hatfield, USA) attached to each gas exit port. The composition of the evolved biogas was measured with a gas chromatograph (6890 N Network GC System, Agilent Technologies, Santa Clara, USA) equipped with a 5 Å molecular sieve column (length 30 m, I.D. 0.53 megabore, film 25 μm). Ultrapure N_2 was used as carrier gas.

Determination of fermentation parameters

oDM

The dry organic matter content was quantified by drying the biomass at 105°C overnight and weighing the residue giving the dry mass content. Further heating of this residue at 550°C provided the organic dry mass content.

Density measurement

Sample density was measured with a MINIDENS automatic density meter (Grabner Instruments, Wien, Austria).

C/N

To determine C/N, an Elementar Analyzer Vario MAX CN (Elementar Group, Hanau, Germany) was employed. The equipment works using the principle of catalytic tube combustion under an O_2 supply at high temperatures

Table 2 Lysis conditions for total community DNA preparation

	Lysozyme[a] (μL)	10% CTAB[b] (μL)	Genomic CTAB lysis buffer[c] (μL)	Qiagen buffer[d] (μL)	Zymo buffer[e] (μL)
A	-	100	-	100	550
B	250	100	-	100	300
C	250	-	300	200	-

[a]100 mg mL^{-1} (Applichem, Barcelone, Spain). [b]Cetyltrimethylammonium bromide (*w/v*). [c]1 M Tris-HCl 100 mL, 500 mM EDTA 50 mL, 5 M NaCl 300 mL, 10% CTAB, 20% SDS, pH = 8 (Wirth *et al.* [64]). [d]ASL buffer from Qiagen QIAamp DNA Stool miniprep kit (51504, Qiagen, Limburg,Netherlands). [e]From Zymo Research Fecal DNA kit (Zymo Research, D6010).

(combustion temperature: 900°C, postcombustion temperature: 900°C, reduction temperature: 830°C, column temperature: 250°C). The desired components were separated from each other with the aid of specific adsorption columns (containing Sicapent (Merck, Billerica, USA), in C/N mode) and determined in succession with a thermal conductivity detector. Helium served as flushing and carrier gas.

NH_4^+-N

For the determination of NH_4^+ ion content, the Merck Spectroquant Ammonium test (1.00683.0001) (Merck, Billerica, USA) was used. At the beginning of the experiment the $NH_4^+ - N$ was 1,100 mg L^{-1}.

VOAs/TAC

Five grams of fermenter sample was taken for the analysis and diluted to 20 g with distilled water. The subsequent process was carried out with Pronova FOS/TAC 2000 Version 812-09.2008 automatic titrator (Pronova, Berlin, Germany). At the beginning of the experiment the VOAs/TAC ratio was 0.2.

DNA isolation for metagenomic studies

Two-milliliter samples, taken from the reactors, were used for total community DNA isolation. The extractions were carried out with a slightly modified version of the Zymo Research kit (D6010, Zymo Research, Irvine, USA). Parallel samples from each reactor were lysed with three different lysis mixes (Table 2). After lysation (bead beating), the Zymo Research kit protocol was followed. The quantity of DNA was determined in a NanoDrop ND-1000 spectrophotometer (NanoDrop Technologies, Wilmington, USA) and a Qubit 2.0 Fluorometer (Life Technologies, Carlsbad, USA). DNA purity was tested by agarose gel electrophoresis and with an Agilent 2200 Tape Station (Agilent Technologies, Santa Clara, USA).

Next-generation DNA sequencing and data handling

The sample preparation for total metagenome sequencing of the pooled samples was carried out following the recommendations of the Ion Torrent PGM sequencing platform (Life Technologies, Carlsbad, USA). Sequencing was performed with Ion Torrent PGM 316 chips. The reads were analyzed and quality values were determined for each nucleotide. The 150 to 250 nucleotide-long individual sequences were further analyzed by using the MG-RAST software package [74], which is a modified version of Rapid Annotations based on Subsystem Technology (RAST). The MG-RAST server computes results against several reference datasets (protein and ribosomal databases) [75]. The generated matches to external databases were used to compute the derived data [59,76].

Abbreviations

A + S: mixture of *Chlamydomonas* sp., and *Scenedesmus* sp., non-sulfur deprived; AB + S: mixture of *Chlamydomonas* sp., *Scenedesmus* sp., and mutualistic bacteria, non-sulfur deprived; AD: Anaerobic digestion; AE + S: mixture of *Chlamydomonas* sp., *Scenedesmus* sp., and *Escherichia coli* Δ*hypF*, non-sulfur deprived; A-S: mixture of *Chlamydomonas* sp., and *Scenedesmus* sp., sulfur deprived; oDM: organic dry matter; TAC: Total alkaline capacity; TAP: TRIS-acetate-phosphate medium; TP: TRIS-phosphate medium; VOS: Volatile organic acids.

Competing interests

The authors declare that they have no competing interests.

Authors' contributions

RW and ZB developed the DNA extraction protocol, designed and performed the experiments, and contributed to the evaluation of metagenomic data. GL helped in the cultivation of the algal-bacterial biomass. GM organized and performed the DNA sequencing work. KLK and ZB conceived the project and participated in its design. RW and KLK drafted the manuscript. GR critically evaluated the manuscript. ÉK analyzed and revised the manuscript. JM and KN analyzed the data and participated in the writing of the manuscript. All authors read and approved the final manuscript.

Authors' information

RW is a postdoctoral fellow at the Department of Biotechnology, University of Szeged, Szeged, Hungary. GL is a PhD student at the Department of Biotechnology, University of Szeged and Institute of Biochemistry, Hungarian Academy of Sciences, Szeged, Hungary. GM is Head of the Metagenomics Laboratory, Institute of Biochemistry, Biological Researh Center, Hungarian Academy of Sciences, Szeged, Hungary. ZB is an Assistant Professor at the Department of Biotechnology, while JM and KN are Full Professors at the Department of Oral Biology and Experimental Dentistry, University of Szeged. ÉK is Full Member of the Hungarian Academy of Sciences and a Senior Scientific Advisor at the Biological Research Center, Hungarian Academy of Sciences. RG is an Associate Professor, Department Chairman, and Director of the Environmental Research Institute at the University of Szeged. KLK is a Full Professor both at the Department of Biotechnology and at the Department of Oral Biology and Experimental Dentistry, University of Szeged; he is a Senior Scientific Adviser at the Institute of Biophysics, Biological Research Center, Hungarian Academy of Sciences. KLK serves as President of the Hungarian Biogas Association.

Acknowledgements

The authors thank Ms. Netta Bozóki for technical assistance. This work was supported by the domestic grants GOP-1.1.1-11-2012-0128, TÁMOP-4.2.2.A-11/1/KONV-2012-0007 and PIAC_13-1-2013-0145 and by the EU projects H2020-LCE-2014-3 646533 BIOSURF and 'SYMBIOTICS' ERC AdG to Éva Kondorosi.

Author details

[1]Department of Biotechnology, University of Szeged, Közép fasor 52, H-6726 Szeged, Hungary. [2]Institute of Biochemistry, Biological Research Center, Hungarian Academy of Sciences, Temesvári krt. 62, H-6726 Szeged, Hungary. [3]Institute of Biophysics, Biological Research Center, Hungarian Academy of Sciences, Temesvári krt. 62, H-6726 Szeged, Hungary. [4]Department of Oral Biology and Experimental Dental Research, University of Szeged, Tisza L. krt. 64, 6720 Szeged, Hungary.

References

1. McKendry P. Energy production from biomass (Part2-3): conversion technologies. Biores Technol. 2002;83:47–63.
2. Angelidaki I, Ellegaard L. Codigestion of manure and organic wastes in centralized biogas plants: status and future trends. Appl Biochem Biotechnol. 2003;109:95–105.
3. Santosh Y, Sreekrishnan TR, Kohli S, Rana V. Enhancement of biogas production from solid substrates using different techniques—a review. Biores Technol. 2004;95:1–10.
4. Goyal HB, Seal D, Saxena RC. Bio-fuels from thermochemical conversion of renewable resources: a review. Renew Sust Energy Rev. 2008;12:504–17.

5. Schenk PM, Thomas-Hall SR, Stephens E, Marx UC, Mussgnug JH, Posten C, et al. Second generation biofuels: high efficiency microalgae for biodiesel production. Bioenergy Res. 2008;1:20–43.
6. Johanson D, Azar CA. Scenario based analysis of land competition between food and bioenergy production in the US. Climate Change. 2007;82:267–91.
7. Richmond A. Handbook of Microalgal culture: Biotechnology and Applied Phycology. Oxford: Blackwell Science; 2004.
8. Rodolfi L, Zitteli GC, Bassi N, Padovani G, Biondi N, Bonini G, et al. Microalgae for oil: strain selection, induction of lipid synthesis and outdoor mass cultivation in a low-cost photobioreactor. Biotechnol Bioeng. 2009;102:100–12.
9. Edward M. The algal industry survey - a white paper by Dr. Mark Edward. In: Association with the Centre for Management Technology. 2009.
10. Singh J, Gu S. Commercialization potential of microalgae for biofuels production. Renew Sust Energy Rev. 2010;14:2596–610.
11. Guccione A, Biondi A, Sampietro G, Rodolfi L, Bassi N, Tredici MR. *Chlorella* for protein and biofuels: from strain selection to outdoor cultivation in green wall panel photobioreactor. Biotechnol Biofuels. 2014;7:84. doi:10.1186/1754-6834-7-84.
12. Posten C, Schaub G. Microalgae and terrestrial biomass as a source for fuels – a process view. J Biotechnol. 2009;142:64–9.
13. Harun R, Singh M, Forde MG, Danquah KM. Bioprocess engineering of microalgae to produce a variety of consumer products. Renew Sust Energy Rev. 2010;14:1037–47.
14. Chen P, Min M, Chen Y, Wang L, Li Y, Chen Q, et al. Review of biological and engineering aspects of algae to fuels approach. Int J Agric Biol Eng. 2009;2:1.
15. González-Delgado AD, Kafarov V. Microalgae based biorefinery: issues to consider. A review. CT&F - Ciencia Tecnología y Futuro. 2011;4:5–22.
16. Yen H-W, Hu I-C, Chen C-Y, Ho S-H, Lee D-J, Chang J-S. Microalgae based biorefinery - from biofuels to natural products. Biores Technol. 2013;135:166–74.
17. Becker EW. Microalgae: Biotechnology and Microbiology. Cambridge, UK: Cambridge University Press; 1994.
18. Golueke CG, Oswald WJ, Gotaas HB. Anaerobic digestion of algae. Appl Microbiol. 1957;5:47–55.
19. Uziel M, Oswald WJ, Golueke CG. Solar energy fixation and conversion with algal-bacterial system. Washington, D.C: U.S. National Science Foundation Rep. No. NSF-RA-N-74-195, NSF; 1974.
20. Keenan JD. Bioconversion of solar energy to methane. Energy. 1977;2:365.
21. Binot R, Martin D, Nyns EJ, Naveau H. Digestion anaerobic d'algues cultivees dans les eaux de refroidissement industrielles. Martigues, France: Proc. Heliosynthese aquaculture Semin; 1977.
22. Samson R, Le Duy A. Biogas production from anaerobic digestion of *Spirulina maxima* algal biomass. Biotechnol Bioeng. 1982;24:1919.
23. Becker EW. The production of microalgae a source of biomass. Biomass Util. 1983;67:205.
24. Hernández EPS, Córdoba LT. Anaerobic digestion of *Chlorella vulgaris* for energy production. Res Con Recyc. 1993;9:127–32.
25. Mussgnug JH, Klassen V, Schlüter A, Kruse O. Microalgae as a substrates for fermentative biogas production in a combined biorefinery concept. J Biotechnol. 2010;150:51–6.
26. De Schamphelaire L, Verstraete W. Revival of the biological sunlight to biogas energy conversion system. Biotechnol Bioeng. 2009;103:296–304.
27. JT, Tramp C, Sims RC, Miller CD. Characterization of a methanogenic community within an algal fed anaerobic digester. ISRN Microbiol. 2012. doi:10.5402/2012/753892.
28. Melis A, Happe T. Hydrogen production. green algae as a source of energy. Plant Physiol. 2001;127:740–8.
29. Zhang L, Happe T, Melis A. Biochemical and morphological characterization of sulfure-deprived and H_2- producing *Chlamydomonas reinhardtii* (green alga). Planta. 2002;21:552–61.
30. Fouchard S, Hemscheimer A, Caruana A, Pruvost J, Legrand J, Happe T, et al. Autotrophic and mixotrophic hydrogen photoproduction in sulfur-deprived *Chlamydomonas reinhardtii* cells. Appl Environ Microbiol. 2005;10:6199–205.
31. Keshtacher-Liebso E, Hadar Y, Chen Y. Oligotrophic bacteria enhance algal growth under iron – deficient conditions. Amer Soc Microbiol. 1995;61:2411–39.
32. Watanabe K, Takihana N, Aoyagi H, Hanada S, Watanabe Y, Ohmura N, et al. Symbiotic association in *Chlorella* culture. FEMS Microbiol Ecol. 2005;51:187–96.
33. Nikolaev YA, Plakunov YK, Voronina NA, Nemtseva NV, Platnikov AO, Gogoleva OA, et al. Effect of bacterial satellites on Chlamydomonas reinhardtii in an algo-bacterial community. Microbiology. 2008;77:78–83.
34. Amin SA, Green DH, Hort MC, Küpper FC, Sunda WG, Carrano JC. Photolysis of ion – siderophore chelates promotes bacteria – algal mutualism. Proc Natl Acad Sci U S A. 2009;106:17071–6.
35. Rivas MO, Vargas P, Riquelme CE. Interactions of *Botryococcus braunii* cultures with bacterial biofilms. Microb Ecol. 2010;60:628–35.
36. Kazamia E, Czesnick H, Nguyen TTV, Croft MT, Sherwood E, Sasso S, et al. Mutualistic interaction between vitamin B-12 dependent algae and heterotrophic bacteria exhibit regulation. Environ Microbiol. 2012;14:1466–76.
37. Xie B, Bishop S, Stessman D, Wright D, Spalding MH, Halverson LJ. *Chlamydomonas reinhardtii* thermal tolerance enhancement mediated by mutualistic interaction with vitamin B_{12}-producing bacteria. ISME J. 2013;7:1544–55.
38. Chwenk D, Nohynek L, Rischer H. Algae bacteria association inferred by 16S rDNA similarity in established microalgae cultures. Microbiol. 2014;3:356–68.
39. Kim B-H, Ramanan R, Cho D-H, Oh H-M, Kim H-S. Role of *Rhizobium*, a plant growth promoting bacterium, in enhancing algal biomass through mutualistic interaction. Biomas Bioenergy. 2014;69:95–105.
40. Wu S, Li X, Yu J, Wang Q. Increased hydrogen production in co-culture of *Chlamydomonas reinhardtii* and *Bradyrhizobium japonicum*. Biores Technol. 2012;123:184–8.
41. Lakatos G, Deák ZS, Vass I, Rétfalvi T, Rozgonyi Sz, Rákhely G, et al. Bacterial synbionts enhance photo-fermentative hydrogen evolution of Chlamydomonas algae. Green Chem. 2014. doi:10.1039/C4GC00745J.
42. Maier T, Binder U, Böck A. Analysis of the *hydA* locus of *Escherichia coli*: two genes (*hydN* and *hypF*) involved in formate and hydrogen metabolism. Arch Microbiol. 1996;165:333–41.
43. Melis A, Zhang L, Seibert M. Sustained photobiological hydrogen gas production upon reversible inactivation of oxygen evolution in the green alga *Chlamydomonas reinhardtii*. Plant Physiol. 2000;122:127–36.
44. Ghirardi ML, Zhang L, Lee JW, Flynn T, Seibert M, Greenbaum E, et al. Microalgae: a green source of renewable H_2. Trends Biotechnol. 2000;18:506–11.
45. Kosourov SN, Batyrova KA, Petushkova EP, Tsygankov AA, Ghirardi ML, Seibert M. Maximizing the hydrogen photoproduction yields in *Chlamydomonas reinhardtii* cultures: the effect of the H_2 partial pressure. Int J Hydrogen Energy. 2012;37:8850–8.
46. Ward AJ, Lewis DM, Green FB. Anaerobic digestion of algae biomass: a review. Algal Res. 2014. doi:10.1016/j.algal.2014.02.001.
47. Amon T, Gruber W, Hoffstede U, Jäger P, Jäkel K, Kaiser F, et al. Gasausbeute in landwirtschaftichen biogasanlagen. KTBL; 2010. ISBN:978-3-941583-49-9. BOKU University, Wien, Austria 2010
48. McGhee TJ. A method for approximation of the volatile acid concentrations in anaerobic digesters. Water Sewage Works. 1968;115:162–6.
49. Nordmann W. Die Überwachtung der Schlammfaulunk. KA-Informationen für das Betriebspersonal, Beilage zur Korrespondenz Abwasser. 1977. 3/77.
50. Alexander M. Biodegradation of organic chemicals. Environ Sci Technol. 1985;19:106–11.
51. Chen Y, Cheng JJ, Creamer KS. Inhibition of anaerobic digestion process: a review. Biores Technol. 2008;99:4044–64.
52. Nielsen HB, Angelidaki I. Strategies for optimizing recovery of the biogas process following ammonia inhibition. Biores Technol. 2008;99:7800–995.
53. Parkin GF, Owen WF. Fundamental of anaerobic-digestion of wastewater sludge. J Environ Eng. 1986;112:867–920.
54. Yadvika S, Sreekrishnan TR, Kohli S, Rana V. Enhancement of biogas production from solid substrates using different techniques - a review. Biores Technol. 2004;95:1–10.
55. Olsson J, Shadiimam MA, Nehrenheim E, Thorin E. Co-digestion of cultivated microalgae and sewage from municipal waste water treatment. International Conference on Applied Energy ICAE 2013 Jul. 1–4. 2013, Pretoria, South Africa, Paper ID: ICAE2013-518.
56. Yen H-W, Brune DE. Anaerobic co-digestion of algal sludge and waste paper to produce methane. Biores Technol. 2007;98:130–4.
57. Makarova K, Slesarev A, Wolf Y, Sarokin A, Mirkin B, Koonin E, et al. Comparative genomics of the lactic acid bacteria. Proc Natl Acid Sci USA. 2006;103:15611–6.
58. Yamada Y, Yukphan P. Genera and species in acetic acid bacteria. Int J Food Microbiol. 2008;125:15–24.

59. Wirth R, Kovács E, Maroti G, Bagi Z, Rakhely G, Kovacs KL. Characterization of a biogas-producing microbial community by short-read next generation DNA sequencing. Biotechnol Biofuels. 2012;5:1–16.
60. Daniel RM, Smith M, Phillip AD, Ratcliffe HD, Drozd JW, Buel AT. Anaerobic growth and denitrification by *Rhizobium japonicum* and other *Rihzobia*. J Gen Microbiol. 1980;120:517–21.
61. Tjepkema J, Evans HJ. Nitrogen fixation by free-living Rhizobium in a defined liquid medium. Biochem Biophys Res Commun. 1975;65:625–8.
62. Schlüter A, Bekel T, Diaz NN, Dondrup M, Eichenlaub R, Gartemann KH, et al. The metagenome of a biogas-producing microbial community of a production-scale biogas plant fermenter analyzed by the 454-pyrosequencing technology. J Biotech. 2008;136:77–90.
63. Krause L, Diaz NN, Edwards RA, Gartemann K-H, Krömeke H, Neuwger H, et al. Taxonomic composition and gene content of a methane-producing microbial community isolated from a biogas reactor. J Biotech. 2008;136:91–101.
64. Kröber M, Bekel T, Diaz NN, Goesmann A, Sebastian J. Phylogenetic characterization of a biogas plant microbial community integrating clone library 16S-rDNA sequences and metagenome sequence data obtained by 454-pyrosequencing. J Biotech. 2009;142:38–49.
65. Stantscheff R, Kuever J, Rabenstein A, Seyfarth K, Dröge S, König H. Isolation and differentiation of methanogenic *Archaea* from mesophilic corn-fed on-farm biogas plants with special emphasis on the genus *Methanobacterium*. Appl Environ Biotechnol. 2014;98:5719–35.
66. Ziganshina EE, Bagmanova AR, Khilyas IV, Ziganshin AM. Assessment of biogas-generating microbial community in a pilot-scale anaerobic reactor. J Biosci Bioeng. 2014;117:730–6.
67. Sirohi SK, Pandey N, Singh B, Puniya AK. Rumen methanogens: a review. Indian J Microbiol. 2010;50:253–62.
68. Rastogi G, Ranade DR, Yeole TY, Patole MS, Houche YS. Investigation of methanogen population structure in biogas reactor by molecular characterization of methyl-coenzyme M reductase A *(mcrA)* genes. Biores Technol. 2008;99:5317–26.
69. Lee C, Kim J, Hwang K, O'Flaherty V, Hwang S. Quantitative analysis of methanogenic community dynamics in three anaerobic batch digesters treating different wastewaters. Water Res. 2009;43:157–65.
70. Blume F, Bergmann I, Nettmann E, Schelle H, Rehde G, Munkdt K, et al. Methagenomic population dynamics during semi-continuous biogas fermentation and acidification by overloading. J Appl Microbiol. 2010;109:441–50.
71. Sialve B, Bernet N, Bernard O. Anaerobic digestion of microalgae as a necessary step to make microalgal biodiesel sustainable. Biotechnol Adv. 2009. doi:10.1016/j.biotechadv.2009.03.001.
72. Lakaniemi A-M, Hulatt CJ, Thomas DN, Tuovinen OH, Puhakka JA. Biogenic hydrogen and methane production from *Chlorella vulgaris* and *Dunaliella tertiolecta* biomass. Biotechnol Biofuels. 2011;4:34.
73. Kovács KL, Ács N, Kovács E, Wirth R, Rákhely G, Strang O, et al. Improvement of biogas production by bioaugmentation. BioMed Res Internat. 2013. http://dx.doi.org./10.1155/2013/482653.
74. Meyer F, Paarmann D, D'Souza M, Olson R, Glass EM, Kubal M, et al. The metagenomics RAST server – a public resource for the automatic phylogenetic and functional analysis of metagenomes. BMC Bioinformatics. 2008;9:386.
75. MG-RAST manual for version 3.3.6 revision 9. ftp://ftp.metagenomics.anl.gov/data/manual/mg-rast-manual.pdf.
76. Kovács E, Wirth R, Maróti G, Bagi Z, Rákhely G, Kovács KL. Biogas production from protein-rich biomass: fed-batch anaerobic fermentation of casein and pig blood and associated changes in microbial community composition. PLoS One. 2013;8(10), e77265.

Study of liquid-phase molecular packing interactions and morphology of fatty acid methyl esters (biodiesel)

Paula Berman[1], Nitzan Meiri[1], Luiz Alberto Colnago[2], Tiago Bueno Moraes[3], Charles Linder[4], Ofer Levi[5], Yisrael Parmet[5], Michael Saunders[6] and Zeev Wiesman[1*]

Abstract

Background: ^{1}H low field nuclear magnetic resonance (LF-NMR) relaxometry has been suggested as a tool to distinguish between different molecular ensembles in complex systems with differential segmental or whole molecular motion and/or different morphologies. In biodiesel applications the molecular structure versus liquid-phase packing morphologies of fatty acid methyl esters (FAMEs) influences physico-chemical characteristics of the fuel, including flow properties, operability during cold weather, blending, and more. Still, their liquid morphological structures have scarcely been studied. It was therefore the objective of this work to explore the potential of this technology for characterizing the molecular organization of FAMEs in the liquid phase. This was accomplished by using a combination of supporting advanced technologies.

Results: We show that pure oleic acid (OA) and methyl oleate (MO) standards exhibited both similarities and differences in the ^{1}H LF-NMR relaxation times (T_2s) and peak areas, for a range of temperatures. Based on X-ray measurements, both molecules were found to possess a liquid crystal-like order, although a larger fluidity was found for MO, because as the temperature is increased, MO molecules separate both longitudinally and transversely from one another. In addition, both molecules exhibited a preferred direction of diffusion based on the apparent hydrodynamic radius. The close molecular packing arrangement and interactions were found to affect the translational and segmental motions of the molecules, as a result of dimerization of the head group in OA as opposed to weaker polar interactions in MO.

Conclusions: A comprehensive model for the liquid crystal-like arrangement of FAMEs in the liquid phase is suggested. The differences in translational and segmental motions of the molecules were rationalized by the differences in the ^{1}H LF-NMR T_2 distributions of OA and MO, which was further supported by ^{13}C high field (HF)-NMR spectra and ^{1}H HF-NMR relaxation. The proposed assignment allows for material characterization based on parameters that contribute to properties in applications such as biodiesel fuels.

Keywords: ^{1}H low field nuclear magnetic resonance relaxometry, Biodiesel physical properties, Methyl oleate, Molecular packing, Oleic acid, Segmental motion

* Correspondence: wiesman@bgu.ac.il
[1]The Phyto-Lipid Biotechnology Laboratory, Departments of Biotechnology, Energy and Environmental Engineering, Ben-Gurion University of the Negev, P.O. Box 653, Beer-Sheva 84105, Israel
Full list of author information is available at the end of the article

Background

Biodiesel production has increased dramatically over the last decade, raising the need for new rapid and non-destructive analytical tools and technologies. Several ^1H low field nuclear magnetic resonance (LF-NMR) applications have been suggested by the authors for the field of biodiesel, including characterization of new alternative biodiesel resources by direct analysis of raw material sources, monitoring of the biodiesel transesterification reaction, and quality evaluation of the final product [1-6]. ^1H LF-NMR relaxometry involves the measurement of relaxation constants, T_1 and T_2, as a consequence of interactions between nuclear spins and their surroundings and among nuclear spins [6]. It was also suggested that the application of a novel numerical optimization method for analyzing ^1H LF-NMR data [6,7] provides better resolved relaxation time distributions and more accurate solutions compared with those shown by existing numerical tools. For example, using this optimization method, the relaxation time distributions of rapeseed oil and biodiesel samples revealed four and three peaks, respectively, compared to a broad bimodal distribution and a single wide peak distribution for the same samples analyzed using WinDXP software [8].

^1H LF-NMR spin-spin (T_2) relaxometry has been suggested as a tool to distinguish between molecular populations in complex systems with differential mobilities and/or microscopic compartmentalization [9-13]. Still, the peaks in the analyzed T_2 distributions of lipids have not been hitherto assigned to the appropriate molecular population arrangements with certainty. Marigheto *et al.* [14] speculated that the analyzed bimodal T_2 distribution of an avocado oil sample arises from molecules of differing mobility, such as the oleic and palmitic constituents, or from nonequivalent proton pools of different mobility, such as those on methyl and olefinic groups. Adam-Berret *et al.* [15] found a similar two-peak distribution for tricaprin in the melt state, and suggested that this may be due to inhomogeneous relaxation rates for the protons along the side chains, or inhomogeneous organization of the triglycerides (TGs) in the liquid with intermolecular interactions. Of course, these hypotheses are interrelated, since different mobilities along the side chains of the TGs in the liquid phase are characteristic of different organizations and vice versa. Callaghan [16] studied the molecular motion of tristearin in the melt and found different T_2s along the chains, which in turn were used to explain the tuning fork molecular configuration.

The liquid morphological structures of lipids, as opposed to crystal structures, have not been significantly studied because their experimental determinations are very difficult and require a combination of different investigation methods. Three models have been previously suggested for the arrangement of TGs in the melt, the smectic, nematic, and discotic liquid crystal models, as reviewed by Iwahashi and Kasahara [17]. However, a conclusive structure of liquid TGs still requires further research. Fatty acids (FAs) are significant building blocks of most lipids, including TGs. Short-range order was also found to exist between the aliphatic chains in the liquid state of FAs. This was attributed by Small [18] to a relatively small volume increase occurring during the melting of the crystalline chains to liquids, and was strengthened by the fact that X-ray scattering showed that domains of layered structures, with one dimension roughly equivalent to the lengths of the molecules, are present in the liquid.

The group of Iwahashi has thoroughly studied the self-organization of FAs in the neat liquid state [17,19-25]. They concluded, using near-infrared spectroscopy and vapor pressure osmosis on various FAs, that these exist mostly as dimers, even at high temperature, where the dimers are the units in their intra- or intermolecular movements. The dimers were found to aggregate to form clusters possessing the structure of a quasi-smectic liquid crystal, where the long-chained FA dimers arrange longitudinally and alternately to make an interdigitated structure in the clusters, with the tail of two dimers near the interacting head groups of the adjacent one. This has been determined from measurements of viscosity, density, high field (HF)-NMR, and X-ray diffraction [26].

The physical properties of FAs and their derivatives are largely determined by the length of the hydrocarbon chain and the degree of unsaturation, which affect the different degrees of molecular packing. In fully saturated compounds, free rotation around each carbon-carbon bond gives the hydrocarbon chain greater flexibility; the most stable conformation is the fully extended form, in which the steric hindrance of neighboring atoms is minimized. These molecules can pack together tightly in nearly crystalline arrays, with atoms all along their lengths in van der Waals contact with the atoms of neighboring molecules. In unsaturated FAs, a cis double bond forces a kink in the hydrocarbon chain. FAs with several such kinks cannot pack together as tightly as one-kink or fully saturated FAs, and their intermolecular interactions are therefore weaker. FAs of the same chain length have lower melting points as the degree of unsaturation is increased, as it takes less thermal energy to disorder them [27].

Surprisingly, the liquid structure of fatty acid methyl esters (FAMEs), which are derivatives of FAs, has attracted very little attention in the literature. FAMEs, the basic molecules that constitute biodiesel, can be achieved by transesterification of TGs using methanol in the presence of a catalyst. Their molecular organization in the melt is of high importance to the field of biodiesel, as it determines physicochemical properties of the fuel, including flow properties, operability during cold weather, blending, and more.

Knothe [28] analyzed the cetane number, heat of combustion, cold flow, oxidative stability, viscosity, and lubricity of common FAMEs, and showed that oleic acid methyl ester (methyl oleate) is the best FAME for high quality biodiesel. Furthermore, there are several genetically modified oil seeds available in the market (soybean, sunflower, peanut) with high methyl oleate content (80%).

It was therefore the objective of this work to explore the potential of ^{1}H LF-NMR spin-spin relaxometry technology to study the molecular details and aggregation of FAs and their FAME lipid derivatives in the liquid phase, using oleic acid and oleic acid methyl ester as a model biodiesel. This was accomplished by using a combination of supporting advanced technologies, including ^{1}H LF-NMR diffusiometry, X-ray diffraction, and ^{13}C and ^{1}H HF-NMR. As will be shown, this new application of ^{1}H LF-NMR is of high importance to the field of biodiesel characterization and also to other research and applied disciplines.

Results

^{1}H LF-NMR T_2 distributions

The combined ^{1}H LF-NMR T_2 distributions of oleic acid (OA) and methyl oleate (MO) at different temperatures are presented in Figure 1A and B, respectively. The intrinsic T_2 values and relative contributions of each peak are marked on each plot. For example, in Figure 1A at 288 K, the intrinsic T_2 values for peaks 1 and 2 are 103 and 251 ms, respectively, and the relative contributions are 57 and 43% for the same peaks. In both figures, the T_2 distributions exhibit two distinct peaks at different T_2 values. As suggested before [14,15] for TGs, the two peaks may be the result of two distinct mobility populations of the protons on the chain, or inhomogeneous structural organizations with two different packing densities and intermolecular interaction intensities or types.

Furthermore, pronounced alterations in the T_2 distributions are found. When comparing two distributions for the same temperature, a shift in the T_2 of the peaks of MO is observed towards higher values. Another interesting result is related to the change in the relative contribution of the peaks in the distribution as the temperature changes, which is mostly pronounced for MO. A closer look into the effect of temperature on the intrinsic T_2 values and relative areas of each peak can be seen in Figure 2A-D. A highly linear increase with temperature of intrinsic T_2 values can be observed for all peaks ($R^2 \geq 0.98$, Figure 2A and B). This suggests an increase in the

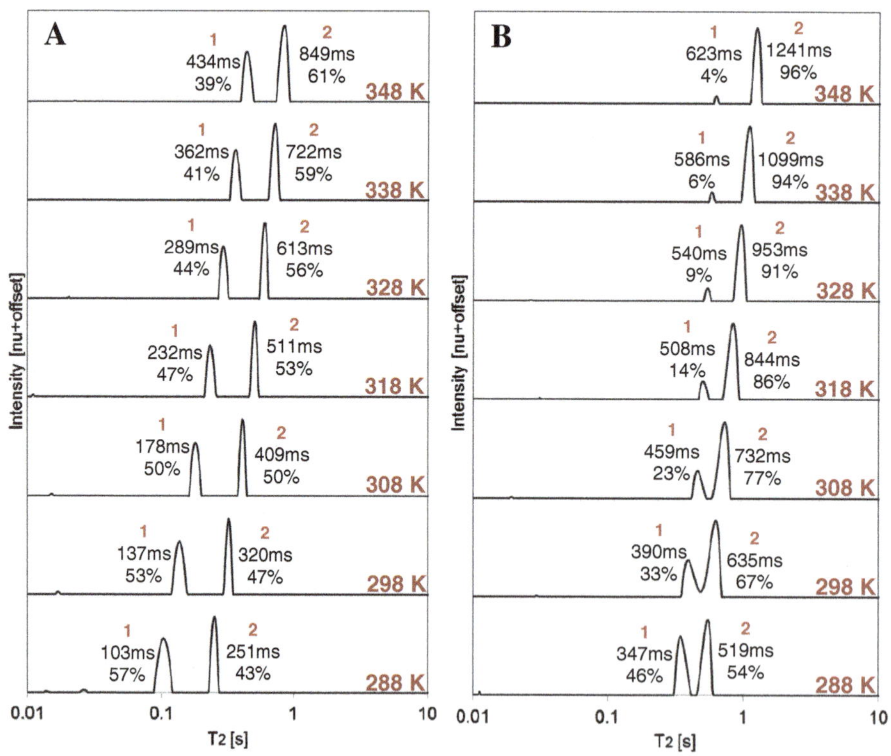

Figure 1 Combined ^{1}H low field (LF)-NMR T_2 distributions of (A) OA and (B) MO at different temperatures. The relative contributions of each peak, in relation to other peaks and intrinsic T_2 values, are shown on each plot. For the same temperature, the peaks in MO have larger T_2 values compared to OA. As temperature increases, a shift in the T_2 of the peaks is observed towards higher values, and the relative concentration of the peaks changes, especially for MO.

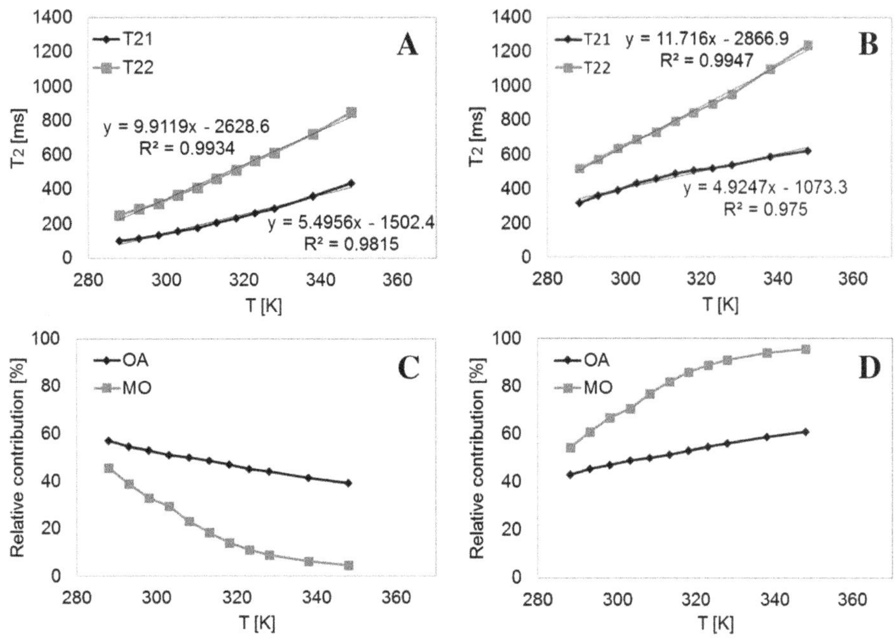

Figure 2 Variation with temperature of intrinsic T_2 values and relative contributions of OA and MO peaks. Comparison of intrinsic T_2 values of **(A)** OA and **(B)** MO; and relative contributions of **(C)** short (T_{21}) and **(D)** long (T_{22}) peaks of OA and MO.

mobility of different protons along the chain, or a change in the molecular organization with a change in intermolecular interactions towards the higher mobility peak (or population). Interestingly, the higher mobility peak (T_{22}) is more affected by temperature, as it exhibits larger slopes compared to the low mobility peak (T_{21}).

These differences are induced by small changes in the chemical structure of the lipid materials. Since both molecules (OA and MO) consist of the same tail (same length and position of double bond), the differences in this case are attributed to a methyl ester versus a carboxylic head group, which is responsible for a major intermolecular interaction of the chain with its neighbor.

As described in the introduction, the molecular organization of OA in the liquid has been extensively studied by Iwahashi's team ([17] and references therein). The stability of the interdigitated structure of OA was mainly attributed to their arrangement in a head-to-head conformation, driven by strong intermolecular hydrogen bonding of the carboxyl groups. In the case of esters, the arrangement of the heads is not obvious and depends on the polarity of the head group. Malkin [29] stated that methyl esters in the solid state behave in a weaker degree like the acids, crystallizing as dimer molecules with the polar groups together. It has also been suggested that a head-to-tail arrangement, where the polar groups in all layers have the same direction, can only be formed in cases of extremely weak polar forces, as in ethyl stearate [30]. In a study to determine end-to-end distances of liquid alkanes, Brady *et al.* [31] substituted one or two ends of the molecules with bromine atoms. They found that with a single substitution the chains lined up end to end so that the Br atoms seek maximum contact with each other, and concluded that this was due to stronger interactions between two polar Br than those between Br and a hydrocarbon chain end. This can be considered as analogous to the polar interactions of two ester groups. In a study of the crystal structure of methyl stearate, the authors suggested that the molecules form double sheets like the acids, probably due to polar forces between oxygen and carbon atoms [32].

We therefore assume the head-to-head conformation for the MO molecules in the liquid. Based on the crystal structure recently suggested for ethyl acetate [33], we would like to propose the configuration shown in Figure 3. Here the resonance structures of the ester group result in weak interactions between the polarized hydrogen of the methyl carbon and the oxygen of the carbonyl on the opposite MO molecule. This molecular arrangement for MO would maximize the polar interactions. The effect of this weaker interaction on their structure and mobility will be further discussed.

X-ray measurements

In order to investigate whether short-range order exists for MO, and compare to the already-determined X-ray bands of OA, X-ray measurements were performed. X-ray patterns of liquids consist of one or several broad rings where the position of the maximum $2\theta_{max}$ corresponds with high accuracy to the average intermolecular spacing, d. Iwahashi *et al.* [21] performed X-ray diffraction

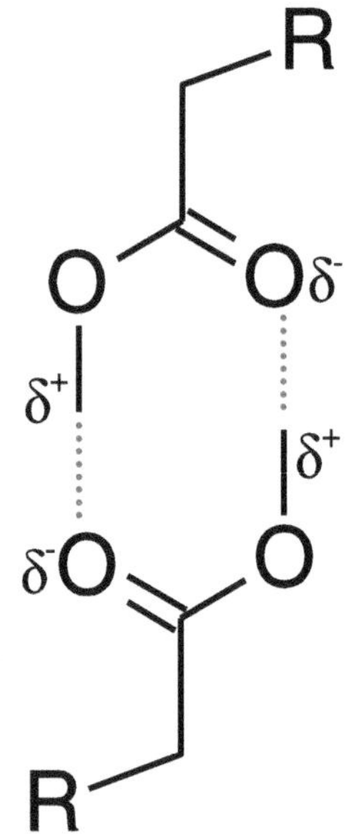

Figure 3 Proposed configuration for the ester groups of two opposite MO molecules interacting with each other. The resonance structures of the ester group result in weak interactions between the polarized hydrogen of the methyl carbon and the oxygen of the carbonyl on the opposite MO molecule. This molecular arrangement for MO would maximize the polar interactions.

(XRD) measurements on several types of FAs, and found similar spectra for all the materials, consisting of mainly a large and sharp band around 0.14 nm^{-1} and a small and broad band around 0.03 nm^{-1}. They suggested that the band around 0.14 nm^{-1} gives a measure of the spacing between adjacent molecules (short range), and the small band at around 0.03 nm^{-1} provides information regarding the long spacing of the plane made by the aligned molecules. In this study, XRD measurements on both materials showed only a single broad peak at around 0.14 nm^{-1}, probably due to instrumentation differences. The 0.03 nm^{-1} peak was therefore measured using small angle X-ray scattering (SAXS) technology.

Figure 4A and B show the X-ray spectra for OA and MO at 298 K using XRD and SAXS instruments, respectively. As shown, the peak around 0.14 nm^{-1} is sharp for both materials, whereas the peak around 0.03 nm^{-1} is very broad and difficult to resolve, especially for the MO sample. The broadness of the peaks suggests a lower degree of order, especially in the long-range spacing. The short- and long-range spacing values, d, derived from the corresponding spectra are summarized in Table 1.

The short- and long-range spacings for the OA material were found to be in excellent agreement with those previously reported [21]. The short-range spacing of both materials was similar, probably due to similarities in the structure of the tails, leading to close interchain interactions. On the other hand, a larger long-range spacing was found for MO compared to OA, even when considering the distance from melting point (2.383 nm for OA at 298 K versus 2.517 nm for MO at 263 K). The difference in the long-range spacing originates from a larger distance between repeating planes. Considering the model proposed in this work for the head-to-head configuration of two opposite MO molecules (Figure 3), it would be fair to suggest that this is the result of the weak CH--O hydrogen bridges leading to larger distances between head groups. To be more specific, MO heads interact through weak polar interactions that form octagons (with a larger long-range spacing), while OA heads interact through hydrogen bonding, which makes hexagons (with a shorter long-range spacing).

The longer spacing between two MO molecules is a clear explanation for the reduced density and lower melting point of MO compared to OA (0.874 g/cm^3 [34] and 253.1 K [35] versus 0.891 g/cm^3 and 286 K (both taken from [19]) for MO and OA, respectively, with densities reported for 293 K). Both density and melting point parameters suggest a less efficient packing of MO.

To explore the effect of temperature on the short- and long-range spacings, further X-ray measurements were performed (Figure 5A and B). Iwahashi *et al.* [21] found that the long spacing in OA is constant regardless of temperature, whereas the short spacing increases with temperature. Our measurements show a similar increase

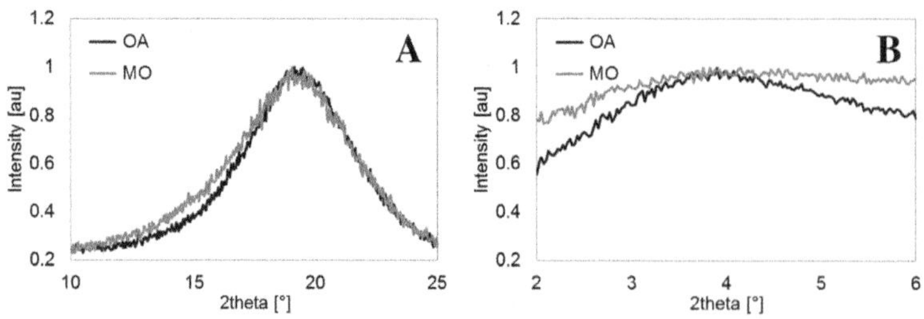

Figure 4 X-ray spectra of OA and MO measured using (A) XRD and (B) SAXS at 298 K. The peak at around 0.14 nm^{-1} ($2\theta \approx 19.8°$) is sharp for both materials, whereas the peak at around 0.03 nm^{-1} ($2\theta = 4.2°$) is very broad and difficult to resolve, especially for the MO sample. SAXS: small angle X-ray scattering; XRD: X-ray diffraction.

Table 1 Short- and long-range spacing, d, of OA and MO at 298 K

	d_{OA} [nm]	d_{MO} [nm]
Short spacing (XRD)	0.459	0.460
Long spacing (SAXS)	2.383	2.531
		2.517[a]

[a]Long spacing measured at 263 K.
d_{OA} and d_{MO} are the short- and long-range spacings of OA and MO, respectively.
SAXS: small angle X-ray scattering; XRD: X-ray diffraction.

in the short spacing with temperature for both OA and MO. The long spacing for MO also increased with temperature and showed a moderate, close-to-linear increase from a temperature close to the melting point until 343 K. These results imply a larger fluidity of MO compared to OA, because MO molecules separate both longitudinally and transversely from one another as the temperature is increased. Still, the appearance of both peaks for MO suggests that a degree of order, though on a smaller scale, is maintained even at high temperatures.

Self-diffusion coefficients

The molecular structure of the molecules is expected to affect their translational movement. We therefore measured the self-diffusion coefficients, D, of OA and MO at various temperatures (Figure 6). As presented, MO has larger D values compared to OA for all the temperatures, meaning the translational movement of the ester is considerably larger than the acid. Both materials exhibit Arrhenius dependence of the form:

$$D = D_0 exp\left(-\Delta E_{app}/RT\right) \tag{1}$$

with apparent activation energies, ΔE_{app}, of 27.0 and 19.5 KJ/mol for the OA and MO molecules, respectively. ΔE_{app} of OA is in fair agreement with the value reported by Yamamoto *et al.* [25] for the same molecule, and interestingly also with the value reported for liquid TGs [36]. These results suggest that OA molecular movement within its bulk liquid is considerably reduced compared with MO, and requires an equivalent amount of energy as a TG to initiate diffusion.

Dynamic viscosity measurements at different temperatures were performed on the MO sample (Table 2). These were then used to calculate the apparent hydrodynamic radius, r, from the Stokes-Einstein formula under a slip boundary condition (Equation (2) [17]):

$$r = \frac{kT}{4\Pi\eta D} \tag{2}$$

where k is Boltzmann's constant, D is the self-diffusion coefficient, and η the dynamic viscosity. The apparent hydrodynamic radius of MO follows a very moderate linear ($R^2 = 0.80$) increase with temperature (Table 2). This result agrees with the moderate increase with temperature of the short- and long-range spacings (Figure 5A and B). The r values of OA were also calculated using the dynamic viscosities given in [19].

Despite the large difference in ΔE_{app} of OA and MO, their apparent hydrodynamic radii were found to be very close (Table 2).

Iwahashi *et al.* [22] found that nonanoic acids in the liquid state remain as dimers even at 363 K. Based on this finding, they concluded that for normal FAs, dimers are the units in their intra- or intermolecular movements. Furthermore, they calculated the hydrodynamic radius of several normal FAs in the range of C8 to C18, and found that it decreases very slightly with increasing hydrocarbon chain. This suggested that the rotational (end-over-end) as well as transverse motion of each dimer is severely restricted, and that only a longitudinal translation (translational movement along molecular axes) is allowed.

Since the hydrodynamic radii r of OA and MO are very close, the large difference in D, and consequently ΔE_{app}, can be attributed to their viscosity differences. It is well known that for long hydrocarbons, viscosity increases with number of carbons, due to a higher number of van der Waals interactions with adjacent molecules. Since both molecules have similar tails, one may rationalize that

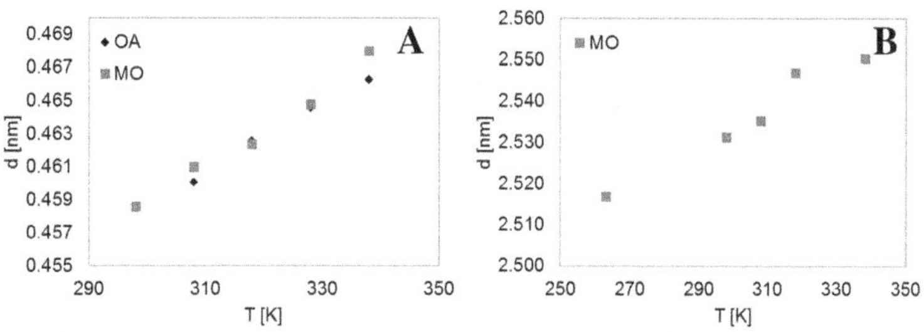

Figure 5 Response of (A) short- and (B) long-range spacings, d, with temperature using X-ray measurements. A similar increase in the short spacing with temperature for both OA and MO was observed. The long spacing for MO also increased with temperature and showed a moderate, close-to-linear increase from a temperature close to the melting point until 343 K.

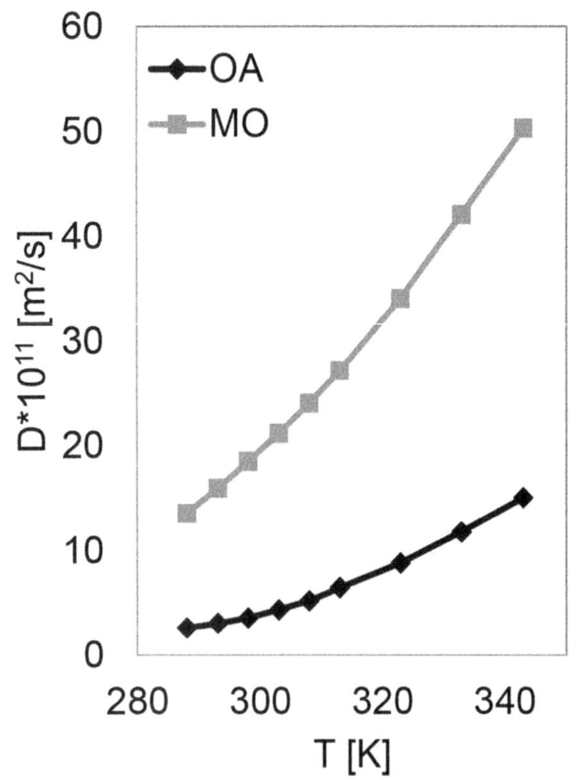

Figure 6 Self-diffusion coefficient, D, of OA and MO at various temperatures. MO exhibits larger D values compared to OA for all the temperatures, meaning that the translational movement of the ester is considerably larger than that of the acid.

Table 2 Dynamic viscosity, η, of MO and apparent hydrodynamic radius, r, of MO and OA according to temperature

T [K]	η_{MO} [mPa s]	r_{MO} [nm]	r_{OA}[a] [nm]
288	6.97	0.339	
293	6.03	0.332	0.315
298	5.27	0.336	
303	4.66	0.339	0.337
308	4.14	0.340	
313	3.71	0.341	0.330
318	3.34		
323	3.03	0.344	0.337
328	2.76		
333	2.52	0.345	0.345
338	2.32		
343	2.15	0.348	0.364
348	1.99		
353	1.85		
358	1.72		

[a]Calculated using the dynamic viscosities given in [19].
η_{MO} is the dynamic viscosity of MO; r_{MO} and r_{OA} are the apparent hydrodynamic radii of MO and OA, respectively.

for the temperature range applied in this study, unlike the case of OA, which occurs as dimers, single MO molecules are the units in translational diffusion. Thus, for free motion, we would expect a larger difference in hydrodynamic radii r for single versus dimerized molecules (18 carbons versus 36 carbons). However, as stated before, the motion for long rod-like molecules is restricted to linear molecular movement; therefore, similar r values for the OA and MO molecules are to be expected.

Segmental motion

The spin lattice, T_1, is likely to be correlated to the movement of the carbon atoms, i.e., segmental motion (specifically rotational tumbling and to a lesser extent translational and internal motion) in the molecule. The segmental movements at the end and near the end of the molecule are probably most important for the OA and MO molecules to find the spaces for their translational diffusion [21]. Of course, their close molecular arrangement and intermolecular interactions will work to hinder the segmental motion of some carbons. Segmental motion through the reciprocal of the effective correlation time, $1/\tau_c$, of each carbon can be calculated from T_1, measured by ^{13}C HF-NMR. Figure 7 shows the $1/\tau_c$ values of OA at 298 K and MO at 298, 318, 338, and 358 K. Assignment of ^{13}C chemical shifts to the appropriate peaks was performed according to [37]. The results of $1/\tau_c$ values of OA presented in this work are in excellent agreement with the results presented elsewhere [21].

A very close pattern can be seen when comparing the segmental motion of the different carbons on the OA and MO chains at 298 K. For both molecules, the rotational movement of the double bond carbons is considerably restricted due to stronger intermolecular interactions between the pi electrons of the double bond, and increases towards the end of the chains.

Nevertheless, two main differences are observed, which are both associated with the dissimilarities of head-to-head intermolecular interactions. The first is that all $1/\tau_c$ values of the carbons along the chain of OA are smaller compared to the $1/\tau_c$ of the carbons at the same position on MO. This is analogous to the shift of the entire distribution of MO to higher T_2 values (Figure 1B), and can be attributed to the lower viscosity of MO. The second involves the rotational movements of the carbons closest to the head (C2 to C4). As suggested before, at a temperature of 298 K, OA molecules are almost entirely dimerized. This dimerization restricts the rotational movement, leading to pronounced rigidity compared to the methyl end. The head in MO, on the other hand, has a higher freedom of movement and is not tightly bonded, since $1/\tau_c$

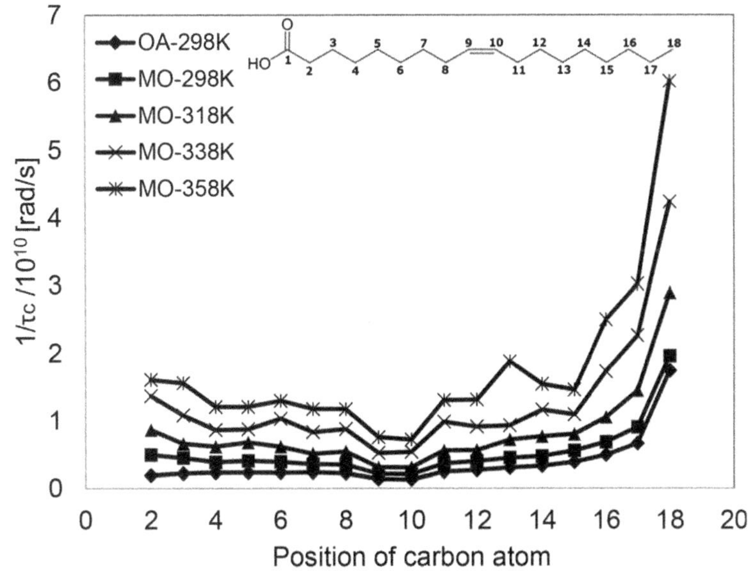

Figure 7 Segmental motion of OA and MO at different temperatures. Segmental motion through the reciprocal of the effective correlation time, $1/\tau_c$, of each carbon can be calculated from T_1, measured by ^{13}C high field (HF)-NMR. OA was measured at 298 K, and MO at 298, 318, 338, and 358 K. The structure of OA along with designation of carbon numbers is shown for reference.

values decrease from the second carbon towards the double bond.

Nevertheless, there is a pronounced difference in $1/\tau_c$ between the two ends of MO. Even though no strong hydrogen bonding exists in this ester, the polar interactions are sufficiently strong to limit the rotation of the head, so that it does not behave like the tail. This strengthens our initial assumption of polar interactions between the ester head groups.

As stated before, translational diffusion is probably initiated by the ends of the molecules. In the case of OA, dimers of two hydrogen-bonded molecules would move by the flipping of both tails on the dimer. MO molecules, on the other hand, would find available spaces for translational movement by very vigorous rotation of the tail, but also by wagging of the head.

$1/\tau_c$ values of all carbons in the MO chain increase with temperature, while maintaining the pattern described before. Interestingly, the $1/\tau_c$ values of all the carbons on MO at 358 K (apart from C9 and C10) are higher than the $1/\tau_c$ value of C17 at 298 K. This implies that the entire MO molecule moves around in the same vigorous manner as the tail does at 298 K. This is not the case for OA, judging by the values presented in [21]. Owing to the dimerization of the head, even at comparable high temperatures, the head and adjacent carbons do not reach the same degree of motion as the tail at 298 K.

^1H and ^{13}C HF-NMR chemical shift analyses

HF-NMR is a useful tool for identifying non-covalent interactions, as chemical shift is a sensitive measure of local chemical environment. Intermolecular interactions can be identified by changes in chemical shift. Therefore, to learn about the intermolecular interactions and their changes with temperature, we observed the change in chemical shifts of each proton and carbon on OA and MO, acquired using ^1H and ^{13}C HF-NMR, respectively.

Unless otherwise stated, all chemical shifts moved downfield with increasing temperature. This can be explained by the increase of bond length with temperature, which requires less energy to cause nuclei inversion. In contrast, the carbon and proton on the carboxyl and hydroxyl groups, respectively, of the OA showed a varying behavior with increasing temperature, with maxima at approximately 320.5 and 327 K, respectively (Figure 8A and B for the carbon and proton, respectively). Iwahashi *et al.* [19] found that a discontinuous change takes place in some physical properties of the liquid OA around 328 K, and concluded that the liquid structure or molecular conformation changes into a more highly disordered one. This may serve as an explanation for the upfield increase of chemical shifts above this approximate temperature. Another possible explanation is the possibility that above a given temperature the H-bonding interactions between the carboxyl and hydroxyl groups is reduced by the mobility, so that the polar-electrostatic interaction in H-bonding is reduced, inducing an upfield shift. This is similar to the upfield shift observed on C1 of OA, as a function of pH [38], and on the OH group of octanoic acid as a function of temperature [26].

The ^1H and ^{13}C chemical shifts for all other protons and carbons show a linear increase with temperature with

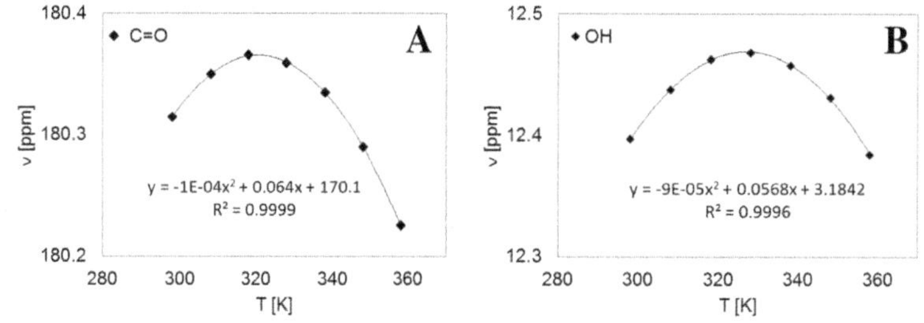

Figure 8 HF-NMR chemical shift, v, of the OA head (A) ^{13}C and (B) ^{1}H at different temperatures. The carbon and proton on the carboxyl and hydroxyl groups, respectively, of OA, showed a varying behavior with increasing temperature, with maxima at approximately 320.5 and 327 K, respectively.

different slopes. While the slopes for the different protons along the chain are very similar, those for the carbons differ to a great extent depending on their position (Figure 9). It is likely that the carbons with larger $1/\tau_c$, meaning high rotational motion (Figure 7), are less affected by the changes in temperature and therefore exhibit small slopes. This is true for the carbons in the tail side of both OA and MO molecules. The double bond carbons C9 and C10, and the neighbor carbons, which show smaller $1/\tau_c$ and low mobility, consequently show larger slopes. The C2 and C3 carbons on the head side of the molecules do not follow this behavior. For MO, although these carbons show a moderate increase in mobility with temperature (Figure 7), they show larger slopes than the carbons on the tail side. This apparent anomaly could be explained by the deshielding effect of the carboxyl group with temperature. The effect is induced through the C-C σ-bond, and is more effective in the α carbon than the β carbon, so that the slope of C2 > C3.

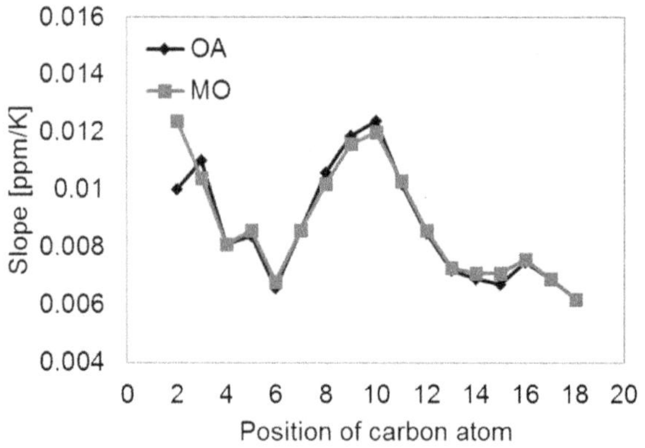

Figure 9 The slopes for the ^{13}C HF-NMR chemical shift, v, as a function of temperature for each carbon. The slopes were calculated using linear regression of the change in ^{13}C chemical shifts with temperature. Apart from C2, OA and MO have very similar slopes. On the other hand, the slopes differ to a great extent depending on their position.

Conversely, in OA, the slope of C2 < C3. This can be explained by the varying behavior observed for the carboxyl group, which shows an upfield increase in chemical shift with temperature. Similar to MO, this shielding effect propagates trough the C-C σ-bonds, and is more intense in carbon α than β, explaining the C2 < C3 slope.

^{1}H HF-NMR T_1 and T_2

In order to assign T_2 values of different protons along the chains of OA and MO, the perfect echo sequence [39] was applied on a Bruker AVANCE III 600 MHz NMR spectrometer, at 298, 318, and 338 K. Peaks were assigned according to [40]. All protons exhibited a monoexponential behavior. Figure 10A and B show the monoexponential T_2 values for the resolved protons according to the position of the attached carbon. The T_2 values fit very well with the segmental motion presented in Figure 7, with the exception of the double bond protons. For all temperatures, the protons on MO have larger T_2 values compared to the equivalent ones on OA. All the protons show a linear increase in T_2 values with temperature. Interestingly, the T_2 values of MO show a greater response with temperature, compared to OA. This is possibly comparable to the larger increase with temperature of the higher mobility peak (T_{22}) and relative peak contributions of MO as seen in the T_2 distributions (Figure 2A-D).

^{1}H T_1 values were also measured for OA and MO, on the 600 MHz NMR spectrometer, at 298 K. Table 3 shows the T_1/T_2 ratio of protons according to the position of the attached carbon for the two samples at 298 K. In the case of $2\Pi\nu_0\tau_c < 1$, where ν_0 is the Larmor frequency and τ_c is the correlation time, $T_1 \approx T_2$ with T_2 slightly smaller than T_1. This is the case for non-viscous small molecules that exhibit fast rotation and tumbling. For larger molecules, rotation and tumbling are hindered, and an increase in T_1 along with a decrease in T_2 is expected. Based on this designation, MO exhibits liquid-like behavior, while OA shows a more hindered nature, with the highest rigidity presented for the head and double bond protons (Table 3).

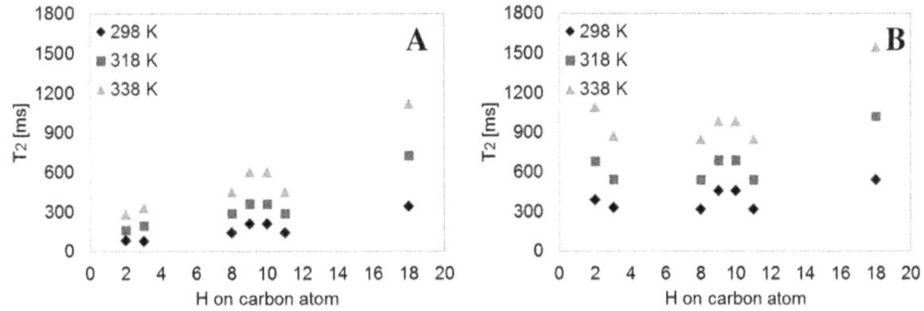

Figure 10 Monoexponential T_2 values of the resolved protons according to the position of the attached carbons. Measurements of **(A)** OA and **(B)** MO were performed at 298, 318, and 228 K using ^{1}H HF-NMR spectrometer.

Discussion

The only primary molecular difference between MO and OA is the methyl group on the acid moiety. This significantly affects the morphological structure of the material above its melting point. A key question at this point is whether MO in the liquid at the temperature range of this study has a molecular arrangement resembling that of OA. A molecular arrangement in this case would resemble to some extent that of a liquid crystal (as suggested for OA [19]), meaning molecules within this arrangement diffuse much like those in a liquid, while maintaining some degree of orientational and sometimes positional order as a function of the specific morphology. More specifically, this would mean that at least one molecular axis tends to point along a preferred direction as the molecules undergo diffusion [41].

Following this classification, it is suggested that both OA and MO possess a liquid crystal-like order, according to the apparent hydrodynamic radius, r, since for both molecules the rotational, as well as transverse, motion is to some extent restricted, and only a longitudinal translation is allowed. Still, the dimerization due to hydrogen bonding of the head groups in OA leads to a greater molecular rigidity compared to MO, and to a more efficient packing. Considering the head-to-head model interaction suggested in this paper, it is also reasonable to assume the quasi-smectic liquid crystal-like structure for MO, where heads are aligned next to tails in an interdigitated structure, similarly to the case of OA [19]. Evidence can be seen from the long-spacing (Table 1) of lateral planes, which is roughly equivalent to the length of a single MO molecule. In the case of adjacent heads the long-spacing is expected to be approximately double the one reported. However, the broader small angle peak acquired for MO (Figure 4B) suggests a less ordered structure in this respect.

It is also important to consider that the self-organization of these long rod-like molecules is temperature dependent, and may vary to a great extent as temperature increases further away from the melting point. Phase transition temperatures can give an estimation of intermolecular interactions and the degree of molecular order. Higher melting points are related to more efficient packing in the solid. A boiling temperature is required to break all intermolecular interactions present in the liquid. Crystallization is affected by the degree of order present in the liquid, such that more ordered molecules will crystallize at higher temperatures. The crystallization, melting, and boiling points of OA and MO according to the literature are shown in Table 4 (for OA the lower melting point for the α polymorph is referenced). The higher crystallization, melting, and boiling points of OA are an indication of the higher degree of intermolecular interactions and molecular packing arrangements compared to MO, both in the solid and in the liquid. Based on their chemical structure, this is clearly the effect of hydrogen bonding of the head groups in OA versus weaker polar interactions for MO.

Interestingly, a significantly larger temperature hysteresis between crystallization and melting points exists for MO compared to OA (20.6 versus 8.3 K, respectively, calculated from Table 4). This suggests a smaller degree of order in MO, thus requiring a substantially reduced temperature to crystallize, since crystallization requires

Table 3 ^{1}H T_1/T_2 ratio of OA and MO measured using HF-NMR spectrometer at 298 K

	^{1}H T_1/T_2	
	OA	MO
C2	8.50	2.54
C3	7.59	2.65
C8/C11	4.48	2.23
C9/C10	5.87	2.71
C18	3.95	2.66

Table 4 Comparison of phase transition points of OA and MO

	Crystallization point [K]	Melting point [K]	Boiling point[a] [K]
OA	277.7 [14]	286.0[b] [14]	496.0 [42]
MO	232.5 [43]	253.1 [35]	474.0 [42]

[a]Measured at 1.333 kPa.
[b]The melting point for the α polymorph is referenced.

good molecular packing to reach an adequate entropy level.

Lastly, the task of assigning the molecular basis of the two peaks in the T_2 distributions remains (Figure 1A and B). Based on the acquired information, we now refer to the original suggestions for the two peaks and offer more informed explanations.

The two peaks are the result of inhomogeneous organizations with two different packing densities and intermolecular interaction intensities or types

Badmaev *et al.* [44] suggested a cluster model by which any liquid presents a micro-inhomogeneous medium, consisting of two dynamic components: ordered areas (clusters) and an inhomogeneous disorder matrix. Hernqvist [45] also proposed a dynamic model for liquid tristearin that consists of a lamellar liquid crystalline phase, where the size and orientation of the units vary with diffusion rates of the molecules and therefore with changes in temperature. This can be described as a transiently structured liquid with centers of organized structures forming and dissolving continuously, thus forming an equilibrium structure.

Following the cluster model, a reasonable assignment for the T_2 distributions of OA and MO would be that the first peak (T_{21}) consists of the molecules in the liquid crystal clusters and the second peak (T_{22}) would be the result of the more liquid-like molecules. The molecular structure described before would therefore be responsible for the organization within the liquid crystal clusters, whereas the other liquid molecules in the amorphous morphology volumes would diffuse randomly throughout the sample volume with the molecular axes in rapid rotary-like movement. This kind of microstructural organization can be the result of structural memory coming from the solid structure, and can very well explain the two peaks.

Following this model, T_{21} would stand for the average T_2 value for all protons inside the liquid crystal cluster, and T_{22} the average T_2 value for the free more mobile ones. Since a very small difference exists between the two groups, T_{21} and T_{22} have close values at each temperature, although due to a more ordered nature and closer packing, T_{21} is smaller. As temperature increases, molecules from the cluster break up and transfer to the disordered matrix, as can be seen for the change in the relative contributions of the peaks. This pattern is more observed for MO, in the range of temperatures tested in this study. As suggested before, due to dimerization of the heads in OA, it has a denser arrangement and requires higher activation energy to initiate diffusion. As a result, the exchange of molecules between the two groups is considerably slower than that of MO, and the peak loss of T_{21} in OA is significantly less than in MO as the temperature is increased.

A very interesting phenomenon that supports this model was observed in the T_2 distributions, whose possible explanation may be monitoring of the exchange of molecules between amorphous and liquid crystal environments until stabilization. This can be seen for MO heated from 193 to 288 K, the temperature of measurement (Figure 11). Measurements were taken at five increasing times (t1 to t5 according to the order of measurement) until final stabilization was achieved at t4. As shown, the relative contribution of the peaks changes in favor of the second peak as time progresses, until reaching a steady state.

Pulse NMR has long been suggested as a tool for measuring solid fat content in partially crystallized fats. In the food industry, solid fat content values measured at different temperatures can be used to help predict important attributes such as mouth-feel and hardness. In this application the signals of both the solid (crystallized) fat and liquid oil are acquired. Possibly, with the current method, it is also conceivable to measure the ratio between the more rigid (clusters) and looser parts of the

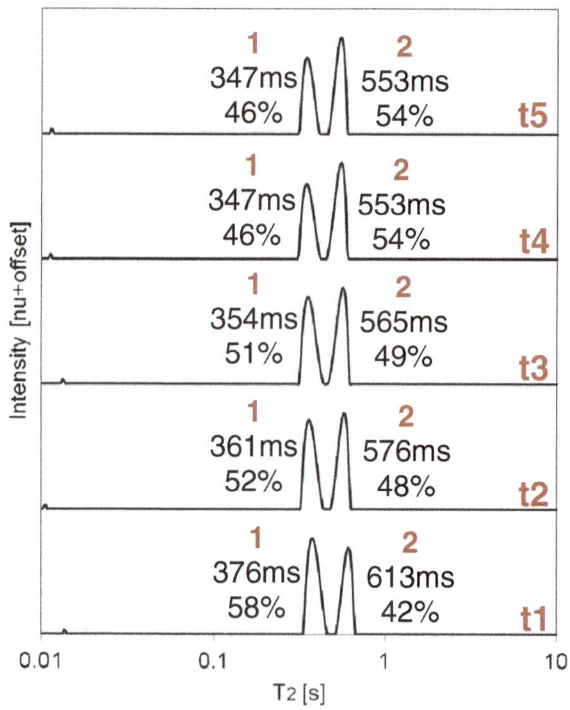

Figure 11 Monitoring of the exchange of molecules between populations (possibly amorphous and liquid crystal environments) until stabilization. Combined ^{1}H low field (LF)-NMR T_2 distributions of MO heated from 193 to 288 K, the temperature of measurement, at five increasing times (t1 to t5 according to the order of measurement). The first measurement (t1) was performed following 1 h of stabilization at 288 K. The relative contributions of each peak, in relation to other peaks and intrinsic T_2 values, are shown on each plot. The relative contribution of the peaks changes in favor of the second peak as time progresses, until reaching a steady state.

liquid. NMR and especially LF-NMR relaxometry would therefore be an excellent tool for monitoring changes in weak morphologies and/or interactions. This may be due to the low energy required to excite spin systems. The low frequency relaxation process in liquids concerns low energy processes, and can be explained by interaction of clusters.

The two peaks are the result of two distinct mobility populations of the protons on the chain

Since temperature influences the self-organization of OA and MO, as suggested before, comparison of distributions would be made relative to the temperature of melting of each compound. Therefore, the comparison would be performed by subtracting the melting point of each compound (about 288 and 253 K for OA and MO, respectively) from the temperature of measurement. Following this rule, the MO distribution acquired at 288 K should be compared with the OA distribution acquired at 318 K, and so on. It can be seen (Figure 1A and B) that T_{22} is almost similar for the two materials (most pronounced for high temperatures), whereas T_{21} of OA is shorter than that of MO. This can indicate two populations: T_{22} is very close for both materials and can be assigned as the less restricted parts of the molecules; T_{21}, on the other hand, is the more rigid part of the molecules, where OA is more restricted than MO (due to smaller values). Rigidity will lead to differences in intermolecular interactions, such that the more rigid parts have a close neighbor to interact with, leading to lower T_2. The relative contribution of the peaks changes with temperature towards the less rigid peak. This is more pronounced for MO, since its head is freer to move compared to OA, as shown by the segmental motion and T_2 values measured on a 600 MHz ^{1}H HF-NMR spectrometer (Figures 7 and 10A, B, respectively). Peak assignments would therefore be as follows:

For OA, at intermediate temperature, the hydrogens close to the head, from C2 to C10 (18 hydrogens) are less mobile (T_{21} group) than the hydrogens from OH and from the tail, C11 to C18 (18 hydrogens, T_{22} group). The 1:1 ratio between these two groups is observed at 308 K. At higher temperatures, part of the head (C2 to C10 hydrogens) develops a mobility similar to that of the tail. At lower temperatures part of the tail hydrogens are in the head signal.

The assignment for MO is similar. At low temperatures, the C11 to C18 and OCH_3 (20 hydrogens) are grouped in T_{22} and C2 to C10 (16 hydrogens) in T_{21}. This is the ratio at 288 K. At higher temperatures, the T_2 of part of the hydrogens from the head group "jumps" from the short to long T_2. At 338 K, apparently, only the two olefinic hydrogens are part of T_{21} (6/94%).

In order to rule out the possibility that the large change in relative contributions of the peaks of MO, in response to an increase of temperature, is due to a loss of structural organization at temperatures far above the melting point, additional measurements were carried out at 258, 268, and 278 K (Figure 12, this is an extension of Figure 1B). Two peaks were once again observed for all temperatures, with similar trends of increasing T_2 values and relative contributions with increasing temperatures, meaning the change in T_2 values of the peaks and distribution between populations is constant from the melting point and above. This strengthens our assumption that the different response to temperature of OA and MO is due to differences in their chemical composition (head interactions). Following the comparison rule proposed before, the T_2 distributions near melting should be compared (288 and 258 K for OA and MO molecules, respectively). Based on the peak assignment suggested before, it appears that at 258 K 75% of MO is rigid, though still more mobile than OA (both T_{21} and T_{22} have larger values than for OA).

Both theories for the assignment of peaks in the T_2 distributions can be sustained by logical reasoning. However, the supporting data presented, especially ^{13}C and ^{1}H HF-NMR measurements, suggest that the two peaks are the result of two distinct mobility populations of the protons on the chain. An additional possible experiment to get a more conclusive choice between the models could be the use of partially deuterated OA and MO, for example, deuterating the carbons in the tail end (C11 to C18). If the T_2 distribution remains unchanged, it would support the first model. However, a large reduction in

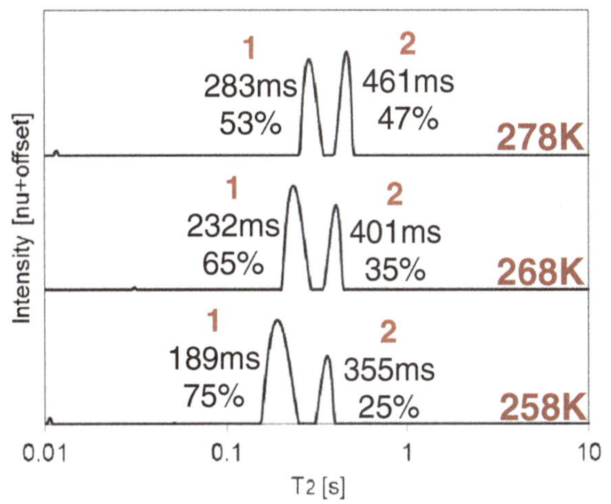

Figure 12 Combined ^{1}H low field (LF)-NMR T_2 distributions of MO at low temperatures (just above melting point). This figure is an extension of Figure 1B. The relative contributions of each peak, in relation to other peaks and intrinsic T_2 values, are shown on each plot. The change in T_2 values of the peaks and distribution between populations is constant from the melting point and above.

the relative contribution and T_2 value of the second peak would support the second model. Additional possible experiments would also involve measurement of several other standard FA and FAME materials of various chain lengths and degrees of unsaturation. Our research group has already initiated a thorough research design comprising these experiments. However, these are beyond the scheme of the current work and will be discussed in a separate publication.

The goal of this work was to explore the potential of ^{1}H LF-NMR relaxometry for characterizing the molecular organization of lipids in the liquid phase. Although the T_2 distributions acquired may well explain the cluster model of a microstructure arrangement and exchange of molecules between domains, the supporting measurements performed suggest otherwise. Still, bearing in mind that the mobility of the molecules is the direct outcome of their morphological structure, the differences in the molecular arrangement of OA and MO can be proposed by monitoring the differences in T_2 distributions and peak area in response to a gradient of temperatures. This can be observed from the similarities in T_2 distributions in relation to melting point. In this way, the large change in relative contribution of the peaks for MO suggests a less dense packing compared to OA. Ultimately, the dimerization of the head in OA compared to weaker polar interactions of the head can be clearly deduced from the increased intrinsic T_2 values of MO.

Compared to other spectroscopic methods used in this study, ^{1}H LF-NMR was found to be highly susceptible to low energy, weak intermolecular interactions, and aggregation of molecules. This technology is therefore suggested as a potentially important tool for the field of biodiesel. Several disadvantages have been suggested for the use of biodiesel in unmodified diesel engines, including cold weather operability and flow properties. In addition, a major problem exists with oxidative stability and the stability of biodiesel-petrodiesel blends, especially during prolonged storage. These physico-chemical properties are the direct outcome of the mobility and molecular structure of the fuel in the liquid. A possible application could be to test different additives to reduce dipolar interaction between MO heads. A small amount of additive could increase mobility and reduce viscosity, which would be easily observed by ^{1}H LF NMR. The same could be applied to test the effect of biodiesels prepared from different feedstocks, meaning different FA profiles, and/or biodiesel-petrodiesel blends, and their differences in mobilities and packing. By reducing the temperature, going from the liquid down to the crystallization point, this technology may also be applied to test the initiation of crystallization and to monitor its progress. ^{1}H LF-NMR can also be used for monitoring oxidation of biodiesels. We found an indication for this application in the course of our experiments on a sample of methyl linolenate. An additional peak at lower than usual T_2 values was observed following prolonged storage.

Conclusions

We have shown that OA and MO have both similarities and differences as seen in the ^{1}H LF-NMR relaxation times and relative contributions according to temperature. These were attributed to a degree of liquid crystal-like order of both molecules that is temperature dependent, and that affects their ability to interact with close neighbors. Two suggestions have been made for the peaks: two distinct mobility populations of the protons on the chain; or a microstructural organization of the liquid with ordered areas and inhomogeneous disorder matrix. These properties allow for the material characterization based on parameters that contribute to important material properties in applications such as biodiesel fuels. This new application is of high potential to the field of biodiesel, and to other research and applied disciplines wherein relative weak interaction forces play an important part in physicochemical characteristics.

Materials and methods

Materials

High purity (≥99%) OA and MO lipid standards were purchased from Sigma-Aldrich. The standards were used as received without further purification. Both standards were kept at 253 K between measurements, and heated from this temperature to the required temperature of measurement.

^{1}H LF-NMR relaxometry and diffusiometry

All ^{1}H LF-NMR measurements were performed on a 20-MHz minispec bench-top pulsed NMR analyzer (Bruker Analytik GmbH, Rheinstetten, Germany), equipped with a permanent magnet and a 10-mm temperature-controlled probe head. Measurements were first performed in the temperature range of 288 to 348 ± 0.03 K in 5 K steps. Prior to measurement, the samples were heated for minimum 1 hour and then allowed to equilibrate inside the instrument for 5 minutes. All measurements were performed on liquid standards (above melting point). For each temperature and sample the following parameters were optimized: receiver gain, magnetic field offset, detection angles, P90, P180, and homogeneity.

For MO, comparison of the data acquired at several temperatures starting from 193 or 253 K was performed, and similar distributions were found for the same temperature. However, the time to achieve stable results from 193 K was very long (several hours, especially for the higher temperatures). Therefore, MO was kept at 253 K between measurements, and heated from this temperature to the required temperature of measurement.

Determination of spin-spin relaxation (T_2) was performed using a CPMG (Carr, Purcell, Meiboom, and Gill [46,47]) pulse sequence. Values of τ of 0.4 and 1.125 s and recycle delays of 4 and 6 s were used for the OA and MO samples, respectively. Additional CPMG measurements of MO were performed at close to melting temperatures (258 to 278 ± 0.03 K in 10 K steps). Acquisition at low temperatures for MO was performed with a τ of 0.4 to 0.5 s and a recycle delay of 5 s. For all the analyses, 32 scans were accumulated and 8,000 echoes were acquired. Data was acquired in magnitude mode due to better repeatability and stability of results, and further analyzed using the primal-dual interior method for convex objectives (PDCO) optimization algorithm with $\alpha_2 = 0.5$, as described in [7]. CONTIN (software application for inverse Laplace transformation of LF-NMR relaxometry data available in minispec) was also used for analyzing the acquired CPMG data, in order to compare distributions with PDCO. The mathematical formulation of CONTIN is described in [48]. Both methods exhibited very similar T_2 distributions, although better repeatability and stability was found with PDCO analyses.

The self-diffusion coefficient, D, was determined by the pulsed-field gradient spin echo (PFGSE) method [49]. The PFGSE sequence was used with 16 scans, a τ of 7.5 ms, and a recycle delay of 6 s. Typical gradient parameters were Δ of 7.5 ms, δ of 0.5 ms, time between the 90° pulse to the first gradient pulse of 1 ms, and g of 1.6 T/m. A water sample (1.25 g/L $CuSO_4$) was used for calibration. The D values of water were taken from [50]. Each reported value is the average of a minimum of ten measurements.

High field ^1H and ^{13}C-NMR chemical shift, spin-spin, and spin-lattice relaxation time

The high field ^1H and ^{13}C-NMR measurements were performed on a BRUKER AVANCE III operating at 600 MHz for ^1H nuclei and 150 MHz for ^{13}C. Prior to measurement, samples were heated for a minimum of 10 minutes and added in a 5-mm NMR tube. For the lock signal, a closed 1-mm capillary tube, filled with D_2O, was added to the sample. The non-spinning samples were allowed to equilibrate inside the instrument for 15 minutes after reaching the set temperature. Before each measurement, shimming was optimized using automated and manual procedures. The chemical shifts, in parts per millions (ppm), were obtained without a reference signal. The ^1H and ^{13}C NMR spectra were obtained using 4 and 8 scans and recycle delays of 30 s and 120 s, respectively.

The ^1H and ^{13}C longitudinal relaxation times, T_1, were measured using the inversion recovery method [51]. The ^1H transverse relaxation time, T_2, was measured using a modified (perfect echo) CPMG pulse sequence (PROJECT - Periodic Refocusing of J Evolution by Coherence Transfer) that resulted in spectra without J modulation [39]. The calculations of T_1 and T_2 were carried out with the subroutines included in the TOPSPIN 3.2 software package.

^{13}C HF-NMR spin-lattice relaxation of a protonated carbon is overwhelmingly dominated by dipole-dipole interactions with the attached protons [21]. T_1 is therefore related to the number of directly bonded hydrogens, N, and the effective correlation time, τ_c, for the rotational movement of the carbon atoms in the object molecule. Thus, T_1 is approximately given in terms of N and $1/\tau_c$:

$$T_1 = \frac{r_{CH}^6}{N\hbar^2\gamma_C^2\gamma_H^2}\left(\frac{1}{\tau_C}\right) \quad (3)$$

where ħ is Planck's constant and γ_C and γ_H are the gyromagnetic ratios of ^{13}C and ^1H, respectively. Here, r_{CH} is the C-H distance, usually about 0.109 nm, and the reciprocal of the effective correlation time, $1/\tau_c$, represents the magnitude of the segmental rotation for the carbon atom at a different position.

X-ray methods

In this study, XRD and SAXS techniques were used for measuring the short- and long-range spacing between adjacent molecules, respectively.

XRD data was collected on a Panalytical Empyrean Powder Diffractometer equipped with a position-sensitive (PSD) X'Celerator detector using Cu K_α radiation (λ = 0.154 nm) and operated at 40 kV and 30 mA. The usual Bragg-Brentano $\theta/2\theta$ geometry was employed. $\theta/2\theta$ scans were run for 15 minutes in a 2θ range of 2 to 35° with step equal to about 0.0167°. Measurements were performed at a range of 298 to 338 K in 10 K steps.

SAXS measurements were performed on a SAXSLAB GANESHA 300-XL (Skovlunde, Denmark) instrument. Cu K_α ($\lambda = 0.154$ nm) radiation was generated by a Genix 3D Cu source (operated at 47 mV and 0.55 mA) with integrated monochromator, 3-pinhole collimation, and a two-dimensional Pilatus 300 K detector. The distance between the sample and detector was 350 mm. The q range was between 0.0012 to 0.067 nm^{-1}. OA was measured at 298 K, for comparison with the literature, and MO was measured at several temperatures, including 263 K, 298 to 318 K in 10 K steps, and 338 K.

Dynamic viscosity

Dynamic viscosity measurements of MO were performed on an AR 2000 Rheometer (TA Instruments), on a double-gap Peltier cylinder system in steady-state flow mode, in the temperature range of 288 to 358 K. For each temperature, 10 points were acquired in the range shear rate 10 to 300 s^{-1}, and the average was reported.

Abbreviations

CPMG: Carr, Purcell, Meiboom, and Gill; FA: fatty acid; FAME: fatty acid methyl ester; HF-NMR: high field nuclear magnetic resonance; LF-NMR:

low field nuclear magnetic resonance; MO: methyl oleate; OA: oleic acid; PFGSE: pulsed-field gradient spin echo; SAXS: small angle X-ray scattering; TG: triglyceride; XRD: X-ray diffraction.

Competing interests
The authors declare that they have no competing interests.

Authors' contributions
This work is part of the doctoral thesis of PB, and she is the main author. PB designed most of the experiments, analyzed results, contributed to the establishment and validation of the new PDCO algorithm, and drafted the manuscript. PB and NM performed the ^{1}H LF-NMR measurements and analyzed the results. LAC and TBM performed the ^{1}H HF-NMR measurements and analyzed the results. OL, YP, and MS formulated, designed, and established the PDCO algorithm. ZW, LAC, and CL contributed to the development of the experimental design and proofread the manuscript. ZW led and coordinated the overall project. All authors read and approved the final manuscript.

Acknowledgements
PB acknowledges support from the Women in Science scholarship of the Israel Ministry of Science and Technology. NM acknowledges support from the Substitutes for Oil Transportation fellowship of the Israel Ministry of Science and Technology. MS acknowledges support from the National Institute of General Medical Sciences of the National Institutes of Health (award U01GM102098). The content is solely the responsibility of the authors and does not necessarily represent the official views of the funding agencies. The authors would like to thank Dr. Dimitri Mogiliansk, Dr. Sharon Hazan, and Dr. Mark Karpasas from the Ilse Katz Institute for Nanoscale Science and Technology at Ben Gurion University for performing the X-ray and dynamic viscosity measurements; and the Phyto-Lipid Biotechnology Lab (PLBL) members at Ben Gurion University of the Negev for their contribution to this work.

Author details
[1]The Phyto-Lipid Biotechnology Laboratory, Departments of Biotechnology, Energy and Environmental Engineering, Ben-Gurion University of the Negev, P.O. Box 653, Beer-Sheva 84105, Israel. [2]Embrapa Instrumentação, Rua 15 de Novembro 1452, São Carlos, SP 13560-970, Brazil. [3]Instituto de Física de São Carlos, Universidade de São Paulo, Av. Trabalhador Sao-Carlense 400, São Carlos, SP 13566-590, Brazil. [4]Zuckerberg Center for Water Sciences and Technology and Department of Biotechnology, Ben-Gurion University of the Negev, P.O. Box 653, Beer-Sheva 84105, Israel. [5]Department of Industrial Engineering and Management, Ben-Gurion University of the Negev, P.O. Box 653, Beer-Sheva 84105, Israel. [6]Department of Management Science and Engineering, Stanford University, Stanford, CA, USA.

References
1. Colnago LA, Engelsberg M, Souza AA, Barbosa LL. High-throughput, non-destructive determination of oil content in intact seeds by continuous wave-free precession NMR. Anal Chem. 2007;79:1271–4.
2. Prestes RA, Colnago LA, Forato LA, Vizzotto L, Novotny EH, Carrilho E. A rapid and automated low resolution NMR method to analyze oil quality in intact oilseeds. Anal Chim Acta. 2007;596:325–9.
3. Berman P, Nizri S, Parmet Y, Wiesman Z. Large-scale screening of intact castor seeds by viscosity using time domain NMR and chemometrics. J Am Oil Chem Soc. 2010;87:1247–54.
4. Willson RM, Wiesman Z, Brenner A. Analyzing alternative bio-waste feedstocks for potential biodiesel production using time domain (TD)-NMR. Waste Manag. 2010;30:1881–8.
5. Cabeça LF, Marconcini LV, Mambrini GP, Azeredo RBV, Colnago LA. Monitoring the transesterification reaction used in biodiesel production, with a low cost unilateral nuclear magnetic resonance sensor. Energy Fuel. 2011;25:2696–701.
6. Berman P, Leshem A, Etziony O, Levi O, Parmet Y, Saunders M, et al. Novel ^{1}H low field (LF)-NMR applications for the field of biodiesel. Biotechnol Biofuels. 2013;6:55.
7. Berman P, Levi O, Parmet Y, Saunders M, Wiesman Z. Laplace inversion of LR-NMR relaxometry data using sparse representation methods. Concepts Magn Reson A. 2013;42:72–88.
8. WinDXP Toolbox. Software release version 1.8, Distributed EXPonential analysis. Witney, UK: Oxford Instruments; 2006.
9. Silva RC, Carneiro GF, Barbosa LL, Lacerda V, Freitas JC, Castro EV. Studies on crude oil-water biphasic mixtures by low-field NMR. Magn Reson Chem. 2012;50:85–8.
10. Barros CN, Arêas EP, Figueiredo EN, Arêas JA. Low-resolution ^{1}H spin-spin relaxation of n-decane/water emulsions stabilized by β-casein. Colloid Surface B. 2006;48:119–27.
11. Bertram HC, Lars W, Jacob HN, Henrik JA. Direct measurement of phase transitions in milk fat during cooling of cream - a low-field NMR approach. Int Dairy J. 2005;15:1056–63.
12. Chatakanonda P, Chinachoti P, Sriroth K, Piyachomkwan K, Chotineeranat S, Tang HR, et al. The influence of time and conditions of harvest on the functional behaviour of cassava starch - a proton NMR relaxation study. Carbohydr Polym. 2003;53:233–40.
13. Hills BP, Le Floc'h G. NMR studies of non-freezing water in cellular plant tissue. Food Chem. 1994;51:331–6.
14. Marigheto N, Duarte S, Hills B. NMR relaxation study of avocado quality. Appl Magn Reson. 2005;29:687–701.
15. Adam-Berret M, Boulard M, Riaublanc M, Mariette F. Evolution of fat crystal network microstructure followed by NMR. J Agric Food Chem. 2011;59:1767–73.
16. Callaghan PT. The use of ^{13}C spin relaxation to investigate molecular motion in liquid tristearin. Chem Phys Lipids. 1977;19:56–73.
17. Iwahashi M, Kasahara Y. Dynamic molecular movements and aggregation structures of lipids in a liquid state. Curr Opin Colloid Interface Sci. 2011;16:359–66.
18. Small DM. Lateral chain packing in lipids and membranes. J Lipid Res. 1984;25:1490–500.
19. Iwahashi M, Yamaguchi Y, Kato T, Horiuchi T, Sakurai I, Suzuki M. Temperature dependence of molecular conformation and liquid structure of cis-9-octadecenoic acid. J Phys Chem. 1991;95:445–51.
20. Iwahashi M, Suzuki M, Czarnecki MA, Liu Y, Ozaki Y. Near-IR molar absorption coefficient for the OH-stretching mode of cis-9-octadecenoic acid and dissociation of the acid dimers in the pure liquid state. J Chem Soc Faraday Trans. 1995;91:697–701.
21. Iwahashi M, Kasahara Y, Matsuzawa H, Yagi K, Nomura H, Terauchi H, et al. Self-diffusion, dynamical molecular conformation and liquid structures of n-saturated and unsaturated fatty acids. J Phys Chem B. 2000;104:6186–94.
22. Iwahashi M, Kasahara Y, Minami H, Matsuzawa H, Suzuki M, Ozaki Y. Molecular behaviors of n-fatty acids in liquid state. J Oleo Sci. 2002;51:157–64.
23. Iwahashi M, Takebayashi S, Umehara A, Kasahara Y, Minami H, Matsuzawa H. Dynamical dimer structure and liquid structure of fatty acids in their binary liquid mixture: dodecanoic and 3-phenylpropionic acids system. Chem Phys Lipids. 2004;129:195–208.
24. Iwahashi M, Umehara A, Wakisaka K, Kasahara Y, Minami H, Matsuzawa H, et al. Effect of cholesterol and other additives on viscosity, self-diffusion co-efficient, and intramolecular movements of oleic acid. J Phys Chem B. 2007;111:740–7.
25. Yamamoto S, Matsuda H, Kasahara Y, Iwahashi M, Takagi T, Baba T, et al. Dynamic molecular behavior of semi-fluorinated oleic, elaidic and stearic acids in the liquid state. J Oleo Sci. 2011;61:649–57.
26. Matsuzawa H, Tsuda M, Minami H, Iwahashi M. Dynamic molecular behavior and cluster structure of octanoic acid in its liquid and CCl_4 solution. Food Nutr Sci. 2013;4:25–32.
27. Nelson DL, Cox MM. Lehninger principles of biochemistry. 5th ed. New York: WH Freeman; 2009.
28. Knothe G. "Designer" biodiesel: optimizing fatty ester composition to improve fuel properties. Energy Fuel. 2008;22:1358–64.
29. Malkin T. Alternation in properties of long chain carbon compounds. Nature. 1931;127:126–7.
30. Larsson K. Molecular arrangement in glycerides. Fette Seifen Anstrichmittel. 1972;74:136–42.
31. Brady GW, Wasserman E, Wellendorf J. Structure studies in liquid alkanes including direct determination of end-to-end distances. J Chem Phys. 1967;47:855–6.
32. Aleby STIG, von Sydow E. The crystal structure of methyl stearate. Acta Crystallogr. 1960;13:487–92.
33. Boese D, Kirchner M, Echeverria GA, Boese R. Ethyl acetate: x-ray, solvent and computed structure. Chem Phys Chem. 2013;14:799–804.

34. Ramírez-Verduzco LF, Rodríguez-Rodríguez JE, Jaramillo-Jacob ADR. Predicting cetane number, kinematic viscosity, density and higher heating value of biodiesel from its fatty acid methyl ester composition. Fuel. 2012;91:102–11.
35. Knothe G, Dunn RO. A comprehensive evaluation of the melting points of fatty acids and esters determined by differential scanning calorimetry. J Am Oil Chem Soc. 2009;86:843–56.
36. Callaghan PT, Jolley KW. Translational motion in the liquid phases of tristearin, triolein and trilinolein. Chem Phys Lipids. 1980;27:49–56.
37. Gunstone FD, Pollard MR, Scrimgeour CM, Vedanayagam HS. Fatty acids: part 50. ^{13}C nuclear magnetic resonance studies of olefinic fatty acids and esters. Chem Phys Lipids. 1977;18:115–29.
38. Forato LA, Yushmanov VE, Colnago LA. Interaction of two prolamins with 1-^{13}C oleic acid by ^{13}C NMR. Biochemistry. 2004;43:7121–6.
39. Aguilar JA, Nilsson M, Bodenhausen G, Morris GA. Spin echo NMR spectra without J modulation. Chem Commun. 2012;48:811–3.
40. Gunstone FD. The lipid handbook. 2nd ed. Gunstone FD, Harwood JL, Padley FB, editors. London: Chapman & Hall; 1995.
41. Collings PJ, Hird M. Introduction to liquid crystals: chemistry and physics. Taylor & Francis Inc; 1997.
42. Cermak SC, Evangelista RL, Kenar JA. Distillation of natural fatty acids and their chemical derivatives. Distillation - advances from modeling to applications. Zereshki D, editor. ISBN: 978-953-51-0428-5; 2012. http://www.intechopen.com/books/distillation-advances-from-modeling-to-applications/distillation-of-natural-fatty-acids-and-their-chemical-derivatives.
43. Dunn RO. Crystallization behavior of fatty acid methyl esters. J Am Oil Chem Soc. 2008;85:961–72.
44. Badmaev BB, Dembelova TS, Damdinov BB. The investigation of shear elasticity of liquids at low-frequent shear influence. J Mol Liq. 2006;127:163–7.
45. Hernqvist L. On the structure of triglycerides in the liquid state and fat crystallization. Fette Seifen Anstrichmittel. 1984;86:297–300.
46. Carr HY, Purcell EM. Effects of diffusion on free precession in nuclear magnetic resonance experiments. Phys Rev. 1954;94:630–8.
47. Meiboom S, Gill D. Modified spin-echo method for measuring nuclear relaxation times. Rev Sci Instrum. 1958;29:688–91.
48. Provencher SW. A constrained regularization method for inverting data represented by linear algebraic or integral equations. Comput Phys Commun. 1982;27:213–27.
49. Stejskal EO, Tanner JE. Spin diffusion measurements: spin echoes in the presence of a time-dependent field gradient. J Chem Phys. 1964;42:288–92.
50. Holz M, Heil SR, Sacco A. Temperature-dependent self-diffusion coefficients of water and six selected molecular liquids for calibration in accurate ^{1}H NMR PFG measurements. Phys Chem Chem Phys. 2000;2:4740–2.
51. Abragam A. The principles of nuclear magnetism. Clarendon: Oxford University Press; 1961.

17

New generation biofuel: continuous acidogenesis of sucrose-raffinose mixture simulating vinasse is promoted by γ-alumina pellets

Katerina Lappa[1], Panagiotis Kandylis[1], Nikolaos Bastas[2], Stavros Klaoudatos[2], Nikolaos Athanasopoulos[3], Argyro Bekatorou[1], Maria Kanellaki[1] and Athanasios A Koutinas[1*]

Abstract

Background: This investigation comprises a contribution on the production of a new generation biofuel using the industrial liquid waste of bioethanol distilleries, known as vinasse. This study focuses on the exploitation of vinasse as an acidogenesis substrate for volatile fatty acids and simultaneous ethanol production. These products can be used for ester production, which is the new generation biofuel. Therefore, the aims of the present study are (i) to examine any promotional effect of γ-alumina on acidogenesis of a sucrose-raffinose mixture simulating vinasse, (ii) to study the operational stability of the continuous acidogenesis of sucrose and raffinose and subsequently of vinasse, and (iii) to determine the volatile fatty acid chemical composition and ethanol formation.

Results: Batch acidogenesis of sucrose and raffinose mixtures showed that γ-alumina promoted fermentation leading to an increase in the volatile fatty acid yield factor from 40% to 95% compared to free cells. The application of the system in continuous mode for more than 3 months showed that the continuous volatile fatty acid productivity obtained was higher than 7 g/L/day. Lactic acid was the predominant acid when sucrose and raffinose were used while butyric acid in the case of vinasse. The highest volatile fatty acid concentration reached was 19 g/L for vinasse.

Conclusions: A promoting effect of γ-alumina in acidogenic fermentation of sucrose-raffinose and vinasse is reported. Continuous acidogenesis of sucrose-raffinose mixtures and vinasse using γ-alumina with immobilized cells showed high operational stability (more than 3 months). These findings result in easy scale up of the process that will produce a high annual added value of $11,000,000 in a daily bioethanol production plant of 50,000 L.

Keywords: γ-Alumina, Acidogenic fermentation, Butyric acid, Lactic acid, Ethanol, Volatile fatty acids, Vinasse, Raffinose

Background

In the European Union, about 29% of the total ethanol produced comes from industries using molasses [1]. After ethanol recovery, a liquid residue known as vinasse is produced (9 to 14 L per liter of ethanol) with a very high organic load [2,3]. The treatment and disposal of vinasse are very important since it may cause serious environmental problems due to its high chemical oxygen demand (COD).

Several methods have been proposed for the treatment of vinasse prior its disposal. Among them, fertirrigation (fertilization and irrigation) is the method with most applications [4], but several problems related to soil and groundwater contamination have been reported [5,6]. Other methods of vinasse disposal, but too expensive, are vinasse recirculation and vinasse concentration for volume reduction [7].

A biorefinery concept for vinasse exploitation includes its use as a substrate for fermentation and the production of value-added co-products, such as volatile fatty acids and ethanol, for application in novel biofuel production. This approach could reduce the problem of vinasse disposal and enhance the sustainability of sugarcane-to-ethanol plants. The use of volatile fatty acids and ethanol for the production of esters is very promising as a novel biofuel. In the frame of this possibility, it was reported that

* Correspondence: a.a.koutinas@upatras.gr
[1]Food Biotechnology Group, Department of Chemistry, University of Patras, 26500 Patra, Greece
Full list of author information is available at the end of the article

in the acidogenesis of glucose in a batch mode using anaerobic mixed culture, volatile fatty acids and ethanol were formed [8]. This gives the possibility for simultaneous production of the two chemicals that are necessary for the esterification reaction. The use of such esters in a homogeneous charge compression ignition engine gave promising results for use of such alternative liquid biofuels produced by anaerobic treatment of waste materials [9]. Furthermore, the porous γ-alumina was found to be an effective support for cell immobilization [10] and promoter of methane [11] and alcoholic [12] fermentations. Based on these properties of γ-alumina, it was recently found that this material promoted anaerobic acidogenic fermentation of glucose as a model compound both in batch and continuous modes, leading to accumulation of mainly volatile fatty acids (VFAs) and ethanol [13,14].

Therefore, the aims of the present study are (i) to examine any promotional effect of γ-alumina on the acidogenesis of a sucrose-raffinose mixture simulating vinasse, the liquid waste of alcohol distillery; (ii) to study the operational stability of the continuous acidogenesis of sucrose and raffinose and subsequently of vinasse; and (iii) to determine the VFA chemical composition and ethanol formation.

Results and discussion

Rationale of the investigation

The last decades, research efforts have been started, aiming to produce a new generation biofuel based on ester production from organic acids of the acidogenic fermentation using anaerobic mixed culture in the fermentation of glucose, resulting also in the simultaneous production of ethanol [8]. Furthermore, we discovered the promotional effect of γ-alumina in the acidogenesis of glucose [13] and under conditions favoring butyric acid production in the same fermentation of glucose medium [14]. These works open the way for the production of the new generation biofuel from wastes of the food production industry. Vinasse, the liquid waste of ethanol distilleries with high production capacity, contains sucrose and raffinose as residual sugar. Therefore, a study of acidogenesis of a model mixture of sucrose and raffinose is necessary to be carried out, as a precursor of vinasse treatment. This investigation's objectives are to show any promotional effect by γ-alumina, as compared with free cells in the acidogenesis of sucrose and raffinose mixtures; to examine the chemical composition of VFA and ethanol formation in continuous processing; and to study the effectiveness of a γ-alumina-supported biocatalyst in the continuous acidogenesis of vinasse after adaptation of biocatalyst in a synthetic medium containing sucrose and raffinose found in sugarcane and sugar-beet vinasse. To achieve the objectives, batch and continuous acidogenic fermentations were performed and samples were analyzed for ethanol, acetic acid, propionic acid, butyric acid, lactic acid, succinic acid, malic acid, and valeric acid.

Effect of γ-alumina on VFA and ethanol production using sucrose and raffinose mixtures

Batch acidogenesis was performed at 37°C and a pH value of 7 as the optimum conditions were obtained. Mixtures of sucrose and raffinose (20 g/L) were used in order to simulate vinasse. The residual sugar in vinasse is less than 10 g/L containing sucrose and raffinose. However, the total organic material in vinasse is higher due to amino acid content, natural colorings, and so on. This is estimated to be a total organic mass of more than 20 g/L. Firstly, sucrose was used and then raffinose was added in various concentrations in order to obtain adaptation to the sugars before vinasse treatment, which contains these carbohydrates. The simulation of vinasse was attempted with carbohydrates and yeast extract addition which is rich in amino acids. γ-Alumina increased the VFA yield factor in all mixtures of sucrose and raffinose from 40% to 95% as compared with free cells (Figure 1). The lower improvement was for sucrose, while the higher was for raffinose. Table 1 shows results of batch acidogenesis of sucrose and raffinose mixtures using immobilized cells on porous γ-alumina. VFAs formed in all repeated batch fermentations, proving the suitability of the biocatalyst for continuous mode operation. All sucrose and raffinose have been bioconversed in all batches and all mixtures. The yield factor varies from 0.20 to 0.42 g/g using an initial sugar concentration of 20 g/L. VFA concentration was from 4 to 8.49 g/L while ethanol formation did not exceed 1.4 mL/L. VFAs that were formed were acetic acid, propionic acid, butyric acid, lactic acid, and valeric acid. Lactic acid had a higher concentration in comparison with other acids. These results encouraged the study of continuous acidogenesis by immobilized cells on γ-alumina in order to achieve the aforementioned objectives.

VFA and ethanol production in continuous acidogenesis of sucrose and raffinose mixtures

Continuous acidogenesis was performed using successively sucrose, sucrose-raffinose mixtures, and vinasse (Figure 2). Lactic acid was produced at the highest concentration as compared with acetic acid, butyric acid, succinic acid, and malic acid. Lactic acid, in continuous fermentation of sucrose and sucrose-raffinose mixtures, accounts for 65% of total VFAs while butyric acid was not produced at all. Figure 3 illustrates that the VFA yield factor increased as the time proceeds and as the acidogenesis proceeds from sucrose to sucrose-raffinose mixtures. The ethanol yield factor is reduced as the continuous acidogenesis proceeds. The operational stability

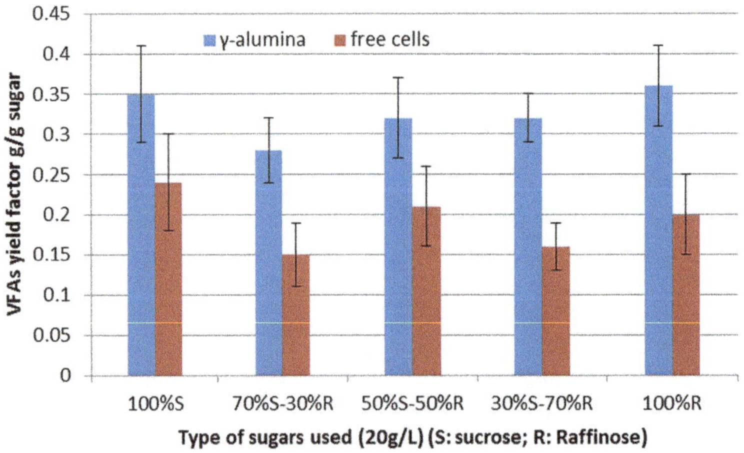

Figure 1 VFA yield factor during batch acidogenesis. Comparison of VFA yield factor during batch acidogenesis using γ-alumina-supported mixed anaerobic culture versus composition of sugars at 37°C and initial pH 7; suc: sucrose, raf: raffinose, vinasse diluted from concentrate, VFAs: volatile fatty acids.

of the continuous mode acidogenesis remains constant and increases for a long period of time as indicated by the stability of VFA productivity in Figure 4. Conversion varies from 34% up to 78.8% (Tables 2 and 3). The residual sugar that remains makes it possible for the effluent from the bioreactor to pass through γ-alumina-supported yeast cells to increase ethanol content which is necessary for the esterification reaction.

VFA production in continuous acidogenesis of alcohol distillery's liquid waste

After more than 2 months of operation, vinasse was used as influent in the continuous system and the results are summarized in Table 3. Vinasse was prepared from concentrate after dilution. Firstly, a dilution ratio of 1:5 was used simulating the content of fresh vinasse. The use of vinasse caused an increase in the total VFA content with a mean value of 14 g/L (varying from 7 to 17 g/L) and changed the VFA composition (Table 3 and Figure 2). More specifically, the concentration of lactic acid decreased to a mean value of 7 g/L (accounting for 50% of total VFA content), and subsequently, butyric acid and succinic acid concentrations increased to 4 g/L for each one (accounting for 25% of total VFAs each). The system continued its operation using 1:5 vinasse for 15 days, and then a more diluted vinasse was used. This time, the dilution ratio was 1:10 from the concentrated vinasse which is similar to the 1:2 dilution from fresh vinasse. This led to a further increase in total VFA concentration to a mean value of 16 g/L reaching at the last fermentations even up to 19 g/L. Furthermore, the lactic acid concentration continued to decrease reaching a value of 2 g/L, and the corresponding mean values for butyric and succinic acids were 7 and 4.5 g/L, respectively. The percentage of each acid in the total VFA content was 63% for butyric acid, 26% for succinic acid, and 10% for lactic acid. The results of VFA, lactic acid, and butyric acid productivities versus fermentation time are presented in Figure 4, showing the decrease of lactic acid and subsequently increase of butyric acid productivity. The VFA productivity continuously increases, and especially when vinasse is used, it reaches values of more than 7 g/L/day.

Scientific and technological consideration

γ-Alumina is a porous material with numerous applications in catalysis. Furthermore, it has been used in several fermentation systems such as methane production and alcoholic fermentation and has been proven to be a very promising promoter. Recently, γ-alumina had promotional effect on acidogenic fermentations of glucose showing improvement of ethanol and VFA production [13,14]. The present investigation also proves the promotional effect of γ-alumina on the acidogenic fermentation of sucrose and raffinose sugars which are contained in sugarcane and sugar-beet vinasse. Sucrose-raffinose mixture was used as a model substrate, simulating vinasse, and γ-alumina is also used as an immobilization support for the mixed anaerobic culture leading to increased cell densities in the bioreactor facilitating the repeated batch and continuous processing. Repeated batch processing of sucrose-raffinose mixtures shows the suitability of biocatalyst for continuous processing. However, the batch process gave a lower VFA yield factor and VFA concentration as compared with the continuous process, which had an operational stability for a long time. The adaptation of biocatalyst in sucrose-raffinose mixtures accommodates a rapid start-up for vinasse acidogenic fermentation. Vinasse treatment leads to an important VFA yield factor, and in the mixtures of VFAs produced, lactic acid and butyric acid predominate.

Table 1 Batch acidogenesis of sucrose and raffinose mixtures using γ-alumina-supported anaerobic mixed culture

Initial sugar composition	Repeated batch	Final pH	Ethanol (mL/L)	Acetic acid (g/L)	Propionic acid (g/L)	Butyric acid (g/L)	Lactic acid (g/L)	Succinic acid (g/L)	Malic acid (g/L)	Valeric acid (g/L)	Total VFAs (g/L)	Residual sugar (g/L)	Sugar conversion (%)	VFA yield (g/g)
100% sucrose	1	5.6	0.55	1.21	0.29	1.31	1.23	-	-	-	4.04	0	100	0.20
	2	5.1	0.44	1.73	1.15	1.29	1.88	-	-	2.80	8.85	0	100	0.44
	3	4.7	0.34	1.50	0.79	1.05	3.43	-	-	1.43	8.20	0	100	0.41
70% sucrose 30% raffinose	1	5.6	0.74	1.44	0.69	2.56	0.63	-	-	-	5.32	0	100	0.27
	2	5.2	0.40	1.73	1.24	1.96	0.70	-	-	-	5.63	0	100	0.28
	3	4.9	0.20	0.99	0.64	0.94	0.81	-	-	2.33	5.71	0	100	0.29
50% sucrose 50% raffinose	1	5.4	1.04	1.19	-	1.36	1.45	-	-	-	4.00	0	100	0.20
	2	5.0	0.85	1.43	0.53	1.51	1.36	-	-	2.64	7.47	0	100	0.37
	3	4.7	0.30	1.04	0.47	1.01	3.77	0.86	0.94	-	8.09	0	100	0.40
30% sucrose 70% raffinose	1	5.5	0.98	1.53	0.35	1.77	1.09	0.86	-	-	5.60	0	100	0.28
	2	5.1	0.47	1.49	0.58	1.87	0.98	-	-	-	4.92	0	100	0.25
	3	4.8	0.29	1.39	0.68	1.55	2.36	-	-	2.32	8.30	0	100	0.42
100% raffinose	1	5.2	1.4	2.63	1.55	3.20	1.11	-	-	-	8.49	0	100	0.42
	2	4.9	0.6	1.11	0.51	1.34	1.36	-	-	2.83	7.15	0	100	0.36
	3	4.5	0.1	0.32	0.37	0.75	4.53	-	-	-	5.97	0	100	0.30

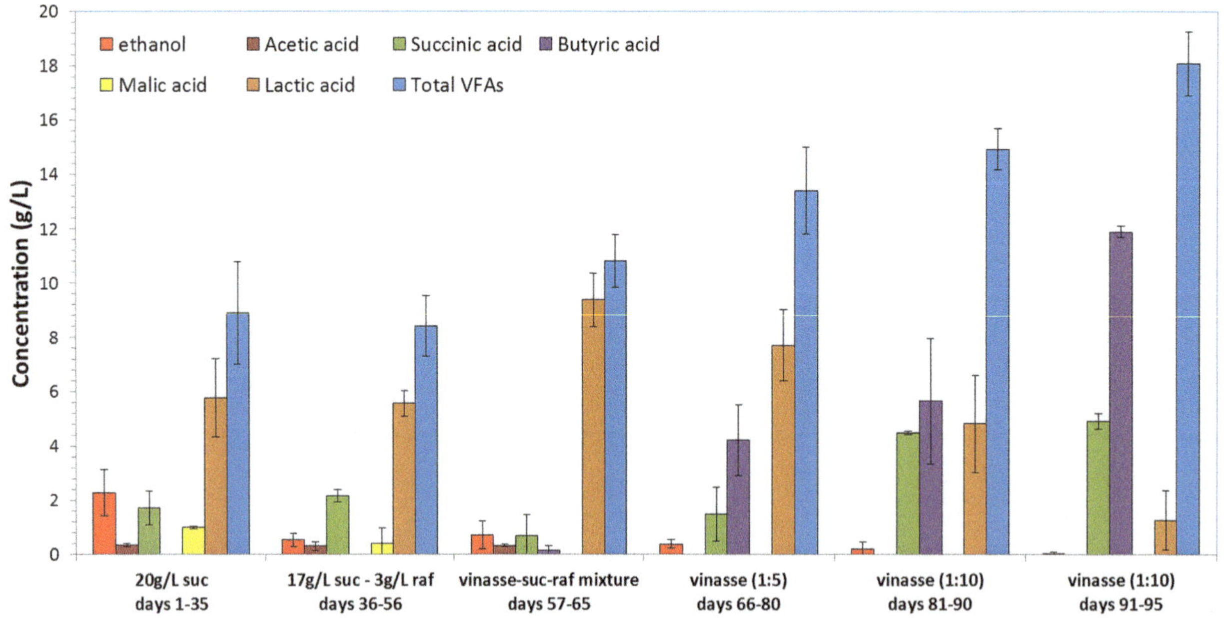

Figure 2 VFA and ethanol production during continuous acidogenesis. Ethanol and VFA composition of the effluent during the continuous acidogenesis using γ-alumina-supported mixed anaerobic culture versus fermentation time and different compositions of influent at 37°C and initial pH 7; suc: sucrose, raf: raffinose, vinasse diluted from concentrated, VFAs: volatile fatty acids.

The last is of technological importance due to their esters with ethanol which have increased boiling and ignition points and viscosity, simulating better gasoline and therefore being a more suitable fuel to be added in biodiesel. Moreover, the results clearly showed that lactic acid is the main VFA in the acidogenesis of sucrose, sucrose-raffinose mixtures, and sucrose-raffinose-vinasse mixtures, while butyric acid is the main VFA when only vinasse was used. The formation of butyric acid at high concentrations in continuous acidogenesis of vinasse complies with the result of favoring butyric acid formation of a previous study [14]. Ethanol is necessary for the esterification of VFAs and can be produced after mixing of the effluent containing VFAs with a liquid waste containing glucose, sucrose, or lactose and supplying it through a second continuous bioreactor for ethanol formation in the same solution with VFAs. The second bioreactor can contain γ-alumina-supported yeast [10] or kefir yeast cells [15]. Based on the results of this investigation, one ethanol distillery with 50,000 L daily ethanol production can produce

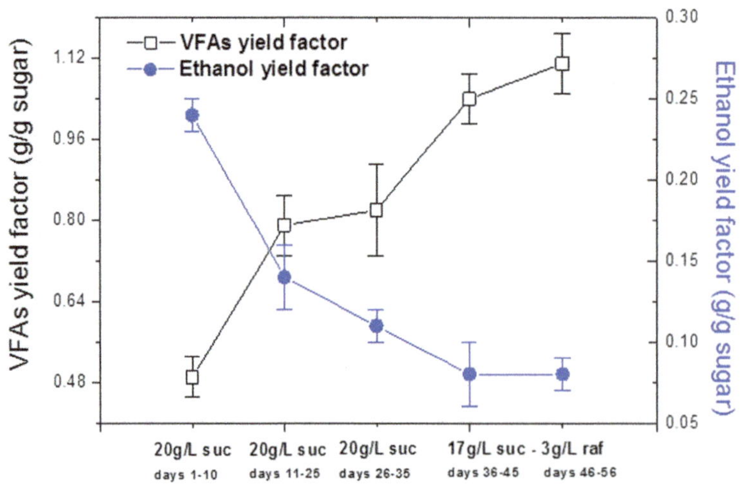

Figure 3 VFA and ethanol yield factor during continuous acidogenesis. Yield factors during the continuous acidogenesis using γ-alumina-supported mixed anaerobic culture versus fermentation time and different compositions of influent at 37°C and initial pH 7; suc: sucrose, raf: raffinose, VFAs: volatile fatty acids.

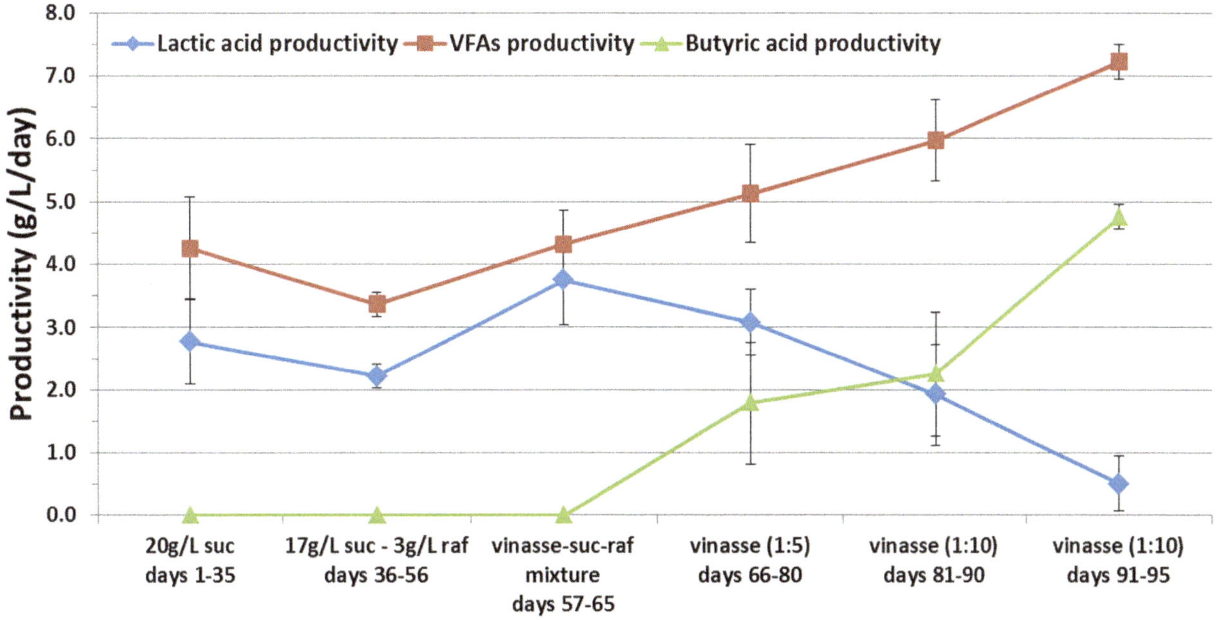

Figure 4 VFA, lactic acid, and butyric acid productivities versus time in continuous acidogenesis. Productivities during the continuous acidogenesis using γ-alumina-supported mixed anaerobic culture versus fermentation time and different compositions of influent at 37°C and initial pH 7; suc: sucrose, raf: raffinose, vinasse diluted from concentrate, VFAs: volatile fatty acids.

from its liquid waste vinasse a maximum of 15,000 L of esters daily as a new generation biofuel. Therefore, the annual added value is estimated to be $11,000,000. More specifically, 50,000 L of ethanol will produce 575,000 L of vinasse, which will result, after acidogenesis, to 11,500 kg of acids daily. After esterification, these acids will result to 15,000 L of esters per day and therefore 5,475,000 L annually. If this replaces gasoline, it means an added value of almost $11,000,000. This added value will reduce bioethanol production cost by about 30%.

Conclusions

The acidogenic fermentation of sucrose and raffinose is promoted by γ-alumina. Batch acidogenesis of sucrose-raffinose mixtures gave a lower concentration and yield factor of VFAs than the continuous process. Continuous acidogenesis of sucrose-raffinose mixtures using γ-alumina with immobilized cells has high operational stability, and the adaptation of a biocatalyst results to a rapid scale up for the acidogenesis of vinasse. Lactic acid is the predominant VFA when synthetic medium is used, and butyric acid predominates in VFAs produced when vinasse is used. The added value produced from the new generation biofuel as a bioproduct of an ethanol distillery will reduce its production cost by about 30%.

Materials and methods

Culture and growth media

Mixed bacterial anaerobic culture was obtained from an upflow anaerobic sludge blanket (UASB) bioreactor and was inoculated in a medium containing 50 g/L glucose, aqueous NH_3 and 50% H_3PO_4 (to achieve a COD:N:P ratio of 100:5:1), $NaHCO_3$ 4 g/L, and yeast extract 4 g/L, without pH adjustment [11]. Cell growth was carried out at 37°C. The medium was sterilized by autoclaving at 120°C for 10 min.

Batch acidogenic fermentation of sucrose and raffinose mixtures

For the repeated batch fermentations, conical flasks were used containing immobilized mixed anaerobic culture on 100 g cylindrical γ-alumina and 100 mL of synthetic medium containing 20 g/L sugars. Fermentations were carried out at 37°C with an initial pH of 7. Synthetic medium with five different compositions of sugars was used containing 20 g/L sucrose, 14 g/L sucrose and 6 g/L raffinose, 10 g/L sucrose and 10 g/L raffinose, 6 g/L sucrose and 14 g/L raffinose, and 20 g/L raffinose. Each medium also contained aqueous NH_3 and 50% H_3PO_4 (to achieve a COD:N:P ratio of 100:5:1), $NaHCO_3$ 4 g/L, and yeast extract 4 g/L, with pH adjustment at 7 [11]. The media were sterilized by autoclaving at 120°C for 10 min.

Cell immobilization and continuous anaerobic acidogenic fermentation of sucrose

The mixed culture was fixed on porous cylindrical γ-alumina pellets (1.40 m^2/g surface area) to increase rate of fermentation and enable continuous processing. The experimental apparatus used for cell immobilization and for continuous acidogenic fermentation consisted of

Table 2 Continuous acidogenesis of synthetic media using γ-alumina-supported anaerobic mixed culture

Influent				Effluent												
Feed	Initial sugar (g/L)	Flow rate (L/day)	Operation time (days)	pH	Ethanol (ml/L)	Acetic acid (g/L)	Lactic acid (g/L)	Succinic acid (g/L)	Malic acid (g/L)	Butyric acid (g/L)	Total VFAs (g/L)	Residual sugar (g/L)	Sugar conversion (%)	VFA yield (g/g)	VFA productivity (g/L/day)
Sucrose	20	0.2	1	4.9	7.7	0.18	3.83	1.32	1.01	-	6.34				2.54
	20	0.2	2	4.8	8.9	0.21	4.21	1.36	1.01	-	6.79	5.44	72.8	0.47	2.72
	20	0.4	3	4.7	10.5	0.23	4.41	1.34	1.08	-	7.06				5.65
	20	0.4	4	4.8	5.7	0.33	3.89	1.01	1.01	-	6.24				4.99
	20	0.4	6	4.8	2.9	0.39	3.35	0.84	1.01	-	5.59	7.06	64.7	0.43	4.47
	20	0.3	7	4.7	1.1	0.38	3.32	0.91	1.00	-	5.61				3.37
	20	0.3	8	4.7	1.0	0.35	3.83	0.99	0.99	-	6.16				3.70
	20	0.3	9	4.7	1.0	0.34	4.60	1.09	0.99	-	7.02	7.60	62.0	0.57	4.21
	20	0.25	10	4.7	1.94	0.32	5.35	1.17	0.97	-	7.81				3.91
	20	0.25	11	4.6	1.96	0.35	6.61	1.26	0.97	-	9.18				4.59
	20	0.25	12	4.5	1.85	0.32	6.91	1.29	0.95	-	9.48	4.24	78.8	0.60	4.74
	20	0.25	13	4.6	2.11	0.40	7.59	1.34	0.97	-	10.29				5.15
	20	0.25	14	4.6	1.85	0.38	7.37	1.34	0.96	-	10.05				5.03
	20	0.25	15	4.5	1.75	0.38	8.16	1.47	0.96	-	12.96	6.88	65.6	0.99	6.48
	20	0.25	16	4.7	2.18	0.34	6.58	1.30	0.96	-	9.19				4.60
	20	0.25	17	4.7	2.33	0.38	6.70	1.34	0.98	-	9.39				4.70
	20	0.25	18	4.7	2.4	0.38	7.02	1.40	0.77	-	9.57	6.99	65.0	0.74	4.79
	20	0.25	21	4.7	2.28	0.37	6.19	1.38	0.99	-	8.93				4.47
	20	0.2	23	4.5	4.38	0.37	9.04	2.30	0.97	-	12.67	5.06	74.7	0.85	5.07
	20	0.2	24	4.6	3.46	0.41	7.51	2.59	0.99	-	11.50				4.60
	20	0.2	25	4.6	3.43	0.42	6.10	2.61	1.01	-	10.15				4.06
	20	0.2	26	4.6	2.96	0.42	5.82	2.58	1.01	-	9.83	8.12	59.4	0.83	3.93
	20	0.2	27	4.6	6.04	0.43	5.99	2.54	1.02	-	9.98				3.99
	20	0.2	28	4.6	5.78	0.38	5.67	2.34	1.02	-	9.41				3.76
	20	0.2	29	4.6	5.26	0.36	5.61	2.24	1.02	-	9.23	7.74	61.3	0.75	3.69
	20	0.2	31	4.7	4.28	0.37	5.50	2.12	1.02	-	9.01				3.60
	20	0.2	32	4.7	1.89	0.33	5.62	2.29	1.03	-	9.28				3.71
	20	0.2	33	4.7	1.27	0.36	5.57	2.37	1.03	-	9.33	7.67	61.6	0.76	3.73
	20	0.2	34	4.7	1.40	0.25	5.22	2.53	1.05	-	9.05				3.62
	20	0.2	35	4.7	1.46	0.33	5.58	2.70	1.06	-	9.67	9.79	51.0	0.95	3.87
		0.2	37	4.7	0.60	0.24	6.38	2.52	1.17	-	10.31	10.3	51.5	1.06	4.12

Table 2 Continuous acidogenesis of synthetic media using γ-alumina-supported anaerobic mixed culture *(Continued)*

Sucrose and raffinose	17 g sucrose + 3 g raffinose														
		0.2	38	4.7	1.07	-	5.57	2.24	1.06	-	8.87				3.55
		0.2	39	4.7	1.29	-	6.41	2.66	1.18	-	10.26				4.10
		0.2	40	4.7	1.09	0.18	6.36	2.62	1.18	-	10.15	11.9	40.5	1.25	4.06
		0.2	41	4.7	0.78	0.34	6.12	2.37	1.17	-	10.00				4.00
		0.2	43	4.7	0.84	0.34	5.52	2.22	1.08	-	9.16				3.66
		0.2	45	4.8	0.74	0.34	5.21	2.03	-	-	7.59	10.8	46.0	0.82	3.04
		0.2	46	4.7	0.78	0.33	5.32	2.07	-	-	7.72				3.09
		0.2	47	4.6	0.54	0.31	5.35	2.05	-	-	7.71				3.08
		0.2	48	4.7	0.83	0.25	5.39	2.05	-	-	7.69				3.08
		0.2	50	4.7	0.41	0.28	5.05	2.08	-	-	7.41				2.96
		0.2	51	4.7	0.83	0.33	5.13	2.13	-	-	7.59	13.2	34.0	1.12	3.04
		0.2	52	4.6	0.45	0.79	5.10	1.97	-	-	7.86				3.14
		0.2	53	4.7	0.48	0.35	5.34	1.97	-	-	7.66	13.1	34.5	1.11	3.06
		0.2	54	4.7	0.19	0.31	5.45	1.99	-	-	7.75				3.10
		0.2	55	4.7	0.22	0.38	5.48	2.04	-	-	7.90				3.16
		0.2	56	4.8	0.32	0.35	5.13	2.00	-	-	7.49				3.00

Table 3 Continuous acidogenesis of vinasse using γ-alumina-supported anaerobic mixed culture

Influent			Effluent							
Feed	Flow rate (L/day)	Operation time (days)	pH	Ethanol (ml/L)	Acetic acid (g/L)	Lactic acid (g/L)	Succinic acid (g/L)	Butyric acid (g/L)	Total VFAs (g/L)	VFA productivity (g/L/day)
25% vinasse (1:5)	0.2	58	5.2	0.35	0.36	6.33	1.80	-	8.49	3.40
	0.2	59	5.3	1.09	0.38	6.41	1.07	-	7.86	3.14
50% vinasse (1:5)	0.2	60	5.4	0.49	0.28	9.28	0.61	-	10.79	4.32
	0.2	61	5.4	1.96	0.31	13.09	-	-	14.03	5.61
75% vinasse (1:5)	0.2	64	5.5	0.70	0.31	11.72	-	0.78	12.81	5.12
100% vinasse (1:5)	0.2	66	5.3	0.56	-	10.45	-	2.36	12.81	5.12
	0.2	67	5.3	0.34	-	9.33	-	8.22	17.55	7.02
	0.2	68	5.2	0.22	-	8.60	-	7.53	15.79	6.32
	0.2	69	5.4	0.48	-	7.06	-	2.36	9.42	3.77
	0.2	71	5.4	0.28	-	6.73	1.56	7.78	16.07	6.43
	0.2	72	5.3	0.44	-	6.81	-	-	6.81	2.72
	0.2	73	5.5	0.36	-	7.37	-	-	7.37	2.95
	0.2	76	5.4	0.76	-	6.46	4.40	2.36	13.22	5.29
	0.2	78	5.3	0.84	-	7.38	4.49	2.69	14.56	5.82
	0.2	80	5.5	0.61	-	6.81	4.34	3.45	14.60	5.84
100% vinasse (1:10)	0.2	81	6.1	0.89	-	7.66	4.60	4.10	16.36	6.54
	0.2	82	5.8	0.40	-	6.98	4.47	3.50	14.95	5.98
	0.2	84	6.0	0.16	-	6.93	4.52	4.02	15.47	6.19
	0.2	85	6.2	-	-	5.27	4.43	4.66	14.36	5.74
	0.2	86	6.3	0.17	-	4.45	4.44	5.75	14.64	5.86
	0.2	88	6.3	0.24	-	2.37	4.47	7.56	14.40	5.76
	0.2	89	6.4	-	-	-	4.44	9.85	14.29	5.72
	0.2	90	6.2	0.13	-	-	4.60	12.14	16.74	6.70
	0.2	91	6.5	-	-	1.88	4.93	11.71	18.52	7.41
	0.2	94	6.4	-	-	1.91	5.17	11.86	18.94	7.58
	0.2	95	6.3	-	-	2.01	5.07	12.16	19.24	7.70

a 1.25-L glass tower reactor. The bioreactor was filled with 700 g of γ-alumina pellets and equal volumes of 20 g/L sucrose medium, and anaerobic culture was added. The bioreactor was placed in an incubator set at 37°C, and it was left to ferment for 2 days without feeding in order to achieve cell immobilization. Subsequently, fresh sucrose medium was pumped continuously into the bioreactor using a high-accuracy peristaltic pump, and a steady state was obtained in a few days. The bioreactor was pumped continuously and successively with sucrose, sucrose and raffinose mixture, and vinasse as described previously. Samples were taken at various time intervals and analyzed for volatile fatty acids, ethanol, and residual sugar by GC and HPLC as described below.

Media for continuous acidogenic fermentation

During the continuous anaerobic fermentation, media of various sugar substrates were used successively. At the beginning, 20 g/L sucrose was used for 35 days, then a mixed medium of 17 g/L sucrose and 3 g/L raffinose was tested for 21 days and then concentrated vinasse (sugar-beet) after dilution with water (1:5 dilution for 15 days and 1:10 dilution for 15 days) was the last medium used. Each medium also contained aqueous NH_3 and 50% H_3PO_4 (to achieve a COD:N:P ratio of 100:5:1), $NaHCO_3$ 4 g/L, and yeast extract 4 g/L, with pH adjustment at 7 [11]. The media were sterilized by autoclaving at 120°C for 10 min.

GC analysis for ethanol determination

Ethanol was determined on a Shimadzu GC-8A system (Shimadzu Corporation, Kyoto, Japan), with a Teknokroma HAYE SEP Q 80/100 column, a C-R6A Chromatopac integrator, a He as carrier gas (40 mL/min), and a FID detector. The injection port and detector temperature was 210°C. The column temperature was 130°C. Samples of 2 μL were injected directly into the column. Determinations were done by means of standard curves.

HPLC analysis for VFA and residual sugar determination

Residual sugar was determined by high-performance liquid chromatography, using a Shimadzu chromatograph with a Nucleogel Ion 300 OA column, a LC-9A pump, a CTO-10A oven at 30°C, and a RID-6A refractive index detector. H_2SO_4 0.008 N was used as mobile phase with a flow rate of 0.8 mL/min, and propanol-1 was used as an internal standard. A volume of 0.25 mL of the sample and 0.625 mL of a 1% (*v*/*v*) solution of propanol-1 were diluted to 25 mL. Then, 60 μL of the final solution was injected directly to the column. Residual sugar and concentrations were calculated using standard curves.

Volatile fatty acids were determined by high-performance liquid chromatography, using a Jasco chromatograph LC-2000 Plus (Jasco Analytical Instruments, Tokyo, Japan) with a Bio-rad Aminex HPX-87H column (7.8 mm ID × 300 mm, 9 μm particle size), a PU-2089 plus quaternary gradient pump, a Jasco CO-2060 Plus oven at 50°C, a MD-2018 plus photodiode array detector operated at 210 nm, and an autosampler AS 2050 plus. H_2SO_4 0.008 N was used as mobile phase with a flow rate of 0.6 mL/min. The samples were filtered by a membrane filter of 0.45 nm, and all the data were processed with the ChromNav program.

Abbreviations

COD: chemical oxygen demand; Raf: raffinose; Suc: sucrose; UASB: upflow anaerobic sludge blanket; VFAs: volatile fatty acids.

Competing interests

The authors declare that they have no competing interests.

Authors' contributions

KL performed the experiments presented herein. PK supervised the experiments, assisted in the experimental design, and drafted the manuscript. NB, SK, and NA are industrial partners who designed the experiment, advised academics, and performed the experiments in the continuous system using vinasse. AB and MK supervised the work, designed the experiments, and drafted the manuscript. AAK conceived of the study, supervised the work, designed the experiments, and finally drafted the manuscript. All authors participated in the evaluation of the data and read and approved the final manuscript.

Acknowledgements

The research project was co-financed by the European Union (European Regional Development Fund - ERDF) and Greek national funds through the Operational Program 'Competitiveness and Entrepreneurship' of the National Strategic Reference Framework (NSRF) 2007-2013 - National Action 'Cooperation 2011: Partnerships of Production and Research Institutions in Focused Research and Technology Sectors' of the General Secretariat for Research and Technology (Contract Nr. 11SYN-8-839).

Author details

[1]Food Biotechnology Group, Department of Chemistry, University of Patras, 26500 Patra, Greece. [2]B.G. Spiliopoulos S.A, 26500 Patra, Greece. [3]I. Athanasopoulos and Co, 106b Lontou Str, 26224 Patra, Greece.

References

1. Balat M, Balat H. Recent trends in global production and utilization of bio-ethanol fuel. Appl Energy. 2009;86:2273–82.
2. Goldemberg J, Coelho ST, Guardabassi P. The sustainability of ethanol production from sugarcane. Energy Policy. 2008;36:2086–97.
3. España-Gamboa EI, Mijangos-Cortés JO, Hernández-Zárate G, Domínguez Maldonado JA, Alzate-Gaviria LM. Methane production by treating vinasses from hydrous ethanol using a modified UASB reactor. Biotechnol Biofuels. 2012;5:82.
4. Wheals AE, Basso LC, Alves DMG, Amorim HV. Fuel ethanol after 25 years. Trends Biotechnol. 1999;17:482–7.
5. Smeets E, Junginger M, Faaij A, Walter A, Dolzan P, Turkenburga W. The sustainability of Brazilian ethanol - an assessment of the possibilities of certified production. Biomass Bioenergy. 2008;32:781–813.
6. Navarro AR, Sepúlveda MC, Rubio MC. Bio-concentration of vinasse from the alcoholic fermentation of sugar cane molasses. Waste Manage. 2000;20:581–5.
7. Moura AG, Medeiros JR. Applying consistent technology for fuel ethanol production. Sugar Tech. 2008;10:20–4.
8. Ren NQ, Wang BZ, Huang JC. Ethanol-type fermentation from carbohydrate in high rate acidogenic reactor. Biotechnol Bioeng. 1997;54(5):428–33.
9. Contino F, Foucher F, Mounaim-Rousselle C, Jeanmart H. Combustion characteristics of tricomponent fuel blends of ethyl acetate, ethyl propionate, and ethyl butyrate in homogeneous charge compression ignition (HCCI). Energy Fuel. 2011;25:1497–503.
10. Kana K, Kanellaki M, Papadimitriou A, Psarianos C, Koutinas AA. Immobilization of *Saccharomyces cerevisiae* on γ-alumina pellets and its ethanol production in glucose and raisin extract fermentation. J Ferment Bioeng. 1989;68:213–5.
11. Koutinas AA, Toutoutzidakis G, Kana K, Kouinis I. Methane fermentation promoted by gamma-alumina pellets. J Ferment Bioeng. 1991;72:64–7.
12. Kanellaki M, Koutinas AA, Kana K, Nicolopoulou M, Papadimitriou A, Lycourghiotis A. Ethanol production by *Saccharomyces cerevisiae* promoted by γ-alumina. Biotechnol Bioeng. 1989;34:121–5.
13. Syngiridis K, Bekatorou A, Kallis M, Kandylis P, Kanellaki M, Koutinas AA. γ-Alumina as a process advancing tool for a new generation biofuel. Bioresour Technol. 2013;132:45–8.
14. Syngiridis K, Bekatorou A, Kandylis P, Larroche C, Kanellaki M, Koutinas AA. Favouring butyrate production for a new generation biofuel by acidogenic glucose fermentation using cells immobilized on γ-alumina. Bioresour Technol. 2014;161:118–23.
15. Athanasiadis I, Boskou D, Kanellaki M, Kiosseoglou V, Koutinas AA. Whey liquid waste of the dairy industry as raw material for potable alcohol production by kefir granules. J Agric Food Chem. 2002;50(25):7231–4.

Genome-wide analysis of the endoplasmic reticulum stress response during lignocellulase production in *Neurospora crassa*

Feiyu Fan[1,2], Guoli Ma[1,2], Jingen Li[1,2], Qian Liu[1], Johan Philipp Benz[3], Chaoguang Tian[1*] and Yanhe Ma[1]

Abstract

Background: Lignocellulolytic fungal cells suffer endoplasmic reticulum (ER) stress during lignocellulase synthesis; however, an understanding of this integrated process on a genome-wide scale remains poor. Here, we undertook a systematic investigation of this process in *Neurospora crassa* (*N. crassa*) using transcriptomic analysis coupled with genetic screens.

Results: A set of 766 genes was identified as the ER stress response targets (ESRTs) in *N. crassa* under cellulose utilization conditions. Among these, the expression of 223 and 186 genes showed dependence on IRE-1 and HAC-1, respectively. A total of 527 available mutants for ESRT genes were screened, 249 of which exhibited ER stress susceptibility, including 100 genes with unknown function. Disruption of *ire-1* or *hac-1* in *N. crassa* did not affect transcriptional induction of lignocellulase genes by cellulose but severely affected secretion of the corresponding enzymes. A global investigation of transcription factors (TFs) discovered three novel regulators (RES-1, RES-2, RRG-2) involved in lignocellulase secretion. Production of lignocellulases in *Δres-1* increased by more than 30% in comparison to wild type (WT), while secretion decreased by nearly 30% in strains *Δres-2* and *Δrrg-2*. Transcriptional profiling of the three TF mutants suggests they are deeply involved in lignocellulase secretion and ER stress response.

Conclusions: Here, we determined the transcriptional scope of the ER stress response during lignocellulase synthesis in the model cellulolytic fungus *N. crassa*. Through genome-wide mutant screening and analysis, dozens of novel genes were discovered to be involved in the process. The findings of this work will be useful for strain improvement to facilitate lignocellulase and biomass-based chemical production.

Keywords: ER stress, UPR, Transcription factor, Lignocellulase secretion, RNA-seq, RESS

Background

Saprophytic fungi evolved a highly efficient capability to secrete enzymes into their extracellular matrix to synergistically depolymerize biomass. This natural property has been exploited for lignocellulase production in industry [1]. Although *Trichoderma reesei* (*Hypocrea jecorina*) and *Aspergillus spp.* have been successfully developed as systems for lignocellulase production, a clear genetic basis underpinning the requirements of effective secretion has not yet been elucidated [2]. Early electron microscopy analysis revealed that endoplasmic reticulum (ER) proliferation occurred in the *T. reesei* hyper-secretion mutant RUT-C30 compared with its parental strain, QM6a [3,4]. Further characterization of the RUT-C30 strain revealed the transcript levels of ER-resident chaperone genes, such as *pdi1* and *bip1*, were induced in response to increased production of cellulase proteins [5]. These data suggested that the intracellular secretory pathway tightly regulates lignocellulase secretion in particular components of the ER.

The ER acts as the central hub where membrane and secretory proteins are properly folded and matured. When the ER encounters high protein flux or intractable heterologous proteins, its folding capacity could be transiently saturated, thus leading to a secretory pathway traffic jam and causing ER stress [6,7]. Cells activate multiple pathways to respond to ER stress, one of the best characterized being the unfolded protein response

* Correspondence: tian_cg@tib.cas.cn
[1]Key Laboratory of Systems Microbial Biotechnology, Tianjin Institute of Industrial Biotechnology, Chinese Academy of Sciences, Xiqi Dao32, Tianjin Airport Economic Area, Tianjin 300308, China
Full list of author information is available at the end of the article

(UPR). In *Saccharomyces cerevisiae* and filamentous fungi, the UPR mainly depends on an evolutionarily conserved signaling cascade which is mediated by ER-resident transmembrane kinase/endoribonuclease IRE1 (Inositol-requiring enzyme-1) and the basic-leucine zipper (bZIP) transcription factor HAC1 (homologous to ATF/CREB1) [8]. Although several aspects of ER stress response and the UPR have been intensively unraveled using the *S. cerevisiae* model, filamentous fungi are capable of much higher protein production levels, suggesting different or additional response mechanisms might exist between the two systems. For instance, filamentous fungi utilize a feedback mechanism termed repression under secretion stress (RESS) [9-11] which selectively down-regulates transcription of genes encoding extracellular enzymes upon ER stress and thus helps to reduce ER load. However, the detailed mechanism behind RESS is not clear at the moment.

The Ascomycete *Neurospora crassa* is a common inhabitant of burnt plant material in nature and has recently become a model system to address lignocellulose deconstruction and utilization [12,13]. Previous transcriptome profiling of *N. crassa* grown on *Miscanthus* and cellulose revealed that the expression of most lignocellulase genes is significantly induced at early time points (16 h) but rapidly declines thereafter [14], implying RESS exists in *N. crassa* and might be a limiting step of lignocellulase synthesis. Moreover, recent work reported *hac-1* activation by unconventional mRNA splicing when *N. crassa* was grown on cellulose [15], suggesting the UPR cascade is needed to coordinate the increased lignocellulase flux and the cellular folding capacity. *N. crassa* has the unique advantage of a nearly complete genome deletion strain collection [16], which makes it an ideal resource for a systematic study of the ER stress response and lignocellulase synthesis. To our knowledge, studies on the cross-talk of these two processes have not been performed so far.

In this work, by using high-throughput mRNA sequencing technology (RNA-Seq) coupled to genetic screens, we explored the connection between fungal lignocellulase secretion and ER stress on a genome-wide level. We thus defined the regulon of the ER stress response and identified dozens of genes as well as three novel transcription factors: RES-1 (NCU03699), RES-2 (NCU02724), and RRG-2 (NCU02413) to be involved in this integrated process.

Results

The experimental design setup for the analysis of mild and acute ER stress during cellulase production in *N. crassa*

To trigger ER stress while eliminating any non-specific drug response noise, the *N. crassa* wild-type (WT; FGSC#2489, *Mat A)* was treated with two kinds of ER stress-causing agents: either dithiothreitol (DTT; prevents disulfide bond formation,) or tunicamycin (TM; inhibits *N*-linked glycosylation). Cells can dynamically activate distinct physiological processes, such as adaptation, alarm, or apoptosis, depending on the ER stress intensity [7,17]. In order to maximize the detection of targets which contribute to ER stress resistance and minimize apoptotic effects, three experiments were initially employed to determine the appropriate treatment (Figure 1). When the sensitivity of *N. crassa* WT to DTT and TM was measured using concentration gradients (Figure 1A), we observed a range of growth phenotypes from mild inhibition at low concentrations (1 to 3 mM DTT or 0.5 to 10 μg/mL TM), to significant impairment at moderate concentrations (5 to 11 mM DTT or 20 to 80 μg/mL TM) and complete inability to grow when the conidia were exposed to 13 mM DTT or 120 μg/mL TM, respectively. We considered these results to reflect lethal, acute, and mild stress levels. We next checked the IRE-1-mediated mRNA splicing of the unfolded protein response marker *hac-1* [15] by treating *N. crassa* cells with a similar concentration gradient of either DTT or TM during growth on cellulose. The results shown in Figure 1B suggest that a treatment with 0.1 mM DTT or 5 μg/mL TM is sufficient to achieve mild stress levels. Finally, to determine the impact of distinct stress levels on lignocellulase synthesis, we measured the expression levels of two well-known UPR maker genes, *grp-78* (NCU03928) and *pdi-1* (NCU09223), as well as two major cellobiohydrolase genes, *cbh-1* (NCU07340) and *cbh-2*(NCU09680), by quantitative PCR (qPCR) (Figure 1C). We found the folding machinery to be activated at all tested stress levels, while *cbh-1* and *cbh-2* were dramatically down-regulated at concentrations exceeding 0.1 mM DTT or 5 μg/mL TM, respectively, suggesting an induction of RESS to render cells more resistant against acute ER stress. Thus, based on growth phenotype, the splicing pattern of *hac-1* mRNA and RESS mediated down-regulation of cellulase genes, four stress conditions were finally chosen for further analysis: mild ER stress by treatment with 0.1 mM DTT or 5 μg/mL TM; and acute ER stress by treatment with 5 mM DTT or 40 μg/mL TM.

Defining the ER stress response targets in wild-type *N. crassa* during cellulase production

Comparative analyses of the transcriptional profiling data of WT under four ER stress conditions (mild ER stress and acute ER stress by treatment with DTT or TM as described above) *versus* the no drug condition were performed. Of the 9,730 annotated genes in the *N. crassa* genome, we detected 1,322 (13.59%), 1,349 (13.86%), 1,810 (18.60%), and 1,882 (19.34%) genes with altered expression which determined by NOISeq (for detail of cut-off,

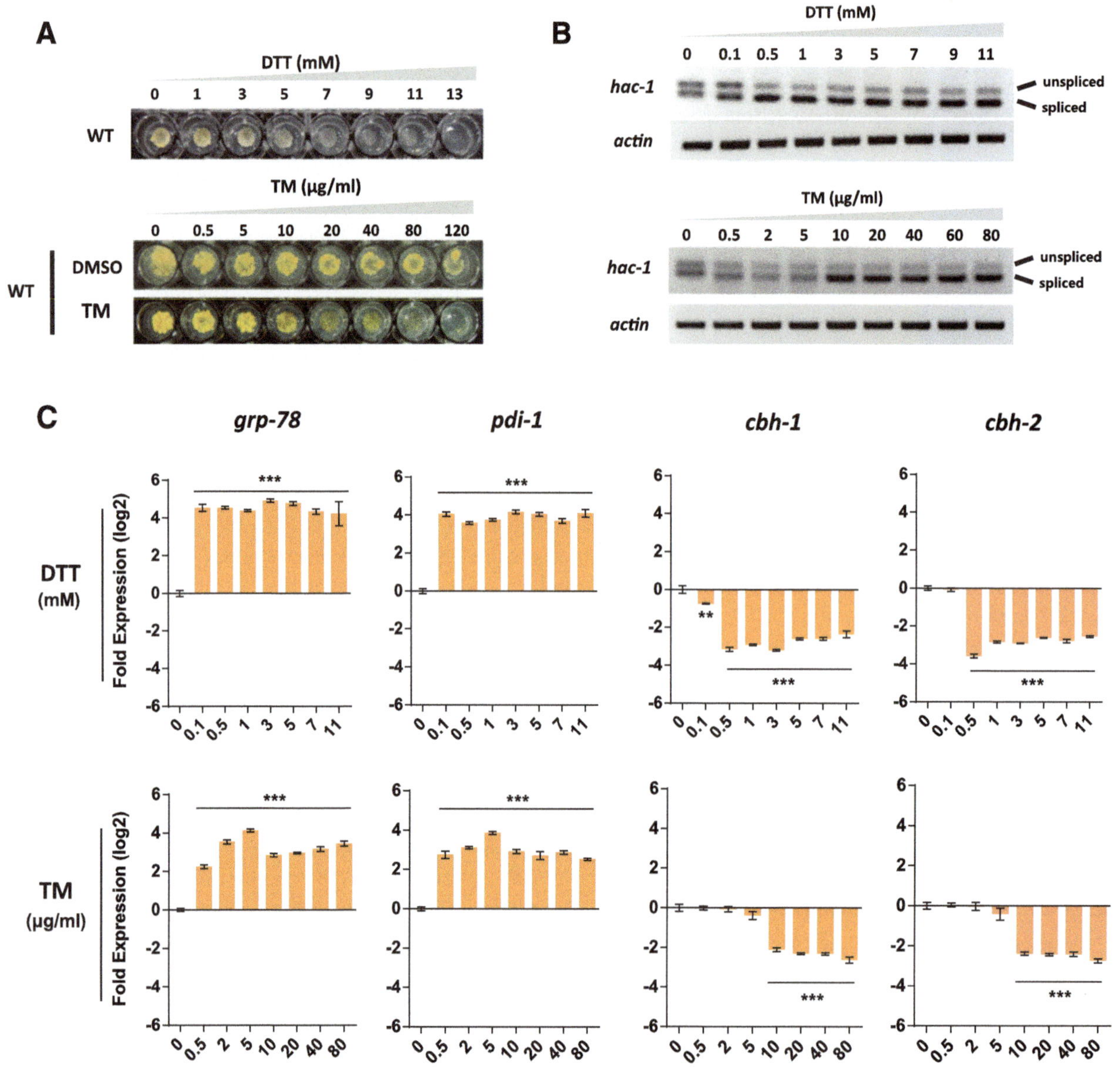

Figure 1 Defining applicable ER stress stimulation conditions. ER stress was induced by incubating *N. crassa* in the presence of concentration gradients of dithiothreitol (DTT) or tunicamycin (TM), and several criteria were used to define ER stress levels, including (**A**) sensitivity of growth to the drugs. WT conidia were inoculated onto solid GFS agar in a multi-well plate containing the indicated concentrations of DTT or TM and incubated at 25°C for 3 days. (**B**) Monitoring *hac-1* non-canonical splicing using semi-quantitative RT-PCR. WT conidia were inoculated into Vogel's medium with Avicel as carbon source for 36 h, then either DTT or TM was added at different concentrations as indicated for an extra 1 h before harvesting. The product sizes representing *hac-1* spliced and un-spliced forms were 201 bp and 224 bp, respectively. The gene *actin* (NCU04173) was used as a loading control. (**C**) Monitoring UPR and RESS activation by qPCR. Culture conditions were as in (**B**). The data are normalized to expression on mock = 1 and re-calculated to log_2-fold (standard error of the mean (SEM), $n = 3$). The asterisk indicates a significant difference from mock (**$P < 0.01$, ***$P < 0.001$) via one-way ANOVA.

see the 'Materials and methods' section) for each of the four conditions, respectively (Additional file 1: Table S1 p1 and Additional file 2: Table S2 p1-4). The global regulation patterns during mild and acute ER stresses were graphed separately (Figure 2A). Elevated expression levels in a subset of 766 (7.87%) genes induced by both DTT and TM at either mild or acute ER stress levels were deemed to be the result of true ER stress rather than non-specific effects and therefore defined as the ER stress response targets (ESRTs: 277 mild-specific targets, 285 acute-specific targets, and 204 genes generally up-regulated under both types of stress) (Figure 2B, Additional file 3: Table S3 p1, p3). Several well-established UPR targets [18] were found to be included in our ESRT dataset (Figure 2A). This result was confirmed

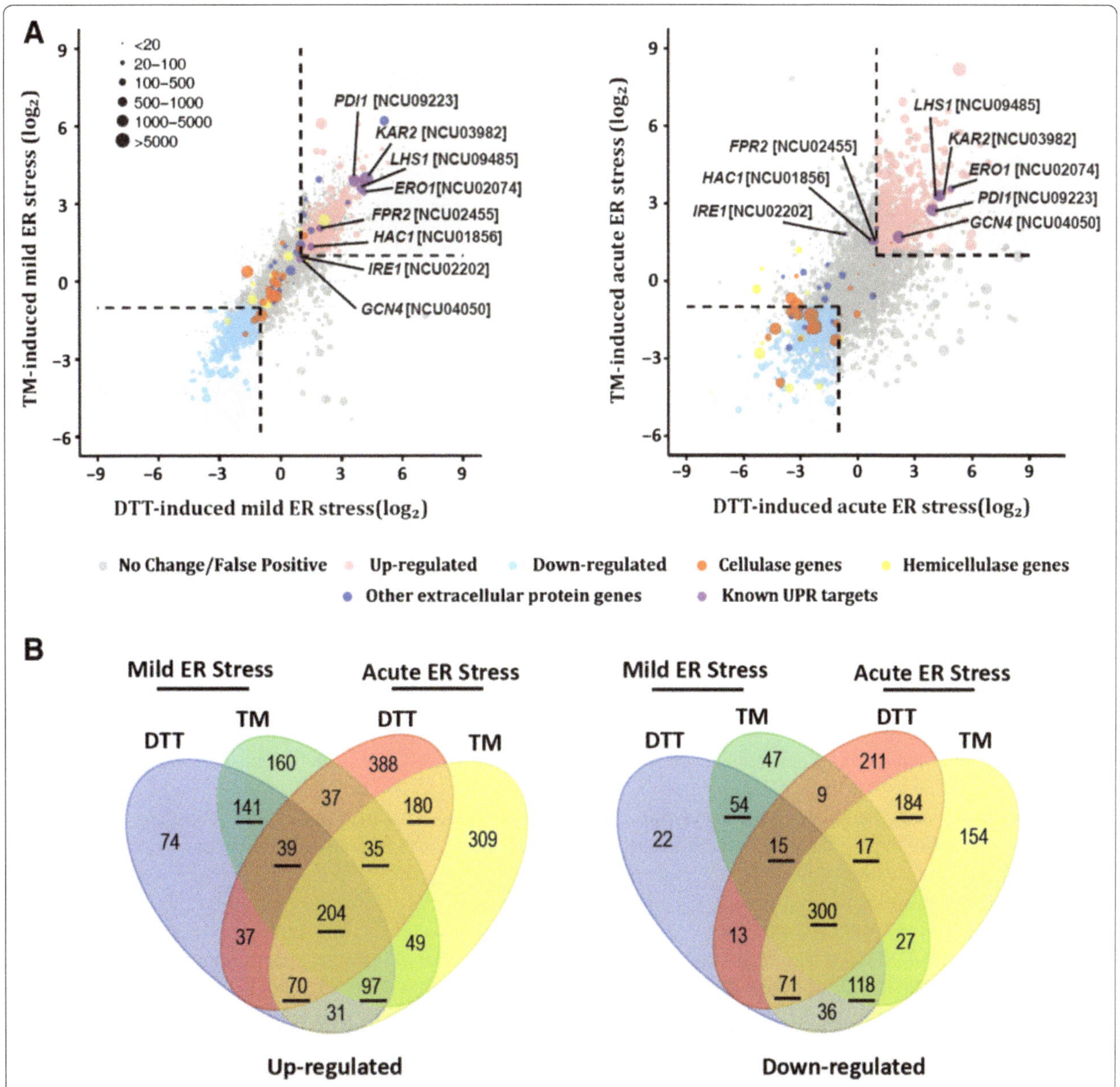

Figure 2 Panorama of ER stress response in *N. crassa*. (**A**) The pattern of intensity-dependent ER stress response (left: mild stress induced by 0.1 mM DTT or 5 μg/mL TM; right: acute stress induced by 5 mM DTT or 40 μg/mL TM) was plotted by calculating the $\log_2$ normalized fold changes in wild-type cells treated with DTT or TM *versus* the mock treatment. Each point represents a transcript with a detected expression level. Points filled with light-pink or light-blue represent genes expression with increased or decreased abundance (*q* value ≥0.95) upon ER stress. Gray points represent genes with no differences in the expression or false positives (see the 'Materials and methods' section for detail). Cellulases (red points), hemicellulases (yellow points), and extracellular proteins (dark blue points) previously identified by LC-MS [14] are indicated to show the RESS. The black dotted line represents a manually defined threshold ($|\log_2 \text{ratio}| = 1$). The mean of *D* value (the difference of RPKM for each gene between two samples in each comparison) of two comparisons (DTT and TM) from NOISeq results were plotted and the size of dots reflects a given level of transcript abundance. The locations of eight known UPR targets in the diagram are indicated to demonstrate that UPR was elicited successfully. (**B**) Four-way Venn diagram pooling genes according to their expression response to the four different treatments. Marked (black bar) are the 766 ESRTs up-regulated by both DTT and TM at either mild or acute ER stress levels (left) and the 759 down-regulated targets upon ER stress (right).

by qPCR assays using independent biological samples, showing consistency with the RNA-seq data (Additional file 4: Figure S1). Furthermore, we also collected 759 (7.80%) genes (Figure 2B) with decreased abundance upon ER stress (Figure 2B and Additional file 3: Table S3 p4, p6). Including cellobiohydrolase-1 (*cbh-1,* NCU07340) and cellobiohydrolase-2 (*cbh-2,* NCU09680), a total of 23 major lignocellulase genes were found to belong to this dataset and be specifically down-regulated under acute ER stress (Figure 2A and Additional file 3: Table S3 p7), implying that a rapid correction of intracellular transcript pools is needed to tolerate acute stress.

Further functional categorization of 462 (60.31%) annotated genes of the 766 ESRTs using FunCat [19] in addition to manual classification based on Pfam domain annotations and BLAST hits showed most of them were linked to protein secretion or the biogenesis of secretory organelles (Additional file 3: Table S3 p2, p3). Overrepresented categories found were for example as follows: protein folding, modification, destination (155 genes, $P = 1.49e\text{-}07$), membrane lipid metabolism (19 genes, $P = 3.89e\text{-}03$), transport facilities and transport routes (166 genes, $P = 1.86e\text{-}09$), as well as cell growth/morphogenesis (41 genes, $P = 4.52e\text{-}03$). Extending previous observations, the expression pattern of certain functional gene groups were found to depend on stress intensity. For example, genes encoding members of the p24 family (for example, NCU01342 [*ERV25* in *S. cerevisiae*], NCU03800 [*ERP3*], and NCU04003 [*ERP1*]), which are involved in ER to Golgi transport, were specifically up-regulated upon mild stress levels. In contrast, genes encoding most components of COPII transport vesicles involved in ER to Golgi transport were generally up-regulated under any stress levels. In addition, the expression levels of half of the genes engaged in *N*-linked or *O*-linked glycosylation (13 of 26) were enhanced solely under mild stress conditions whereas most genes related to phospholipid metabolism (9 of 13) were only up-regulated upon acute stress intensity (Additional file 3: Table S3 p3). Furthermore, our ESRT dataset included several complete complexes within the secretory pathway, such as the SEC61 translocon complex (NCU08897 [*SEC61* in *S. cerevisiae*], NCU08379 [*SBH1/SBH2*], and NCU04127 [*SSS1*]), the SEC63 complex (NCU00169 [*SEC63*], NCU06333 [*SEC62*], NCU02681 [*SEC66*], and NCU07746 [*SEC72*]), as well as the SNARE complex involved in vesicle-mediated Golgi to ER retrograde trafficking (NCU06708 [*SEC22*], NCU00953 [*SEC20*], NCU07939 [*UFE1*], and NCU00184 [*USE1*]). Given that around 40% of genes in the *N. crassa* genome are without clear function, the ESRTs identified here will aid in detecting novel component function in secretory pathway of filamentous fungi.

Genome-wide overview of the ESRT deletion mutants in *N. crassa*

To provide a basic understanding of the function of ESRT genes discovered by transcriptome analysis, we phenotypically screened 527 mutants of the total 766 ESRT targets (including 526 homokaryotic gene knockout strains and one single-base-pair deletion mutant for *cpc-1* which results in a nonfunctional, truncated polypeptide of this amino acid regulator [20]) and 249 genes that appear to be pivotal for ER stress resistance in the WT (Figure 3) for their sensitivity to ER stress (Additional file 5: Table S4). Most mutants were able to survive in the presence of 1 mM DTT, except for strain *ΔNCU10762*. In *S. cerevisiae*, the homolog of NCU10762 is *ALG7*, which encodes the UDP-N-acetyl-glucosamine-1-P transferase (GPT) that transfers the first N-acetylglucosamine to dolichyl-phosphate in the first step of the dolichol pathway of *N*-linked protein glycosylation [21]. Moreover, we found another 248 genes whose deletion caused the *N. crassa* cells to become DTT-sensitive, with 44, 84, 82, and 38 of them sensitive to 3, 5, 7, and 9 mM DTT, respectively. As anticipated, many of these genes had roles in the secretory pathway, such as genes involved in protein translocation into the ER (for example, NCU02681 [*SEC66* in *S. cerevisiae*], NCU08379 [*SEC61*], protein folding (for example, NCU00813 [*MPD1*], NCU09485 [*LHS1*], NCU 11102 [*SCJ1*], NCU00968 [*SIL1*]), ER-associated degradation (ERAD) (for example, NCU00146 [*DER1*], NCU 01296 [*DER1*], NCU01268 [*UBC6*]), protein glycosylation (for example, NCU06386 [*ALG5*], NCU06166 [*KTR7*]), and protein trafficking (for example, NCU03819 [*SEC16*], NCU 01342 [*ERV25*], NCU05514 [*YIP1*], NCU06708 [*SEC22*]). However, many mutants for genes with pivotal functions as unfolded protein response targets could not be assayed because they lacked homokaryotic mutants, such as *grp-78* (NCU03982) and *pdi-1* (NCU09223), as well as nearly half of the genes engaged in protein translocation, glycosylation, and trafficking (Additional file 3: Table S3 p3). It is likely that these genes are essential for cell viability.

Importantly, our screen provided functional data that 100 (13.05%) formerly unclassified genes could be directly or indirectly involved in ER stress. To test whether these ESRTs also contribute to cellulase secretion, we next checked the cellulase production of 18 mutants sensitive to 3 mM DTT and 31 mutants with sensitivity to 5 mM DTT in liquid batch culture. Deletion of five genes, namely NCU07135, NCU08986, NCU04360, NCU01886, and NCU00918, decreased cellulase secretion titers by more than 30% when compared with the WT (Additional file 4: Figure S2), suggesting the products of those five genes might function in the secretory pathway. The function of the other candidates remains unknown, although the DTT-sensitive phenotype suggests they might be involved in the ER stress response.

Identification of ER stress response targets that are regulated by the canonical IRE-1/HAC-1 UPR pathway

As in *S. cerevisiae*, no clear homologs to ER stress sensors such as PKR-like ER kinase (PERK) and activating transcription factor-6 (ATF6) were found in *N. crassa*, implying the IRE-1 (encoded by NCU02202) and HAC-1 (NCU01856) mediated pathway might be the only route to trigger the UPR cascade. In comparison to the homologs from other species [22-24], the *N. crassa hac-1*

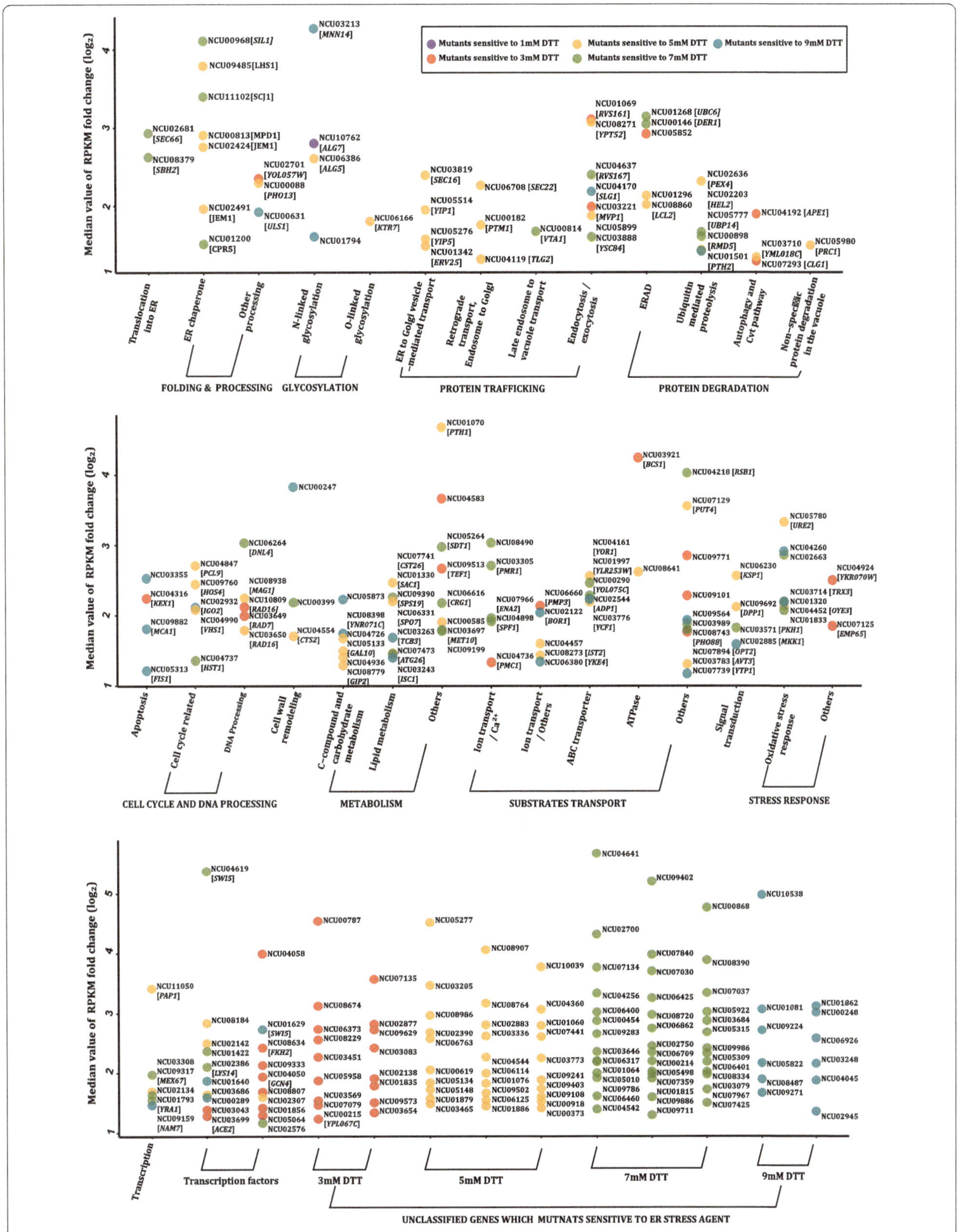

Figure 3 (See legend on next page.)

(See figure on previous page.)
Figure 3 Expression induction and functional categorization of 249 *N. crassa* KO mutants which exhibited ER stress susceptibility. Each point on the scatter plot represents the median value of fold changes of WT on mild DTT, mild TM, acute DTT, and acute TM (log_2 normalized, the full data see Additional file 3: Table S3, p3) and functional category of an ESRT gene whose KO mutants exhibited ER stress susceptibility. DTT-sensitive phenotypes are plotted using different colors, from strongest to weakest along a gradient of dark-purple (sensitive to 1 mM DTT), red (3 mM), orange (5 mM), green (7 mM) to cyan (9 mM).

mRNA contains an atypical intron (23 nt) [15] similar in size to higher eukaryotes, such as *Arabidopsis thaliana* (23 nt), *Drosophila melanogaster* (23 nt), and *Homo sapiens* (26 nt), rather than *S. cerevisiae* (252 nt) (Additional file 4: Figure S3A), revealing an intimate phylogenetic link between higher eukaryotes and filamentous fungi. To explore how the IRE-1/HAC-1 mediated UPR pathway affects ER stress response and lignocellulase production, a KO mutant for *hac-1* was constructed while *Δire-1* was purified by sexual crossing (Additional file 4: Figure S3B,C,D). Consistent with previous observations in *Aspergillus fumigatus* [25] and other eukaryotes [22,23], a loss of IRE-1 was sufficient to impair the unconventional splicing of *hac-1* mRNA in *N. crassa* (Additional file 4: Figure S3E). In comparison to their parental strains, both *Δire-1* and *Δhac-1* mutants showed a marked reduction (approximately 32%) in radial growth on solid minimal medium (Figure 4A). Deletion of either *ire-1* or *hac-1* did not visibly affect the width (approximately 5 μm) of the middle intercalary cell of conidia in *N. crassa* (Additional file 4: Figure S4A); however, during germination, huge vacuoles accumulated within young hyphae of both mutants, probably caused by a vesicle trafficking malfunction (Additional file 4: Figure S4B). In liquid batch culture using microcrystalline cellulose (Avicel) as the sole carbon source, more than 97% of both *Δire-1* and *Δhac-1* conidia failed to germinate (Figure 4B), suggesting UPR is a key factor for the cellulase secretion capacity of *N. crassa*.

We further explored which ESRTs were regulated by the UPR pathway through transcriptome profiling of both *Δire-1* and *Δhac-1* strains (Additional file 1: Table S1 p2 and Additional file 2: Table S2 p5-10). Because these strains are unable to use cellulose, sucrose pre-grown mycelia were used for the profiling analysis instead of directly inoculating cultures with conidia. Bioinformatic analysis (Additional file 6: Supporting information) revealed that 223 and 186 genes out of the 766 ESRTs were regulated by IRE-1 and HAC-1, respectively (Additional file 7: Table S5 p5). As expected from previous observations in *S. cerevisiae* [18,26] and other model systems [11,25,27-29], the IRE-1/HAC-1 governed UPR regulon mainly functions to facilitate protein folding and to remodel the secretory pathway. Major enriched functional categories (FunCat) were protein folding and stabilization ($P = 4.44e-06$ for IRE-1, $P = 3.72e-06$ for HAC-1), glycosylation ($P = 1.15e-$

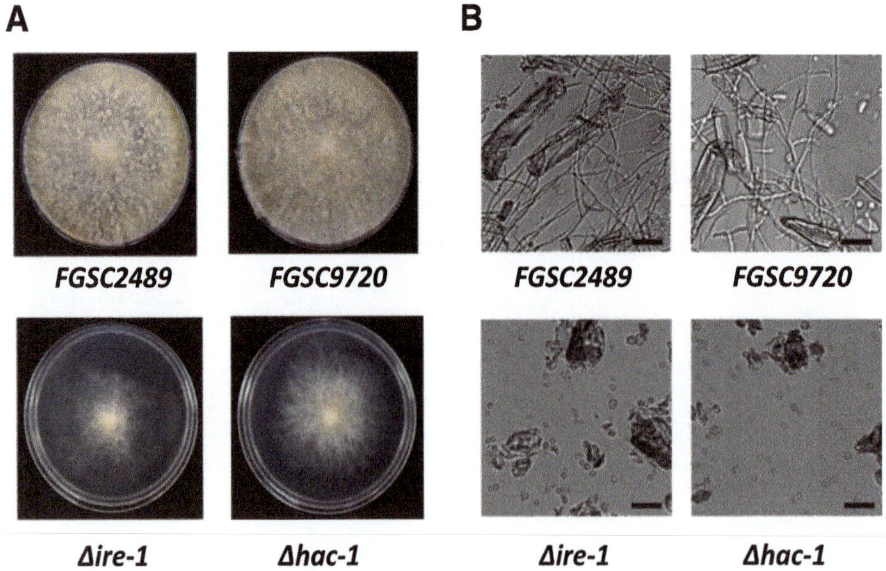

Figure 4 Deletion of *ire-1* and *hac-1* affects cell growth and causes cellulose utilization defects. (**A**) Comparison of radial growth phenotypes of *ire-1* and *hac-1* with their parental strain. Conidia (around 10^3) were dotted on minimal medium (MM) plates, grown for 72 h at 25°C. (**B**) The conidia of *Δire-1* and *Δhac-1* mutants fail to germinate under cellulolytic induction conditions (presence of Avicel as sole C-source). Pictures were taken after 36 h of culture; the irregular blocks in the images represent Avicel particles. The black bar indicates a length of 20 μm.

11 for IRE-1, $P = 7.07e-13$ for HAC-1) as well as vesicular transport (Golgi network, etc.) ($P = 6.35e-24$ for IRE-1, $P = 5.11e-22$ for HAC-1). Although *Δire-1* and *Δhac*-1 failed to grow on Avicel, lignocellulase genes were found to be induced to similar levels compared with their parental strains (Additional file 1: Table S1 p2), suggesting the UPR pathway does not directly regulate the induction of hydrolytic enzyme genes by cellulose. In addition, the expression levels for most lignocellulase genes were markedly down-regulated when acute ER stress emerged in both the *Δire-1* and *Δhac-1* backgrounds (Additional file 1: Table S1 p2 and Additional file 2: Table S2 p6, p8), implying RESS is independent of the UPR pathway. Furthermore, the analysis of intracellular cellulase synthesis and secretory pathway protein loading in *Δire-1* and *Δhac*-1 revealed that the cellulase synthesis in both mutants was similar to the parental strains (after a culture switch from sucrose to cellulose for 4 h and 36 h). Therefore, the observed reduction in secreted cellulases implies a reduced secretion capacity within this time frame. At a later stage (96 h), the amount of intracellular cellulases was diminished, which might be a result from a negative feedback-loop of cellulase hypo-secretion (Additional file 4: Figure S5A,B). These observations demonstrate that IRE-1 and HAC-1 play an important role in the *N. crassa* cellulase secretion system.

Systematic characterization of TFs required for ER stress response during cellulase synthesis

For both ER stress response and cellulase synthesis, transcription factor (TF) mediated orchestration of gene expression is crucial. We systematically searched for TFs with significantly altered expression patterns in our recorded differential expression data based on acute/mild ER stress (Figure 2). Of the 266 predicted TFs present in the *N. crassa* genome (based on Broad version 7) [30] (Additional file 8: Table S6 p1-2), 33 were up-regulated in the presence of ER stress, while 6 TFs were down-regulated (Figure 5A and Additional file 8: Table S6 p3).

Of the 33 TFs with increased abundance upon ER stress, five (15.15%) belonged to the basic-region leucine zipper (bZIP) family, 11 (33.33%) to the fungal Zn(2)-Cys(6) binuclear cluster family, nine (27.27%) to the Cys2His2 (C2H2)-type zinc fingers, and eight to miscellaneous families (24.24%) (Additional file 8: Table S6 p3). We next screened 30 available homokaryotic deletion mutants out of 33 TFs for DTT-sensitivity; compared with WT, 20 TFs exhibited significant DTT-sensitive phenotypes, including the UPR regulator *hac-1* (NCU01856) (Figure 5B). Additionally, six TFs were found to be regulated in an IRE-1/HAC-1 mediated manner; including four TFs (NCU09333, NCU03043, NCU03686, NCU02386) affected by IRE-1 and two TFs (NCU01422, NCU06095) that were dependent on HAC-1 (Additional file 7: Table S5 p5).

To assess whether those TFs regulated by ER stress also contribute to cellulase secretion, liquid batch cultures with (Figure 5C) or without (Additional file 4: Figure S6) a rich nitrogen source were performed. The *Δire-1* and *Δhac-1* strains, along with their parental strains, were included as controls. Consistent with previous observations [31], the *Δcre-1* strain displayed an increase of approximately 50% in the amount of secreted protein compared to the WT strain. A loss of *ire-1* led to a 50% decrease of the secreted protein levels in comparison to the WT strain, while the *Δhac-1* KO mutant almost completely lost its ability to produce lignocellulases. Using ±30% of secreted proteins produced by WT as a threshold (refer to the *Δcre-1* published data [31]) and considering the data from both nitrogen conditions, we found *ΔNCU03699* could significantly promote lignocellulase secretion similar to *Δcre-1*, while *ΔNCU02413* (*Δrrg-2*) and *ΔNCU02724* exhibited repressive effects. The intracellular cellulase loading of the secretory pathway was further examined in these mutants using switch experiments of sucrose pre-grown strains shifted to Avicel. While the intracellular cellulase production was not affected until after 96 h (Additional file 4: Figure S5A,B), extracellular cellulase accumulation was substantially increased (*ΔNCU03699*) or reduced (*ΔNCU02413* (*Δrrg-2*) and *ΔNCU02724*) (Additional file 4: Figure S7A,B), implying that the secretion capacity rather than cellulase synthesis is affected in these strains. These findings suggest that the hyper- or hypo-secretion phenotype of these mutants is linked to cellulase secretion (Additional file 4: Figure S7A,B). To the best of our knowledge, this is the first report providing strong functional data indicating that these TFs are involved in lignocellulase production and ER stress, although NCU02413 (*rrg-2*) had previously been suggested to play a role in the oxidative stress response in *N. crassa* [32]. Based on their involvement in the overall process, we named the TFs encoded by NCU03699 and NCU02724 RES-1 and regulator for ER stress response (RES-2), respectively.

Of the six TFs with decreased abundance upon ER stress, three belong to the fungal Zn(2)-Cys(6) binuclear cluster domain family, including the two core regulators of plant cell wall degradation: CLR-2 [33] and XLR-1 [34]. Because most lignocellulase genes that are down-regulated in the presence of acute ER stress belong to the CLR-2 or XLR-1 regulons, it seems likely that RESS results to a great extent from the down-regulation of these core regulators. Two of the other four TFs that were found to be down-regulated under ER stress (NCU09549, NCU04848, NCU08923, and NCU07430 had available mutants that allowed us to test their enzyme production. Slightly decreased lignocellulase production was recorded for ΔNCU09549 when compared to WT, while no obvious defect was observed in the ΔNCU04848 mutant (Figure 5C).

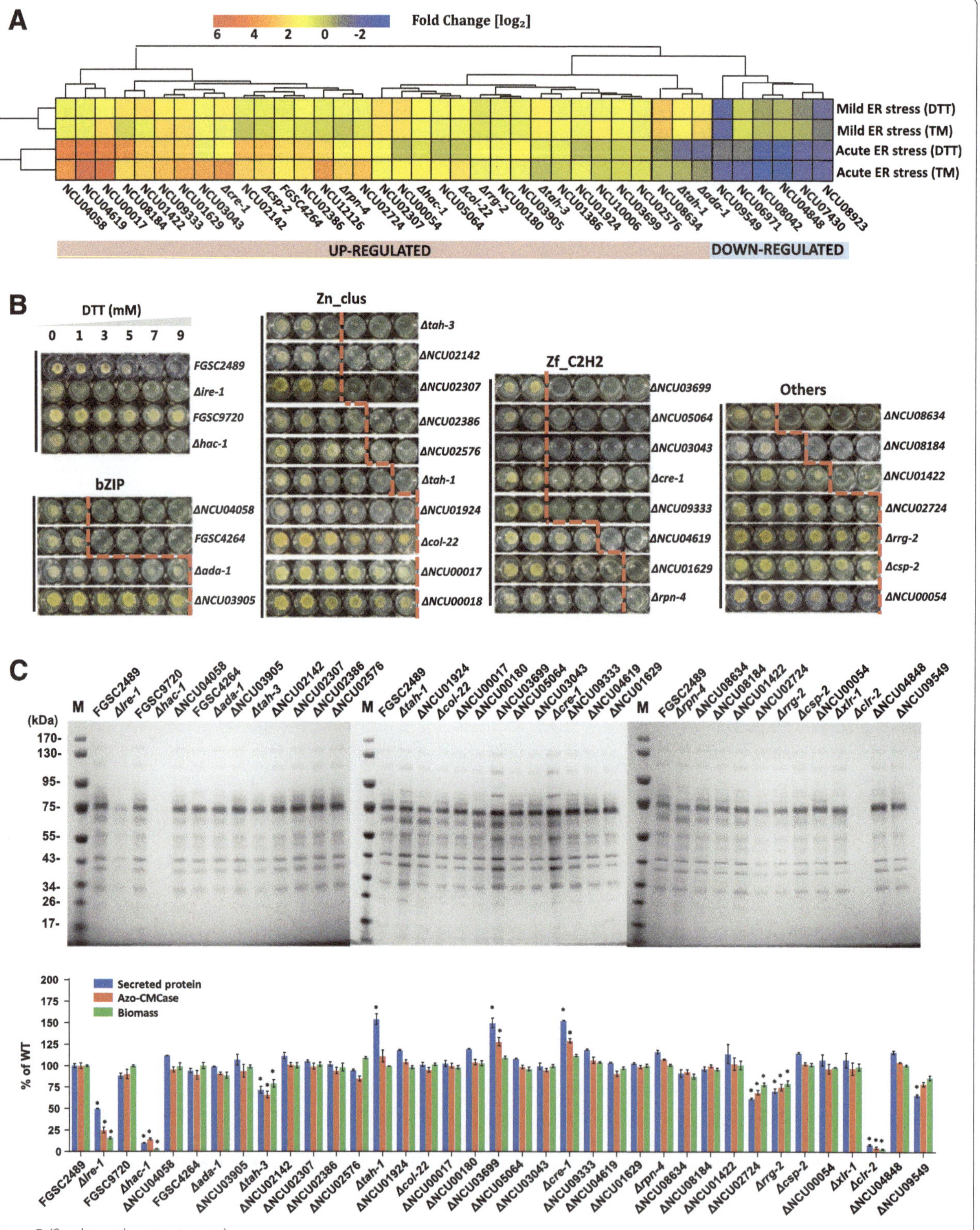

Figure 5 (See legend on next page.)

(See figure on previous page.)
Figure 5 Screen for transcription factors involved in ER stress response and cellulase synthesis. (**A**) Heatmap of 39 transcription factors (TFs) found to be differentially expressed in response to ER stress under distinct stress intensities: 33 up-regulated and 6 down-regulated. The degree of expression abundance was normalized by RPKM-fold change (log_2). Names of TF DNA-binding domains according to Pfam annotation were indicated in square brackets. (**B**) Screening of DTT susceptibility of 30 available ESRT TF KO mutants. The red line indicates the growth to no growth threshold and is an indicator for sensitivity to ER stress elicited by DTT. (**C**) Batch cultures of ESRT TF KO mutants were prepared to discover targets affecting cellulase secretion. After 7 days of culture with rich nitrogen source, the typical secretome of each TF candidate is shown on SDS-PAGE gel (upper panel), while total extracellular protein concentration, CMCase activity, and biomass (as total protein) were measured and displayed after normalization to the WT control (lower panel) by percentage (standard error of the mean (SEM), $n = 3$, the asterisk (*) indicates a significant difference from WT with an unadjusted P value of <0.001 using a one-way ANOVA).

RES-1, RES-2, and RRG-2 are required for ER stress response and cellulase synthesis

Since RES-1, RES-2, and RRG-2 likely function to bridge the ER stress response to lignocellulase expression, transcriptional profiling of the three corresponding TF deletion strains was performed under ER stress conditions. Deletion of *res-1* (encoded by NCU03699) resulted in up-regulation of 583 genes in comparison to WT in cells grown on Avicel for 36 h. Consistent with the protein secretion data (Figure 5C), FunCat analysis of the 583 potential *res-1* targets showed 38 genes belonging to the 'polysaccharide metabolism' group, with many lignocellulase genes being enriched ($P = 6.11e\text{-}09$) (Figure 6A and Additional file 9: Table S7 p1). For example, the gene expression of several *AA9* family members such as NCU08760 (*AA9-5*), NCU03328 (*AA9-6*), NCU00836 (*AA9-7*), NCU03000 (*AA9-8*) as well as some other glycosyl hydrolases such as NCU05121 (*gh45-1*, endoglucanase V), NCU06861 (*gh43-3*), NCU07326 (*gh43-6*), and NCU04952 (*gh3-4*, β-glucosidase) showed a significant increase in the *Δres-1* strain compared to WT. The expression levels of NCU09664 and NCU04870, encoding acetylxylan esterases, increased more than 33- and 7-fold, respectively, and a gene encoding the extracellular feruloyl esterase B (NCU09491) increased over 30-fold upon loss of RES-1. Surprisingly, the expression of

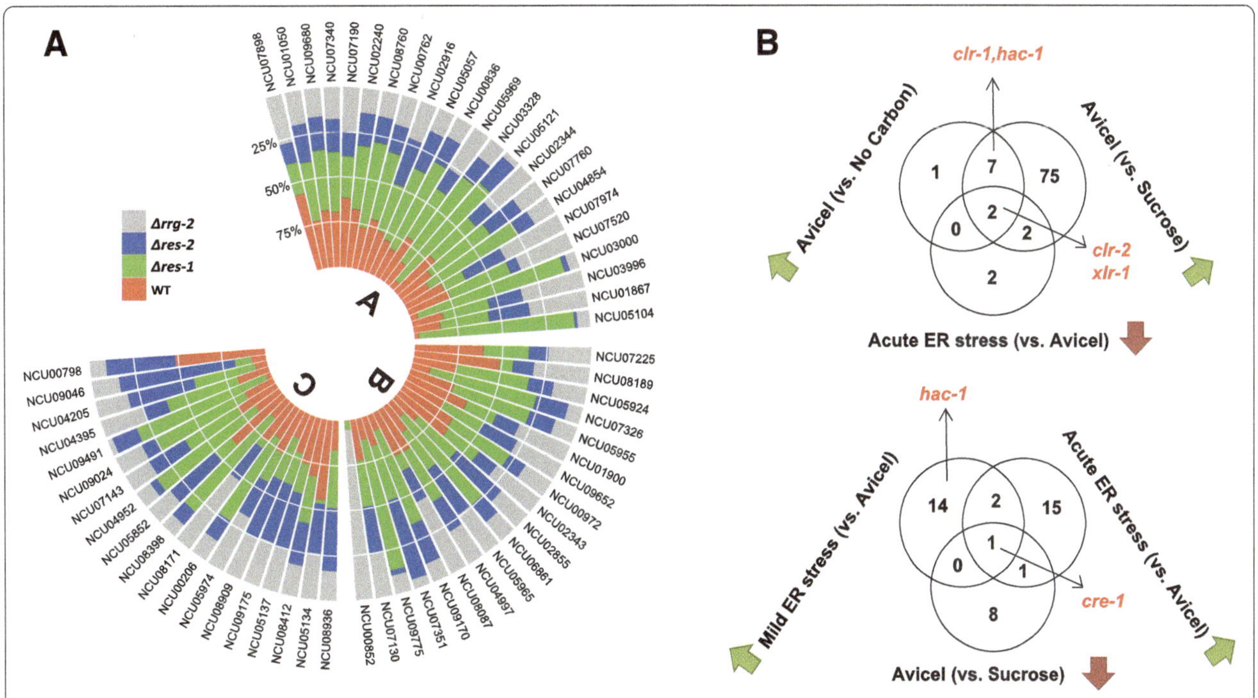

Figure 6 RES-1, RES-2, and RRG-2 indirectly affect lignocellulase gene expression. (**A**) Comparison of lignocellulase expression levels in *Δres-1*, *Δres-2*, and *Δrrg-2* with WT. A total of 62 genes were grouped into three clusters: [A] cellulase genes, [B] hemicellulase genes, and [C] other genes encoding extracellular proteins. For each target, the relative proportion of transcript abundance (RPKM) is indicated as a percentage of the total. (**B**) Three-way Venn diagrams generated to search for TFs, which are coordinately expressed with lignocellulase genes. Top diagram: TFs induced (green arrow) by Avicel *versus* no carbon (top left) and *versus* sucrose condition (top right) and repressed (red arrow) under acute ER stress (bottom) were reasoned to potentially have a positive effect on lignocellulase gene expression. Bottom diagram: In addition, TFs being repressed when exposed to Avicel *versus* sucrose treatment while being induced upon ER stress could likely exert a negative effect on lignocellulase gene expression.

several major cellulase genes, such as NCU07340 (*cbh-1*), NCU09680 (*cbh-2*), and NCU00762 (*gh5-1*, endoglucanase 3), was WT-like in *Δres-1* after growth on Avicel for 36 h, although significant hyper-secretion was observed after 7 days of culture (Figure 5C), suggesting RES-1 could affect lignocellulase production not only at the transcriptional but also at the post-transcriptional level as protein, thus the latter effect is exerted indirectly. In good agreement with this hypothesis, we found that a deletion of *res-1* resulted in 76 ESRTs failing to be up-regulated upon exposure to DTT (Additional file 9: Table S7 p2-3), especially genes functioning in glycosylation ($P = 2.8e-03$, for example, NCU02541 and NCU03995, which separately encode the oligosaccharyltransferase alpha and gamma subunit-like proteins) as well as lysosomal and vacuolar protein degradation ($P = 5.38e-05$, for example, NCU01955 [autophagy-related protein 3], NCU02515 [dipeptidyl aminopeptidase]).

The *Δres-2* and *Δrrg-2* deletion strains exhibited repressive effects on lignocellulase secretion. Transcriptome profiles showed deletion of *res-2* specifically down-regulated 491 genes compared with WT when cells were grown on Avicel, including 87 genes involved in C-compound and carbohydrate metabolism ($P = 1.62e-06$) (Figure 6A and Additional file 9: Table S7 p4). For instance, hemicellulase genes such as NCU07225 (*gh11-2*, endo-1, 4-beta-xylanase 2), NCU08189 (*gh10-2*, endo-1,4-beta-xylanase), and NCU00972 (*gh53-1*, arabinogalactan endo-1,4-β-galactosidase) decreased more than three- to fourfold under Avicel conditions in the *Δres-2* mutant as compared to WT. We noticed the major cellulase genes such as NCU07340 (*cbh-1*), NCU09680 (*cbh-2*), and NCU00762 (*gh5-1*) also showed a 1.6 ~ 1.8-fold decrease in expression under Avicel conditions in the *Δres-2* strain, although the respective data did not pass the *Q* value cut-off. In addition, 118 ESRTs were shown to be regulated by the TF RES-2, including targets involved in cellular export and secretion ($P = 7.68e-03$), cellular import ($P = 5.49e-03$), enzyme inhibitors ($P = 7.10e-03$), as well as several targets functioning in autophagy and cytoplasm to vacuole-targeting (Cvt) pathways, such as NCU01955 (autophagy-related protein 3), NCU04192 (vacuolar aspartyl aminopeptidase Lap4), and NCU01545 (autophagy protein 8) (Additional file 9: Table S7 p5-6).

Similar to *res-2*, a deletion of *rrg-2* down-regulated 102 genes compared with WT when cells were grown on Avicel, including 25 genes involved in C-compound and carbohydrate metabolism ($P = 6.11e-05$) (Figure 6A and Additional file 9: Table S7 p7), such as lignocellulase genes NCU09680 (*cbh2*), NCU02240 (*AA9-1*, endoglucanase II), and NCU05955 (xyloglucanase). The expression of genes encoding glucoamylases was also found to be down-regulated (for example NCU01517 and NCU08746). In addition, *rrg-2* also appeared to regulate 115 ESRTs related to lipid, fatty acid, and isoprenoid metabolism ($P = 4.90e-03$), glycosylation ($P = 1.97e-03$), ER to Golgi transport ($P = 9.05e-03$), etc. (Additional file 9: Table S7 p8-9).

Lignocellulase genes can be induced significantly when *Neurospora* is exposed to cellulose; however, even in the presence of a cellulosic inducer, these genes can also be rapidly down-regulated when suffering RESS. Therefore, TFs with a similar expression pattern to lignocellulase genes under both physiological conditions are likely to be involved in cellulase regulation. We re-analyzed published transcriptome profiling data of *N. crassa* grown on sucrose, Avicel, and no carbon (starvation) and compared these with our ER stress response data [35] (Additional file 8: Table S6 p3-4; Figure 6B). Ten and 86 TFs were significantly induced by Avicel compared to either the no carbon or sucrose controls, respectively (Figure 6B, upper panel); nine of them were up-regulated in both conditions. Besides *clr-1*, *clr-2*, and *xlr-1*, *hac-1* was also included, suggesting lignocellulase induction and UPR activation act synergistically. Filtering additionally for TFs with decreased abundance under acute ER stress (six total; potentially due to RESS), only *clr-2* and *xlr-1* overlapped, implying CLR-2 and XLR-1 maybe the only direct activators of the lignocellulase regulon. Trying to find the key repressor of cellulase genes, we reasoned that such a regulator might be up-regulated under ER stress (mild as well as acute) and down-regulated under standard cellulolytic induction (Avicel only) compared to the repression condition on sucrose. Only *cre-1* was identified to match these criteria (Figure 6B, lower panel) and is therefore confirmed to have a major influence on lignocellulase gene expression. However, this result does not imply that CRE-1 is the only repressor for cellulose genes, and additional repressors that are not necessarily co-regulated might still be involved. The expression pattern for *res-1*, *res-2*, and *rrg-2* did not correlate with major cellulase genes or core regulators such as *clr-2*, *xlr-1*, and *cre-1*, suggesting their influence on cellulase expression may occur through another way.

Discussion

In the present study, we systematically analyzed the ER stress response of *N. crassa* during lignocellulase induction and secretion by experiments at both mild and acute stress levels. We found 766 genes to be specifically up-regulated in response to this process (ESRTs), of which 223 and 186 genes were regulated by UPR pathway core components IRE-1 and HAC-1, respectively. Screening 527 deletion strains for ESRTs lead to the identification of 249 genes that appear to be pivotal for resistance to ER stress in the WT. Of these, 100 targets were previously annotated as 'unknown' and are reported here for the first time to be involved in the response to ER stress. Interestingly, we found 357 ESRTs to be highly conserved, with orthologs in higher eukaryotes such as human

(Additional file 3: Table S3 p3). For example, the ESRT NCU09101 encodes a hypothetical protein and has orthologs in both *S. cerevisiae* (YBR220C) and *H. sapiens* (SLC33A1/AT-1). The function of *S. cerevisiae* YBR220C is unknown; however, *H. sapiens* SLC33A1/AT-1 encodes an ER membrane transporter that was recently reported to regulate the induction of autophagy downstream of the IRE1/XBP1 pathway [36]. Our screening data showed *ΔNCU09101* was sensitive to DTT (Additional file 5: Table S4), consistent with previous reports, validating that this target is related to ER stress. To our knowledge, systematic information on ER stress at a cellular level has only been performed in three model systems to date, including screening of KO mutants in *S. cerevisiae* [37] and screening of natural variation of human B cells [38] and *Drosophila* lines [39]. Thus, our data not only fill the knowledge gap in filamentous fungi but also contribute to the general understanding of the mechanisms mediating this process in higher eukaryotes.

In this study, we further addressed how TF networks coordinate the signal flow and gene expression during the ER stress response and cellulase synthesis. Thirty-three TFs were up-regulated upon ER stress under conditions of cellulase synthesis, including the well-characterized UPR regulator HAC-1 (NCU01856) as well as CPC-1 (NCU04050). We found *N. crassa* HAC-1 to act as an important factor for lignocellulase secretion while not mediating the RESS feed-back loop (profiling data mentioned above are presented in Additional file 1: Table S1 p2). In addition, we found that hyphal growth is severely impaired in *hac-1* deletion strains of *N. crassa*, which is consistent with previous observations in *Aspergillus niger* [40], suggesting that HAC-1 plays an important role in hyphal polarized growth and development, as it was shown in *Candida albicans* [28]. Interestingly, though the observed phenotype of Δ*ire-1* was slightly weaker than the phenotype changes in Δ*hac-1*, both mutants showed a pronounced defect in their conidial germination and cellulase secretion capacity compared to their parental strains. Moreover, an un-spliced *hac-1* transcript is present in the *N. crassa* Δ*ire*-1 strain. While in *S. cerevisiae* an un-spliced *hac-1* mRNA cannot be translated into a functional protein [41], previous studies could show that mammalian cells are indeed capable of translating un-spliced *hac-1/Xbp-1* transcripts [42]. Whether such an event also occurs in filamentous fungi and in which way these proteins would contribute to certain physiological functions such as hyphal polarized growth remain unclear. The TF CPC-1 has been previously characterized as a regulator associated with amino acid biosynthesis in *N. crassa* [43]. In metazoans, ATF4 as the functional homolog of CPC-1 has been proven to act in the PERK mediated UPR branch and it regulates those UPR genes involved in redox response and apoptosis [6,7,11]. The fact that CPC-1/CpcA is specifically induced upon ER stress has been observed in several independent fungal systems [11,44,45] and could be confirmed by our data. Here, we found additionally that a deletion of *cpc-1* renders *N. crassa* cells sensitive to DTT (Figure 5B) while significantly increasing extracellular protein secretion under limited nitrogen conditions (Figure S5). These observations suggest that CPC-1 may play a critical role in *N. crassa* ER stress response and cellulase secretion, although its regulation does not seem to be controlled by the IRE-1/HAC-1 mediated UPR cascade (Additional file 7: Table S5 p5). With NCU04058, a novel putative bZIP TF was identified, whose expression levels were markedly elevated upon acute ER stress. Moreover, the fact that the corresponding deletion strain displayed an apparent DTT-sensitive phenotype implies that it might also function as a stress response regulator.

The TF genes *res-1* (NCU03699) and NCU03043 are the homologs of *A. fumigatus zfpA* and *flbC*, respectively. It has been reported that both *A. fumigatus zfpA* and *flbC* transcripts can be induced by exposing cells to calcium [46,47], suggesting RES-1 and the TF encoded by NCU03043 may respond to intracellular calcium disturbances induced by ER stress. Moreover, we saw that loss of RES-1 results in a boost of lignocellulase secretion, probably owing to the de-repressive effect of several auxiliary cellulose-degrading enzymes on transcription (for example, several AA9 family members) which contribute to promote the efficiency of classical hydrolytic enzymes (cellulases). Deletion of *res-2* (NCU02724) and *rrg-2* (NCU02413) on the other hand, significantly decreased lignocellulase expression and secretion in *N. crassa*, although probably in an indirect manner. In this work, FunCat characterization of the RES-2 regulon revealed that genes involved in cellular export and secretion ($P = 7.68e\text{-}03$) as well as cellular import (endocytosis) ($P = 5.49e\text{-}03$) were enriched. Moreover, endocytosis has been reported to contribute to hyphal tip growth and tip secretion [48], suggesting RES-2 might regulate secretory pathways of *N. crassa*. In addition, the ortholog of RES-2 in *A. fumigates*, Afu1g09670, has been reported to be induced by calcium treatment [46]. While ER stress is always associated with a disturbance in the intracellular calcium homeostasis, changes in calcium levels are also reported to influence the hyphal tip growth [49,50] and thus might in turn affect tip secretion. Therefore, it is feasible, that RES-2 is involved in the ER stress response and affects lignocellulase secretion.

The *S. cerevisiae* SKN7, a homolog of the *N. crassa* RRG-2 (Identity: 57.5%, *E* value: 1E-38), controls the oxidative stress response in yeast cells together with YAP1 [51]. It has been reported that over-expression of *YAP1* contributes to modified intracellular redox conditions and enhances recombinant protein secretion [52]. However, whether SKN7 is also involved in protein secretion remains to be elucidated. In this work, deletion of RRG-2/SKN7 in *N. crassa* leads to a hypo-secretion phenotype, suggesting that RRG-2/SKN7 maybe has similar functions as YAP1 in the protein secretion process. It is also known that ER stress and oxidative stress are closely linked events, and therefore antioxidants, which can diminish oxidative stress [53], could additionally decrease ER stress and improve protein secretion. Hence, it can be concluded that RRG-2 might be involved in the regulation of the redox balance to affect cellulase secretion. The TF TAH-3 (NCU03686) was previously found to be required for fungal tolerance to harsh non-thermal plasma treatment in *N. crassa* [54]. The counterpart of TAH-3 in *S. cerevisiae* is UPC2, which functions by regulating sterol biosynthetic gene expression [55]. Potentially, disturbed sterol biosynthesis could trigger ER stress [56]. We found that extracellular protein secretion was reduced by circa 30% in *Δtah-3* compared to the WT, suggesting TAH-3 might be involved in ER stress resistance by maintaining intracellular sterol homeostasis.

The TF encoded by NCU09333 was identified as the homolog of the cellulase repressor ACE1 in *T. reesei* [57] and the stress responsive factor StzA in *Aspergillus nidulans* [58]. In *N. crassa*, NCU09333 was up-regulated upon ER stress while showing only a limited effect on cellulase secretion, suggesting NCU09333 functions more like StzA rather than ACE1 in *N. crassa*. Finally, the expression level of the carbon catabolite repression (CCR) regulator CRE-1 (NCU08807) was found to be up-regulated upon ER stress. The *Δcre-1* strain moreover showed a DTT-sensitive phenotype, suggesting the CCR pathway might also cross-talk with the ER stress response.

By monitoring TFs with a decreased abundance upon ER stress, we detected CLR-2 and XLR-1. This result was confirmed by qPCR (Additional file 4: Figure S8), suggesting that the RESS-mediated repression of cellulase genes was likely a direct result of the down-regulation of these two regulators. The expression levels of some carbohydrate transporter genes, which are belonging to the *clr-2* and *xlr-1* regulons, such as *cdt-1* (NCU00801) and *lat-1* (NCU02188), were also decreased, supporting our hypothesis. It was intriguing to find that the transcription level of another key regulator, cellulose degradation regulator 1 (CLR-1), was not significantly down-regulated in the RESS condition (Additional file 4: Figure S8). CLR-1 has been shown to be necessary for expression of *clr-2* [33] and is up-regulated when fungal cells are exposed to cellulose, suggesting the expression of *clr-1* may be directly regulated by biopolymer inducers while *clr-2* transcription is not only controlled by CLR-1 but also regulated by some other players in the ER stress response. With the discovery mentioned above, a new model for the lignocellulase induction and synthesis of *N. crassa* is presented (Figure 7).

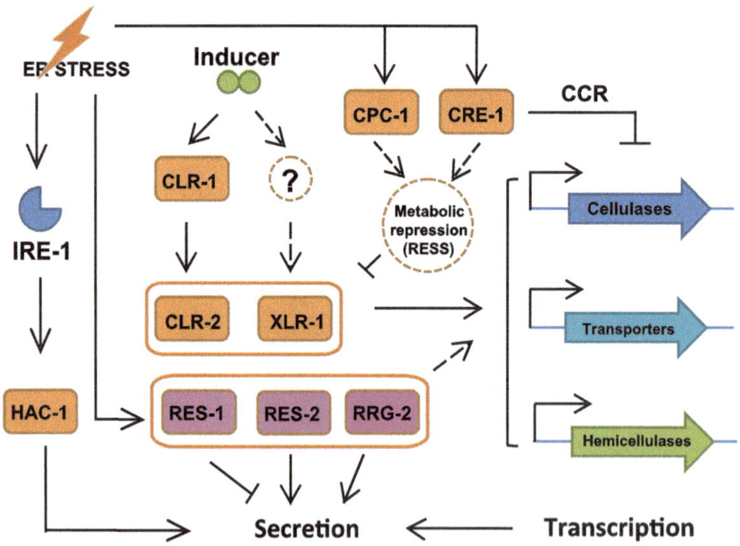

Figure 7 Schematic model illustrating the coordinated action of ER stress and cellulase synthesis. ER stress affects cellulase gene expression and secretion via the core lignocellulase regulators. Acute ER stress specifically down-regulates *clr-2* and *xlr-1* rather than *clr-1* for rapid adjustment of lignocellulase transcript abundance. The CRE-1-mediated CCR pathway activated by ER stress might also contribute to this process. The newly identified regulators RES-1, RES-2, and RRG-2 indirectly affect lignocellulase production. However, enzyme synthesis was increased in *Δres-1* when compared with WT, whereas *Δres-2* and *Δrrg-2* showed repressive effects, suggesting RES-1 likely acts as repressor while RES-2 and RRG-2 act as activators for the synthesis and secretion of extracellular enzymes. The IRE-1/HAC-1-mediated UPR pathway also acts indirectly on lignocellulase production by mainly regulating protein folding, modification (such as glycosylation), and transport. The finding of RESS being independent of UPR pathway, raise the possibility that metabolic repression resulting from ER stress might contribute to this process.

RESS was shown to be activated independently of the IRE-1/HAC-1 mediated UPR pathway in this work (profiling data on Additional file 1: Table S1 p2 and Figure 7). How RESS is triggered by ER stress remains to be elucidated. Notably, a previous study in *T. reesei* demonstrated that the intracellular concentrations of some free amino acids increased upon acute ER stress [44]. Moreover, we found *cre-1* expression levels to be elevated when cells suffered ER stress, suggesting that the intracellular glucose homeostasis may also be perturbed under these conditions. These observations raise the possibility that metabolic repression that results from internal fluctuations of intracellular nutritional cues, such as simple sugars or free amino acids, might have an impact on RESS, although more data need to be acquired to support this hypothesis.

Conclusions

Here, we deeply investigated the complicated cross-talk between the ER stress response and lignocellulase synthesis in *N. crassa* by combining transcriptional profiling, mutant screening, and a detailed regulatory network analysis. The genome-wide screening uncovered hundreds of genes that might be involved in this pathway, including 100 formerly hypothetical genes. Our data furthermore allowed connecting the function of three novel transcription factors to cellulase expression and the ER stress response, shedding new light on the molecular mechanism of this complicated cross-talk in fungi. Since filamentous fungi are important systems for biotechnological enzyme production, the findings of the present study should have an impact on industrial fungal strain development for both lignocellulase production as well as synthesis of biomass-based chemicals.

Materials and methods

Strains

The *N. crassa* strains used in the study were obtained from the Fungal Genetics Stock Center (http://www.fgsc.net/) [59], including the WT strain FGSC#2489 (*Mat A*), FGSC#9720 (*mus-52::bar; his-3, Mat A*), the *cpc-1* mutant (CD15 allele mutation, FGSC#4264) and a set of 526 homokaryotic deletion strains individually knocked-out in non-essential genes encoding proteins playing potential roles in ER stress response. The *hac-1* gene was deleted following standard methods provided by the *Neurospora* Functional Genomics Project (http://www.fgsc.net/ neurosporaprotocols/KO_Protocols.pdf, for details see Additional file 6: Supporting Information), while a homokaryotic *ire-1* KO mutant was generated by a sexual cross of the heterkaryon FGSC#21727 with wild-type strain FGSC#2489.

Liquid cultures for ER stress stimulation

For strains able to utilize cellulose, such as WT, *Δres-1*, *Δres-2*, and *Δrrg-2*, liquid cultivations were carried out on a defined cellulose medium (1× Vogel's salts, 1% *w/v* crystalline cellulose (Avicel PH-101; Sigma-Aldrich, St. Louis, USA), 0.2% *w/v* NH_4NO_3) according to published data [60]. Ten-day-old conidia were inoculated into a 50-mL medium at 10^6 conidia/mL final concentrations and grown at 25°C in constant light and agitation (200 rpm). After 36 h, the cultures were treated with either dithiothreitol (DTT) or tunicamycin (TM) with indicated doses for one more hour to induce ER stress. Since both strains, *Δire-1 and Δhac-1*, are unable to utilize cellulose, both mutants and their parental strains were pre-grown in minimal medium (1× Vogel's salts, 2% *w/v* sucrose) for 16 h, and then young hyphae were harvested and transferred into cellulose medium as mentioned above for another 4 h to induce cellulase gene expression. After that, 5 mM DTT were added into each culture to stimulate ER stress for one more hour.

ER stress sensitivity screening

Fresh conidia were harvested from 10-day-old slants for each strain and suspended in sterilized water to a final concentration around 3×10^5 per mL. Three microliters of a conidia suspension (approximately 1,000 conidia) were spotted onto modified solid GFS medium [61] (1× Vogel's salts, 2% *w/v* sorbose, 0.05% *w/v* glucose, 0.05% *w/v* fructose, 1.5% *w/v* agar) including the indicated dose of DTT or TM. Colonial growth was performed at 25°C for 3 days. The antibiotic Hygromycin B was added into the medium at a final concentration of 200 μg/mL for all knock-out mutants, while L-histidine was added with a final concentration of 100 μg/mL for *Δhac-1* and FGSC#9720.

Total RNA extraction and real-time qPCR analysis

Total RNA from frozen samples was isolated using Zirconia/Silica beads (0.5 mm diameter; Biospec Products, Bartlesville, USA) and a Mini-Beadbeater (Biospec Products, Bartlesville, USA) with 1 mL TRIzol reagent (Invitrogen, Grand Island, USA) according to the manufacturer's protocol. An additional clean-up including the on-column DNase I treatment was performed by using the RNeasy mini kit (QIAGEN, Valencia, USA). For qPCR analysis, reverse transcription was performed using iScript™ cDNA synthesis kit (Bio-Rad, Hercules, USA). The qPCR was performed using iQ™ SYBR Green Supermix (Bio-Rad, Hercules, USA) on the CFX connect™ Real-Time system (Bio-Rad, Hercules, USA). PCR reaction setup was according to the manufacturer's instructions. The relative transcript level of each gene was calculated by the 2-ΔΔCt (comparative Ct) method. Data were normalized to the endogenous control *actin* (NCU04173) with expression on untreated control as the reference sample.

RNA sequencing and data analysis

The total RNA of two biological replicate samples was extracted separately, mixed together after measuring the quality of each sample, and used for high-throughput RNA sequencing. The 13 49-nt single-end RNA-seq libraries were generated commercially at Beijing Genomics Institute (BGI, Shenzhen, China) while eight 90-nt paired-end RNA-seq libraries were generated at GENEWIZ (Biotechnology Co. Ltd, Suzhou, China) by using Illumina's HiSeq™ 2,000 platform (Illumina, San Diego, USA) (see summary of Additional file 1: Table S1 for details). Most of the data in the present study were generated by sequencing only one sample. However, as a reference, we performed parallel sequencing in this study for WT to make sure the sequencing data we got were technically reliable. The data of libraries 01 and 14 (Additional file 1: Table S1) come from biological replicates (WT no drug control), similar as the data from libraries 03 and 15 (Additional file 1: Table S1, WT treated with 5 mM DTT). The correlations between the two RNA-seq replicate data were analyzed and are shown in Additional file 4: Figure S9.

The sequencing data are deposited in the Gene Expression Omnibus database (GEO, http://www.ncbi.nlm.nih.gov/geo/) with accession number GSE61949. Filtered clean eads were then aligned to the latest *N. crassa* OR74A genome (version 12) [62] (http://www.broadinstitute.org/annotation/genome/neurospora/MultiHome.html) with splicing-aware aligner TopHat2 (version 2.0.12) [63]. Abundance for each transcript was calculated using the reads per kilobase per million (RPKM) [64]. Genes with altered expression was performed by using R package NOISeq (version 2.6.0) [65] (Q value ≥0.95 or 0.90 used as threshold for single-end or paired-end libraries, respectively, which approximately corresponds to a $|\log_2 \text{ratio}| \geq 1$). To discover significantly expression changes between the conditions tested, only the genes with relative high transcriptional abundance (RPKM values above 20 or 15 in at least one condition for single-end or paired-end libraries, respectively [66]) were deemed as positives and went into further analysis. MIPS FunCat online tools (http://mips.helmholtz-muenchen.de/funcatDB/) were used to functional classification of genes with altered expression. Meanwhile, manual annotation was assigned to these genes by referring to Blast hits and Pfam domain annotation (Broad version 7) to prove or verify FunCat results (for detail, see Additional file 6: Supporting Information).

Growth, lignocellulase secretion, and enzyme activity assays

Liquid cultures for cellulase production were carried out either in cellulose medium with rich nitrogen source (1× Vogel's salts, 2% *w*/*v* Avicel, 0.75% *w*/*v* yeast extract) or poor nitrogen source (1× Vogel's salts, 2% *w*/*v* Avicel). The 10-day-old conidia of tested strains were inoculated into 100 mL liquid media at 10^6 conidia/mL final concentration and grown at 25°C in constant light and shaking (200 rpm) over the course of 7 days. Since the deletion strain from the TF NCU03043 had severe defects in conidiation, young hyphae were pre-grown in minimal medium for 24 h to accumulate biomass, which was then shifted into rich/poor nitrogen source medium and fermented for 6 days to produce lignocellulases. For switching experiments with the hyper- or hypo-secretion mutants (such as Δ*res-1*, Δ*res-2*, and Δ*rrg-2*), 10-day-old conidia were inoculated into 100 mL minimal medium at 10^6 conidia/mL final concentration and grown at 25°C in constant light and agitation (200 rpm). After 16 h, young hyphae were harvested and transferred into cellulose medium with rich nitrogen source for the indicated times to produce cellulases. Relative enzyme activity of culture supernatants was assayed with Remazol brilliant Blue R-conjugated CMC kits (Megazyme, Wicklow, Ireland). The amount of secreted protein was determined using the Bradford protein assay (Bio-Rad, Hercules, USA). For protein gel electrophoresis, 4-μl unconcentrated culture supernatants were loaded on SDS-PAGE (Novex® NuPAGE® Pre-cast Protein Gels, Thermo fisher scientific, Waltham, USA). Directly measuring the dry weight of fungal biomass was replaced by measuring the total protein content in order to mitigate the interference of Avicel on the test. Assays were performed as previously described [67]. To check the actual protein loading of the secretory pathway of certain mutants (for example, Δ*hac-1*, Δ*ire-1*, Δ*res-1*, Δ*res-2*, and Δ*rrg-2*), total protein from the same amount of biomass for each strain grown on sucrose or Avicel were extracted as previously described [67]. Unconcentrated extractions (4 μl) for each sample were loaded on SDS-PAGE.

Microscopy and image processing

The growth phenotype of Δ*ire-1*/Δ*hac-1* as well as their parental strains was monitored by inoculating around 1,000 conidia for each strain into the center of petri dishes containing minimal medium and growth for 3 days. For recording conidial and mycelial differential interference contrast (DIC) images, cultures were performed in liquid minimal medium and samples collected at indicated time points. DIC images were recorded by using an Olympus BX51 Microscope (Olympus, Tokyo, Japan). Image Pro Plus (v6.0) was used for image processing.

Statistical significance tests and data plotting

Unless otherwise noted, all experiments were performed in triplicate and statistical tests for significance determined via one-way ANOVA by using R (version 3.1.1). R Packages include pheatmap (version 0.7.7) [68], ggplot2 (version 0.9.3.1) [69], VennDiagram (version 1.6.9) [70], and phenotypicForest (version 0.2) [71] used to data plotting.

Additional files

Additional file 1: Table S1. RNA-seq read profiles mapped to the genome of *N. crassa*.

Additional file 2: Table S2. Differential expression analysis.

Additional file 3: Table S3. Functional groups of genes with altered expression in response to ER stress with different stress intensities.

Additional file 4: Figures S1-S9. Supplementary figures referred to in the main text.

Additional file 5: Table S4. Phenotypical screen of 527 ESRT KO mutants under DTT-induced ER stress.

Additional file 6: Supporting information. Additional methods referred to in the main text.

Additional file 7: Table S5. Genes regulated by either IRE-1 or HAC-1 in *N. crassa*.

Additional file 8: Table S6. TFs associated with ER stress response and cellulase synthesis in *N. crassa*.

Additional file 9: Table S7. Genes regulated by the TFs RES-1, RES-2, or RRG-2 in *N. crassa*.

Abbreviations

Bp: base pair; *clr-1/clr-2*: cellulose degradation regulator 1/2; *cpc-1*: cross pathway control protein 1; ER: endoplasmic reticulum; ESRTs: ER stress response targets; *hac-1*: homologous to ATF/CREB1; *ire-1*: inositol-requiring enzyme-1; Nt: nucleotide; ORF: open reading frame; *res-1/res-2*: regulator for ER stress response 1/2; RESS: repression under secretion stress; RPKM: reads per kilobase per million; *rrg-2*: response regulator 2; TF: transcription factor; UPR: unfolded protein response; *xlr-1*: xylan degradation regulator 1.

Competing interests

The authors declare that they have no competing interests.

Authors' contributions

FF, GM, CT, and YM designed the research; FF and GM performed the research; FF, JL, JPB, and QL analyzed the data; and FF, GM, JPB, and CT wrote the paper. All authors read and approved the final manuscript.

Acknowledgements

This work was supported by a grant from the 973 Program of China (2011CB707403 and 2011CBA00800), the 863 Project (2012AA022203D and SS2014AA021300), and the NSF of China (31171207 and 31471186). The authors wish to thank Mr. Wenliang Sun and Ms. Lixian Wang for the help in the experiments and Dr. Carolyn Rasmussen, Dr. William Beeson and Mr. Nils Thieme for critical reading of the manuscript.

Author details

[1]Key Laboratory of Systems Microbial Biotechnology, Tianjin Institute of Industrial Biotechnology, Chinese Academy of Sciences, Xiqi Dao32, Tianjin Airport Economic Area, Tianjin 300308, China. [2]University of Chinese Academy of Sciences, Yuquan Road, Beijing 100049, China. [3]Holzforschung München, TUM School of Life Sciences Weihenstephan, Technische Universität München, Hans-Carl-von-Carlowitz-Platz 2, Freising, Germany.

References

1. Himmel ME, Ding SY, Johnson DK, Adney WS, Nimlos MR, Brady JW, et al. Biomass recalcitrance: engineering plants and enzymes for biofuels production. Science. 2007;315:804–7.
2. Yan S, Wu G. Secretory pathway of cellulase: a mini-review. Biotechnol Biofuels. 2013;6:177–88.
3. Ghosh A, Al-Rabiai S, Ghosh BK, Trimiño-Vazquez H, Eveleigh DE, Montenecourt BS. Increased endoplasmic reticulum content of a mutant of Trichoderma reesei (Rut-C30) in relation to cellulase synthesis. Enzyme Microb Technol. 1982;4:110–3.
4. Ghosh A, Ghosh BK, Trimiño-Vazquez H, Eveleigh DE, Montenecourt BS. Cellulase secretion from a hypercellulolytic mutant of *Trichoderma reesei* Rut-C30. Arch Microbiol. 1984;140:126–33.
5. Pakula TM, Salonen K, Uusitalo J, Penttila M. The effect of specific growth rate on protein synthesis and secretion in the filamentous fungus Trichoderma reesei. Microbiology. 2005;151:135–43.
6. Ron D, Walter P. Signal integration in the endoplasmic reticulum unfolded protein response. Nat Rev Mol Cell Biol. 2007;8:519–29.
7. Hetz C. The unfolded protein response: controlling cell fate decisions under ER stress and beyond. Nat Rev Mol Cell Biol. 2012;13:89–102.
8. Hollien J. Evolution of the unfolded protein response. Biochim Biophys Acta. 2013;1833:2458–63.
9. Pakula TM, Laxell M, Huuskonen A, Uusitalo J, Saloheimo M, Penttila M. The effects of drugs inhibiting protein secretion in the filamentous fungus Trichoderma reesei. Evidence for down-regulation of genes that encode secreted proteins in the stressed cells. J Biol Chem. 2003;278:45011–20.
10. Al-Sheikh H, Watson AJ, Lacey GA, Punt PJ, MacKenzie DA, Jeenes DJ, et al. Endoplasmic reticulum stress leads to the selective transcriptional downregulation of the glucoamylase gene in *Aspergillus niger*. Mol Microbiol. 2004;53:1731–42.
11. Carvalho ND, Jørgensen TR, Arentshorst M, Nitsche BM, van den Hondel CA, Archer DB, et al. Genome-wide expression analysis upon constitutive activation of the HacA bZIP transcription factor in *Aspergillus niger* reveals a coordinated cellular response to counteract ER stress. BMC Genomics. 2012;13:350–66.
12. Glass NL, Schmoll M, Cate JH, Coradetti S. Plant cell wall deconstruction by ascomycete fungi. Annu Rev Microbiol. 2013;67:477–98.
13. Znameroski EA, Glass NL. Using a model filamentous fungus to unravel mechanisms of lignocellulose deconstruction. Biotechnol Biofuels. 2013;6:6–12.
14. Tian C, Beeson WT, Iavarone AT, Sun J, Marletta MA, Cate JH, et al. Systems analysis of plant cell wall degradation by the model filamentous fungus *Neurospora crassa*. Proc Natl Acad Sci U S A. 2009;106:22157–62.
15. Benz JP, Chau BH, Zheng D, Bauer S, Glass NL, Somerville CR. A comparative systems analysis of polysaccharide-elicited responses in *Neurospora crassa* reveals carbon source-specific cellular adaptations. Mol Microbiol. 2014;91:275–99.
16. Colot HV, Park G, Turner GE, Ringelberg C, Crew CM, Litvinkova L, et al. A high-throughput gene knockout procedure for Neurospora reveals functions for multiple transcription factors. Proc Natl Acad Sci U S A. 2006;103:10352–7.
17. Xu C, Bailly-Maitre B, Reed JC. Endoplasmic reticulum stress: cell life and death decisions. J Clin Invest. 2005;115:2656–64.
18. Travers KJ, Patil CK, Wodicka L, Lockhart DJ, Weissman JS, Walter P. Functional and genomic analyses reveal an essential coordination between the unfolded protein response and ER-associated degradation. Cell. 2000;101:249–58.
19. Ruepp A, Zollner A, Maier D, Albermann K, Hani J, Mokrejs M, et al. The FunCat, a functional annotation scheme for systematic classification of proteins from whole genomes. Nucleic Acids Res. 2004;32:5539–45.
20. Paluh JL, Plamann M, Kruger D, Barthelmess IB, Yanofsky C, Perkins DD. Determination of the inactivating alterations in two mutant alleles of the Neurospora crassa cross-pathway control gene cpc-1. Genetics. 1990;124:599–606.
21. Barnes G, Hansen WJ, Holcomb CL, Rine J. Asparagine-linked glycosylation in *Saccharomyces cerevisiae*: genetic analysis of an early step. Mol Cell Biol. 1984;4:2381–8.
22. Lin JH, Li H, Yasumura D, Cohen HR, Zhang C, Panning B, et al. IRE1 signaling affects cell fate during the unfolded protein response. Science. 2007;318:944–9.
23. Nagashima Y, Mishiba K, Suzuki E, Shimada Y, Iwata Y, Koizumi N. Arabidopsis IRE1 catalyses unconventional splicing of bZIP60 mRNA to produce the active transcription factor. Sci Rep. 2011;1:29.
24. Hooks KB, Griffiths-Jones S. Conserved RNA structures in the non-canonical Hac1/Xbp1 intron. RNA Biol. 2011;8:552–6.
25. Feng X, Krishnan K, Richie DL, Aimanianda V, Hartl L, Grahl N, et al. HacA-independent functions of the ER stress sensor IreA synergize with the canonical UPR to influence virulence traits in Aspergillus fumigatus. PLoS Pathog. 2011;7, e1002330.

26. Kimata Y, Ishiwata-Kimata Y, Yamada S, Kohno K. Yeast unfolded protein response pathway regulates expression of genes for anti-oxidative stress and for cell surface proteins. Genes Cells. 2006;11:59–69.
27. Shen X, Ellis RE, Sakaki K, Kaufman RJ. Genetic interactions due to constitutive and inducible gene regulation mediated by the unfolded protein response in C. elegans. PLoS Genet. 2005;1, e37.
28. Wimalasena TT, Enjalbert B, Guillemette T, Plumridge A, Budge S, Yin Z, et al. Impact of the unfolded protein response upon genome-wide expression patterns, and the role of Hac1 in the polarized growth, of Candida albicans. Fungal Genet Biol. 2008;45:1235–47.
29. Wakasa Y, Oono Y, Yazawa T, Hayashi S, Ozawa K, Handa H, et al. RNA sequencing-mediated transcriptome analysis of rice plants in endoplasmic reticulum stress conditions. BMC Plant Biol. 2014;14:101.
30. Shelest E. Transcription factors in fungi. FEMS Microbiol Lett. 2008;286:145–51.
31. Sun J, Glass NL. Identification of the CRE-1 cellulolytic regulon in Neurospora crassa. PLoS One. 2011;6, e25654.
32. Banno S, Noguchi R, Yamashita K, Fukumori F, Kimura M, Yamaguchi I, et al. Roles of putative His-to-Asp signaling modules HPT-1 and RRG-2, on viability and sensitivity to osmotic and oxidative stresses in Neurospora crassa. Curr Genet. 2007;51:197–208.
33. Coradetti ST, Craig JP, Xiong Y, Shock T, Tian C, Glass NL. Conserved and essential transcription factors for cellulase gene expression in ascomycete fungi. Proc Natl Acad Sci U S A. 2012;109:7397–402.
34. Sun J, Tian C, Diamond S, Glass NL. Deciphering transcriptional regulatory mechanisms associated with hemicellulose degradation in Neurospora crassa. Eukaryot Cell. 2012;11:482–93.
35. Znameroski EA, Coradetti ST, Roche CM, Tsai JC, Iavarone AT, Cate JH, et al. Induction of lignocellulose-degrading enzymes in Neurospora crassa by cellodextrins. Proc Natl Acad Sci U S A. 2012;109:6012–7.
36. Pehar M, Jonas MC, Hare TM, Puglielli L. SLC33A1/AT-1 protein regulates the induction of autophagy downstream of IRE1/XBP1 pathway. J Biol Chem. 2012;287:29921–30.
37. Jonikas MC, Collins SR, Denic V, Oh E, Quan EM, Schmid V, et al. Comprehensive characterization of genes required for protein folding in the endoplasmic reticulum. Science. 2009;323:1693–7.
38. Dombroski BA, Nayak RR, Ewens KG, Ankener W, Cheung VG, Spielman RS. Gene expression and genetic variation in response to endoplasmic reticulum stress in human cells. Am J Hum Genet. 2010;86:719–29.
39. Chow CY, Wolfner MF, Clark AG. Using natural variation in Drosophila to discover previously unknown endoplasmic reticulum stress genes. Proc Natl Acad Sci U S A. 2013;110:9013–8.
40. Mulder HJ, Nikolaev I. HacA-dependent transcriptional switch releases hacA mRNA from a translational block upon endoplasmic reticulum stress. Eukaryot Cell. 2009;8:665–75.
41. Ruegsegger U, Leber JH, Walter P. Block of HAC1 mRNA translation by long-range base pairing is released by cytoplasmic splicing upon induction of the unfolded protein response. Cell. 2001;107:103–14.
42. Yoshida H, Oku M, Suzuki M, Mori K. pXBP1(U) encoded in XBP1 pre-mRNA negatively regulates unfolded protein response activator pXBP1(S) in mammalian ER stress response. J Cell Biol. 2006;172:565–75.
43. Tian C, Kasuga T, Sachs MS, Glass NL. Transcriptional profiling of cross pathway control in Neurospora crassa and comparative analysis of the Gcn4 and CPC1 regulons. Eukaryot Cell. 2007;6:1018–29.
44. Arvas M, Pakula T, Lanthaler K, Saloheimo M, Valkonen M, Suortti T, et al. Common features and interesting differences in transcriptional responses to secretion stress in the fungi Trichoderma reesei and Saccharomyces cerevisiae. BMC Genomics. 2006;7:32.
45. Wang B, Guo G, Wang C, Lin Y, Wang X, Zhao M, et al. Survey of the transcriptome of Aspergillus oryzae via massively parallel mRNA sequencing. Nucleic Acids Res. 2010;38:5075–87.
46. Soriani FM, Malavazi I, da Silva Ferreira ME, Savoldi M, Von Zeska Kress MR, de Souza Goldman MH, et al. Functional characterization of the Aspergillus fumigatus CRZ1 homologue CrzA. Mol Microbiol. 2008;67:1274–91.
47. Kwon NJ, Garzia A, Espeso EA, Ugalde U, Yu JH. FlbC is a putative nuclear C2H2 transcription factor regulating development in Aspergillus nidulans. Mol Microbiol. 2010;77:1203–19.
48. Wendland J, Walther A. Tip growth and endocytosis in fungi. In: Samaj J, Baluska F, Menzel D, editors. The plant endocytosis 2005. Heidelberg, Germany: Springer; 2005. p. 293–310.
49. Silverman-Gavrila LB, Lew RR. Calcium gradient dependence of Neurospora crassa hyphal growth. Microbiology. 2003;149:2475–85.
50. Bowman BJ, Abreu S, Margolles-Clark E, Draskovic M, Bowman EJ. Role of four calcium transport proteins, encoded by nca-1, nca-2, nca-3, and cax, in maintaining intracellular calcium levels in Neurospora crassa. Eukaryot Cell. 2011;10:654–61.
51. Lee J, Godon C, Lagniel G, Spector D, Garin J, Labarre J, et al. Yap1 and Skn7 control two specialized oxidative stress response regulons in yeast. J Biol Chem. 1999;274:16040–6.
52. Delic M, Rebnegger C, Graf AB, Mattanovich D, Gasser B. Overexpression of the redox transcription factor Yap1 modifies intracellular redox conditions and enhances recombinant protein secretion. Microbial Cell. 2013;11:376–86.
53. Malhotra JD, Miao H, Zhang K, Wolfson A, Pennathur S, Pipe SW, et al. Antioxidants reduce endoplasmic reticulum stress and improve protein secretion. Proc Natl Acad Sci U S A. 2008;105:18525–30.
54. Park G, Ryu YH, Hong YJ, Choi EH, Uhm HS. Cellular and molecular responses of Neurospora crassa to non-thermal plasma at atmospheric pressure. Appl Phys Lett. 2012;100:063703.
55. Lewis TL, Keesler GA, Fenner GP, Parks LW. Pleiotropic mutations in Saccharomyces cerevisiae affecting sterol uptake and metabolism. Yeast. 1988;4:93–106.
56. Pineau L, Colas J, Dupont S, Beney L, Fleurat-Lessard P, Berjeaud JM, et al. Lipid-induced ER stress: synergistic effects of sterols and saturated fatty acids. Traffic. 2009;10:673–90.
57. Aro N, Ilmen M, Saloheimo A, Penttila M. ACEI of Trichoderma reesei is a repressor of cellulase and xylanase expression. Appl Environ Microbiol. 2003;69:56–65.
58. Chilton IJ, Delaney CE, Barham-Morris J, Fincham DA, Hooley P, Whitehead MP. The Aspergillus nidulans stress response transcription factor StzA is ascomycete-specific and shows species-specific polymorphisms in the C-terminal region. Mycol Res. 2008;112:1435–46.
59. McCluskey K. The fungal genetics stock center: from molds to molecules. Adv Appl Microbiol. 2003;52:245–62.
60. Yazdi MT, Woodward JR, Radford A. Cellulase production by Neurospora-crassa - the enzymes of the complex and their regulation. Enzyme Microb Technol. 1990;12:116–9.
61. Davis RH, de Serres FJ. Genetic and microbiological research techniques for *Neurospora crassa*. Methods Enzymol. 1970;17A:79–143.
62. Galagan JE, Calvo SE, Borkovich KA, Selker EU, Read ND, Jaffe D, et al. The genome sequence of the filamentous fungus Neurospora crassa. Nature. 2003;422:859–68.
63. Kim D, Pertea G, Trapnell C, Pimentel H, Kelley R, Salzberg SL. TopHat2: accurate alignment of transcriptomes in the presence of insertions, deletions and gene fusions. Genome Biol. 2013;14:R36.
64. Mortazavi A, Williams BA, Mccue K, Schaeffer L, Wold B. Mapping and quantifying mammalian transcriptomes by RNA-Seq. Nat Methods. 2008;5:621–8.
65. Tarazona S, Garcia-Alcalde F, Dopazo J, Ferrer A, Conesa A. Differential expression in RNA-seq: A matter of depth. Genome Res. 2011;21:2213–23.
66. McIntyre LM, Lopiano KK, Morse AM, Amin V, Oberg AL, Young LJ, et al. RNA-seq : technical variability and sampling. BMC Genomics. 2011;12:293.
67. Brown NA, de Gouvea PF, Krohn NG, Savoldi M, Goldman GH. Functional characterisation of the non-essential protein kinases and phosphatases regulating Aspergillus nidulans hydrolytic enzyme production. Biotechnol Biofuels. 2013;6:91.
68. Kolde R. pheatmap: Pretty Heatmaps. 2011. R package version 0.5.1. [http://CRAN.R-project.org/package=pheatmap].
69. Ginestet C. ggplot2: elegant graphics for data analysis. J R Stat Soc Ser A-Stat Soc. 2011;174:245.
70. Chen H, Boutros PC. VennDiagram: a package for the generation of highly-customizable Venn and Euler diagrams in R. Bmc Bioinformatics. 2011;12:35.
71. Ladroue C. phenotypicForest. 2012. R package version 0.2 [http://chrisladroue.com/phorest/].

Permissions

All chapters in this book were first published in BFB, by BioMed Central; hereby published with permission under the Creative Commons Attribution License or equivalent. Every chapter published in this book has been scrutinized by our experts. Their significance has been extensively debated. The topics covered herein carry significant findings which will fuel the growth of the discipline. They may even be implemented as practical applications or may be referred to as a beginning point for another development.

The contributors of this book come from diverse backgrounds, making this book a truly international effort. This book will bring forth new frontiers with its revolutionizing research information and detailed analysis of the nascent developments around the world.

We would like to thank all the contributing authors for lending their expertise to make the book truly unique. They have played a crucial role in the development of this book. Without their invaluable contributions this book wouldn't have been possible. They have made vital efforts to compile up to date information on the varied aspects of this subject to make this book a valuable addition to the collection of many professionals and students.

This book was conceptualized with the vision of imparting up-to-date information and advanced data in this field. To ensure the same, a matchless editorial board was set up. Every individual on the board went through rigorous rounds of assessment to prove their worth. After which they invested a large part of their time researching and compiling the most relevant data for our readers.

The editorial board has been involved in producing this book since its inception. They have spent rigorous hours researching and exploring the diverse topics which have resulted in the successful publishing of this book. They have passed on their knowledge of decades through this book. To expedite this challenging task, the publisher supported the team at every step. A small team of assistant editors was also appointed to further simplify the editing procedure and attain best results for the readers.

Apart from the editorial board, the designing team has also invested a significant amount of their time in understanding the subject and creating the most relevant covers. They scrutinized every image to scout for the most suitable representation of the subject and create an appropriate cover for the book.

The publishing team has been an ardent support to the editorial, designing and production team. Their endless efforts to recruit the best for this project, has resulted in the accomplishment of this book. They are a veteran in the field of academics and their pool of knowledge is as vast as their experience in printing. Their expertise and guidance has proved useful at every step. Their uncompromising quality standards have made this book an exceptional effort. Their encouragement from time to time has been an inspiration for everyone.

The publisher and the editorial board hope that this book will prove to be a valuable piece of knowledge for researchers, students, practitioners and scholars across the globe.

List of Contributors

Lee R Lynd and Mark Laser
Thayer School of Engineering, Dartmouth College, Hanover, NH, USA

Mariam Sow, Mosad Elmissiry and Ibrahim A Mayaki
New Partnership for Africa's Development (NEPAD), Johannesburg, South Africa

Annie FA Chimphango
Department of Process Engineering, University of Stellenbosch, Stellenbosch, South Africa

Luis AB Cortez and Luiz AH Nogueira
Faculty of Agricultural Engineering, University of Campinas, Campinas, Brazil

Carlos H Brito Cruz
São Paulo Research Foundation São Paulo, São Paulo, Brazil
Physics Institute, University of Campinas, Campinas, Brazil

Marcia AFD Moraes
Department of Applied Economics, University of São Paulo, ESALQ, Piracicaba, Brazil

Gideon M Wolfaardt
Department of Microbiology, University of Stellenbosch, Stellenbosch, South Africa
Water Institute, University of Stellenbosch, Stellenbosch, South Africa

Jeremy Woods
Centre for Environmental Policy, Imperial College London, London, UK

Willem H van Zyl
Department of Microbiology, University of Stellenbosch, Stellenbosch, South Africa

Sudhanshu S Pawar, Thitiwut Vongkumpeang and Ed WJ van Niel
Division of Applied Microbiology, Lund University, Getingevägen 60, PO Box 124, SE-221 00 Lund, Sweden

Carl Grey
Department of Biotechnology, Lund University, Getingevägen 60, PO Box 124221 00 Lund, Sweden

Jinlan Cheng
Jiangsu Provincial Key Lab of Pulp and Paper Science and Technology, Nanjing Forestry University, Nanjing, China USDA Forest Service, Forest Products Laboratory, Madison, WI 53719, USA

Shao-Yuan Leu
USDA Forest Service, Forest Products Laboratory, Madison, WI 53719, USA
Department of Civil and Environmental Engineering, Hong Kong Polytechnic University, Kowloon, Hong Kong

JY Zhu and Rolland Gleisner
USDA Forest Service, Forest Products Laboratory, Madison, WI 53719, USA

Balázs Frankó, Mats Galbe and Ola Wallberg
Department of Chemical Engineering, Lund University, PO Box 124, Getingevägen 60, Lund SE-221 00, Sweden

Sandra Cerqueira Pereira
Embrapa Instrumentation, Rua XV de Novembro 1452, 13560-970 São Carlos, SP, Brazil

Larissa Maehara and Cristiane Sanchez Farinas
Embrapa Instrumentation, Rua XV de Novembro 1452, 13560-970 São Carlos, SP, Brazil
Graduate Program of Chemical Engineering, Federal University of São Carlos, Rodovia Washington Luiz, km 235, 13565-905 São Carlos, SP, Brazil

Cristina Maria Monteiro Machado
Embrapa Agroenergy, Parque Estação Biológica s/n°, 70770-901 Brasília, DF, Brazil
Jun Niu, Jiyong An, Xinyu Hou, Zheng Xiang, Sufan Zhou, Zhixiang Zhang, Xinyi Feng and Shanzhi Lin
College of Biological Sciences and Biotechnology, College of Nature Conservation, National Engineering Laboratory for Tree Breeding, Key Laboratory of Genetics and Breeding in Forest Trees and Ornamental Plants, Ministry of Education, Beijing Forestry University, Beijing 10083, China

Libing Wang and Haiyan Yu
Research Institute of Forestry, Chinese Academy of Forestry, Beijing 10091, China

Chengliang Fang, Denglong Ha, Lin Qiu and Haiyan Zhao
Jigongshan National Nature Reserve, Xingyang 464133, China

Chengyu Fu
Liaocheng Food and Drug Administration, Liaocheng 252000, Shandong, China

Bang Wang, Pengli Cai and Jingen Li
Key Laboratory of Systems Microbial Biotechnology, Tianjin Institute of Industrial Biotechnology, Chinese Academy of Sciences, Tianjin 300308, China
University of Chinese Academy of Sciences, Beijing 100049, China

Wenliang Sun, Chaoguang Tian and Yanhe Ma
Key Laboratory of Systems Microbial Biotechnology, Tianjin Institute of Industrial Biotechnology, Chinese Academy of Sciences, Tianjin 300308, China

Nasir Mehmood, Eric Husson, Catherine Sarazin and Isabelle Gosselin
Unité Génie Enzymatique et Cellulaire, FRE-CNRS 3580, Université de Picardie Jules Verne, 33 rue Saint-Leu, 80039 Amiens Cedex, France

Cédric Jacquard
Unité de Recherche Vignes et Vins de Champagne—UPRES-EA 4707, Université de Reims Champagne-Ardenne, BP103951687 Reims Cedex 2, France

Sandra Wewetzer and Jochen Büchs
AVT—Biochemical Engineering, RWTH Aachen University, Aachen, Germany

Zemin Fang, Xiaoman Liu, Xuecheng Zhang, Wei Fang and Yazhong Xiao
School of Life Sciences, Anhui University, 111 Jiulong Road, Hefei, Anhui 230601, China
Anhui Provincial Engineering Technology Research Center of Microorganisms and Biocatalysis, 111 Jiulong Road, Hefei, Anhui 230601, China

Liyuan Chen, Yu Shen and Xiaoming Bao
The State Key Laboratory of Microbial Technology, School of Life Sciences, Shandong University, 27 Shanda Nanlu, Jinan, Shandong 250100, China

Xiaotang Wang
Department of Chemistry & Biochemistry, Florida International University, Miami, FL 33199, USA

Prachand Shrestha, Timothy M Szaro, Thomas D Bruns and John W Taylor
Department of Plant and Microbial Biology, University of California, Berkeley, CA 94720-3102, USA

Ana B Ibáñez and Stefan Bauer
Energy Biosciences Institute, University of California, Berkeley, CA 94720, USA

Sydney I Glassman
Department of Environmental Science Policy and Management, University of California, Berkeley, CA 94720, USA

Ranjita Biswas
Biosciences Division, Oak Ridge National Laboratory, Oak Ridge, TN 37830, USA
BioEnergy Science Center, Oak Ridge National Laboratory, Oak Ridge, TN 37830, USA
Centre for Rural Development and Technology, Indian Institute of Technology Delhi, Hauz Khas, New Delhi 110016, India

Tianyong Zheng, Daniel G Olson and Lee R Lynd
BioEnergy Science Center, Oak Ridge National Laboratory, Oak Ridge, TN 37830, USA
Thayer School of Engineering at Dartmouth College, Hanover, NH 03755, USA

Adam M Guss
Biosciences Division, Oak Ridge National Laboratory, Oak Ridge, TN 37830, USA
BioEnergy Science Center, Oak Ridge National Laboratory, Oak Ridge, TN 37830, USA
One Bethel Valley Road, Oak Ridge, TN 37831-6038, USA

Juan Mangas-Sánchez and Patrick Adlercreutz
Department of Biotechnology, Lund University, P.O. Box 124, Lund SE-221 00, Sweden

Patricia J Slininger, Maureen A Shea-Andersh, Stephanie R Thompson, Bruce S Dien, and Michael A Cotta
Bioenergy Research Unit, National Center for Agricultural Utilization Research, USDA-ARS, 1815 N. University, Peoria, IL 61604, USA

Cletus P Kurtzman
Bacterial Foodborne Pathogens and Mycology Research, National Center for Agricultural Utilization Research, USDA-ARS, 1815 N. University, Peoria, IL 61604, USA

Venkatesh Balan, Leonardo da Costa Sousa, Nirmal Uppugundla and Bruce E Dale
DOE Great Lakes Bioenergy Research Center, Michigan State University, East Lansing, MI 48824, USA

Haoyu Wang
State Key Laboratory of Urban Water Resource and Environment, Harbin Institute of Technology, 150090 Harbin, China
Section of Sanitary Engineering, Department of Water Management, Delft University of Technology, 2628 CN, Delft, The Netherlands
UNESCO-IHE Institute for Water Education, 2601 DA, Delft, The Netherlands

Yu Tao
State Key Laboratory of Urban Water Resource and Environment, Harbin Institute of Technology, 150090 Harbin, China
Section of Sanitary Engineering, Department of Water Management, Delft University of Technology, 2628 CN, Delft, The Netherlands

Margarida Temudo, Margot Schooneveld, Henk Bijl and Joris Kloek
DSM Biotechnology Center, 2600 MA, Delft, The Netherlands

Nanqi Ren
State Key Laboratory of Urban Water Resource and Environment, Harbin Institute of Technology, 150090 Harbin, China

Monika Wolf, Cornelia Heine, Anne Foerster and Vincent Pelenc
Biopract GmbH, 12489 Berlin, Germany

Jules B van Lier
Section of Sanitary Engineering, Department of Water Management, Delft University of Technology, 2628 CN, Delft, The Netherlands
UNESCO-IHE Institute for Water Education, 2601 DA, Delft, The Netherlands

Merle de Kreuk
Section of Sanitary Engineering, Department of Water Management, Delft University of Technology, 2628 CN, Delft, The Netherlands

Roland Wirth and Zoltán Bagi
Department of Biotechnology, University of Szeged, Közép fasor 52, H-6726 Szeged, Hungary

Gergely Lakatos, Gergely Maróti and Éva Kondorosi
Institute of Biochemistry, Biological Research Center, Hungarian Academy of Sciences, Temesvári krt. 62, H-6726 Szeged, Hungary

János Minárovics and Katalin Nagy
Department of Oral Biology and Experimental Dental Research, University of Szeged, Tisza L. krt. 64, 6720 Szeged,
Hungary

Gábor Rákhely
Department of Biotechnology, University of Szeged, Közép fasor 52, H-6726 Szeged, Hungary
Institute of Biophysics, Biological Research Center, Hungarian Academy of Sciences, Temesvári krt. 62, H-6726 Szeged, Hungary

Kornél L Kovács
Department of Biotechnology, University of Szeged, Közép fasor 52, H-6726 Szeged, Hungary
Institute of Biophysics, Biological Research Center, Hungarian Academy of Sciences, Temesvári krt. 62, H-6726 Szeged, Hungary
Department of Oral Biology and Experimental Dental Research, University of Szeged, Tisza L. krt. 64, 6720 Szeged, Hungary

Paula Berman, Nitzan Meiri and Zeev Wiesman
The Phyto-Lipid Biotechnology Laboratory, Departments of Biotechnology, Energy and Environmental Engineering, Ben-Gurion University of the Negev, P.O. Box 653, Beer-Sheva 84105, Israel

Luiz Alberto Colnago
Embrapa Instrumentação, Rua 15 de Novembro 1452, São Carlos, SP 13560-970, Brazil

Tiago Bueno Moraes
Instituto de Física de São Carlos, Universidade de São Paulo, Av. Trabalhador Sao-Carlense 400, São Carlos, SP 13566-590, Brazil

Charles Linder
Zuckerberg Center for Water Sciences and Technology and Department of Biotechnology, Ben-Gurion University of the Negev, P.O. Box 653, Beer-Sheva 84105, Israel

Ofer Levi and Yisrael Parmet
Department of Industrial Engineering and Management, Ben-Gurion University of the Negev, P.O. Box 653, Beer-Sheva 84105, Israel

Michael Saunders
Department of Management Science and Engineering, Stanford University, Stanford, CA, USA

Katerina Lappa, Panagiotis Kandylis, Argyro Bekatorou, Maria Kanellaki and Athanasios A Koutinas
Food Biotechnology Group, Department of Chemistry, University of Patras, 26500 Patra, Greece

Nikolaos Bastas and Stavros Klaoudatos
B.G. Spiliopoulos S.A, 26500 Patra, Greece

Nikolaos Athanasopoulos
I. Athanasopoulos and Co, 106b Lontou Str, 26224 Patra, Greece

Feiyu Fan, Guoli Ma and Jingen Li
Key Laboratory of Systems Microbial Biotechnology, Tianjin Institute of Industrial Biotechnology, Chinese Academy of Sciences, Xiqi Dao32, Tianjin Airport Economic Area, Tianjin 300308, China
University of Chinese Academy of Sciences, Yuquan Road, Beijing 100049, China

Qian Liu, Chaoguang Tian and Yanhe Ma
Key Laboratory of Systems Microbial Biotechnology, Tianjin Institute of Industrial Biotechnology, Chinese Academy of Sciences, Xiqi Dao32, Tianjin Airport Economic Area, Tianjin 300308, China

Johan Philipp Benz
Holzforschung München, TUM School of Life Sciences Weihenstephan, Technische Universität München, Hans-Carl-von-Carlowitz-Platz 2, Freising, Germany